D1061673

**METHODS IN**

# MICROBIOLOGY

# Methods in Microbiology



A complete list of the books in this series is available from the publishers on request.

# METHODS IN

# MICROBIOLOGY

## Volume 22
## Techniques in
## Microbial Ecology

Edited by

**R. GRIGOROVA**

*Institute of Microbiology, Bulgarian Academy of Sciences,*
*Acad. G. Bontchev, Sofia, Bulgaria*

and

**J. R. NORRIS**

*Lord Zuckerman Research Centre, University of Reading, Reading, UK*

**ACADEMIC PRESS**
*Harcourt Brace Jovanovich, Publishers*
London San Diego New York Boston
Sydney Tokyo Toronto

ACADEMIC PRESS LIMITED
24/28 Oval Road
London NW1 7DX

*United States Edition published by*
ACADEMIC PRESS INC.
San Diego, CA 92101



This book is printed on acid-free paper



*British Library Cataloguing in Publication Data*
*is available*
*ISBN 0-12-521522-3*

Typeset by Page Bros (Norwich) Ltd
Printed and bound in Great Britain by The University Press, Cambridge

# CONTRIBUTORS

**M. Bazin,** Department of Microbiology, King's College, University of London, Campden Hill Road, London W8 7AH, UK

**L. Boddy,** School of Pure & Applied Biology, University of Wales, College of Cardiff, PO Box 915, Cardiff CF1 3TL, UK

**R. Campbell,** Department of Botany, University of Bristol, Bristol BS8 1UG, UK

**J. W. Costerton,** Biology Department, University of Calgary, Calgary, Alberta T2N 1N4, Canada

**D. J. Dickinson,** Department of Pure & Applied Biology, Imperial College, London SW7 2BB, UK

**J. Dighton,** Institute of Terrestrial Ecology, Merlewood Research Station, Grange-over-Sands, Cumbria LA11 6JU, UK

**M. Fletcher,** Center of Marine Biotechnology, Maryland Biotechnology Institute, The University of Maryland, 600 East Lombard Street, Baltimore, Maryland 21202, USA

**J. C. Frankland,** Institute of Terrestrial Ecology, Merlewood Research Station, Grange-over-Sands, Cumbria LA11 6JU, UK

**J. C. Fry,** School of Pure & Applied Biology, University of Wales, College of Cardiff, PO Box 915, Cardiff CF1 3TL, UK

**J. C. Gottschal,** Department of Microbiology, Biological Centre, University of Groningen, Kerklaan 30, 9751 NN Haren, The Netherlands

**T. R. G. Gray,** Department of Biology, University of Essex, Colchester, Essex, CO4 3SQ, UK

**R. Grigorova,** Institute of Microbiology, Bulgarian Academy of Sciences, Sofia 1113, Bulgaria

**M. P. Greaves,** Department of Agricultural Sciences, University of Bristol, AFRC Institute of Arable Crops Research, Long Ashton Research Station, Bristol BS18 9AF, UK

**G. H. Hall,** Institute of Freshwater Ecology, The Ferry House, Far Sawrey, Ambleside, Cumbria LA22 0LP, UK

**R. A. Herbert,** Department of Biological Sciences, University of Dundee, UK

**J. E. Hobbie,** Marine Biological Laboratory, Woods Hole, MA 02543, USA

**J. G. Jones,** Institute of Freshwater Ecology, The Ferry House, Far Sawrey, Ambleside, Cumbria LA22 0LP, UK

**C. W. Kaspar,** University of Maryland, Department of Microbiology, College Park, MD USA (Present address: San Labs, 405 Eighth Avenue, S.E., Cedar Rapids, Iowa 52401, USA)

**T. I. Ladd,** Biology Department, Saint Mary's University, Halifax, Nova Scotia B3H 3C3, Canada (Present address: Department of Biology, Millersville University, Millersville, PA 17551, USA)

**J. F. Levy,** Department of Pure & Applied Biology, Imperial College, London SW7 2BB, UK

**M. F. Madelin,** Department of Botany, University of Bristol, Bristol BS8 1UG, UK

**A. J. McCarthy,** Department of Genetics & Microbiology, University of Liverpool, Liverpool L69 3BX, UK

**A. Menell,** Department of Microbiology, King's College, University of London, Campden Hill Road, London W8 7AH, UK

**D. J. W. Moriarty,** CSIRO Marine Laboratories, PO Box 120, Cleveland, Queensland 4163, Australia

**R. W. Pickup,** Institute of Freshwater Ecology, The Ferry House, Far Sawrey, Ambleside, Cumbria LA22 0LP, UK

**F. G. Priest,** Department of Biological Sciences, Heriot Watt University, Riccarton, Edinburgh EH14 4AS, UK

**B. M. Simon,** Institute of Freshwater Ecology, The Ferry House, Far Sawrey, Ambleside, Cumbria LA22 0LP, UK

**C. Tartera,** University of Maryland, Department of Microbiology, College Park, MD, USA (Present address: Department of Microbiology, Faculty of Biology, University of Barcelona 08071 Barcelona, Spain)

**S. E. Underhill,** Schools of Chemical Engineering, University of Bradford, West Yorkshire BD7 1DP, UK

**S. T. Williams,** Department of Genetics & Microbiology, University of Liverpool, Liverpool L69 3BX, UK

# PREFACE

Volume 22 of *Methods in Microbiology* deals with the important subject of Microbial Ecology. We have considered this topic from time to time over the past few years but we delayed embarking on this volume until we felt that the subject had reached a stage in its development when an attempt to pull together the central research methods would be valuable. Before starting on the project we approached many of the workers in the field for their comments and suggestions. They responded with a great deal of enthusiasm and thorough advice. We are most grateful to the many people who helped in the early stages to define topics and identify authors.

It is clear from the start that the formula developed for other volumes in the Series would not work for microbial ecology. There were already several well constructed publications dealing with the subject. To ignore them would lead to much quite unjustified repetition of material already in print. The subject was also unusual in that it could be tackled in two ways; as a study of habitats or as a study of individual microbial types. We recognised that there was a need for both approaches since some research workers are mainly interested in microbial systems in habitats and others focus their attention on individual groups of organisms.

We decided that the habitat was the best way to develop techniques as a theme, concentrating on the modifications to central methods required to adapt them for different applications. We came to the conclusion that it would be far too big a task, and for too repetitive, to try to deal separately with all of the microbial groups that have been studied ecologically. Nevertheless we felt that it would be useful to include a few chapters describing the methods used to study particular groups, both as examples of approaches to be taken in such studies and because the groups chosen were of special significance.

We accept that there is a certain amount of repetition in the methods described. This is inevitable when dealing with so many different habitats, and we have found it interesting and valuable to compare the ways in which central techniques have been adapted by different authors to meet their own needs. We have accepted in this volume more of a review approach than is usual in the Series. The subject matter, we feel, lends itself to this treatment and it also helps to overcome the problem of needless repetition of previously published material.

The first two chapters set the scene by describing the methods for enumerating microorganisms in their natural environments and for determining biomass. Chapters on continuous culture methods and mathematical models then lead in to a study of aquatic environments, methods for studying adhesion and attachment and biofilm formation. Soil and forest litter habitats are then dealt with both in general terms and with more specific focus on particular groups and biochemical functions. This leads naturally on to the rhizosphere and the study of timber and forest products. A chapter on food concentrates on new and future techniques and the important actinomycetes and spore-forming bacteria are described as models of group orientated studies.

We hope that this volume will serve as a useful reference for workers in, or entering, the ecology field. We believe that the subject is emerging as one of central importance as biologists begin to tackle the major environmental problems now challenging mankind. We would like to thank the authors for their kindness and patience in dealing with our often naive questions and, as always, the staff of Academic Press for their careful and professional attention throughout the project.

R Grigorova
J. R. Norris

*August, 1990*

# CONTENTS

# 1
# Methods for Enumerating Microorganisms and Determining Biomass in Natural Environments

R. A. HERBERT
*Department of Biological Sciences, University of Dundee, UK*



METHODS IN MICROBIOLOGY
VOLUME 22 ISBN 0-12-521522-3

## I. Introduction

A fundamental aspect of many ecological investigations is the need for a reliable quantitative estimation of the microbiological populations that are present. Whilst this may initially appear a relatively simple objective to achieve, it is in practice fraught with problems. Ideally the counting method used should enable all the organisms present to be enumerated and allow the differentiation of viable and non-viable cells. In practice this ideal has rarely been achieved and in most instances a compromise has to be accepted. The principal reason for this inability to estimate microbial populations accurately is that natural environments contain an extremely diverse assemblage of microorganisms which occupy many different ecological niches, each of which presents its own methodological difficulties. Planktonic organisms generally present the least difficulty since the majority are freely suspended in water although a proportion may be attached to particulate matter. Depending upon the proportion of cells attached to particulate matter the water samples may be subjected to varying degrees of pretreatment prior to enumeration of the resident microflora. Jones (1979) considered that for unpolluted water samples vigorous shaking by hand 20 times for 30 s was more than sufficient to disrupt bacterial clumps and microorganisms attached to particulate matter. In contrast, water samples from eutrophic ponds and lakes often contain substantial numbers of attached microorganisms (1–13%) whilst in highly turbid habitats such as estuaries where levels of suspended solids may reach concentrations of 1000 mg $l^{-1}$ the majority of the microbial cells are attached to particulate matter (Goulder *et al.*, 1981; Plummer *et al.*, 1987). In order to enumerate the total microbial populations in water samples taken from these habitats a more rigorous pretreatment is required. Sonication is commonly used to detach microorganisms from particles. As a general guide three 1-min bursts of sonication are used with 30-s intervals between each treatment. It is essential to cool the samples in ice during sonication to avoid overheating. Equally the lowest amount of power and period of sonication commensurate with effective release of the attached cells should be used to minimize disruption of the cells. The proportion of free-living and attached microbial populations can be estimated if differential filtration techniques are employed. Dodson and Thomas (1964) developed a simple but effective system for separating attached from planktonic microorganisms consisting of a Perspex cylinder to one end of which is secured by gluing or clamping a membrane or other suitable filter. Typically a 3–5 μm pore size membrane is used and the operation is carried out by inverting the Perspex tube into a vessel containing the water sample. A hydrostatic head of 25–30 cm between the outside and inside of the tube is sufficient to allow planktonic microorganisms to pass through the filter and collect within the tube. Recovery of the attached microorganisms can be achieved by backwashing the filter with sterile water together with mechanical agitation of the filter surface. The advantage of this system is that it provides a gentle yet effective means of separating the two populations and the speed of filtration can be controlled by regulating the head of water and filter area used. By employing different types and pore sizes of filter, fractionation of particular size classes of microorganisms can be achieved. Equally, the system is eminently suitable for concentrating microbial populations when cell densities are low, e.g. oligotrophic waters. This approach is very effective for rapidly concentrating several litres of water down to a few millilitres.

To enumerate benthic populations of microorganisms more rigorous methods for detachment of microbial cells are required. Sonication is again widely used to detach microbial cells from sediment particles and the protocol described previously in this review is commonly used. Alternatively the sediment sample can be homogenized in a high-speed blender such as a Sorvall Omnimixer, Wareing blender or similar homogenizer. Dale (1974) homogenized sediment samples at 23 000 r.p.m. for 5 min but any speed above 10 000 r.p.m. is effective. Microorganisms can also be efficiently detached from sediment particles using a Meikle tissue homogenizer. Balanced pairs of specimen tubes containing 10 ml of sediment and 10 g of sterile glass Ballotini glass beads (Grade 11, 0.176–0.249 mm diameter) are shaken in the homogenizer at 50 $H_3$ for 1–5 min. The Braun MSK homogenizer works on a similar principle to the Meikle and the sample bottles containing sediment and Ballotini beads are oscillated with a rotary motion at 4000–5000 r.p.m. for 10–30 s. The efficiency of homogenization of flocculant material, particularly sewage and sewage sludge, can be increased by diluting the samples in either 5 mg $l^{-1}$ sodium tripolyphosphate (Pike *et al.*, 1972) or in a solution containing 0.01% each of sodium pyrophosphate and Lubrol W (Gayford and Richards, 1970). The addition of surfactants not only increases desorption of cells from particulate matter but also prevents re-aggregation. Gibson *et al.* (1987) showed that the addition of 0.0001% w/v cetyl trimethylammonium bromide (CTAB) significantly increased the recovery of viable *in situ* populations of sulphate-reducing bacteria from estuarine sediments.

Removal of epiphytic bacteria from plants is best achieved by using a stomacher, e.g. Colworth Stomacher-400. Fry and co-workers (Fry *et al.*, 1985) reported that small quantities of plant material (10–15 leaves of *Elodea*) should be added to 50–150 ml of sterile distilled water and treated in a stomacher for 5 min. Using this technique these workers reported that 41% of the epiphytic bacteria could be removed from plant leaves. To remove epilithic bacteria from stones Goulder (1987) recommended scrubbing with a sterile toothbrush. Fry (1988) considers that the best system is to put 100 ml of membrane-filtered autoclaved water into a sterile stomacher bag and then scrub the stone(s) within the bag to remove the epilithic bacteria. The suspended material is then treated in the stomacher for 5 min. Since stones may cause abrasion and puncturing of the stomacher bag during processing, Fry (1988) recommended that a second stomacher bag should be used to enclose the first bag.

A necessary prerequisite to enumerating microbial populations in natural environments is the need to collect representative sample material. It is outwith the remit of this review to provide a detailed discussion of the rationale and planning of sampling programmes and the types of sampling equipment available. Suffice to say that if meaningful and valid interpretation of the data obtained is to be made then due evaluation of the sampling procedures to be employed must be undertaken. For details on sampling equipment the reader should consult reviews by Holme and Mackintyre (1971), Collins *et al.* (1973), Williams and Gray (1973), Mackereth *et al.* (1978), Sieburth (1979) and Herbert (1988). Equally, whilst statistical analysis is an essential component of any quantitative ecological investigation, it is beyond the scope of this review to provide detailed descriptions of statistical methods. Several excellent texts on methods of statistical analysis are available and for detailed information readers should consult Snedocor and Cochran (1967), Sokal and Rohlf (1969), Elliot (1977), Jones (1979) and Kirchman *et al.* (1982).

Methods used to enumerate microbial populations fall into two major divisions, direct and indirect counting methods. Direct counting methods can be further subdivided. Those which attempt to count all the microorganisms of a particular type, e.g. bacteria, fungi, algae and protozoa, in a given sample give a total count of all the organisms present irrespective of whether they are living or dead. Alternatively, viable counting methods can be used to differentiate between living and dead cells by assessing their ability to grow either in liquid media, on solid media or on membrane filters. Since there is no universal growth medium on which all microorganisms will grow, it is therefore inevitable that viable counting methods substantially underestimate the true microbial populations present. For example, in marine waters Kogure *et al.* (1979) estimated that the viable counting methods estimated between 0.001 and 0.1% of the total microbial count whilst in fresh waters the proportion was between 0.9 and 22% (Sorokin, 1972). Viable count data for natural microbial populations should therefore be treated with considerable caution except when other supporting information such as total count data and estimates of biomass are available or when specific physiological groups are being enumerated.

Indirect counting methods in contrast do not require visual or cultural examination of the microorganisms and rely upon the presence of a specific chemical component that is only present in the living cells and which is rapidly degraded upon death of the organism. A crucial aspect of indirect methods is that the chemical component to be determined should have a known and constant ratio to cell biomass and not be influenced by factors such as growth rate and nutrient status of the cells.

## II. Total counting methods

The direct counting method selected is dependent upon the type of organism to be enumerated in the samples. Eukaryotes are normally enumerated using counting chambers and a variety of different types are available.

### A. Counting chambers for algae, protozoa and fungi

Sedimentation chambers are widely used to count phytoplankton since they provide a gentle but efficient system to concentrate the low numbers of cells that are usually present in the water column. The construction of such systems is well documented (Lund *et al.*, 1958; Vollenwieder, 1974; Jones, 1979) but basically they comprise a glass tube (15–20 mm diameter), the volume of which can be adjusted to suit the population density of algae anticipated. To one end of the glass tube is cemented, using an epoxy resin adhesive, a thin glass disc (coverslip or similar thickness glass). The sedimented cells which collect on the thin glass disc can then be viewed and counted using an inverted microscope. The usual method of sample preparation is to add preservative, usually Lugol's iodine (10 g $I_2$, 20 g KI, 200 ml distilled water to which is added 20 ml glacial acetic acid prior to use; the fixative is then stored in the dark), in the proportion of 1 part fixative to 100 parts of sample. Not only does this fix the phytoplankton cells but it also speeds up the sedimentation process. As an alternative fixative Jones (1979) recommends the addition of 0.2 ml each of 5% v/v Tween 80, 40% v/v formaldehyde and 20% w/v copper sulphate to each water sample. It is essential that an adequate period of time is allowed for the cells to sediment out and typically this is in the order of 3 h for each centimetre of column height. The sedimented cells are counted using an inverted microscope fitted with a low-power objective and normally the whole area of the chamber is scanned. Details of the counting procedures used are given by Lund *et al.* (1958) and in an abbreviated form by Jones (1979).

An alternative counting chamber for algae and protozoa is the Sedgewick–Rafter cell. The counting chamber comprises a slide with 1-mm raised margins such that when a coverslip is placed on top the volume enclosed is 1 $cm^3$. The chamber is divided up into 1000 1-$mm^2$ units and, to ensure that the volume is 1 $cm^3$, the chamber should be weighed empty and when full. Great care should be taken when filling the chamber to ensure that its volume is 1 ml. Jones (1979) recommended that the best way to fill the chamber was to place the coverslip diagonally across the cell and introduce the sample with a piston pipette (Finnpipette or similar model) through the gap at one corner allowing air to be expelled from the opposite corner. Finally, the coverslip is gently eased across to completely cover the sample and the cells then counted using bright-field microscopy. The Sedgewick–Rafter cell is particularly useful for counting larger microorganisms such as the slow-moving protozoa. Woelkerling *et al.* (1976) examined the errors incurred when using the Sedgewick–Rafter chamber for counting phytoplankton. They concluded that a settling time of at least 15 min was required and this was dependent on the type of preservative used. Counting random fields was recommended rather than strip counting across the chamber and greater accuracy was obtained by making fewer counts on a larger number of cells.

In addition to the two systems described here there are a variety of algal counting chambers that can be constructed simply in the laboratory. These include a moat chamber comprising a moat which is machined into a clear Perspex circular block which has a barrier at one point along its length. The Perspex disc is attached by a clamp at its centre point so that it can be clamped onto the stage of a binocular or dissecting microscope. The sample is pipetted into the moat, the chamber is rotated about its central axis and the organisms are counted along the moat commencing at one side of the barrier and finishing at the other. Depending upon the size of microorganisms to be counted, the depth and hence the volume of the moat can be modified to suit individual needs. This type of chamber is particularly useful for counting large ciliates.

Whilst the moat chamber system is appropriate for large-celled microorganisms the Lund counting chamber (Lund, 1959) is more appropriate for counting microalgae. Jones (1979) describes the construction of the Lund chamber in considerable detail and therefore only the minimum of constructional information is given here. The chamber comprises a $22 \times 50$ mm coverslip along the two long edges of which are cemented, with epoxy resin, two thin strips of coverslip of the same length. The coverslip is then cemented onto a microscope slide thereby forming a chamber between the two strips of coverslip. The chamber can then be filled by capillary action from a Pasteur pipette. The depth of the chamber can be varied depending on the thickness of the supporting coverslips and the volume can be calculated by weighing the chamber before and after filling with a water sample.

Trolldenier (1973) developed a simple but effective system for counting microorganisms in soil suspensions but this method is equally applicable to benthic material. The method involves marking a known area, usually a circle on a glass slide. A sample of known volume is then pipetted onto this area and allowed to dry. To facilitate even distribution of particulate matter in the sample, sodium hexa-metaphosphate should be added to give a final concentration of 0.001% w/v. The microorganisms in the dried sample are then counted by either bright-field or phase-contrast microscopy. A number of variations in this technique are available. For example, Edmonson (1974) describes the enumeration of algae by covering a known sample volume on a microscope slide with a coverslip and counting transects. Benthic ciliates may be counted in drops of known volume on microscope slides (Goulder, 1971; Finlay *et al.*, 1979). By including an internal standard of fungal spores a quantitative estimate of the microorganisms present can be obtained using these simple counting systems. The procedure is to mix known volumes of sample and standard fungal spore suspension and count the number of organisms and fungal spores per microscope field. The number of microorganisms $ml^{-1}$ of the original sample can be derived from the equation of Jones (1979):

$$N_0 = \frac{O_f \cdot [SP]V_{sp}}{SP_f \cdot V_0}$$

where $N_0$ = number of microorganisms $ml^{-1}$ of sample
$O_f$ = number of microorganisms/microscope field
$SP_f$ = number of fungal spores/microscope field
$V_0$ = volume of sample (ml)
$V_{sp}$ = volume of standard fungal suspension (ml)

## B. Counting chambers for bacteria and microalgae

A number of different counting chambers are available for the enumeration of bacteria and microalgae but of these the Improved Neubauer and Helber systems are most widely used. They all consist of a glass slide, a section of which is precisely ground to a depth of either 0.02 mm or 0.1 mm below the surface and this is surrounded by a moat. An area of 1 $mm^2$ on the platform is marked with a Thoma-type grating of engraved lines into 400 small squares (≡2500 $\mu m^2$) and the volume of each square is $5 \times 10^{-5}$ µl. The chamber is usually closed with a thick optically flat coverslip but this may be replaced with thinner coverslips when using short working distance objectives. The bacterial suspension is fixed by adding 2 to 3 drops of 40% v/v formaldehyde and mixing thoroughly. If the suspension is too dense it should be diluted to give between 40 and $200 \times 10^6$ cells $ml^{-1}$. The sample should be carefully pipetted onto the ruled area using a Finnpipette or similar piston action pipette and the coverslip lowered evenly until 'Newton rings' are seen uniformly distributed over the areas of contact. The preparation should then be examined by phase-contrast microscopy using a ×40 or ×60 objective. The average number of bacteria/square is calculated from counts made in sufficient squares (typically 100) to yield a significant total number of bacteria (e.g. 100 to 1000 and preferably >300). Counts are best made in preparations containing between 2 and 10 bacteria/square. The average count of cells/square multiplied by $20 \times 10^6$ and any dilution factor used gives the bacterial count $ml^{-1}$ in the original sample. Two further preparations of the same sample should be counted and the mean value of the three results determined. When cell numbers are $<40 \times 10^6$ bacteria $ml^{-1}$ a counting chamber with a depth of 0.1 mm should be used so as to obtain a significant count in fewer squares.

In my laboratory Helber counting chambers are routinely used since they are thinner in section than the Neubauer system and thus allow the use of short working distance objectives. All counting chambers of this type are prone to error and these have been discussed in detail by Jones (1979). The major sources of error are due to variations in the depth and hence volume of the sample in the chamber, especially when thinner flexible coverslips are employed. To avoid this problem the depth of the chamber should be measured each time the chamber is used. This can be achieved by noting the difference in micrometer readings on the fine focus control when focusing on the lower surface of the counting chamber and under surface of the coverslip (Pike *et al.*, 1972). To obtain reproducible results considerable care should be exercised when filling the chamber, and piston-type pipettes are to be preferred. Motility of the organisms also presents problems and Jones (1979) recommends that 0.5% w/v $NiSO_4$ be added to water samples to slow down protozoa, whereas 4% w/v polyvinyl alcohol or methyl cellosolve is more appropriate for bacteria. When it is not essential to differentiate viable from non-viable cells samples can be preserved in formaldehyde to give a final concentration of 1% w/v. Alternatively, cells can be stained by adding freshly filtered methylene blue (0.2 μm pore size) to give a final concentration of 0.1% w/v.

### C. Direct counting by epifluorescence microscopy

Whilst counting chambers such as those described in the foregoing section are still widely used, the application of epifluorescence microscopy is now generally acknowledged to be the best method currently available for counting aquatic microorganisms. Like many experimental techniques it is basically a simple method but there are several essential precautions that must be observed if reproducible and meaningful results are to be obtained. The method involves staining the microorganisms in the sample with a fluorochrome, collecting the cells on a membrane filter, followed by counting using epifluorescence microscopy. For detailed descriptions of the development of epifluorescence microscopy as a technique for counting microorganisms the reader should consult the excellent reviews by Jones and Simon (1975), Hobbie *et al.* (1977), Zimmerman (1977), Daley (1979) and Fry (1988). It cannot be overemphasized that all reagents used for epifluorescence microscopy should be free of bacteria and particulate matter. Fry (1988) recommends that all reagents used for epifluorescence direct counts should be prepared by filtering through 0.2-μm membrane filters and then autoclaved. As an additional precaution against contamination this author recommends the addition of formaldehyde to give a final concentration of 2% v/v, and storage of all reagents at 4°C. Samples from the reagents should be routinely checked for contamination and if present the contents should be immediately discarded and new stocks prepared.

In the early phase of development of epifluorescence direct counting methods cellulose acetate tortuous path membrane filters were commonly used. However, it is now accepted that when using 0.45-μm pore size membranes substantial underestimates of the total microbial count occur since many of the smaller bacterial cells will penetrate as much as 40–100 μm into the membrane and cannot be counted. To overcome this problem 0.2-μm pore size cellulose membranes can be used since Jones (1979) reported that bacteria rarely penetrated more than 10 μm depth in these filters. However, polycarbonate membranes with 0.2-μm pore size have now superceded cellulose acetate filters since they confer a number of advantages over tortuous path filters. They are much thinner (~10 μm thick), more robust and can therefore be rigorously backwashed, size fractionation of microbial populations is more precise since the pores are of a uniform diameter, and they have a shiny smooth surface which ensures that the cells lie in a single focal plane and are therefore easier to count. Before use polycarbonate membranes must be stained black to minimize autofluorescence when illuminated with U.V. light. The most widely used method is to stain the membranes for 5 min in Irgalan Black ($2\,g\,l^{-1}$) dissolved in membrane-filtered 2% v/v acetic acid (Hobbie *et al.*, 1977). Alternatively, the staining method of Jones and Simon (1975) can be used with dyes such as Dylon No. 8, Ebony Black or that of Fry and Davies

(1985) using Lanasyn Brilliant Black. Which staining method is chosen is a matter of personal preference as in the author's laboratory they are all equally effective. In the past problems occasionally arose with hydrophobic patches on polycarbonate membranes and if this problem should develop they should be pretreated with detergent according to the method of Hobbie *et al.* (1977), but this can result in an unacceptably intrusive background fluorescence. Zimmerman and co-workers (1978) used an ethanolic solution of Sudan black B to stain polycarbonate membranes which overcame the problem of hydrophobic patches without increasing the background fluorescence. Staining can be carried out in sterile Petri dishes and the filters should be thoroughly rinsed in membrane-filtered water prior to use. Jones (1979) recommends that the washed membranes should be placed on filter paper to absorb water and then dried over silica gel in a vacuum desiccator. Filters which have been stored in this way should be moistened with sterile water prior to use.

The experimental procedure is that the fluorochrome [acridine orange or 4′-6-diamidino-2-phenylindole (DAPI)] is added to the water sample as a $1\ g\ l^{-1}$ aqueous solution to give a final concentration of $10\ mg\ l^{-1}$ for a contact time of not more than 3 min (acridine orange) or 5 min (DAPI). The sample is then gently filtered through a 0.2-μm pore size 25-mm diameter black polycarbonate membrane. The filter unit used may be either an all-glass unit (Millipore or equivalent type) fitted to a Buchner funnel or alternatively a Luer syringe fitting filter assembly (Swin-Lok or equivalent). Jones (1979) reported that ensuring an even distribution of the microbial cells on the membrane was a more common problem with the funnel units compared with syringe fitting filters. Jones and Simon (1975) reported that a volume of 6 ml was the minimum that could be used with a 25-mm all-glass filter unit. This can readily be achieved by diluting the sample with 0.2-μm membrane-filtered water. Whilst disc filter systems are less prone to an uneven distribution of microorganisms due to their greater effective filtration area they do suffer from the dead volume of liquid trapped in the filter holder. The individual investigator must therefore choose the filtration system most appropriate to his needs. Following filtration of the sample the membrane should be rinsed with a volume of filtered water equivalent to the total volume of the sample. Rinsing removes the excess stain and reduces background fluorescence. The membrane should be left on the filtration unit until all the surface water has been removed before placing on top of a small drop of non-fluorescent immersion oil (Fractoil or equivalent) on a microscope slide and a further drop added on top. A coverslip is then gently lowered onto the membrane so that the mountant evenly covers the surface. Wynn-Williams (1985) recommends the use of the photofading retardant Citifluor as mountant since it enhances colour contrast of the fluorescing cells and reduces fading of the image for up to 1 h.

A wide range of epifluorescence microscopes is now available. The brilliance of the image will depend upon the light source, the filter combinations selected, the numerical aperture of the objective, type of membrane used and the fluorochrome and its concentration. Fry (1988) recommends the use of both high magnification and high numerical aperture objectives since this combination provides the brightest image and highest resolution. Acridine orange requires excitation by blue light with a wavelength of 470–475 nm whereas DAPI requires ultraviolet light with a wavelength of 355–365 nm. As a consequence fluorite lenses must be used with DAPI-stained preparations whereas with acridine orange glass or fluorite lenses may be used. In the author's experience the expense in buying the best quality objectives and eyepieces that are available is fully justified if epifluorescence counting is to be a routine method. Acridine orange-stained bacteria fluoresce green whilst particulate matter stains either red, orange or yellow whereas with DAPI the cells stain blueish-white and debris appears yellow. DAPI gives a substantially brighter fluorescence and does not fade as rapidly as acridine orange. However, DAPI cannot be used with sediments as it is difficult to discriminate microbial cells from particulate matter. Fry (1988) recommends that counting should be performed using a $10 \times 10$ graticule in a focussing eyepiece. The number of squares that need to be counted is dependent upon the number of cells present on the membrane. Because of the problems associated with fading of the image it is better to count a large number of fields of view with few bacteria present than large numbers in a fewer number of fields. As with the Helber counting chambers the number of cells counted should be between 100 and 1000 and preferably be greater than 300. The number of bacteria present in the original sample is calculated from the mean cell number/graticule area used, the volume of sample filtered and the effective filtration area which is given by the following formula (Jones, 1979)

$$n = \frac{YAd}{av}$$

where $n$ = number of cells $ml^{-1}$ of samples
$Y$ = mean cell count/graticule area used
$A$ = effective area of filtration
$d$ = dilution factor (if any)
$a$ = graticule area
$v$ = volume of sample filtered

A recent development in epifluorescence microscopy is the linking of image analysing systems to an epifluorescent microscope. At the time of writing this review three commercially available systems have been used and these are the Quantimet 800 (Cambridge Instruments), Artek 810 (Artek Systems) and the IBAS system (Zeiss Kontron). The use of image analysers coupled to epifluorescence microscopy is still at an early stage of development and readers not familiar with such systems should consult articles by Fry and Davies (1985), Bjornsen (1986), Sieracki *et al.* (1985) and Fry (1988).

Before the advent of epifluorescence microscopy, membrane filters were stained with erythrosine and examined by bright-field microscopy (Jannasch, 1958; Rodina, 1972; Sorokin and Overbeck, 1972). The technique is to stain the dried membrane filter by placing on an absorbent pad saturated with a 5% w/v solution of erythrosine in 5% w/v phenol and allowing the stain to be taken up for 1–24 h. The membrane filter is then decolourized by placing on an absorbent pad soaked in distilled water and the process repeated until the filter is a pale pink colour. The membrane is dried at room temperature over silica gel and then placed over a drop of immersion oil on a slide. A second drop of immersion oil is then added and a coverslip gently lowered over the membrane surface to give an even oil film. When viewed by bright-field microscopy the cells appear pink on a colourless background. This method works well with large-celled bacteria but Jones (1979) reported difficulties in differentiating small cells from detritus.

Epiphytic bacteria on the leaves of submerged aquatic plants can be readily counted using the staining method developed by Hossell and Baker (1979). Whole leaves are stained in filtered aniline blue dissolved in phenol–acetic acid (20 ml glacial acetic acid, 3.75 g phenol and 80 ml distilled water) for 2 min. The stained leaves are mounted on microscope slides and examined by bright-field microscopy. Bacteria appear as dark blue cells and can be counted using a similar technique to that described for epifluorescence microscopy. Hossell and Baker (1979) reported that the leaves of some submerged macrophytes, e.g. *Ranunculus*, absorbed the aniline blue stain. These workers recommended that leaves from these plants should be pretreated with 40% v/v formaldehyde for 30 min, washed in distilled water followed by staining with Eosin yellow ($0.2\ g\ l^{-1}$) for 60 min and finally rinsed in distilled water prior to staining with aniline blue.

Fry and Zia (1982) developed a novel yet simple and effective system for counting unstained planktonic bacteria using an agar slide technique. The method involves the coating of thin microscope slides (0.8–1.0 mm thick), marked on one surface with a 1-cm square, with 0.5 ml of 1% w/v purified agar which is then allowed to gel. An aliquot volume (typically 10 µl) of the sample to be counted, which may need a prior concentration step by centrifugation if cell numbers are low, is then carefully spread over the 1-cm square on the agar using a platinum loop. The sample is allowed to dry for 5–10 min, covered by a coverslip and observed by phase-contrast microscopy using a high-quality objective, e.g. a Zeiss Neofluar ×100 objective, N.A. 1.3 or equivalent. Counting can be performed in an analogous manner to that previously described for epifluorescence microscopy.

## D. Counting microorganisms using a Coulter counter

The Coulter counter (Coulter Electronics Inc., Hialeah, Florida 33010) has been used to count planktonic microorganisms in fresh (Mulligan and Kingsbury, 1968; Evans and McGill, 1969) and marine waters (Sheldon and Parsons, 1967; Sieburth, 1979). The theory of the Coulter counter is discussed in depth by Kubitschek (1969) and only the essential details will be described here. The basis of the method is that a suspension of cells in a conductive fluid is passed through a small orifice that has electrodes suspended on either side. A current is passed between the two electrodes and the resistance caused by the orifice is measured. The passage of a cell through the aperture results in an increase in resistance by the displacement of its own volume of electrolyte from that region. The voltage pulse generated by each cell is amplified and registered electronically, giving a count of the number of cells flowing through the aperture. Coulter counting methods are only applicable *in situ* for microorganisms in water samples containing little detrital matter. Usually cells can be counted in the medium in which they are growing which is particularly useful when cell concentrations are low. At low cell concentrations the accuracy of counting is dependent on low background counts and this is particularly true for bacteria which, because of their small size, require a higher detector sensitivity and thus higher electronic noise levels. As a consequence previously undetected foreign particles are counted thus resulting in greater errors in counting. At high cell densities small concentrations of foreign particles can be ignored and accuracy is limited by coincidence counts that arise when more than one cell occupies the counting orifice. This problem can be overcome by dilution of cell suspensions. Care should be taken if saline is used as diluent since it can lead to shrinkage of the cells due to osmotic stress. Changes in cell volume can be prevented by using phosphate buffers or by the addition of formaldehyde to give a final volume of 0.4–4% v/v. Curds *et al.* (1978) consider that for counting protozoa the Coulter counter is the ideal system since it gives accurate and reproducible results. The Coulter counter has also been successfully used to determine growth rates of phytoplankton by determining the increase in cell numbers over a specified time period (Parsons, 1973).

### E. Fluorescent antibody methods for enumerating specific microbial populations

Fluorescent antibody techniques have now been developed which enable specific groups of microorganisms to be counted *in situ*. The technique is widely used in medical microbiology and pathology but has only slowly been introduced into microbial ecology. In principle the fluorescent antibody approach is simple and elegant yet to obtain meaningful data rigorous checks and controls must be included (Schmidt, 1973). The initial stage in the process is to isolate in pure culture the microorganism under investigation and then either whole cells or some subcellular component are injected into a rabbit or similar laboratory animal to raise antibodies. The antibodies are then purified and tagged with an appropriate fluorochrome such as fluorescein isothiocyanate (FITC). The antibody when added to a sample then combines with the antigenic microorganisms, if present, and causes them to fluoresce. The number of fluorescing cells can then be counted using an epifluorescence microscope fitted with a focussing eyepiece and graticule. A serious limitation of the fluorescent antibody technique when counting soil bacterial is the non-specific adsorption of the fluorescent antibody to soil colloids and films. This may be overcome by pretreatment with a gelatin–rhodamine isothiocyanate (RhITC) conjugate to suppress non-specific antibody adsorption as well as serving as a counterstain (Bohlool and Schmidt, 1973). FITC-stained organisms fluoresce green against a red background when this technique is used. The fluorescent antibody technique has been used to enumerate nitrifying bacteria (Belser and Schmidt, 1978), sulphate-reducing bacteria (Smith, 1982) and methanogenic bacteria (Ward and Frea, 1979). This method is particularly useful in autecological studies in that it enables enumeration and identification of the microorganisms in a single step. The major drawback is that it is species specific and in order to broaden the spectrum of this technique it would be necessary to raise antibodies against a range of similar organisms and couple this with FITC as fluorochrome to produce a polyvalent stain. This is clearly a lengthy and tedious process and could be of little value unless all the organisms of the physiological type under study could be isolated in pure culture which is unlikely.

### F. Micro-ELISA technique for enumerating bacteria

A novel recent development has been the application of a micro-ELISA technique for enumerating sulphate-reducing bacteria in natural environments by Bobowski and Nedwell (1978). Antisera were raised by injecting 2-ml volumes of cell-free extracts of different species and strains of sulphate-reducing bacteria in phosphate-buffered saline into rabbits. When the antibody titre reached a maximum the rabbits were bled. A polyvalent antiserum mixture was then prepared in order to maximize detection of as many strains or species of sulphate-reducing bacteria as possible. Bobowski and Nedwell (1978) adapted the ELISA procedure of Engvall and Perlmann (1971) for microtitre plates and used goat anti-rabbit IgG conjugated to horse radish peroxidase as the detection system. Using this technique these workers were able to detect numbers of sulphate-reducing bacteria as low as $10^3$ cells per membrane filter. Comparison of the ELISA technique with conventional most probable number (MPN) methods showed that counts obtained by the former were greater by an order of magnitude. This the authors attributed to the broader specificity of the antiserum cocktail which estimated a more representative part of the total sulphate-reducing bacteria population and secondly to the fact that unlike the MPN method which only detects viable cells the ELISA method also detects cellular antigens of non-viable cells.

## III. Viable counting methods

Despite the problems asociated with viable counting methods they are still widely used in microbiology to enumerate the microbial populations present in a sample. As discussed earlier in this review, interpretation of viable count data should be undertaken with great care since growth and recovery of all the desired organisms rarely occurs and thus *sensuo stricto* they can never be equated with true total counts. The fundamental basis of viable counting methods is that they depend on the ability of the microorganisms present in a sample to grow either in liquid culture media, or media gelled with agar or some other gelling agent or on the surface of a membrane filter. A vast array of growth media have been described in the literature and it is clearly an impossible task to review all the types available. Manufacturers, such as Difco and Oxoid, produce a wide range of selective media and further details can be obtained from the relevant manuals (Oxoid, 1979; Difco, 1984). In addition the reader should consult reviews by Skerman (1967), Collins (1969), Norris and Ribbons (1970), Rodina (1972), Starr *et al.* (1981), Herbert (1982) and Schneider and Rheinheimer (1988). Since new media formulations are continually being described it is important that the investigator selects carefully the media and methods that are most appropriate to his needs and these should be fully tested before a detailed study is undertaken.

### A. Plate counting methods

Plate count methods can be subdivided into (1) the pour plate method and (2) the surface spread method.

#### 1. *Pour plate method*

In the pour plate method a measured volume of the sample (1 ml) is mixed with the appropriate molten agar medium (10 ml) in a sterile Petri dish. After the medium has gelled and incubation at a suitable temperature the number of colonies which develop are counted. As a compromise between sampling and overcrowding errors, counts should be made on plates which contain between 50 and 500 colonies (ideally 200–300). When cell densities in the sample are greater than 500 viable organisms $ml^{-1}$ then serial decimal dilutions of the samples should be made. Nine-ml volumes of an appropriate diluent, e.g. phosphate-buffered saline (Cruickshank *et al.*, 1975), one-quarter strength Ringer's solution (Harrigan and McCance, 1966; Oxoid, 1979) or physiological saline (0.9% w/v NaCl) should be prepared in test tubes or Universal bottles and sterilized by autoclaving at 121°C for 15 min. A 1-ml volume of the sample should be removed with a sterile pipette and transferred to the first tube/bottle of diluent (fill and empty the pipette with the sample to be diluted several times before withdrawing a 1-ml volume). Mix the first dilution by gentle shaking (inversion if using Universal bottles) or on a vortex mixer. With a fresh 1-ml sterile pipette remove a 1-ml volume from the first dilution and carefully transfer to the second dilution tube/bottle. Repeat the procedure until all the dilutions are prepared. Commencing with the greatest dilution aseptically pipette a 0.1-ml volume of each dilution into each of three sterile Petri dishes. Pour into each dish 10 ml of molten agar medium which has been cooled to 45°C. Immediately, mix the contents of each plate by gently moving each plate in a series of circular movements in different directions, ensuring that the contents do not spill out. Allow the agar to gel and incubate the inverted plates at an appropriate growth temperature, e.g. 20°C or 37°C. Count the number of colonies on the plates which have between 50 and 500 colonies. Determine the mean value for the three replicates and multiply by the dilution factor used to obtain the viable count $ml^{-1}$ of the original sample.

Pour plates are rarely used for ecological work since many microorganisms are killed by the thermal shock of being exposed to molten agar.

### 2. *Spread plate method*

In the surface plate method the molten agar medium (45°C) is poured into a series of sterile Petri dishes and allowed to gel. To avoid spreading of the colonies over the agar surface the plates should be dried for at least 2 h at 37°C prior to use. Serial tenfold dilutions of the original sample are prepared as for the pour plate method. Commencing at the highest dilution 0.1-ml volumes from each dilution are pipetted onto the surface of each of three plates. The sample is then spread over the surface of the plate using an ethanol-flamed glass spreader. The plates are then incubated at the appropriate temperature for the organisms under investigation and the number of colonies which develop are counted as for the pour plate method.

### 3. *Miles and Misra method*

Alternatively, the method of Miles and Misra (1938) can be used to determine the microbial populations in samples. The method consists of placing small drops (20 μl) from serial dilutions using a calibrated dropping pipette on the surface of poured agar plates and then counting the colonies which develop after incubation. Agar plates should be prepared and dried as described for the spread plate method. Serial tenfold dilutions of the original sample are then prepared and 20-μl drops from each dilution are allowed to fall from a height of 25 mm onto the medium where they spread over an area of 15–20 mm diameter. For six dilutions, six plates are used for each count, one drop from each dilution being placed on each plate. Allow the drops to be absorbed before inverting and incubating the plates. Counts are made in the drop areas showing the largest number of individual colonies without confluent growth (20 or more). The mean of the six counts gives the viable count/20 μl of the dilution used and should be multiplied to give the number of colonies $ml^{-1}$. The calibrated dropping pipettes are constructed by drawing out a Pasteur pipette such that a capillary is produced which has an internal diameter of 0.91 mm (0.036 inch). This can be confirmed by inserting a standard wire gauge of 0.91 mm diameter into the capillary. Pipettes prepared in this way produce a drop volume of approximately 20 μl when the drops are delivered at a rate of more than one per second with the pipette held vertically. The advantage of the Miles and Misra technique over the spread plate method is its speed and economical use of agar plates. Bousfield *et al.* (1973) introduced a minor modification to the Miles and Misra method by substituting piston pipettes (Finnpipette or equivalent) for the calibrated dropping pipette. These workers found a good correlation using this technique when compared with conventional spread plate methods.

### B. Membrane filter methods

This method is based on the use of highly porous cellulose acetate membranes which are manufactured in different pore sizes. The pore structure of these membranes enables large volumes of water to pass through under pressure but prevents the passage of any bacteria present in the sample. Bacteria are retained on the membrane surface which is then brought into contact with an absorbent pad saturated with suitable nutrients. Membrane filters are manufactured by the Millipore Filter Corporation, Bedford, Massachusetts, USA; Gelman Sciences Inc., Ann Arbor, Michigan 48106, USA; Oxoid Ltd, Basingstoke, Hampshire, UK; Sartorius Gmbh, Gottingen, West Germany; and Nucleopore, Pleasanton, California 94566, USA. Membrane filters with a pore size of 0.43–0.47 μm are usually recommended for bacteriological work. However, in our experience bacteria from river water samples are able to pass through these membranes and we now routinely use 0.2-μm pore size membranes to overcome this problem. It is therefore important that preliminary experiments be undertaken to determine the required pore size of the membranes to be used for a particular application. Jones (1979) reported that he obtained higher viable counts using Millipore HC filters which have a graded pore structure (2.5 μm pore-opening decreasing to 0.7 μm). Thus, whilst these membranes have a larger nominal pore size they may in practice be more efficient in recovering viable microorganisms from water samples. The principal advantage of membrane filtration over conventional plating methods is speed and that large volumes of sample can be processed when cell numbers are low since the concentration and inoculation stages are completed in a single step.

The general procedure is as follows:

(a) Sterilization of equipment. The filter units (47 mm diameter) and Buchner flask are sterilized at 121°C for 15 min. The membrane filters can either be purchased presterilized or alternatively can be sterilized by placing individual filters between sheets of filter paper in glass Petri dishes followed by autoclaving at 121°C for 15 min. Absorbent pads can also be purchased commercially presterilized or again treated like the membranes.

(b) Preparation of absorbent pads. Each sterile pad is placed into a sterile Petri dish and sufficient sterile medium (~1 ml) is added aseptically to achieve saturation without excess liquid being visible.

(c) Sample filtration and incubation of membrane. The sterile filter holder and Buchner flask are attached to a vacuum pump. Using sterile flat-face forceps a sterile membrane is placed, grid side up, on the sintered glass disc of the filter unit. The sample to be filtered is placed into the funnel and gentle vacuum applied. After all the sample has passed through the filter it should be rinsed with sterile one-quarter strength Ringer's solution (2 × 20 ml volumes). The vacuum source is then disconnected and the membrane aseptically removed and transferred to an incubating dish containing the absorbent pad saturated with the appropriate liquid growth medium. It is important that the membrane be placed on the absorbent pad with a rolling action to avoid trapping air bubbles. The membrane is then incubated at an appropriate temperature for the microorganisms under investigation. To avoid drying out of the absorbent pads we normally place the incubation dishes (lids uppermost) in polythene bags to minimize evaporation losses. After incubation the number of colonies are counted and the viable count calculated $ml^{-1}$ of original sample. If the original sample is too concentrated it can be diluted with 0.2-μm membrane-filtered one-quarter strength Ringer's solution.

When there is insufficient contrast between the colonies and the filter surface the membranes can be stained with 0.01% w/v malachite green for 3–10 s. The excess stain is drained off and colonies will appear unstained on a green background. Alternatively, the membrane can be stained with 0.01% w/v methylene blue for 1 min. The membrane is then rinsed with water and the colonies will stain blue on a white background.

### C. Most probable number (MPN) methods

The most probable number (MPN) method provides an estimate of the number of viable microorganisms present in a sample which are capable of growth in a given liquid growth medium. The MPN method is commonly used with highly selective media and a tenfold dilution series of tubes. Serial tenfold dilutions of the sample are prepared and aliquot volumes (1 ml) from each dilution are then inoculated into 9-ml volumes of the growth media. Usually five replicates of each dilution are prepared. The inoculated medium is incubated and the number of positive tubes from each dilution recorded. Since most MPN methods use liquid media a positive tube is one showing turbidity. The degree of dilution of the sample required and the number of positive tubes at different dilutions are then used to determine the MPN of organisms by reference to probability tables (Taylor, 1962; Meynell and Meynell, 1970). The MPN method is particularly useful when it is required to determine low cell numbers in water samples (<1 viable organism $ml^{-1}$). At these low cell densities the plate count method is no longer practicable. Counts obtained using MPN methods are frequently higher than plate count methods and this is usually attributed to the fact that the microorganisms in the sample are not disrupted into individual cells and the interactions which occur *in situ* are therefore continued in the growth medium. The MPN method suffers from a major disadvantage in that it has a very large sampling error. In order to reduce this error, the number of replicates, or alternatively the dilution factor, can be increased. Jones (1979) gives a detailed analysis of the effect of these factors on the confidence limits obtained and the reader is well advised to consult this review before using MPN methods.

The MPN method has been extensively used to determine coliform and faecal streptococci in water samples (Ministry of Housing and Local Government, 1982; American Public Health Association, 1976). It has also been used widely to enumerate bacterial populations in natural environments, e.g. nitrifying bacteria (Alexander, 1965), nitrate-respiring bacteria (Focht and Joseph, 1972; Macfarlane and Herbert, 1984) and sulphate-reducing bacteria (Schroder and Van Es, 1980; Battersby *et al.*, 1985). Battersby *et al.* (1985) elegantly exploited the MPN method to enumerate sulphate-reducing bacteria even when other microbial populations were present. Sulphate-reducing bacteria are strict anaerobes and Battersby *et al.* (1985) used the heterotrophic bacterium *Pseudomonas putida* as a biological reductant to remove residual oxygen from the growth medium. The growth medium comprised: $FeSO_4 \cdot 7H_2O$ 0.3 g, $NH_4Cl$ 0.3 g, $KH_2PO_4$ 0.2 g, bacto-casamino acids 1.0 g, Na ascorbate 1.0 g, Na lactate (70% w/v) 3.6 ml, Resazurin (BDH Chemicals, Poole) 1 tablet, sea water 1 litre. The medium pH is adjusted to 7.5 with 2 M NaOH, dispensed into 25-ml bottles and autoclaved at 121°C for 15 min. After autoclaving the tops of the bottles are tightly sealed. Serial tenfold dilutions of the sample are then made in the medium and 10-, 1-, and 0.1-ml aliquot volumes added to five bottles for each volume used. Aliquot volumes (0.1 ml) of a concentrated cell suspension of *Ps. putida* are then added to each bottle. The bottles are then filled with sterile medium and incubated at 30°C. Bottles showing a black coloration are scored as positive and the MPN of sulphate-reducing bacteria in the original sample calculated from McCrady's probability tables (Cruickshank *et al.*, 1975).

Microsystem MPN methods have been developed based on microtitre plates. Curtis *et al.* (1975) described a MPN microsystem for enumerating nitrifying bacteria in water and sediment. Serial tenfold dilutions were prepared in tubes and then aliquot volumes transferred to a 25-compartment 'repli-plate' for incubation and assay using the MPN tables of

Taylor (1962). Darbyshire *et al.* (1974) developed a micromethod for estimating bacterial and protozoan populations using a 96-well microtitre plate system (12 dilutions, eight replicates per dilution). In the absence of MPN tables, these workers described a statistical method based on the Poisson distribution, to calculate the MPN of bacteria and protozoa present in the original sample. Rowe *et al.* (1977) refined the microtitre plate method for nitrifying bacteria and since it was based on doubling dilutions it has greater accuracy than methods based on decimal dilutions. This method is rapid, occupies little space and requires small volumes of reagents. In the original method Rowe *et al.* (1977) used calibrated loops to prepare the serial twofold dilutions but we routinely use piston pipettes for this operation (Macfarlane and Herbert, 1984).

The basis of the method is that 50 μl of medium is dispensed into each of the 8 × 12 wells of a sterile microtitre plate (Nunclon or equivalent). Aliquot volumes of the sample (50 μl) are pipetted into each of the first eight wells. Twofold dilutions of each sample are then prepared with the result that 12 twofold dilutions are prepared with eight replicates of each dilution. The plates are then incubated at the appropriate temperature for the organisms under study and each well is scored by adding an indicator (0.2 g diphenylamine in 100 ml concentrated $H_2SO$ in the case of ammonia-oxidizing microorganisms). A blue colour reaction indicates a positive reaction. We have used the method for estimating populations of ammonia-oxidizing, nitrite-oxidizing and nitrate-respiring bacteria in estuarine sediments and found it to be a rapid, reliable and accurate method for determining the viable populations of these organisms in marine and estuarine sediments (Macfarlane and Herbert, 1984). To avoid evaporation of water when incubated for prolonged periods the plates can be sealed around the edges with parafilm or adhesive tape. The MPN for the populations present in the original sample can be determined from the MPN tables of Rowe *et al.* (1977).

Since programmable calculators and personal computers are now widely available MPN values for a given microbial population can be calculated using simple computer programs [see Koch (1981), Clarke and Owens (1983) for details]. The use of these programs enables the investigator to calculate the estimate of MPN and confidence limits for any combination of dilution levels, sample volumes and number of replicates.

### D. Enumeration of anaerobes

#### *1. Plate count method*

For the successful enumeration of strict anaerobes on the surface of agar plates it is essential that all oxygen is excluded. This can be most readily and conveniently achieved by pouring and incubating inoculated plates in an anaerobic cabinet (Forma Scientific or equivalent). These are expensive items of equipment to purchase and maintain. A simpler technique is to incubate inoculated plates in an anaerobic jar which has been flushed out with oxygen-free nitrogen (OFN) or 10% $CO_2$/90% $H_2$. When OFN is used the palladium catalyst is ineffective since no $H_2$ is present in the gas phase to remove final traces of $O_2$ which may have entered. The anaerobic jar should be leak-tight and maintained under a positive gas pressure.

The following procedure describes the use of the BTL Anaerobic Jar (Baird and Tatlock Ltd, Chadwell Heath, Essex) which employs the cold catalyst palladium to remove the final traces of oxygen. An indicator solution with the following composition is prepared, dispensed in a Durham tube (25 mm long × 7 mm diameter) and sterilized by heating at 121°C for 15 min: D-glucose 5 g, sodium mercaptoacetate 1 g, 1% w/v methylene blue 0.2 ml, agar 1 g, distilled water 1 litre. The Durham tube is then connected to the side arm of the anaerobic jar with butyl rubber tubing. Reduced methylene blue is colourless whereas in the presence of oxygen the indicator turns blue. The Petri dishes to be incubated should be placed in the anaerobic jar with the medium uppermost and the lid clamped into position. A small quantity of silicon grease should be used to lightly coat the butyl rubber 'O' seal in the lid. One of the two taps in the lid of the jar is attached to a vacuum pump fitted with a vacuum gauge and the jar evacuated to a pressure of 50–100 mm mercury and the tap then closed. The second tap is then connected to a gas mixture consisting of 10% $CO_2$/90% $H_2$ which is allowed to fill the evacuated jar at a constant pressure (0.15–0.2 bar). When the vessel is full the tap is closed and the anaerobic jar evacuated and refilled twice more before incubating at a suitable growth temperature. The most convenient source of $H_2$ and $CO_2$ is a commercially available gas mixture. The indicator should decolorize after a few hours and remain colourless until the jar is opened at the end of the experimental period. Failure of the indicator is indicative of leak in the side arm, gasket failure, faulty taps or poisoning of the catalyst. A number of precautions should be observed when using anaerobic jars. These include preventing the lid of the Petri plate from forming a seal with the dish due to the accumulation of condensation by inserting a paper clip between the two units. Moisture can also adversely affect the activity of the palladium catalyst and this can be removed by heating the whole capsule in a bunsen flame.

Commercially available but more expensive alternatives to the Baird and Tatlock anaerobic jar are the BBL 'Gas pak' system (Baltimore Biological Laboratories) and the Oxoid anaerobic plus system (Oxoid Ltd, Basingstoke). In these systems gas generation is achieved by adding 10 ml of distilled water to an aluminium foil sachet containing sodium borohydride, cobalt chloride, citric acid and sodium bicarbonate. After addition of water the sachet is placed in the jar which is made of clear polycarbonate. These systems require no pumps or cylinders of compressed air and are therefore ideal for field use. The only requirements are for a fresh gas generation pack each time the jar is set up. The gas generation packs can also be used in Baird and Tatlock jars thereby removing the requirement for a vacuum pump and compressed gas mixtures.

### 2. *Hungate roll-tube technique*

The Hungate roll-tube technique depends for its success on the ability of the investigator to exclude air from within the growth chamber by a stream of oxygen-free gas when the vessel is opened and by a butyl rubber stopper when it is closed. The basis of the method has been reviewed in considerable detail by Hungate (1969) and Hungate and Macy (1973) and will not be discussed further in this review. A commercial system is available from Astell Scientific (London). The growth medium is removed from the autoclave as soon as the pressure is down to atmospheric and then held at 45°C under a stream of sterile oxygen-free nitrogen. A suitable system for removing final traces of oxygen from OFN nitrogen is to pass it through an electrically heated column of copper (Hungate, 1969). The medium is then dispensed in 4.5-ml volumes into 25-ml roll-tube bottles (Astell Scientific, London) using either the Hungate technique (Hungate, 1969) or an anaerobic dispenser such as that developed by Herbert and Gilbert (1984). The roll tubes are then sealed with oxygen-impermeable butyl rubber stoppers (Astell Scientific, London) as the gas-flushing needle is carefully withdrawn from each tube. The tubes are then returned to the water bath. A wise precaution is to incorporate 0.0001% w/v resazurin into the growth medium so that any oxidized tube (pink) can be rejected. An aliquot volume (0.5 ml) of the sample is then injected into the first bottle using a sterile syringe, the contents mixed and from this bottle a 0.5-ml volume is withdrawn with a further sterile syringe and inoculated into a second bottle. This procedure is repeated until a decimal dilution series has been prepared. The tubes are then placed on a water-cooled roller (Astell Scientific, London) which coats the inside of each bottle with a thin layer of agar. The bottles are then incubated at an appropriate temperature until discrete colonies are visible which can then be counted using a low-power binocular dissecting microscope. The advantage of this method is that even exacting anaerobes such as methanogens can be enumerated using this technique which requires the minimum of specialized equipment.

### *3. Agar shake method*

The agar shake method has the advantage of being simpler than the Hungate technique and yet is extremely effective for the enumeration and isolation of anaerobes (van Niel, 1931). The growth medium is removed from the autoclave when it has come down to atmospheric pressure and rapidly dispensed in 9-ml volumes in 16 × 150 mm sterile test tubes and the tubes closed with butyl rubber seals (Astell Scientific, London). The tubes are maintained in a water bath at 42–44°C to keep the medium molten. The first tube in the series is inoculated with either 1 ml or 1 g of sample, and after mixing by inversion, a dilution series is prepared by aseptically transferring 1.0 ml from one tube to the next using piston-type pipettes fitted with sterile disposable tips. After mixing, the tubes are rapidly cooled in a bath of cold water and sealed with a plug of paraffin wax (1 part wax:3 parts liquid paraffin). Finally, the tubes are sealed with butyl rubber bungs. As an additional safeguard, to prevent the ingress of oxygen, absorbent cotton wool plugs can be inserted into the tubes and a few crystals of pyrogallol plus 0.5 ml 2 M sodium carbonate added prior to sealing each tube with a rubber bung. We use this method routinely to enumerate viable populations of purple and green photosynthetic bacteria as well as sulphate reducers in estuarine and marine sediments.

## IV. Indirect methods

Indirect methods involve the estimation of a specific component of a microbial cell which is present in direct proportion to the amount of cell biomass. A range of methods has been developed but only those which are commonly used or have potential applications will be discussed.

### A. Measurement of adenosine triphosphate

This is the most commonly used method of biomass estimation and was originally proposed by Holm-Hansen and Booth (1966). ATP fulfils several of the criteria required for a biomass indicator since it is easy to assay, is found as a relatively constant proportion of all living cells and is not present in detritus or dead cells. The basis of the ATP assay is its reaction with luciferin in the presence of the enzyme luciferase.

$$\text{Reduced luciferin} + \text{ATP} + O_2 \xrightarrow{\text{luciferase}} \text{oxidized luciferin} + \text{AMP} + \text{PP} + \text{light}$$

One photon of light is produced for each molecule of ATP hydrolysed and this can be measured using a photometer. In recent years the original method of Holm-Hansen and Booth (1966) has undergone major modifications to improve sensitivity and minimize interference due to the use of crude luciferase preparations and for details the comprehensive review by Karl (1980) should be consulted. It must be stressed that the ATP assay is a measure of total biomass and attempts to fractionate microbial populations into different size classes have met with little success although larger organisms such as 300-μm plankton can be removed by filtering through a 200–400-μm nylon mesh (Fry, 1988). Riemann (1978) has successfully used 1-μm pore size polycarbonate filters to separate algae from bacteria but as Fry (1988) correctly points out this approach is not suitable for waters in which a substantial proportion of the bacterial population is attached to particles. A crucial element of the ATP assay is the necessity to process samples rapidly after collection since the physiological state and hence the ATP levels of the cells change substantially after sampling. Aliquot volumes of the sample are filtered through 0.2-μm pore size cellulose acetate membrane filters, which have been previously rinsed in boiling 20 mM Tris or HEPES buffer, using the minimum of vacuum. Sediment samples require diluting 100–1000-fold to reduce chemical interference at a later stage during the assay. Following filtration the membrane filter is removed and transferred into 5 ml of boiling 20 mM Tris or HEPES buffer (pH 7.8) for 5 min. It is important that the temperature of the buffer is >92°C to ensure that the ATPases released from the bacteria are totally inactivated. The ATP extract is carefully decanted into a 10-ml measuring cylinder and the membrane rinsed with a further 3 ml of boiling buffer for 3 min. The rinsing buffer is then added to the ATP extract and the volume measured. The extract can then be assayed immediately or stored at −20°C for up to 1 month. Crude firefly luciferase (Sigma) is reconstituted with 20 mM Tris buffer (pH 7.8) and stored at 4°C for 12–15 h. The reconstituted enzyme is then centrifuged at 10 000 *g* for 10 min at 4°C to remove particulate matter and the supernatant allowed to equilibrate for 1 h in the dark prior to use. The limit of detection using crude luciferase is ~1–10 ng ATP $ml^{-1}$. If increased sensitivity is required (1–10 pg ATP $ml^{-1}$) then purified luciferase can be used (Deming *et al.*, 1979) or alternatively additional luciferin can be added to the crude enzyme preparation (Jones, 1979). The assay is performed by injecting 0.1 ml of the enzyme preparation (Hamilton or equivalent microlitre syringe) into 0.2 ml of sample in a photometer tube. Following the addition of the enzyme preparation, a flash of light is emitted which has a peak duration of ~3 s, followed by a slow decay. The flash of light is measured using a photometer such as a LKB Luminometer (LKB Instruments, Uppsala). The peak height of the light flash is proportional to the ATP concentration in the sample. Background light emission levels can be determined by injecting the enzyme preparation into 20 mM Tris or HEPES buffer. The concentration of ATP in the sample can be determined by interpolation from a standard ATP calibration curve. Internal standards of pure ATP should also be included at the extraction stage to check on and correct for extraction efficiency in the presence of planktonic and benthic material.

Deming *et al.* (1979) described a simple and yet rapid ATP extraction procedure using 0.1 M $HNO_3$ which in a single step efficiently extracts ATP and inactivates ATPases. Neutralization of the sample is not required provided the luciferase extract is made up in 0.25 M Tris buffer (pH 8.2) and the sample diluted twentyfold with distilled water prior to injection of 0.1-ml aliquot volumes into the photometer. The major advantage of this method is its simplicity and speed. Some ATP assay methods have dispensed with the filtration step (Jones and Simon, 1977) since it has been reported to reduce the amount of ATP measured. However, as Fry (1988) pertinently points out it is essential to separate the microbial cells in the water sample, since dissolved ATP in the water can contribute significantly to the total amount present. For this reason Fry (1988) recommends that use of the direct injection method be discontinued.

Considerable debate has occurred regarding the appropriate conversion factor to convert ATP concentrations to microbial biomass. Data presented by Holm-Hansen and Booth (1966) suggested that a cellular carbon:ATP ratio was ~250:1 for a wide variety of microbial cells although the range was from 140:1 to 2000:1. The general concensus is that a ratio of 250:1 is acceptable in most situations. However, it must be stressed that this value was obtained using the Tris buffer extraction method and any change to this protocol may result in an altered ratio and hence require recalibration.

## B. Lipopolysaccharides

Since all Gram-negative bacteria contain lipopolysaccharide (LPS) it is a good biomass indicator in environments dominated by these bacteria, e.g. freshwater and marine environments, but is of less value in terrestrial habitats where Gram-positive bacteria form a substantial proportion of the resident microflora. The basis of the method is the reaction of amoebocytes from the horseshoe crab (*Limulus polyphemus*) with the lipopolysaccharide of Gram-negative bacteria to yield a turbid suspension. The amount of complex formed is estimated either turbidometrically (Coates, 1977; Watson *et al.*, 1977) or colorimetrically (Nakamura *et al.*, 1977; Maeda and Taga, 1979). The turbidometric assay is more sensitive and Jones (1979) reported that detection limits as low as 500 bacteria per sample could be achieved.

Whilst the turbidometric LPS method is simple it is essential that LPS-free glassware and reagents are used. All glassware should therefore be dry-heat sterilized (160°C for 6 h). The *Limulus* amoebocyte lysate is reconstituted in 5 ml of LPS-free sea water, which has been previously autoclaved (1 l of sea water plus 100 g fresh activated charcoal) and filtered through a 0.2-μm pore size membrane filter. An aliquot volume (1 ml) of the sample is then added to 0.2 ml of reconstituted amoebocyte lysate, mixed gently and then incubated at 37°C for 1 h. The absorbance is then read at 360 nm in 1-cm pathlength cuvettes. The concentration of LPS in the sample is determined by comparing the absorbance with those obtained for standard concentrations of *E. coli* endotoxin (*E. coli* 0127:B8). Since there is considerable variation in the amoebocyte lysate it is essential that calibration curves be prepared each day. To obtain cell biomass the concentration of LPS is multiplied by a conversion factor of 6.35.

## C. Muramic acid

With the exception of the Archaebacteria all prokaryote cells contain muramic acid as a component of the cell wall. In Gram-positive bacteria the peptidoglycan accounts for 40–90% of the wall dry weight whereas in Gram-negative species the proportion is only 5–10%. Cyanobacteria may contain up to 500 times more muramic acid than other bacteria (King and White, 1977) and hence this method is only applicable in samples from which cyanobacteria are absent, if bacterial, as opposed to prokaryote biomass is to be determined. Several methods have been described for estimating muramic acid in sediments and those of Moriarty (1977) and King and White (1977) will be described. Both methods involve the conversion of muramic acid to lactate followed by enzymatic (Moriarty, 1977) or chemical analysis (King and White, 1977) to determine the lactate concentration.

The method of Moriarty (1977) involves the hydrolysis of a sediment or bacterial suspension (250–500 mg or 10–25 mg, respectively) with 1 ml of 3 M HCl in a sealed glass tube at 100°C for 6 h. 0.2 ml of 0.5 M $Na_2HPO_4$ is then added, followed by the addition of 5 M NaOH to neutralize the acid. The neutralized hydrolysate is then clarified by filtration or centrifugation at 6000 *g* for 10 min. The pH of the sample is increased to pH 12.5 with 1 M NaOH and incubated at 35°C for 2 h before readjusting the pH to 8.0 with 1 M HCl followed by centrifugation at 6000 *g* for 10 min. Lactate present in the sample is then assayed by removing a 100-$\mu$l aliquot volume (containing 10–200 ng D-lactate) and adding to this 0.5 ml of enzyme mixture (5 ml 0.3 M glutamate buffer, pH 9.0; 100 μl NAD 33 mg $ml^{-1}$; 10 μl lactate dehydrogenase 5 mg $ml^{-1}$; 5 μl glutamate amino transferase 10 mg $ml^{-1}$). After mixing and incubation at 30°C for 15 min the tubes are transferred to an ice bath. An aliquot volume of this sample (50–100 μl) is then added to a phosphate-luciferase preparation and the light emission measured. The bacterial luciferase solution is prepared by mixing a volume of bacterial luciferase (1.5 mg $ml^{-1}$) with an equivalent volume of bovine serum albumin in phosphate buffer. The phosphate preparation contains 20 ml 0.1 M phosphate buffer, pH 7.5, 1.4 ml 2-mercaptomethanol, 1.0 ml flavin mononucleotide 0.5 mg $ml^{-1}$ and 1 ml ethanol-saturated dodecylaldehyde. The dodecylaldehyde preparation should be centrifuged prior to use. When the sample contains 10–50 ng D-lactate, 100 μl of luciferase solution is added to 2 ml of phosphate preparation whilst for 50–200 mg D-lactate 50 μl is sufficient. The light emission values for the samples are compared with those obtained for standard solutions of D-lactate.

King and White (1977) developed a chemical method to estimate lactate produced from muramic acid after acid hydrolysis. A sediment sample (1–5 g) is hydrolysed by heating in a sealed glass tube with 6 M HCl for 4–5 h at 105°C. The contents of the tube are quantitatively washed through a coarse sintered-glass filter and dried *in vacuo* at 50°C. The residue is then dissolved in acetone: 0.1 M HCl (9:1 v/v) and separated by cellulose TLC with four cycles of solvent (acetone:glacial acetic acid:water 9:1:1 v/v). Material with an $R_f$ range of 0.35–0.7 is carefully eluted with methanol:water (7:3 v/v) and reduced in volume under a stream of $N_2$. A 1-ml volume is transferred to a glass tube and 0.5 ml 1 M NaOH added. The hydrolysis mixture is incubated at 38°C for 30 min before 10 ml concentrated $H_2SO_4$ is added and the tube sealed. The tube is then heated in a boiling water bath for 5 min. After cooling, 0.1 ml of 4% w/v $CaSO_4$ followed by 0.2 ml of 1.5% w/v *p*-hydroxydiphenyl in 95% v/v ethanol are added, the contents mixed, restoppered and incubated at 30°C for 30 min. The absorbance is read in 1-cm pathlength cuvettes at 560 nm and compared with muramic acid standards. The conversion of muramic acid determinations to estimates of bacterial biomass is difficult when dealing with mixed microbial communities for reasons discussed earlier. Moriarty (1977) is of the opinion that if the majority of benthic bacteria are Gram negative then an average value of 12 μg muramic acid per mg of bacterial carbon is applicable.

In an analogous manner to the use of muramic acid to determine bacterial biomass, fungal biomass can be determined by measuring the chitin content of the sample (Swift, 1973a,b). The chitin assay as described by Swift (1973a) is based on the acid hydrolysis of the polymer followed by estimation of glucosamine. The sample (~100 mg) is mixed with 5 ml 6 M HCl in a sealed glass tube for 3 h and then at 80°C for 16 h. The hydrolysate is then filtered and evaporated to dryness at 75°C. The residue (glucosamine content 4–14 g $ml^{-1}$) is dissolved in distilled water and applied to a short column of Dowex 50 cation exchange resin (200–400 mesh) and the glucosamine eluted with 2 M HCl. The eluant is neutralized with NaOH and the glucosamine condensed with acetylacetone (freshly prepared 2% in 1 M $Na_2CO_3$) at 90°C for 45 min. The product is mixed with 1 ml Ehrlich's reagent (2.67% w/v *p*-diaminobenzaldehyde in 50 ml ethanol plus 50 ml concentrated HCl). The pink colour which develops is measured in 1-cm pathlength cuvettes at 530 nm. A standard curve is prepared using D-glucosamine hydrochloride and the glucosamine concentrations of the samples can be obtained by comparison of the absorbance values obtained. The method is relatively insensitive and it is not applicable to the litter layer of forest soils where contamination by insect chitin and free hexosamine will occur. The conversion factor from glucosamine to fungal biomass also presents many problems since the chitin content varies with age and physiological state of the mycelium. At present the biomass conversion factor is only really applicable to monoculture situations and not to field studies.

Chitin can also be determined by enzymatic conversion to *N*-acetylglucosamine. Samples containing fungal mycelium are autoclaved to prevent further growth and then mixed with commercial chitinase (Sigma Ltd) or an enzyme preparation prepared from puff-balls (*Lycoperdon* spp.) made up in 0.08 M sodium acetate–acetic acid buffer, pH 4.8 (Willoughby, 1978). The sample is then incubated at 33°C until a constant *N*-acetylglucosamine figure is achieved (typically 10 days). Aliquot volumes (1 ml) of the sample are mixed with 0.3 ml of saturated sodium borate and placed in a boiling water bath for 7 min and then cooled. The contents are then made up to 10 ml volume with glacial acetic acid and 1 ml of Ehrlich's reagent added (2 g *p*-amino benzaldehyde in glacial acetic acid plus 5 ml concentrated HCl). The pink colour is allowed to develop for 45 min before reading the absorbance at 540 min in 1-cm pathlength cuvettes. A standard calibration curve should be prepared using *N*-acetylglucosamine (0–10 μg $ml^{-1}$). The major disadvantage of this assay is the long incubation period and the probability of interference from a bacterial-derived *N*-acetylglucosamine.

### D. Photosynthetic pigments

Photopigments such as chlorophyll *a* and bacteriochlorophylls have been extensively used to estimate biomass in planktonic and benthic microbial communities. Since only algae and cyanobacteria contain chlorophyll *a* this provides a convenient indicator of oxygenic photosynthesis. Similarily, bacteriochlorophylls are restricted to the phototrophic purple, green and purple non-sulphur bacteria and hence constitute a useful index of anoxygenic photosynthetic activity.

Whilst extraction of chlorophyll *a* and bacteriochlorophyll with organic solvents such as acetone or methanol are satisfactory and well established for pure laboratory cultures they result in large errors when applied to samples obtained from natural environments. Samples from freshwater and marine environments often contain a mixture of different photopigments and their degradation products. The absorption spectra of these photopigments partially overlap and the degree of overlap varies dependent on whether they are measured *in vivo* or in various organic solvents. Recently, Stal *et al.* (1984) have developed a simple yet effective method for simultaneously measuring chlorophyll *a* and bacteriochlorophylls in samples from natural environments. The basis of the method is shown in Fig. 1.The sample (benthic material, water sample or pelleted culture) is extracted by first adding $MgCO_3$ followed by an appropriate volume of methanol (5–10 ml). The extraction is performed in the dark at room temperature for 2 h. The samples are then centrifuged and the supernatants carefully removed before re-extracting the pellet with a further volume of methanol for 2 h. The extracts are then pooled and 3 ml of 0.05% w/v NaCl added to each 10-ml volume of methanol extract. In order to increase the efficiency of extraction the samples may be sonicated or ground with pure quartz sand. Stal *et al.* (1984) reported that two extractions with methanol removed more than 99% of the extractable photopigments from the samples. The addition of NaCl improves the separation of the methanol and hexane phases. After vigorous mixing following the addition of *n*-hexane the extracts are centrifuged at 300 *g*. The *n*-hexane phase (upper layer) is carefully removed with a Pasteur pipette and divided into two equal parts. The absorbance of one fraction is then measured in 1-cm pathlength cuvettes against a pigment-free blank at 600 nm (chlorophyll *a)* and 768 nm (bacteriochlorophyll *a*). The other fraction is acidified by the addition of 100 μl 5 M HCl/5 ml hexane to give the corresponding phaeophytins. After shaking vigorously the extracts are dried with anhydrous sodium sulphate and the absorbance read in 1 cm cuvettes at 660 nm (chlorophyll *a*) and 768 nm (bacteriochlorophyll *a*). The methanol extract is used to determine the absorbance of bacteriochlorophyll *c* (668 nm), bacteriochlorophyll *d* (657 nm) and bacteriochlorophyll *e* (659 nm). The concentration of chlorophyll (bacteriochlorophyll) can be obtained by modifying the equation of Whitney and Darley (1979).

$$\text{Chlorophyll (g l solvent}^{-1}) = \frac{K \times (E_n - E_a)}{A}$$

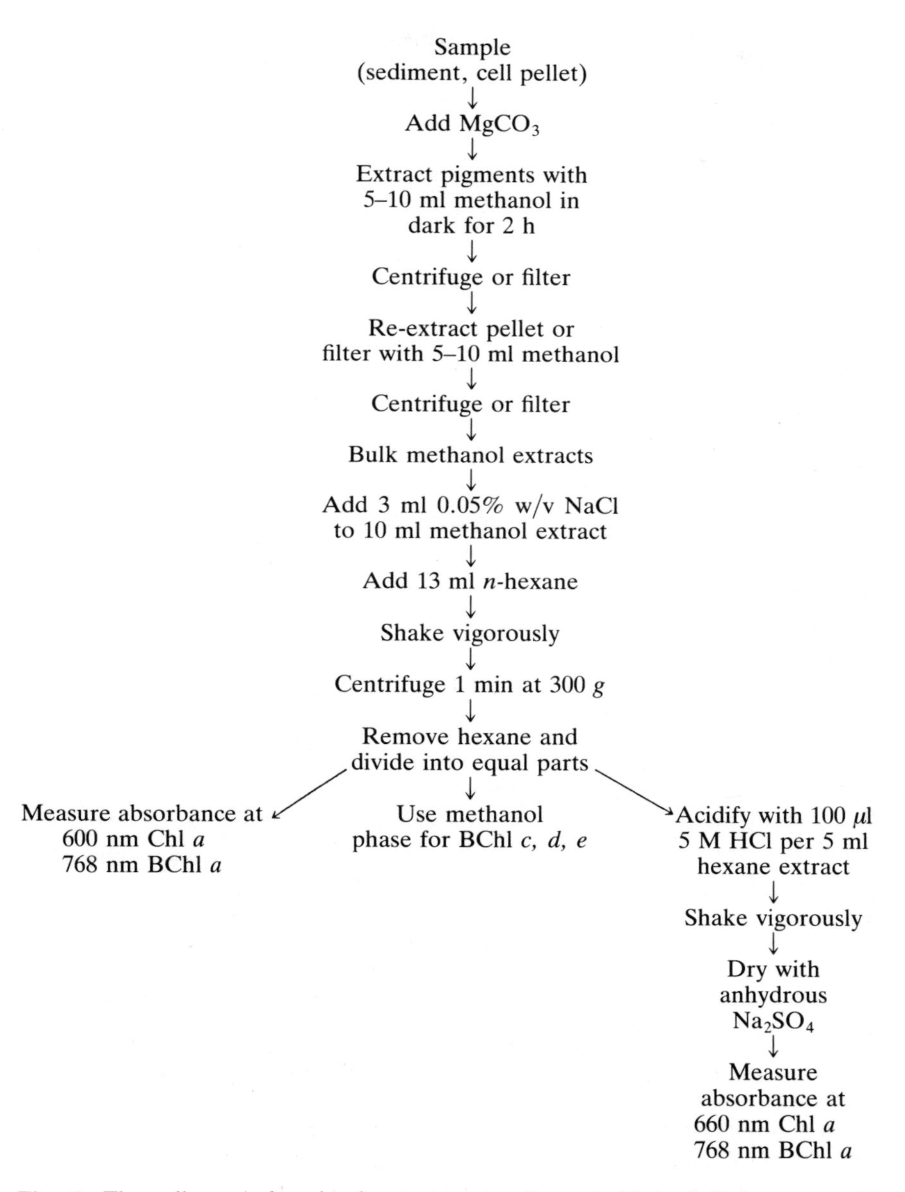


**Fig. 1.** Flow diagram for simultaneous extraction of chlorophyll from natural microbial populations.

**Table I**

Absorption maxima (nm) of photopigments in different organic solvents. Figures in parentheses are absorption coefficients ($g^{-1}$ 1 cm). Data derived from Stal *et al.* (1984)

| Photopigment | Methanol | Solvent<br>Methanol/NaCl | *n*-Hexane |
|---|---|---|---|
| Chlorophyll *a* | 665 (74.5) | 665 (89.4) | 660 |
| (119.9) | | | |
| Phaeophytin *a* | 654 | 657 | 668 |
| Bacteriochlorophyll *a* | 770 (84.1) | 770 (88.1) | 768 |
| (149.5) | | | |
| Bacteriophaeophytin *a* | 747 | 851 | 752 |
| Bacteriochlorophyll *c* | 668 (86.0) | 670 (90.2) | 664 |
| (51.2) | | | |
| Bacteriophaeophytin *c* | 662 | 663 | 670 |
| Bacteriochlorophyll *d* | 657 (82.3) | 658 (86.4) | 650 |
| (89.9) | | | |
| Bacteriophaeophytin *d* | 650 | 651 | 661 |
| Bacteriochlorophyll *e* | 659 (82.3) | 659 (94.8) | 645 |
| (76.3) | | | |
| Bacteriophaeophytin *e* | 656 | 660 | 658 |

$$\text{Phaeophytin (g l solvent}^{-1}) = \frac{K \times F \times (R \times E_a - E_n)}{A}$$

where $A$ = absorption coefficient (g $cm^{-1}$); $E_n$ is the absorbance at the wavelength of maximal absorption for chlorophyll (bacteriochlorophyll); $E_a$ is identical to $E_n$ except it is determined for acidified extracts; $K$ is related to the decrease in absorbance measured following acidification ($K$ for chlorophyll $a = 1.32$ and for bacteriochlorophyll $a = 1.34$); $R = E_n/E_a$ as determined for solutions of pure photopigments ($R$ for chlorophyll $a$ =4.14 and for bacteriochlorophyll $a = 3.91$) and $F$ is a factor to correct for the lower molecular weight of chlorophyll *a* and bacteriochlorophyll *a* ($F$=0.974 for chlorophyll *a* and 0.973 for bacteriochlorophyll *a*). Data presented in Table I show the absorption maxima of photopigments and absorption coefficients $g^{-1}$ 1 cm) in different organic solvents. The amount of chlorophyll (bacteriochlorophyll) present in phototrophic microorganisms is dependent upon their physiological state and hence it is usual practice to express results in terms of photopigment concentration rather than algal or bacterial biomass. Notwithstanding these difficulties a value of 40:1 carbon:chlorophyll *a* content in algal cells has been proposed but conversion factors such as these should be treated with considerable caution.

The method described above simultaneously determines chlorophyll *a* and bacteriochlorophylls in samples from natural environments. Simpler extraction methods are available for chlorophyll *a* which may be more appropriate for laboratory cultures or habitats where algae are dominant and for further information the reader should consult articles by Marker (1972), Jones (1979), Parsons *et al.* (1984) and Fry (1988).

Chlorophyll *a* can also be measured by fluorimetry and this is sufficiently sensitive to be applied directly to water samples. This technique is ideal for continuous monitoring of phytoplankton populations using a flow cell fluorimeter. For further details of this technique the reader should consult articles by Jones (1979) and Parsons *et al.* (1984).

## IV. Miscellaneous methods

### A. DNA as a biomass indicator

DNA, since it is universally present in all living cells, may be used for comparative purposes to provide information on the total biomass present in particular environments. Since it is universally present in all living cells it cannot be used to differentiate bacterial biomass from algal biomass and therefore can only be used as a general biomass indicator in conjunction with other methods.

The method is essentially that of Holm-Hansen *et al.* (1968). A volume of sample which is sufficient for the assay is filtered through a membrane filter. The filter is removed, mixing with 5 ml of acetone in a tube and allowed to stand for 20 min. Following centrifugation at 1500 *g* for 5 min the supernatant is carefully removed. Extraction is repeated with 3-ml aliquot volumes of acetone until all the membrane filter material is removed. The residual precipitate is further extracted with 3 ml 90% w/v acetone, 5 ml 10% w/v trichloroacetic acid at 5°C and finally with 2 × 5 ml 95% v/v ethanol. To the remaining cell pellet is added 1 drop of 0.5% v/v solution of Triton X-100 and the sample dried at 20°C. Diaminobenzoic acid dihydrochloride (DABA · 2HCl) reagent is prepared by dissolving 0.4 g high purity (>99%) DABA · 2HCl in 1 ml of distilled water. 0.1 ml volumes of this reagent are then added to the dried pellets, the tubes covered with parafilm are heated in a water bath at 60°C for 1 h. After 30 min heating the tubes are removed and mixed in a vortex mixer and returned to the water bath. Upon completion of the heating period the tubes are removed and 5 ml 0.6 M perchloric acid added. The contents are thoroughly mixed and then centrifuged at 2000 *g* for 5 min. The concentration of the DABA·2HCl–DNA complex is measured with a fluorimeter with an excitation maximum at 420 nm and maximum emission measured at 520 nm. Fluorescence yield can be optimized by altering the proportion of DABA·2HCl and DNA in the reaction mixture and the individual researcher should investigate different mixtures to optimize fluorescence yield. DNA standards (range 0.04–4 $\mu g\ ml^{-1}$) are prepared by dissolving highly polymerized DNA in 1 M $NH_4OH$. When samples are too concentrated they should be diluted in a solution of 0.1 ml DABA·2HCl (0.4 $ml^{-1}$) in 5 ml 0.6 M perchloric acid.

### B. Extractable lipid phosphate

Phospholipids are major components of the plasma membrane and the content remains remarkably constant under a wide range of growth conditions. White and co-workers (White *et al.*, 1979) proposed that extractable lipid phosphate could be used as an indicator of microbial biomass in marine and estuarine sediments since it correlates accurately with other measures of microbial biomass such as extractable ATP and muramic acid. The method involves the extraction of lipid phosphate from sediments with organic solvents followed by phosphate analysis.

A known weight of sediment (5–15 g) is added to a 250-ml separating funnel and extracted with a solvent mixture of 30 ml 5 mM phosphate buffer (pH 7.4), 75 ml anhydrous methanol and 37.5 ml chloroform. The suspension is vigorously shaken for 2 h on a wrist-action shaker. A further 37.5 ml chloroform and 37.5 ml water are then added and the mixture allowed to separate for 24 h. The upper phase of methanol and water is then carefully removed and the chloroform layer carefully decanted and filtered through a Whatman 2E filter paper. The chloroform is collected in a measuring cylinder. A rotary evaporator is used to remove the chloroform and the extracted lipid is stored under nitrogen at −20°C.

The phosphate content of the dried lipid sample is determined by digestion with 1.5 ml 35% v/v perchloric acid and heated at 180–200°C for 2 h. Considerable care should be taken to ensure that all the chloroform has been removed to avoid the possibility of an explosion occurring during digestion. The digestion mixture should not go to dryness. After cooling 2.4 ml molybdate reagent (4.4 g ammonium molybdate plus 14 ml concentrated $H_2SO_4\ l^{-1}$ distilled water) is added, mixed and 2.4 ml 1-amino-2 naphthol-4 sulphonic acid reagent added (30 g sodium bisulphite, 2 g sodium sulphite, 0.5 g 1-amino-2 napthol-4-sulphonic acid dissolved in 200 ml distilled water, stored at 4°C in the dark and diluted 1:12 with water before use). The mixture is heated in a boiling water bath for 7–10 min, cooled and the absorbance measured at 830 nm. A standard phosphate calibration curve should be prepared from 1–10 μg phosphate (1.097 g $KH_2PO_4$ in 250 ml distilled water, diluted 1:10 to give a working standard of 10 μg phosphate $ml^{-1}$). At present this method is used to indicate the presence of biomass since conversion factors to estimate cellular carbon levels are not available.

### C. Enumeration of methanogens using coenzymes $F_{350}$ and $F_{420}$

This method is based upon the unique fluorescent properties of methanogenic bacteria which are due to the presence of coenzymes $F_{420}$ and $F_{350}$. Factor $F_{530}$ (methanofuran) and Factor $F_{420}$ have been used to count methanogenic bacteria directly as endosymbionts in sapropelic protozoa by fluorescence microscopy (Van Bruggen *et al.*, 1983, 1985). The method involves selecting excitation and barrier filters for fluorescence microscopy which are specific for coenzymes $F_{350}$ and $F_{420}$. Doddema and Vogels (1978) recommend BG3 and KP 425 excitation filters and K 460 barrier filter for fluorescence at 420 nm whereas for excitation at 350 nm a VGI excitation filter together with a K 430 barrier filter should be used. Methanogens fluoresce greenish-yellow at 420 nm and bluish-white at 350 nm. Fluorescence at 420 nm is normally stronger than at 350 nm and both are subject to fading. The method has therefore considerable potential for counting methanogens in natural environments but has less application in biomass determinations since the cellular $F_{420}$ content is very variable.

## References

Alexander, M. (1965). *In* "Methods of Soil Analysis. Part 2 Chemical and Microbiological Properties" (C. A. Black, Ed.), pp. 1467–1472, American Society of Agronomy, Wisconsin.

American Public Health Association (1976). "Standard Methods of the Examination of Water and Wastewater", 14th Edn, American Public Health Association, Washington.

Battersby, N. A., Stewart, D. J. and Sharma, A. P. (1985). *J. Appl. Bacteriol.* **58,** 425–429.

Belser, L. W. and Schmidt, E. L. (1978). *Appl. Environ. Microbiol.* **36,** 584–588.

Bjornsen, P. K. (1986). *Appl. Environ. Microb.* **51,** 1199–1204.

Bobowski, S. and Nedwell, D. B. (1987). *In* "Industrial Microbiological Testing" (J. W. Hopton and E. C. Hill, Eds), Technical Series No. 23, pp. 171–179, Blackwell Scientific Publications, Oxford.

Bohlool, B. B. and Schmidt, E. L. (1973). *Bull. Ecol. Res. Comm. (Stockholm)* **17,** 336–338.

Bousfield, I. J., Smith, G. L. and Trueman, R. W. (1973). *J. Appl. Bacteriol.* **36,** 297–299.

Clarke, K. R. and Owens, N. J. P. (1983). *J. Microbiol. Methods* **1,** 133–137.
Coates, D. A. (1977). *J. Appl. Bacteriol.* **42,** 445–449.
Collins, V. G. (1969). *In* "Methods in Microbiology", Vol. 3B (J. R. Norris and D. W. Ribbons, Eds), pp. 1–52, Academic Press, London.
Collins, V. G., Jones, J. G., Hendrie, M. S., Shewan, J. M., Wynn-Williams, D. W. and Rhodes, M. M. E. (1973). *In* "Sampling—Microbiological Monitoring of Environments" (R. G. Board and D. W. Lovelock, Eds), Soc. Appl. Bact. Tech. Ser. No. 7, pp. 77–110, Academic Press, London.
Cruickshank, R., Duguid, J. P., Marmion, R. H. and Swain, H. A. (1975). "Medical Microbiology", 12th Edn, Churchill Livingstone, Edinburgh.
Curds, C. R., Roberts, D. M. and Wu Chih-Hua (1978). *In* "Techniques for the Study of Mixed Populations" (D. W. Lovelock and R. Davies, Eds), Soc. Appl. Bact. Tech. Ser. No. 11, pp. 165–177, Academic Press, London.
Curtis, E. J., Durrant, C. K. and Harman, M. M. J. (1975). *Water Res.* **9,** 255–268.
Dale, D. G. (1974). *Limnol. Oceanog.* **19,** 509–518.
Daley, R. J. (1979). *In* "Native Aquatic Bacterial: Enumeration, Activity and Ecology", ASTM STP695 (J. W. Costerton and R. R. Colwell, Eds), American Society for Testing and Materials, Philadelphia.
Darbyshire, J. F., Wheatley, M. P., Graves, M. P. and Inkson, R. H. E. (1974). *Rev. Ecol. Biol. Sol.* **11,** 465–475.
Deming, J. W., Picciolo, G. L. and Chappelle, E. W. (1979). *In* "Native Aquatic Bacteria: Enumeration, Activity and Ecology", ASTM STP695 (J. W. Costerton and R. R. Colwell, Eds), pp. 88–98, American Society for Testing and Materials, Philadelphia.
Difco Manual (1984). 10th Edn, Difco Laboratories, Detroit.
Doddema, H. J. and Vogels, G. D. (1978). *Appl. Environ. Microbiol.* **36,** 752–754.
Dodson, A. N. and Thomas, W. H. (1964). *Limnol. Oceanogr.* **9,** 455–456.
Edmonson, W. T. (1974). *In* "A Manual on Methods for Measuring Primary Production in Aquatic Environments" (R. A. Vollenwieder, Ed.), IBP Handbook No. 12, 2nd Edn, pp. 14–16, Blackwell Scientific Publications, Oxford.
Elliot, J. M. (1977). *Sci. Publ. Freshwater Biol. Assoc.* No. 25, 2nd Edn.
Engvall, E. and Perlmann, P. (1971). *Immunochemistry* **8,** 871–874.
Evans, J. H. and McGill, S. M. (1969). *Hydrobiologia* **35,** 401–419.
Finlay, B. J., Laybourn, J. and Strachan, I. (1979). *Oecologia* **39,** 375–377.
Focht, D. D. and Joseph, H. (1972). *Proc. Soil Sci. Soc. Am.* **37,** 698–699.
Fry, J. C. (1988). *In* "Methods in Aquatic Microbiology" (B. Austin, Ed.), pp. 27–72, John Wiley and Sons, Chichester.
Fry, J. C. and Davies, A. R. (1985). *J. Appl. Bacteriol.* **58,** 105–112.
Fry, J. C. and Zia, T. (1982). *J. Gen. Microbiol.* **128,** 2841–2850.
Fry, J. C., Goulder, R. and Rimes, C. (1985). *J. Appl. Bacteriol.* **58,** 113–115.
Gayford, C. G. and Richards, J. P. (1970). *J. Appl. Bacteriol.* **33,** 342–350.
Gibson, G. R., Parkes, R. J. and Herbert, R. A. (1987). *J. Microbiol. Methods* **7,** 201–210.
Goulder, R. (1971). *Oikos* **22,** 199–203.
Goulder, R. (1987). *Letts. Appl. Microbiol.* **4,** 29–32.
Goulder, R., Bent, E. J. and Boak, A. C. (1981). *In* "Feeding and Survival Strategies of Estuarine Bacteria (N. V. Jones and W. J. Wolff, Eds), pp. 1–15, Plenum Press, New York.

Harrigan, W. F. and McCance, M. E. (1966). "Laboratory Methods in Microbiology", Academic Press, London.
Herbert, R. A. (1982) In "Methods in Microbial Ecology" (R. G. Burns and J. H. Slater, eds) pp. 1–22, Blackwell Scientific Publications, Oxford.
Herbert, R. A. (1988). *In* "Methods in Aquatic Bacteriology" (B. Austin, Ed.), pp. 3–25, John Wiley and Sons, Chichester,
Herbert, B. N. and Gilbert, P. D. (1984). *In* "Microbiological Methods of Environmental Biotechnology" (J. M. Grainger and J. M. Lynch, Eds), pp. 235–257, Academic Press, London.
Hobbie, J. E., Daley, R. J. and Jasper, S. (1977). *Appl. Environ. Microbiol.* **33,** 1225–1228.
Holm-Hansen, O. and Booth, C. R. (1966). *Linmol. Oceanogr.* **11,** 510–519.
Holm-Hansen, O., Sutcliffe, W. H. and Sharpe, J. (1968). *Limnol. Oceanogr.* **13,** 507–514.
Holme, N. A. and McIntyre, A. D. (1971). "Methods for the Study of the Marine Benthos", I.B.P. Handbook Vol. 16, Blackwells, Oxford.
Hossell, J. C. and Baker, J. H. (1979). *J. Appl. Bacteriol.* **46,** 63–72.
Hungate, R. E. (1969). *In* "Methods in Microbiology", Vol. 3B (J. R. Norris and D. W. Ribbons, Eds), pp. 117–132. Academic Press, London.
Hungate, R. E. and Macy, J. (1973). *Bull. Ecol. Res. Comm. (Stockholm)* **17,** 123–125.
Jannasch, H. W. (1958). *J. Gen. Microbiol.* **18,** 609–620.
Jones, J. G. (1979). "A Guide to Methods of Estimating Microbial Numbers and Biomass in Fresh Water", Freshwater Biological Association, Windermere.
Jones, J. G. and Simon, B. M. (1975). *J. Appl. Bacteriol.* **39**, 317–329
Jones, J. G. and Simon B. M. (1977). *Freshwater Biol.* **7,** 253–260.
Karl, D. M. (1980). *Microbiol. Rev.* **44,** 739–796.
King, J. D. and White, D. C. (1977). *Appl. Environ. Microbiol.* **33,** 777–783.
Kirchman, D. L., Sigda, J., Kapulscinski, R. and Mitchell, R. (1982). *Appl. Environ. Microbiol.* **44,** 376–382.
Koch, A. L. (1981). *In* "Manual of Methods for General Bacteriology" (P. Gerhardt, Ed.), American Society for Microbiology, Washington.
Kogure, K., Simidu, V. and Taga, N. (1979). *Can. J. Microbiol.* **25,** 415–420.
Kubitschek, H. E. (1969). *In* "Methods in Microbiology", Vol. 1 (J. R. Norris and D. W. Ribbons, Eds), pp. 593–610, Academic Press, London.
Lund, J. W. G. (1959). *Linmol. Oceanogr.* **4,** 57–65.
Lund, J. W. G., Kipling, C. and LeCren, E. D. (1958). *Hydrobiologia* **11,** 143–170.
Macfarlane, G. T. and Herbert, R. A. (1984) *J. Gen. Microbiol.* **130,** 2301–2308.
Mackereth, F. J. H., Heron, J. and Talling, J. F. (1978). Sci. Publ. Freshwater Biol. Assoc. No. 36.
Maeda, M. and Taga, N. (1979). *J. Appl. Bacteriol.* **47,** 175–182.
Marker, M. F. (1972). *Freshwater Biol.* **2,** 361–385.
Meynell, G. G. and Meynell, E. (1970). "Theory and Practice in Experimental Bacteriology", 2nd Edn, University Press, Cambridge.
Miles, A. A. and Misra, S. S. (1938). *J. Hyg. Camb.* **38,** 732–749.
Ministry of Housing and Local Government (1982). "Report on Public Health and Medical Subjects No. 71 Bacteriological Examination of Water Supplies", HMSO, London.
Moriarty, D. J. W. (1977). *Oecologia* **26,** 317–323.

Mulligan, H. F. and Kingsbury, J. M. (1968). *Limnol. Oceanogr.* **13,** 499–506.
Nakamura, S., Morita, T., Iwanaga, S., Niwa, M. and Takahashi, K. (1977). *J. Biochem.* **81,** 1567–1569.
Norris, J. R. and Ribbons, D. W. (1970). "Methods in Microbiology", Vol. 3B, Academic Press, London.
Oxoid Manual (1979) 4th Edn, Oxoid Ltd, London.
Parsons, T. R. (1973). *In* "Handbook of Physiological Methods, Culture Methods and Growth Measurements", pp. 345–358, Cambridge University Press, Cambridge.
Parsons, T. R., Maita, Y. and Lalli, C. (1984). "A Manual of Chemical and Biological Methods for Seawater Analysis", Pergamon Press, Oxford.
Pike, E. B., Carrington, E. G. and Ashburner, P. A. (1972). *J. Inst. Water. Pollut. Control* **6,** 1–23.
Plummer, D. H., Owens, N. J. P. and Herbert, R. A. (1987). *Cont. Shelf. Res.* **7,** 1429–1433.
Riemann, B. (1978). *Vatten* **3,** 187–194.
Rodina, A. G. (1972). "Methods in Aquatic Microbiology", University Park Press, Baltimore.
Rowe, R., Todd, R. and Waide, J. (1977). *Appl. Environ. Microbiol.* **33,** 675–680.
Schmidt, E. L. (1973). *Bull. Ecol. Res. Comm. (Stockholm)* **17,** 67–76.
Schneider, J. and Rheinheimer, G. (1988). *In* "Methods in Aquatic Bacteriology" (B. Austin, Ed.), pp. 73–94, John Wiley and Sons, Chichester.
Schroder, H. G. J. and Van Es, F. B. (1980). *Neth. J. Sea Res.* **14,** 268–287.
Sheldon, R. W. and Parsons, T. R. (1967). "A Practical Manual on the Use of the Coulter Counter in Marine Science", Coulter Electronics (Canada) Ltd.
Sieburth, J. M. (1979). "Sea Microbes", Oxford University Press, New York.
Sieracki, M. E., Johnson, P. W. and Sieburth, J. M. (1985). *Appl. Environ. Microbiol.* **49,** 799–810.
Skerman, V. B. D. (1967). "A Guide to the Identification of the Genera of Bacteria", 2nd Edn, Williams & Wilkins, Baltimore.
Smith, A. D. (1982). *Arch. Microbiol.* **133,** 118–121.
Snedecor, G. W. and Cochran, W. G. (1967). "Statistical Methods", 6th Edn, State University Press, Iowa.
Sokal, R. R. and Rohlf, F. J. (1969). "Biometry", W. H. Freeman & Co., San Francisco.
Sorokin, Y. I. (1972). *In* "Techniques for the Assessment of Microbial Productivity in Aquatic Environments" (R. A. Vollenwieder, Ed.), pp. 128–146, Blackwell Scientific Publications, Oxford.
Sorokin, Y. I. and Overbeck, J. (1972). *In* "Techniques of the Assessment of Microbial Production and Decomposition in Freshwater" (Y. I. Sorokin and H. Kadota, Eds.), Blackwell Scientific Publications, Oxford.
Stal, L. J., van Gemerden, H. and Krumbein, W. E. (1984). *J. Microbiol. Methods* **2,** 295–306.
Starr, M. P., Stolp, H., Turper, H. G., Balows, A. and Schlegal, H. G. (1981). "The Prokaryotes: A Handbook on Habitats, Isolation and Identification of Bacteria", Vols 1 and 2, Springer-Verlag, Berlin.
Swift, M. J. (1973a). *Soil Biol. Biochem.* **5,** 321–322.
Swift, M. J. (1973b). *Bull. Ecol. Res. Comm. (Stockholm)* **17,** 323–328.
Taylor, J. (1962). *J. Appl. Bacteriol.* **25,** 54–76.
Trolldenier, G. (1973). *Bull. Ecol. Res. Comm. (Stockholm)* **17,** 53–59.

Van Bruggen, J. J. A., Stum, C. K. and Vogels, G. D. (1983). *Arch. Microbiol.* **136,** 89–95.

Van Bruggen, J. J. A., Stum, C. K., Zwart, K. B. and Vogels, G. D. (1985). *FEMS Microbiol. Ecol.* **31,** 187–192.

Van Niel, C. B. (1931). *Arch. Mikrobiol.* **3,** 1–112.

Vollenwieder, R. A. (1974). "A Manual for Measuring Primary Productivity in Aquatic Environments", IBP Handbook No. 12, 2nd Edn, Blackwell Scientific Publications, Oxford.

Ward, T. E. and Frea, J. E. (1979). *In* "Methodology for Biomass Determinations and Microbial Activities in Sediments" (C. D. Litchfield and P. L. Seyfried, Eds), pp. 75–86, American Society for Testing and Materials, Philadelphia.

Watson, S. W., Novitsky, T. J., Quinby, H. L. and Valis, F. W. (1977). *Appl. Environ. Microbiol.* **33,** 940–946.

White, D. C., Davies, W. M., Nickels, J. A., King, J. D. and Bobbie, R. J. (1979). *Oecologia* **40,** 51–62.

Whitney, D. E. and Darley, W. M. (1979). *Limnol. Oceanogr.* **24,** 183–186.

Williams, S. T. and Gray, T. R. G. (1973). *In* "Sampling—Microbiological Monitoring of Environments" (R. G. Board and D. W. Lovelock, Eds), pp. 111–121, Academic Press, London and New York.

Willoughby, L. G. (1978). *In* "Techniques for the Study of Mixed Populations" (D. W. Lovelock and R. Davies, Eds), Soc. Appl. Bacteriol. Tech. Ser. No. 11, pp. 31–50, Academic Press, London.

Woelkerling, W. J., Kowal, R. R. and Gough, S. B. (1976). *Hydrobiologica* **58,** 95–107.

Wynn-Williams, D. D. (1985). *Soil Biol. Biochem.* **17,** 739–746.

Zimmerman, R. (1977). *In* "Microbial Ecology of a Brackish Water Environment" (G. Rheinheimer, Ed.), pp. 103–120, Springer-Verlag, New York.

Zimmerman, R., Iturriaga, R. and Becker-Birck, J. (1978). *Appl. Environ. Microbiol.* **36,** 926–935.

# 2
# Direct Methods and Biomass Estimation

JOHN C. FRY

*School of Pure and Applied Biology, University of Wales College of Cardiff, P.O. Box 915, Cardiff CF1 3TL, UK*



## I. Introduction

Direct techniques in microbial ecology are by definition observational and because microorganisms are very small they all use microscopy. This type of approach can be used either qualitatively or quantitatively. As most research relies heavily on quantitative data only this aspect of direct methods will be considered here. This chapter will also concentrate on methods for bacteria because methods for other microorganisms are rather different and have been covered adequately elsewhere (Vollenweider, 1974; Jones, 1979; Burns and Slater, 1982; Cairns, 1982). It was in the early 1970s that epifluorescence microscopy first started to be used in aquatic (Francisco *et al.*, 1973) and terrestrial studies (Babiuk and Paul, 1970; Trolldenier, 1973). Since then this method has become the major technique for direct enumeration of bacteria

METHODS IN MICROBIOLOGY
VOLUME 22 ISBN 0-12-521522-3

in natural communities. Epifluorescence microscopy is used for estimating total counts, sizes, biomass, growth rates and numbers of active bacteria able to grow, take up nutrients and respire. It can even be combined with immunological approaches to count specific bacterial species. For these reasons this chapter will concentrate on epifluorescence approaches. Good accounts of transmitted light and electron microscope methods, which are generally less satisfactory for quantitative studies of natural bacteria, may be found elsewhere (Rodina, 1972; Jones, 1979; Newell *et al.*, 1986; Costerton *et al.*, 1986; Fry, 1988; Baker, 1988).

## II. Sample preparation

Quantitative microscopic observation always demands that the organisms are distributed randomly within the sample. Samples from different habitats require different treatment to ensure this.

Water relatively free of particulate material requires least treatment, thus samples from most marine or unpolluted fresh waters require only vigorous shaking by hand 20 times or for 30 s (Jones, 1979). Water containing a lot of particulate material or sewage, from polluted or very nutrient-rich environments, may require homogenization or sonication to break up clumps of bacteria which would otherwise be very difficult to count. These methods are similar to those used for soil, sediment and sewage and are described below. Estuarine waters often contain very large amounts of particulate material and some studies require that the particle-associated and freely suspended populations are separately assessed. This is best done after separation by differential filtration (Dodson and Thomas, 1964; Goulder, 1977). To do this a 3-μm pore size membrane filter is stuck to the base of a Perspex cylinder which is inverted in a beaker of sample water. A hydrostatic head of 22–24 cm is maintained between the water levels outside and inside the tube by removing the filtrate from within the tube by suction. The concentrated particles must be prevented from clogging the filter by vigorous mixing with a magnetic stirrer during filtration. The freely suspended bacteria are collected with the filtered water and may be studied separately from the whole water.

Soil, sediment and samples from sewage treatment systems normally require pretreatment to break up particles and clumps of bacteria. Some marine sediments are however fine enough so that pretreatment is unnecessary (Rice *et al.*, 1986; Davidson and Fry, 1987). Treatment by homogenization is common for such samples, but the motor should be mounted above the blades to prevent overheating the sample (Fry and Humphrey, 1978). The Waring blender can be used at speeds of up to 23 000 r.p.m.

for periods of 1–16 min (Montagna, 1982; Bakken, 1985). The Ultra-Turrax homogenizer is more vigorous and at 24 000 r.p.m. for 10 min (Meyer-Reil *et al.*, 1978) will give much finer suspensions with very even bacterial distributions. Sonication has also been used, but care is needed not to disrupt bacterial cells. Many different sonication regimes have been used, for example, Wiese and Rheinheimer (1978) and Meyer-Reil (1983) used 20 kHz at 50 W for 5 min and 3 min, respectively, for sandy sediment, whilst Velji and Albright (1986) used a 4-mm probe at 100 W for only 45 s with marine sediment, water and kelp samples.

Soil and sediment samples often need diluting with 0.2-μm pore size membrane-filtered solutions before treatment. Distilled water, natural and artificial sea water, buffers, detergents and deflocullents have all been used for this purpose. Detergents and deflocullents are very useful for preventing reaggregation of particles and bacterial clumps after treatment in all types of sample. A mixture of Cirrasol (ICI Ltd) and sodium pyrophosphate, both at 0.01% (w/v), is generally useful despite being first used for activated sludge (Gayford and Richards, 1970). Sometimes formaldehyde is used to harden the cells and reduce damage during sonication or homogenization. Meyer-Reil *et al.* (1978) used the fixative (0.74%, v/v) alone and Velji and Albright (1986) used formaldehyde (3.7%, v/v) with sodium pyrophosphate for seawater (0.001%) and kelp (0.01%) samples. Much more complex mixtures have been used for some soil studies, for example Bakken (1985) used the following solution: sodium hexametaphosphate (0.22%) adjusted to pH 8.5 with $Na_2CO_3$, sodium pyrophosphate (0.3%), Winogradsky salt solution, bromhexinchloride (0.2%) and Tween 80 (0.5%).

Bacteria are best removed from aquatic plants by the use of a stomacher (Fry *et al.*, 1985; Baker, 1988) which prepares an even suspension with the minimum of plant debris. Epilithic bacterial suspensions are best made by scrubbing with a toothbrush followed by stomaching (Goulder, 1987; Fry, 1988).

It is sometimes very useful to prepare clean suspensions of bacteria from soil or sediment. This only seems to have been done satisfactorily from soil. The procedure (Bakken, 1985) involves up to eight alternate homogenizations and density gradient centrifugations. It worked very well giving clean suspensions containing 10–36% of the original total count of bacteria. There was no difference in either the bacterial size distribution or the proportion of viable cells compared with homogenized soil, so a representative fraction of the population had clearly been obtained.

## III. Reagents and apparatus

All the reagents and apparatus used for direct methods must be sterile and particle free. Thus all solutions must be 0.2-μm membrane filtered before use; it is often prudent to add 2% (v/v) formaldehyde where possible to prevent subsequent bacterial growth. In my experience contamination sometimes occurs despite these precautions, thus in addition heat-stable solutions should be autoclaved and stored at 4°C before use. Control experiments should also be carried out periodically using reagent blanks and if contamination is found all reagents and apparatus currently being used should be discarded or cleaned thoroughly. In the subsequent sections it is assumed that these precautions are taken throughout.

## IV. Techniques for fixed samples

### A. Fixing and storage

When samples are required for total counting or size estimation they should be fixed as soon as possible after sampling, as it is well known that increases in the number and size of bacteria can occur during storage (Christian *et al.*, 1982; Ferguson *et al.*, 1984). Formaldehyde is the most common fixative used, normally at concentrations (v/v) between 0.2% (Fry and Davies, 1985) and 3.7% (Velji and Albright, 1986), however 2% is the most popular concentration (Hobbie *et al.*, 1977). Storage for long periods (3 days to 10 weeks) is possible in such solutions (Fry, 1988) without reductions in total counts or cell shrinkage. Electron microscope-grade glutaraldehyde (2.5%, v/v) has also been used (Clarke and Joint, 1986); in this case storage without loss of numbers for one month was possible. Changes in the number and size of bacteria vary greatly with different storage regimes and sample types so it is best to carry out trials if prolonged storage is anticipated.

### B. Total counts

Aquatic microbiologists use epifluorescence total counting techniques as a major tool in their research, but the technique is less well established in soil microbiology. The most common approach for aquatic samples is to deposit the stained bacteria on a membrane filter before counting. In soil studies smears or other preparations are normally made. The technique is undoubtedly the best available for estimating total numbers of bacteria in natural samples, but its success is dependent on careful attention to many small details. The following sections aim to introduce the novice to the intracacies of these considerations. Trials with different methods are normally needed to get the best results and even the most experienced need to try different approaches with a new type of sample. The principles of the approaches discussed in this section also apply to the other counting methods described in this chapter (Sections IV.E and V).

Some microorganisms can be observed with epifluorescence microscopy without staining because they autofluoresce brightly enough to be observable in natural samples when viewed with the correct filter sets. Appropriate methods for algae, cyanobacteria (Wynn-Williams, 1988) and methanogens (Doddema and Vogels, 1978) have been described.

### *1. Stains*

Table I lists some of the stains and their final working concentrations that have been used successfully. Acridine orange and DAPI are the best stains for aquatic samples and many comparisons have been made in the references quoted in Table I. The position for soil is less clear with several reports providing conflicting results. The intensity of fluorescence varies between the stains and has been ranked as DAPI > mithramycin > ethidium bromide > acridine orange (Milthorpe and Taylor, 1980). Although most of these stains are supposedly specific to DNA, other living and non-living material is commonly visible as well. DAPI and acridine orange are particularly useful because they stain bacteria and other particulate matter or debris differentially. Acridine orange gives the strongest differentiation, with most bacteria appearing pale green and debris orange, yellow or red. However, a small proportion of bacteria, normally <5%, will also appear orange. Bacteria appear pale blue with DAPI and debris looks yellowish. The greater brightness of DAPI-stained bacteria makes counting easier in samples with little debris. In samples containing much particulate matter acridine orange is normally chosen because of the contrast between the green bacteria and red debris. Combinations of stains can also be used, for example Rosser (1980) recommends, for soil bacteria and fungi, a mixture of ethidium bromide which stains cell interiors and FITC which stains the walls.

The concentration of stains used varies greatly (Table I) and trials must again be performed to determine what is best for a particular sample. My experiences from a recent international comparison of methods for measuring sizes of aquatic bacteria will provide a graphic example. A sample of lake water from Germany gave no problems using acridine orange at 5 mg $l^{-1}$ and DAPI at 0.05 mg $l^{-1}$, which are the concentrations normally used in my laboratory. However, a sample of Chesapeake Bay water gave impossibly faint images at these two concentrations and needed 200 mg $l^{-1}$ for acridine orange and 5 mg $l^{-1}$ for DAPI. Oceanic water from

**TABLE I** Some fluorescent stains used for the observation and counting of bacteria from natural environments

| Stain | Final concentration ($mg\ l^{-1}$), contact time (min) | Type of sample | Reference |
|---|---|---|---|
| Acridine orange | 200, 2–5 | Sea water | Bjørnsen (1987) |
| | 100, 1–2 | Lake and sea water | Hobbie *et al.* (1977)[a] |
| | 100, 1–2 | Soil bacteria | Bakken (1985) |
| | 67, 2 | Soil | Trolldenier (1973) |
| | 33–67, 5 | Soil | Wynn-Williams (1985) |
| | 33, 5 | Intertidal sediment | Dale (1974) |
| | 10, 5 | Lake water | Jones and Simon (1975)[a] |
| | 5, 3 | Epiphytic suspension | Fry and Humphrey (1978) |
| | 5, 3 | Whole worms | Harper *et al.* (1981) |
| DAPI | 5, 20 | Lake sediment | Schallenberg *et al.* (1989) |
| | 5, – | Sea water | Sieracki *et al.* (1985) |
| | 0.05, 5 | Lake water | Robarts and Sephton (1981)[a] |
| | 0.01, 5 | Lake water | Porter and Feig (1980) |
| | 0.01, 5 | Marine sediment | Yamamoto and Lopez (1985) |
| Ethidium bromide | 10–500, 3–10 | Soil | Rosser (1980) |
| | 100, 12 | Saltmarsh sediment | Roser *et al.* (1984) |
| FITC | 400, 3 | Soil | Babiuk and Paul (1970)[a] |
| | 400, 3 | Lake water and sediment | Fliermans and Schmidt (1975) |
| | 5, 3 | Lake water | Jones (1974) |
| Euchrysine-2GNX | 10, 5 | Lake water | Jones and Simon (1975)[a] |
| Europium chelate | 2 mM, 90–1080 | Soil | Anderson and Slinger (1975a, b)[a] |
| Mithramycin | 50–100, – | Epiphytic bacteria | Coleman (1980) |

Acridine orange, 3,6-bis(dimethylamino)acridinium chloride; DAPI, 4′,6-diamidino-2-phenylindole; ethidium bromide, 2,7-diamino-10-ethyl-9-phenylphenathridinium bromide; FITC, fluorescein isothiocyanate; euchrysine-2GNX, 3,6-diamino-2,7-dimethyl-9-methyl acridinium chloride; europium chelate, europium(III) thenoyltrifluoroacetonoate; –, data not given.

[a] Key papers with good descriptions of methodology and comparisons of techniques.

the euphotic zone stained well with the lower concentration of DAPI but needed the higher concentration of acridine orange to give a bright image. There is no way in which I could have predicted these difficulties because the lower stain concentrations had proved satisfactory for seawater and sediment samples from several British sites previously.

Both DAPI and acridine orange can be stored for at least 2 months. Acridine orange stores at room temperature but DAPI must be stored in the dark at 4°C.

### 2. *Aquatic methods*

*(a) Membrane filters.* Polycarbonate filters (0.2 μm pore size; 25 mm diameter; Nucleopore Corp.; Hobbie *et al.*, 1977) are far better than the cellulose ones that were originally used (Jones and Simon, 1975). This filter type provides a very flat surface and all the bacteria and particles lie in one focal plane.

The filters have to be stained to prevent background fluorescence. This is most commonly done with irgalan black (Acid Black No. 107; Ciba-Geigy Corp.; 2 g $l^{-1}$ in 2%, v/v, acetic acid); any staining time over 3 min is satisfactory. Some people have had trouble obtaining irgalan black, and Sudan Black B (Merck) can be used instead (67 mg $l^{-1}$ in 50% ethanol for 18–24 h; dissolve dye in absolute ethanol first; Zimmerman *et al.*, 1978). In both cases the filters must be rinsed in particle-free water before use; drying is unnecessary.

*(b) Staining procedures.* In aquatic research most workers stain bacteria prior to filtration, although some have stained them on the membrane filter after filtration (Dale, 1974; Ramsay, 1978). Dilution in particle-free water is normally required for eutrophic and polluted waters or sediments. Dilution should be sufficient so that the filter is not too overcrowded for efficient counting and so that a reasonable number of bacteria are seen in each quadrat counted (see Section IV.B.5).

Diluted samples can be stained in a separate container before filtering through a clean, straight-sided, glass, 25-mm membrane filtration apparatus. The sintered glass support should be kept clean and unblocked by regular boiling in detergent. A cellulose filter (0.45–1.0 μm pore size) can be used as a support to spread the vacuum if distributions of bacteria are uneven. An alternate procedure which minimizes the potential for contamination is to stain the bacteria in the filter funnel, applying vacuum to deposit the bacteria on the filter and to immediately rinse the sintered support when filtration is complete (C. M. Turley, personal communication). The volume filtered affects the evenness of distribution of particles and bacteria on the filter; 2 ml seems to be the minimum for a 25-mm filter but 5–10 ml is probably safer (Jones and Simon, 1975; Fry, 1988).

*(c) Rinsing and mounting.* Rinsing the bacteria on the membrane filter with the most appropriate solution for the sample being examined is essential to obtain a microscope image with a dark background. This is normally done with particle-free pure water using about the same volume of water as filtered. However, some workers have found filtered lake water and natural or artificial (Parsons *et al.*, 1984) sea water to be better. Pure water is normally satisfactory for freshwater samples of planktonic or sediment bacteria, but marine samples are less predictable.

The membrane filter can be mounted on a clean microscope slide immediately after rinsing whilst it is still slightly damp. The filter can be mounted directly without a coverslip or can be mounted in a minimum of non-fluorescent immersion oil or autoclaved paraffin oil under a coverslip. Acridine orange often fades during viewing and if this is a problem mounting in Citifluor AF2 which is glycerol based will probably help (Wynn-Williams, 1985).

*(d) Example procedure: fresh water with acridine orange.*

1. Pour some (about 20 ml) irgalan black solution (2%, v/v, membrane-filtered acetic acid containing 2 g $l^{-1}$ irgalan black) into a Petri dish. Immerse enough polycarbonate membrane filters (25 mm diameter) for a day's work and leave for at least 3 min; overstaining cannot occur. Rinse the membrane filters in another Petri dish containing autoclaved, membrane-filtered 2% formaldehyde (MFF).
2. Put a suitable volume of the water to be examined into a clean, 50-ml container. First try 10 ml for clean water and 1 ml for polluted water. If the volume is less than 10 ml make up to this volume with MFF. Add 415 μl of acridine orange stock solution (membrane-filtered 125 mg $l^{-1}$ acridine orange in MFF) and stain for 3 min.
3. After staining, filter through one of the irgalan black, stained membrane filters and rinse the bacteria on the filter with 10 ml of MFF. Make sure the surface of the membrane is free of water, but that the filter is still damp enough to be flexible.
4. Place the damp filter on the centre of a microscope slide, put a very small drop of autoclaved paraffin oil onto the centre and cover with a clean coverslip. Wait for a few moments for the oil to spread out under the coverslip, pressing the coverslip down gently if necessary.

View the preparation immediately; do not make several and store them.

NB: use 0.2-μm pore size membrane filters throughout. If the bacteria fluoresce too faintly use more stain; a final concentration of 5 mg $l^{-1}$ is used here; up to 200 mg $l^{-1}$ can be used. Adjust the volume of sample water to obtain the correct density of bacteria (see Section IV.B.5).

### *3. Soil methods*

A wide variety of preparation and staining procedures have been used for direct counts of soil bacteria and because there is less literature on this method from soil microbiologists a concensus has not yet emerged. When soil bacteria are well separated from the particulate matter (Bakken, 1985) the approaches used in aquatic studies and described above can be used. However, special techniques must be used in whole soil. Although smears on coverslips (Rosser, 1980) and complex soil–agar films (Anderson and Slinger, 1975a, b; Macdonald, 1980) can be used, simple soil smears on microscope slides would seem the method of choice for most stains (Babiuk and Paul, 1970; Trolldenier, 1973; Wynn-Williams, 1985).

Stained soil smears can be prepared by the following procedure, which is a compilation from the references given above. The blended soil (see Section II) is diluted to between $10^{-2}$ and $10^{-3}$ and allowed to settle for 30 s to 2 min to remove coarse particles. Then 10 μl is spread over 1 $cm^2$ and air dried at room temperature. Trolldenier (1973) recommends using slides with an 11.3-mm engraved circle on them for this purpose, which helps retain the 10-μl drop to the correct area. The drop is likely to contract whilst drying (Wynn-Williams, 1985) and care must be taken to make allowances for this during the calculation of total numbers. The smear is then gently heat fixed. After staining (Table I) the slide must be rinsed thoroughly at least twice, for 2 min with water for acridine orange and with 0.5 M sodium carbonate buffer (pH 9.6; 10 min) and 5% sodium pyrophosphate (2 min) for FITC.

### *4. Viewing*

All reputable microscope manufacturers make good epifluorescence microscopes. I have used those made by Zeiss, Lietz, Olympus and Nikon; all have proved satisfactory.

These manufacturers supply a wide range of filter sets suitable for epifluorescence work. These combinations of exciter filter, dichroic beam splitting mirrors and barrier filters are normally provided in easy-to-change filter blocks. There are three types of filter sets available. The conventional sets consist of coloured glass filters and will not give the brightest image because the exciter filters, although of broad band width, are not very transparent. Selective excitation sets are sold for multiple fluorochrome applications and pass lots of light over a narrow band width. The most suitable filter sets are the high-performance, wide-band pass interference filters which allow maximum excitation light to reach the specimen and so give the brightest fluorescent image.

The excitation peak for acridine orange is about 470 nm in the blue region and for DAPI is in the near ultraviolet (365 nm). The objective lens acts as the condenser in epifluorescence microscopy and glass objectives have poor transmittance in the ultraviolet, so glass only works well for acridine orange, but fluorite objectives are suitable for both stains. It is best to use objectives with very high numerical apertures; these give maximum resolution and focus the maximum amount of exciting light on the specimen. High objective magnification is also required to restrict the area over which this light is spread and so increase the brightness of the fluorescent image. Some glass lenses are now available which combine these features, such as the Olympus S Plan Apo 100×/1.40 and the Nikon N-Plan Apo 100×/1.40 lenses. Two fluorite lenses which are almost as good and rather less expensive are the Leitz Fluotar NPL 100×/1.32 and the Nikon CF Fluor 100×/1.30.

Focusing eyepieces must be chosen which complement these objective lenses and should house the 10 × 10 graticule used for counting. It is a mistake to use eyepieces of too high a magnification as this will reduce the brightness of the image; intermediate magnification changers also have this disadvantage. For routine use 10× eyepieces are suitable, and 16× are the most powerful which should be used only for counting dividing cells (see Section IV.E). I have occasionally used a 20× eyepiece for special photographs but it is very difficult to use.

### 5. *Counting*

*(a) Manual counting with quadrats.* The usual method of enumeration is to count the number of bacteria within squares on an eyepiece graticule which has been calibrated with a slide micrometer. A 10 × 10 graticule is convenient as it allows areas of different sizes to be counted according to the density of bacteria on the membrane. On many microscopes each small square on such a graticule will be about 10–13 μm[8] square with 10× eyepieces and 100× objective. Objects with sharp outlines of typical bacterial shape are scored as bacteria. With DAPI these will stain light blue

and with acridine orange mostly pale green but sometimes orange. The background should be black giving good contrast with the bacteria, but the preparation should not be so dark that the eyepiece graticule is not visible. The whole filter should be covered randomly during counting; the specimen must not be observed while fields are being changed or the counts will not be random. A good method of covering the filter is to select fields along two, central transects at right angles.

Most people have found Poisson distributions of bacteria on membrane filters, so the precision of the count depends on the number of bacteria and not the number of fields counted. The 95% confidence intervals are approximately twice the square root of the number counted, thus are 10% for 400 cells, 20% for 100 cells and 36% for 30 cells (Jones, 1979). For this reason most researchers count 400 bacteria per filter. However, Montagna (1982) carefully studied replication for sediment counts at the level of sediment core, filter and microscope field. He found that if numbers per gram were finally used then more reliable results were obtained by counting five fields of about 30 bacteria on four replicate filters than by counting 20 fields on one filter. This amounted to about 150 bacteria per filter counted. Kirchman *et al.* (1982), in a similar study with water, recommended counting 25 bacteria per field and two replicate filters. Thus it is clearly better to count fewer bacteria on two to four replicate filters than very large numbers of bacteria on one filter per sample. Counting 25–30 bacteria per field is satisfactory with DAPI where fading is not a problem. Acridine orange-stained preparations often fade rapidly, even with fading retardents (see Section IV.B.2(c)), so counting smaller numbers of bacteria (e.g. 10) over many fields is better.

The total number of bacteria per unit volume ($T$) is simply calculated from the mean number of bacteria per graticule area ($N$), filtration area ($A_f$), volume of sample filtered ($V$) and the graticule area ($a$) by the formula

$$T = \frac{NA_f}{aV} \tag{1}$$

The volume filtered must of course be moderated by the degree of dilution if dilution has taken place.

*(b) Effect of particles.* Samples such as estuarine water, saltmarsh water, sediment and soil, which contain a lot of particles, present special problems for counting. Particles will land on the membrane during filtration and cover freely suspended bacteria and other particles. Too many particles will severely reduce counts of both free and attached bacteria as thick layers build up. Schallenberg *et al.* (1989) recommended counting sediment

bacteria at low sediment dilutions and then correcting the counts to take account of the reduction in counted numbers that would have occurred. They claimed that this approach enabled counts to be done with large numbers of bacteria per field, which increased statistical accuracy. However, it is probably safer to count at high dilutions with the proportion of the field of view covered by particles not exceeding 40–70% (Clarke and Joint, 1986; Getliff and Fry, 1990). Bacteria lying behind or on the underside of particles cannot be seen unless the particles are very transparent, which is normally not the case. The problem of hidden bacteria in such fields of view has been overcome in several ways.

It has been proposed (Goulder, 1977) that the total number of visible and hidden bacteria per graticule area ($N_1$) in samples containing particulates can be calculated by

$$N_1 = F + 2R \tag{2}$$

where $F$ = the number of bacteria observed on the filter background and $R$ = the number that appear to lie on particles. This simple formula assumes that on average equal numbers of bacteria fall upon particles during filtration as are covered by the particles as they land on the filter. Thus doubling $R$ takes account of all obscured bacteria.

Clarke and Joint (1986) suggest that the real situation is more complex because the proportion of the filter covered by particles ($P_P$) is also important and propose more complex equations. An estimate of $P_P$ can be made during counting from the proportion of intersection points on the $10 \times 10$ eyepiece graticule which are over particles. They used two models (cases 1 and 2). In case 1 particles could land on top of each other and in case 2 they could not, presumably because they would roll off during filtration. In both cases the formulae calculating the actual number of freely suspended bacteria in the sample ($\beta$) were identical

$$\beta = F/(1 - P_P) \tag{3}$$

The formulae for the actual number of attached bacteria ($\gamma$) were different in cases 1 [$\gamma_1$, Equation (4)] and 2 [$\gamma_2$, Equation (5)].

$$\gamma_1 = 2\left(\frac{-\log_e(1 - P_P)}{P_P}\right)\left\{R - \left[\left(\frac{F}{1 - P_P}\right) \times \left(\frac{P_P}{-\log_e(1 - P_P)} - (1 - P_P)\right)\right]\right\} \tag{4}$$

$$\gamma_2 = 2[R - 0.5FP_P(1 - P_P)^{-1}] \tag{5}$$

The best answer was between these two values so the average can be used.

Hence the total count of all the visible and hidden bacteria from these equations ($N_2$) would be calculated as follows

$$N_2 = \beta + [(\gamma_1 + \gamma_2)/2] \quad (6)$$

These equations worked well for total bacterial counts and estimates of free and attached bacteria in the River Tamar estuary (Clarke and Joint, 1986). However, calculations on test data ($R = 1, 6, 10, 14, 19$; $F = 20 - R$; $P_P = 0.01$, 0.1, 0.3, 0.5, 0.7; Fry, unpublished) show that $\gamma_1$ and $\gamma_2$ are sometimes different and give negative values. These problems tend to occur when $P_P$ is high ($\geq 0.3$) and $F \geq R$ and place doubt on the Clarke and Joint (1986) equations for calculating numbers of free and attached bacteria. With these test data $N_1$ varied between 21 and 39 and $N_2$ was between 13 and 52. In reality $N_1$ and $N_2$ must always be at least 20 because the actual number of visible and hidden bacteria cannot be less than those visible. $N_2$ was only under 20 when there were very low numbers of bacteria visible on particles ($R = 1$; $F = 19$) and $P_P$ was high (0.5, 0.7); $\gamma_1$ and $\gamma_2$ were also negative with these data. These conditions are most unlikely to occur if bacteria do not fall off particles during filtration. For these reasons the Clarke and Joint (1986) equations would seem satisfactory.

It is however possible to calculate a new total count value ($N_3$) which approximates $N_2$ but does not require measurement of $P_P$ by using the following rules: (i) when $P_P \leq 0.4$ then $N_3 = N^3$; (ii) when $P_P \geq 0.4$ then $N_3 = 15.8 + (0.000586 \times N_1)$. Calculations indicate that the values of $N_3$ are always within 10%, and on average within 2%, of $N_2$, except under the most improbable conditions (e.g. $R = 1$; $F = 19$; $P_P = 0.7$). There is so much latitude with these rules around the borderline of 0.4 that $P_P$ can be estimated visually by guesswork. This approach will ensure that the best possible estimates of the total count of bacteria in samples with much particulate matter are obtained with the minimum extra effort.

Kirchman and Mitchell (1982) devised another method for estimating the number of particle-bound bacteria. They filtered particles from pond and saltmarsh water onto 3.0-µm pore size, polycarbonate membrane filters. They then washed the filters in phosphate buffer (2 ml; 0.1 M; pH 7.0), ground them in a tissue homogenizer with a further 1 ml of buffer and rinsed the filter, which was still intact, in another 1 ml of buffer. They then counted the removed bacteria by the normal methods. This method appeared to successfully break up the particulate matter and increased counts of attached bacteria by 46%.

*(c) Other manual methods*. Methods have been suggested to speed up and reduce the tedium of the normal quadrat counting method. These will

probably be more useful when the density of bacteria on the membrane filter is by necessity very low, for example in sparsely populated sediments. However, these methods have not yet been well tried and require justification by further comparative studies before they can be recommended. Two of these methods will be described below.

The third nearest neighbour method (Roser *et al.*, 1984) involves using an eyepiece graticule with a marked centre point and 10 concentric circles each divided into four equal segments (model E44; Graticules Ltd, Tonbridge, UK). Random fields (number of fields viewed = $m$) are observed and the distance ($r$, μm) of third nearest bacterium to the centre point is measured, by reference to the rings, in each of the four segments (number of segments = $k$) for every field. The population density ($d_1$ or $d_2$, number $\mu m^{-2}$) is then calculated from

$$d_1 = \frac{1}{\pi} \cdot \frac{n-1}{m} \sum \left(\frac{1}{r^2}\right) \tag{7}$$

$$d_2 = \frac{1}{\pi} \cdot \frac{n(k-1)}{m} \sum_{i=1}^{m} \frac{k}{\sum_{j=1}^{k} r_{ij}^2} \tag{8}$$

where $n$ is the $n$th nearest neighbour (3 in this case) and the best estimate is $d_1$ when $d_1 > d_2$ and is $(d_1 + d_2)/2$ when $d_1 < d_2$. Roser *et al.* (1984) using 10 fields and the rather low magnification of 625× could obtain counts between 235 bacteria $mm^{-2}$ and $4 \times 10^5$ bacteria $mm^{-2}$. They also obtained a good correlation between quadrat and third nearest neighbour counts ($r = 0.97$), but the latter slightly overestimated the counts below about $10^4$ bacteria $mm^{-2}$.

Another method uses the principle of most probable number (MPN) counting (Roser *et al.*, 1987). This technique uses the particle-sizing British Standard G10 graticule (Graticules Ltd, Tonbridge, UK). This consists of 12 different sized quadrats arranged in a series with each quadrat covering twice the area of the previous one in the series. Randomly selected fields are viewed and all the different sized quadrats are quickly viewed and scored for the presence or absence of bacteria. The average number of quadrats that contain bacteria ($x$) or that are empty ($y$) are then noted. The number of bacteria in the largest sized quadrat ($\lambda$) can then be calculated as follows

$$\log_{10}\lambda = x \log_{10}s - K \tag{9}$$

where $s$ is the size increment used in the series on the graticule (2 in this case) and $K$ is from Table II. Roser *et al.* (1987) recommend that only five fields of view are scored, which gives standard errors comparable to those calculated by Montagna (1982) who used conventional quadrat methods.

**TABLE II**
Values of $K^a$ (equation 9) for the most probable number direct count method (modified from Roser *et al.*, 1987)

| $x$ or $y$ | $K$ when $x$ is smaller | $K$ when $y$ is smaller |
|---|---|---|
| 0.05 | 1.613 | −0.197 |
| 0.1 | 1.324 | −0.108 |
| 0.2 | 1.045 | −0.007 |
| 0.3 | 0.982 | 0.050 |
| 0.4 | 0.789 | 0.101 |
| 0.6 | 0.657 | 0.167 |
| 0.8 | 0.575 | 0.212 |
| 1.0 | 0.521 | 0.245 |
| 1.2 | 0.483 | 0.271 |
| 1.4 | 0.456 | 0.292 |
| 1.6 | 0.438 | 0.309 |
| 1.8 | 0.425 | 0.323 |
| 2.0 | 0.426 | 0.334 |
| 2.5 | 0.405 | 0.356 |
| 3.0 | 0.402 | 0.370 |
| 3.5 | 0.401 | 0.379 |
| 4.0 | 0.401 | 0.386 |
| 4.5 | 0.401 | 0.390 |
| 5.0 | 0.401 | 0.394 |
| 6.0 | 0.401 | 0.397 |
| 7.0 | 0.401 | 0.399 |
| >7.0 | 0.401 | 0.401 |

[a] When $x < y$ use $x$ to select $K$ from column 2 and when $y < x$ use $y$ to select $K$ from column 3.
Only use this table with ⩾11 quadrats in a series where each quadrat is twice the area of the smaller one.

Thus if 42 quadrats contained bacteria from five fields then 8.4 of the quadrats on the graticule ($42/60 \times 12$) would be scored positive. So in this case $x = 8.4$, $y = 3.6$, $a = 2$ and $K = 0.379$, so there would be 141 bacteria in the largest quadrat, which of course must be calibrated for the lenses used with a slide micrometer. This counting procedure takes only about 2 min per filter which can be up to 10 times quicker than conventional methods (Roser *et al.*, 1987)

*Filamentous organisms.* Samples containing filamentous organisms present special counting difficulties. They are easily seen by epifluorescence microscopy and are best counted in sediments using acridine orange staining and membrane filters (Godinho-Orlandi and Jones, 1981). The individual bacteria in chains can be counted as single cells but non-septate filaments may need to be measured. If few are present they will rarely occur in the areas counted and so errors will be minimized. However, if they are abundant they may need to be estimated separately. The length of short filaments can be estimated by reference to the squares of the 10 × 10 eyepiece graticule, but long filaments cannot easily be measured in this way. The total length of long filaments on a filter is best estimated by the grid line intersection method (Giovannetti and Mosse, 1980). To do this fields are observed with a calibrated 10 × 10 eyepiece graticule and the number of intersection points between filaments and the straight lines ($n_i$) on the graticule are counted and number of fields ($f$) viewed is recorded. The total filament length per graticule area ($L$) is then calculated from

$$L = \frac{\pi A_g n_i}{2Hf} \tag{10}$$

where $A_g$ is the area of the whole 10 × 10 graticule and $H$ is the total length of lines on the graticule. As with other methods it is better to count up to 100 intersections on three replicate filters than 300 intersections on one filter; this should give a standard error of about 2% (Giovannetti and Mosse, 1980). Other useful intersection methods are available (Russ, 1986).

*(e) Television image analysis.* Semi-automatic methods for obtaining total counts of natural bacteria will potentially save a great deal of time. Image analysis (see Section IV.C.3 for a full description) has been successfully used to make total counts only in sea and estuarine waters (Sieracki *et al.*, 1985; Bjørnsen, 1986; Bjørnsen *et al.*, 1989; David and Paul, 1989). When large amounts of debris or particles are present the degree of interaction required to provide clean images for automatic counting is prohibitive. So it is unlikely that image analysers will be useful for counting bacteria in soil, sediment or polluted water in the near future.

## C. Size estimation

Electron microscope methods for estimating the sizes of natural bacteria all cause considerable shrinkage and so are not suitable for estimating cell volumes; epifluorescence methods are generally considered better (Fuhrman, 1981; Montesinos *et al.*, 1983; Fry and Davies, 1985). Electron microscopy is also more time consuming than epifluorescence and so it is

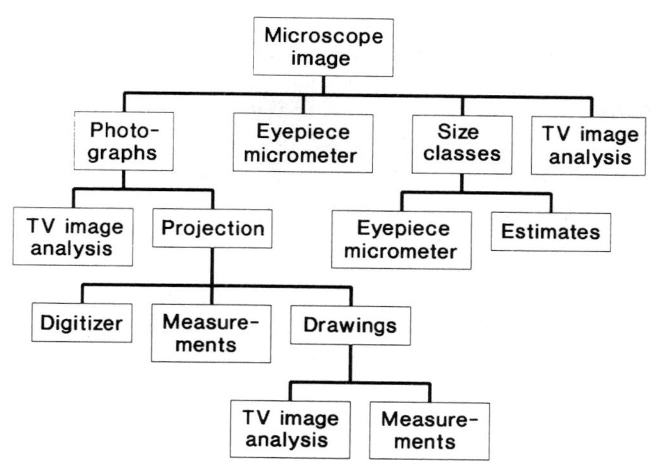


**Fig. 1.** The relationship between the various methods for estimating the volumes of bacteria from fluorochrome-stained epifluorescence images.

not often used. There are a wide variety of different combinations of techniques that can be used for estimating bacterial biovolumes (Fig. 1) so the main elements of these techniques will be discussed here. Sizes of bacteria can be measured either directly from the microscope images or indirectly from photographs of these images.

### *1. Calculating volumes*

When measuring bacteria by eye it is possible to measure the true length of the cell down the central axis and the width at right angles to the length. This approach works well for straight rods and cocci. It is more difficult to measure the true central length of curved rods or spirals but reasonable estimates can be made. From this basic data the volume can be calculated if a suitable model is assumed. The most common model is a straight-sided rod with hemispherical ends. Hence the volume of a single cell ($v$) is calculated by

$$v = (d^2\pi/4)(l - d) + \pi d^3/6 \tag{11}$$

which is simplified to

$$v = (\pi/4)d^2(l - d/3) \tag{12}$$

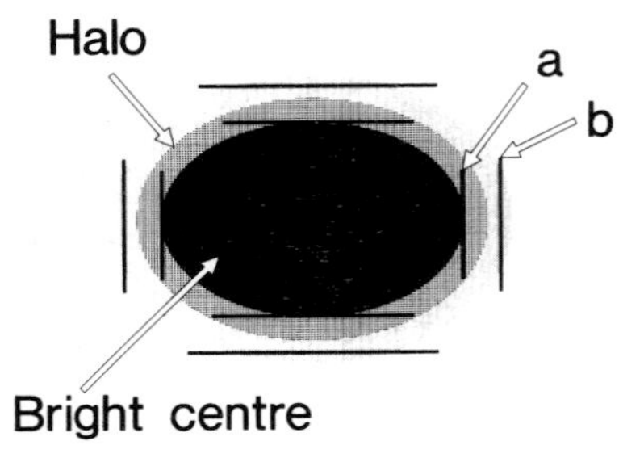


**Fig. 2.** Diagram of the negative image of a bacterium stained with a fluorescent dye showing the 'halo' and the approximate positions at which length and width have been measured by (a) Lee and Fuhrman (1987) and (b) many other workers.

where $d$ = width and $l$ = central length. This formula works for cocci because $l - d$ becomes zero and the formula for the sphere alone is left. Other models have been used, for example Lee and Fuhrman (1987) used the average of this formula and that for an oblate spheroid for rod-shaped bacteria.

It is hard to see the outline of fluorescence images of bacteria due to the so-called 'halo' effect (Lee and Fuhrman, 1987). This halo is caused by a number of factors such as the spread of fluorescent light from the cell interior through the cell wall, optical limitations due to the limit of resolution of the light microscope (*ca* 0.28 μm) and film graininess in photographs or pixel size with image analysers. As yet there is no concensus about how to judge the true edge of the bacterium. Lee and Fuhrman (1987) suggest that the halo should not be measured but many other workers (varied personal communications) tend to measure further towards the edge of the image (Fig. 2). Sieracki *et al.* (1989) used fluorescent microspheres (see Section IV.C.7) to test edge detection and concluded that manual methods often underestimated sizes. This observation supports the use of the outside edge of fluorescent images of bacteria.

Frequency distributions of bacterial lengths, widths and volumes are heavily skewed towards the smaller dimensions in both aquatic (Fuhrman, 1981; Maeda and Taga, 1983; Fry and Davies, 1985; Sieracki *et al.*, 1985) and soil (Bakken and Olsen, 1987; Olsen and Bakken, 1987) populations. Thus only a small number of large bacteria are found. Despite this these organisms can account for most of the biomass; for example Fry and Davies (1985) found that 3.7% of the bacteria could account for 23.6% of the biomass. Large bacteria are often the most active (Rimes and Goulder, 1986; Kirchman and Ducklow, 1987) and can account for most of the viability (Fry and Zia, 1982a, Bakken and Olsen, 1987; Olsen and Bakken, 1987). For these reasons it is vital that the volumes of these rare, large organisms are estimated, so large numbers of individual bacteria must always be measured. Between 50 (David and Paul, 1989) and 800–1000 (Lee and Fuhrman, 1987) bacteria per filter have been used to obtain mean volumes. However, 200 bacteria would seem an adequate number and would ensure that at least some of the rarer large cells are measured. As with total counting techniques replication at the membrane filter level is probably better than estimating more cells on one membrane.

### 2. *Photographs*

Photographic methods are often used for measuring sizes of natural bacteria (Fry, 1988). The basic approach is to take a photograph of the epi-fluorescence image. This can be done with any automatic photographic attachment for the microscope capable of measuring low light levels, because long exposure times are often needed with dim images. A suitable phototube with an Olympus OM2 or OM4 35-mm single lens reflex camera is a satisfactory, low-cost solution. Careful focusing to obtain sharp images is very important. It is best to work from black and white negatives or colour transparencies using a high-speed film (400 ASA). Monochrome film has more exposure latitude and so is more likely to give measureable images in the event of an exposure error. High-contrast films give photographs that are easiest to use. The monochrome films Kodak Tri-X pan (Fry and Davies, 1985) and Ilford HP5 (K. Lochte, personal communication) and the colour films Kodak Ektachrome 400 (Fuhrman, 1981; Meyer-Reil, 1983; Fry and Davies, 1985; Lee and Fuhrman, 1987) and 200 (Bratbak, 1985) have proved suitable for aquatic samples. Good images of cyanobacteria and eubacteria in Antarctic fell field soils were obtained with Kodak Technical Pan 2415 monochrome negatives, Fujichrome 400 colour transparencies and Fujicolor HR400 colour prints (Wynn-Williams, 1986).

The bacteria on the negatives or slides are then measured with an image analyser (see Section IV.C.3) or directly by projection. The projection should be at a magnification of between 10 000× (Lee and Fuhrman, 1987) and 14 400× (Bratbak, 1985). This has been achieved with a slide projector using a final image size of about 70 cm × 48 cm from a whole 35-mm negative (Fry and Davies, 1985), which was equivalent to a magnification of about 10 000×, or an enlarger (Krambeck *et al.*, 1981). Measurements are made directly from the projected image with a ruler or micrometer gauge or indirectly from drawings obtained by outlining the bacterial images on the screen with a pencil or marker. All bacteria giving a sharp image should be measured; this if often up to 95% of the bacteria photographed (Fry and Davies, 1985).

### 3. *Television image analysis*

Television-based image analysis systems are rapidly becoming popular for measuring bacteria in natural samples, because they offer the possibility of very rapid measurement of large numbers of organisms. However, the basic principles of how such instruments can be used are still not widely appreciated so the general reviews by Bradbury (1977, 1979, 1981) will be useful. Although the price of systems able to measure bacteria by epifluorescence microscopy used to be very high, suitable instruments (Getliff and Fry, 1989) now cost about the same as a centrifuge. Other systems (Jackman, 1989), with few advantages, can cost four times this amount. However, many systems in all price ranges are unsuitable, so prospective purchasers should choose carefully.

Figure 3 provides a summary of the main features that are required in an image analysis system for measuring microbial sizes. The video camera is a key part of the system. The standard Newvicon or Chalnicon type of cameras that are successful for counting the brightly fluorescing, acridine orange-stained bacteria in milk (Pettipher and Rodrigues, 1982) are also satisfactory for images of the dense bacterial populations in Antarctic fell field soil crusts (Wynn-Williams, 1986, 1988) and for photographs (Fry and Davies, 1985). Direct viewing of the dimmer epifluorescence images typical of aquatic bacteria requires the TV camera to produce a bright, low-noise image at low light intensities. This necessitates special cameras with image intensifiers incorporated; those used successfully include silicon-intensified target (e.g. Nighthawk SIT 733, Bjørnsen, 1986; Dage MTI SIT 66, David and Paul, 1989) and moonlight (Panasonic WV-1900B, Getliff and Fry, 1989) cameras. Most people have used monochrome cameras because of their high light sensitivity, however some research with colour cameras has been reported (Sieracki and Webb, 1986; Sieracki *et al.*, 1989).

Older image analysers worked with live images (Fry and Davies, 1985; Wynn-Williams, 1986) which is difficult because of fluorochrome dye fading and constantly varying image noise. It is far better to store the original image and to work with that. There are many steps involved in manipulating the stored image before sizes are obtained (Fig. 3) and although Getliff and Fry (1989) describe these details for one image analyser a brief overview will be given here.

The primary stored image will be segmented into up to 256 grey levels and consists of a series of rectangles or pixels. Each pixel will be of one

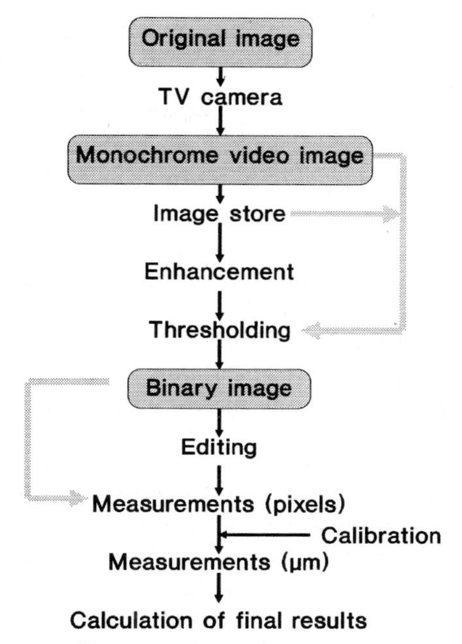


**Fig. 3.** Flow diagram to illustrate the major stages involved in the use of a television image analyser for measuring sizes of bacteria from natural habitats. Stippled arrows show alternative sequences of stages.

grey level and the images used by most modern instruments are 512 × 512 pixels. It is these factors which eventually limit resolution although some instruments can use various mathematical ploys to measure sizes at sub-pixel accuracy. The initial grey level image must be converted into a binary image before measurements can be made. The binary image is created by marking pixels that are considered to be part of an object, and that will eventually be measured, and leaving other pixels unmarked. This process is called segmentation, thresholding or detection. The original images are rarely even enough for thresholding accurately and so image enhancement is used to improve the level of agreement between bacterial objects in the grey image and their binary counterparts. Many different enhancement features are available on the various manufacturers' image analysers. Useful enhancements include shade correction and low or high pass filters to make backgrounds more even, contrast modifications to make selecting bacteria easier and edge improvement routines to reduce halos (see Section IV.C.1). Care should be taken to ensure that enhancement does not change the size of the bacterial images. Thresholding the enhanced image is usually done manually whilst looking at both images superimposed, but recently an automatic technique has been suggested (Sieracki *et al.*, 1989). Such automatic methods take a lot of computing time at present, but might eventually solve the present problem of bacteria with differing brightness in the same field of view requiring different threshold levels.

The binary image can then be edited to remove debris and poorly thresholded or out-of-focus bacteria. Editing can be manual or automatic, but manual editing can be very time consuming and should be avoided if possible. Planktonic samples are normally easy to edit but samples of soil or sediment can be virtually impossible. Erosion (removing a pixel from all around an object) followed by dilation (adding a pixel) has proved a useful (David and Paul, 1989) automatic editing feature. This can separate closely spaced bacteria if rejoining is prohibited in the routine used and can remove noise due to single or small groups of pixels. Most systems have more editing features, both manual and automatic, such as hole-filling and circle-separating routines.

Once the image is edited then measurements can be made, which are either field or feature specific. Field-specific measurements are made in a predefined frame and allow estimates of parameters such as the total number, area ($A_t$), perimeter, length and intercept ($I_t$) of all the objects in a single field of view. Thus sizes of individual organisms are not obtained. This approach has been used successfully for total counts of aquatic bacteria (David and Paul, 1989) and total biovolume ($V_t$) estimates of micro-organisms in soil crusts from Antarctica (Wynn-Williams, 1986) using the following formula

$$V_t = A_t^2/2I_t \tag{13}$$

Feature-specific measurements, which are made on individual objects, are needed to estimate mean cell volumes and are normally only available on the more sophisticated image analysers. Individual cell volumes cannot be estimated directly from two-dimensional images and so other measurements must be made and then converted to volume by calculation. Such calculations are best done within the computer that runs the image analyser (Getliff and Fry, 1989), but raw data can also be transferred to another computer for this purpose (Fry and Davies, 1985; David and Paul, 1989). Image analysers cannot measure central length directly for curved or spiral-shaped organisms. Length can be estimated reasonably accurately for straight rods as the largest feret diameter (caliper diameter at fixed angles; Bradbury, 1977), provided feret diameters at six or more angles are measured. The shortest feret diameter, however, gives a poor estimate of the true width although the feret diameter at right angles to the longest provides a better estimate if large numbers are measured (e.g. 36). For these reasons it is best to use area ($A$) and perimeter ($P$) to estimate central lengths ($l$; equation 15) and widths ($d$; equation 14) and then to use equation 11 or 12 to calculate volume.

$$d = (P - \sqrt{P^2 - 4\pi A})/\pi \tag{14}$$

$$l = P/2 + (1 - \pi/2)d \tag{15}$$

Although the above procedure gives the best estimates of mean cell volume (Fry, 1988) other approaches have been used. Sieracki *et al.* (1985) used area alone to calculate volumes assuming a sphere

$$v = 4/3\sqrt{A^3/\pi} \tag{16}$$

or a prolate spheroid with a length to width ratio of 2

$$v = 0.94\sqrt{A^3/\pi} \tag{17}$$

Bjørnsen (1986, 1989) used a formula that works nearly as well as equations 11, 12, 14 and 15 (Fry, 1988) and which uses the convex perimeter ($P_c$) and area

$$v = 8.5A^{2.5}P_c^2 \tag{18}$$

Feature-specific measurements can be made automatically for all organisms in a field of view at one time or manually one at a time after visual inspection of the thresholded and grey images. Automatic measurements work well when there is even thresholding of all objects and little or no debris present; manual methods are slower, but better, if these criteria do not hold. Larger objects or those of different shape (from calculated form factors; Bradbury, 1977) can sometimes be excluded from automatic measurement by using selection routines provided with the image analyser.

### 4. *Size classification*

Bacterial volumes can be estimated by two types of size class method. In one the sizes and shapes are estimated visually without aids and in the other by comparison with shapes on an eyepiece graticule. Both methods involve recording the number of bacteria in each size category during total counting. These methods have proved popular because they need no specialist equipment and take little more time to perform than the total count.

In the estimation method a predetermined set of shapes and sizes is chosen and the cell volume calculated for each shape and size used. The total and then the mean cell volume of the bacteria in the sample is then calculated from the total count of bacteria in each class. A wide variety of shape and size categories have been used for both soil and aquatic bacteria (Table III). Widely different mean cell volumes can be obtained with this method on the same samples (Lundgren, 1984). These differences are mainly due to the classes selected and can be avoided if classes are chosen which span the likely range of shapes and sizes of bacteria present in the samples. Too few or too many classes should be avoided or classification difficulties will be encountered. At least some bacteria should be found in all the size classes selected. Counts with too many bacteria in one class will give unreliable estimates of mean cell volume. Lundgren (1984) recommends that the relevance of the size classes used should be checked by another method.

The New Porton G12 eyepiece graticule (Graticules Ltd, Tonbridge, UK) has been used to estimate cell volumes for both water (A. Jelmert, personal communication) and soil bacteria (Bakken and Olsen, 1987; Olsen and Bakken, 1987). This graticule is engraved with a series of 11 circles and spheres arranged in a root two progression. Comparison of the fluorescent bacteria with the shapes on the calibrated graticule enables the sizes of the organisms to be estimated. Bratbak (1985) reports that using a 1000× magnification the diameters of the calibrated spheres were 0.20, 0.29, 0.41, 0.57, 0.81, 1.15, 1.63, 2.30, 3.25, 4.60 and 6.50 μm.

### 5. *Other approaches*

Eyepiece graticules can be used to measure directly natural bacteria on fluorochrome-stained membrane filters. The standard measuring graticule with 100 divisions on a linear scale is unsuitable because the divisions are too large. However, it is possible to use a moving hairline graticule (Fry and Davies, 1985) or the New Porton G12 graticule (Bratbak, 1985) to measure lengths and widths and so to calculate volume (equations 11 and 12).

Digitizers have been used to measure sizes of bacteria from SEM negatives (Krambeck *et al.*, 1981) and slides of microscope images (Meyer-Reil, 1983). A digitizer is a graphics tablet attached to a computer loaded with a suitable program. The photograph is then projected onto the graphics tablet and either the outline of the bacteria marked or the lengths and widths indicated with the digitizer's detection pen. Large numbers of bacteria can be measured quickly in this way and the computer program will calculate the bacterial dimensions and convert these into volumes in much the same way as a television-based image analyser does. Digitizers are generally cheaper than television image analysers, but their utility depends on the quality of the software provided.

**TABLE III**

Some examples of size and shape classifications for estimating the biovolume of bacteria in natural communities

| Habitat | Cell shape | Diameter (μm) | Length (μm) | Nominal volume ($\mu m^3$) | Reference |
|---|---|---|---|---|---|
| Soil | Cocci | <0.5 | <0.5 | 0.03 | Lundgren (1982, 1984) |
| | Coccobacilli | ≈0.5 | 0.5–1.0 | 0.22 | |
| | Small rods | ≈0.5 | 1.0–2.0 | 0.34 | |
| | Large rods | >0.5 | >2.0 | 0.75 | |
| | Large cocci | >1.0 | >1.0 | 0.77 | |
| Water | Cocci | 0.3 | <0.4 | 0.014 | Zimmerman (1977) |
| | Cocci | 0.5 | 0.4–0.6 | 0.065 | |
| | Cocci | 0.7 | 0.6–0.8 | 0.180 | |
| | Rods | 0.4 | <0.6 | 0.023 | |
| | Rods | 0.8 | 0.6–1.0 | 0.095 | |
| | Rods | 1.2 | 1.0–1.4 | 0.185 | |
| | Rods | 1.7 | 1.4–2.0 | 0.292 | |
| | Rods | 2.4 | 2.0–2.8 | 0.508 | |
| | Coccoid rods | <0.5 | <0.06 | 0.065 | Palumbo *et al.* (1984) |
| | Cocci | 0.6–1.2 | 0.6–1.2 | 0.320 | |
| | Rods | 0.38–0.44 | 0.6–1.2 | 0.111 | |
| | Rods | 0.44–0.50 | 1.2–1.8 | 0.254 | |
| | Rods | 0.54–0.63 | 1.8–3.0 | 0.574 | |
| | Mini cocci | 0.22–0.57 | 0.90–1.13 | 0.038 | Turley and Lochte (1985) |
| | Mini vibrios | 0.11–0.22 | 0.90–1.13 | 0.038 | |
| | Cocci | 0.57–0.90 | 0.57–0.90 | 0.20 | |
| | Rods | 0.22–0.45 | 1.01–1.24 | 0.11 | |
| | Large rods | 0.45–0.68 | 1.13–2.03 | 0.45 | |
| | Spirilli | 0.22–0.34 | 1.70–2.83 | 0.16 | |
| | Dividing cells | 0.57–0.90 | – | 0.48 | |

### 6. *Choice of method*

Few comparisons have been made between different methods of measuring volumes of natural bacteria. Fry and Davies (1985) found that a moving-hairline eyepiece-graticule grossly overestimated the width (47–65%), and hence volume (77%), of aquatic bacteria observed with phase contrast microscopy. They also found little difference between sizes estimated with image analysis of negatives or slides, and measurements with a ruler taken from these photographs. Bratbak (1985), using acridine orange staining, found little difference between volumes estimated from lengths and widths obtained with the New Porton G12 graticule taken directly and from ruler measurements on projected photographs. Similarly, Lundgren (1984) found no difference between total volumes of soil bacteria in different size categories by either the size class method or direct measurements. One comparison has been made of image analysis estimates of size made directly from negatives and indirectly from outline drawings of the organisms taken from the projected photographs (Fry, 1988). In this case there were significant differences but the differences were not consistent between the water and sediment samples tested.

From these and other considerations discussed earlier it is clear that direct image analysis on epifluorescence microscope images is the best technique for measuring volumes of natural bacteria when particulate matter is low. The more particulate matter that is present, the more difficult the sizes will be to measure and the more editing and separate thresholding will be required. As operator interaction slows down the work considerably, such methods work best on ocean or oligotrophic lake water. However, in my laboratory these techniques have also been successfully used on sewage sludge which contains a lot of finely divided organic matter. When an image analyser and a suitable low-light camera is not available then direct measurements with a ruler on projected photographs are probably equally accurate but are much more time consuming. When a great deal of large particulate matter is present and the bacteria are attached to the particles, as with soil smears or sediment preparations, then direct image analysis from the microscope or photographic images is impossible. In these cases direct measurement of photographs with a ruler or digitizer or image analysis of drawings taken from the projected photographs is probably best (Getliff and Fry, 1990). In all these cases the method of size classing with the aid of a New Porton graticule is probably satisfactory for estimating mean volumes, but is rather more subjective than the other methods. Clearly more comparisons between methods are needed.

### 7. *Calibration*

The usual way of calibrating measurements made by any of the methods discussed above is to use a slide micrometer. These are widely available and usually have a minimum calibration interval of 10 μm, several of which will fit onto the average photograph or field of view. A more recently developed test specimen for incident light microscopy is a 5 mm × 5 mm silicon chip which has been electron beam written with a mesh of squares with a 9.9 μm periodicity (Agar, 1988; Agar Scientific Ltd, Stanstead, UK). This also allows image distortion to be assessed to some degree, which can be particularly important with image analysers.

Uniform, latex, monodispersed, fluorescent microspheres are also available (Fluoresbrite(TM), Polysciences Inc., Warrington, PA, USA). These have been used at various diameters to test measurement methods (0.25, 0.57 and 1.04 μm; Bratbak, 1985; 0.21, 0.63 and 0.70 μm; Lee and Fuhrman, 1987). They would also be useful for calibration and testing image distortion. They are viewed directly once filtered onto stained, polycarbonate membrane filters. They fluoresce very brightly and this had led to gross overestimation of their size (Bjørnsen, 1986; Fry, 1988). To avoid this Lee and Fuhrman (1987) recommend viewing them with non-optimal excitation filters (U.V. exciter filter, 300–400 nm; barrier filter, 515 nm). This cuts the brightness of their fluorescence to a level similar to the green fluorescing bacteria. Despite these problems their diameters have also been underestimated when used to test measurement methods. Lee and Fuhrman (1987) report measurements of 91–92% of the stated diameter. The analogous figures from Bratbak (1985) were 59–97% for ruler measurements from photographs and 66–186% with the New Porton G12 graticule. These problems with fluorescent microspheres indicate that slide micrometers might well be the safest and most reliable calibration method.

### D. Biomass

The bacterial biovolume in a sample can easily be estimated from the product of the total count and mean cell volume. This in turn can be converted to biomass by using a suitable conversion factor. There is a great range between the calibration factors that have been recently estimated experimentally (Table IV). Most investigators have recommended the use of factors that they have calculated, but because of the variation observed it is unclear which factor is really the best to use.

With the exception of that estimated by Bratbak (1985), all the factors obtained with electron microscopy are much smaller than those obtained with other methods (mean = 98 fgC $\mu m^3$; Table IV). This is not unexpected in view of the cell shrinkage known to occur during electron microscopy preparation procedures (see Section IV.C). There is little difference between the average factors obtained from pure cultures (318 fgC $\mu m^3$) and mixtures of natural bacteria (309 fgC $\mu m^3$) using epifluorescence microscopy. In view of the large coefficients of variation obtained in almost all the studies the values obtained with the Elzone particle counter (mean = 209 fgC $\mu m^3$) are probably also similar. The factor of 308 fgC $\mu m^3$ recommended recently by Fry (1988), using a more carefully reasoned set of arguments from less data, is almost exactly the same as the mean values from the epifluorescence studies alone. For these reasons and because most people use epifluorescence microscopy for direct biomass estimation, I recommend that a conversion factor of 310 fgC $\mu m^3$ is used in future.

### E. Frequency of dividing cells

The ratio of numbers of dividing and the total number of bacteria in a sample has often been used in aquatic studies to estimate the growth rate of the population. This ratio expressed as a percentage is called the frequency of dividing cells (FDC) and was first used by Hagstrom *et al.* (1979).

The number of dividing cells in a sample is counted with the total number of bacteria by epifluorescence microscopy on membrane filters as described above. Dividing cells are those with a distinct invagination but without a clear zone between the daughter cells. There are fewer dividing cells than other bacteria and so either larger numbers of fields must be counted or a larger quadrat size used. Investigators often count fewer dividing than total bacteria; most count 30 (Hagstrom *et al.*, 1979; Newell and Christian, 1981; Hanson *et al.*, 1983) but sometimes up to 100 are counted (Meyer-Reil, 1983). These bacteria are difficult to count because each cell must be inspected for an invagination and aquatic bacteria are often very small. FDCs between 1% and 16% have been recorded in water samples (Hanson *et al.*, 1983; Turley and Lochte, 1985) which means between 200 and 3000 bacteria need to be inspected if 30 dividing bacteria are to be counted. A well-set-up microscope with first-class optics makes the task easier.

To estimate growth rate from FDC a calibration equation must be used. Table V gives some of the equations that have been suggested. These equations are obtained by regression of values of specific growth rate ($\mu$) and FDC obtained from experiments. Most of these regressions are satisfactory with high proportions of the variability in the data explained by the equation ($r^2$ = 33–95%) and reasonably low coefficients of variation (7–26%). The experiments have been done either in batch culture (Newell and Christian, 1981; Hanson *et al.*, 1983) or in a chemostat (Hagstrom *et*

**TABLE IV** Some factors used for converting biovolume to biomass that have been estimated by experiment

| Bacteria used | Conversion factor (fgC $\mu m^{-3}$) | Coefficient of variation (%) | Method used for | | Reference |
|---|---|---|---|---|---|
| | | | cell volume | cell C | |
| *Escherichia coli* | 130[a] | – | TEM | CHN | Watson *et al.* (1977) |
| *Arthrobacter globiformis* | 374[a] | 27 | Epi | Combustion | Van Veen and Paul (1979) |
| *Enterobacter aerogenes* | 355[a] | 24 | | | |
| Nine soil isolates | 157[a] | – | Epi | C/D | Bakken and Olsen (1983) |
| *Bacillus subtilis* | 284 | – | – | CHN | Bratbak and Dundas (1984) |
| *E. coli*<br>*Pseudomonas putida* | 164<br>251 | –<br>– | | | |
| *P. putida* | 478[b] | 44 | SEM, Epi + P, Epi + E | CHN | Bratbak (1985) |
| Estuarine | 650[b] | 28 | | | |
| *E. coli* | 90 | – | TEM | CHN | Heldal *et al.* (1985) |
| Lacustrine<br>Estuarine | 307<br>409 | 26<br>33 | Epi + TVIA | Combustion | Bjørnsen (1986) |
| Lacustrine | 106 | 44 | Epi + E | CHN | Nagata (1986) |
| *Alcaligenes* sp.<br>*Pseudomonas* spp.<br>Marine >0.6 μm | 239<br>193<br>196 | 10<br>20<br>37 | Elzone | CHN | Kogure and Koike (1987) |
| Marine | 380 | 36 | Epi + P | CHN | Lee and Fuhrman (1987) |
| Mixed pure cultures and aquatic | 50–100 | – | SEM | X-ray | Norland *et al.* (1987) |
| Lacustrine | 154 | 54 | Epi + P | ICA | Scavia and Laird (1987) |

Epi, epifluorescence microscopy; TEM, transmission electron microscopy; SEM, scanning electron microscopy; Elzone, Elzone particle counter; P, photographs; E, eyepiece micrometer; TVIA, television image analysis; X-ray, energy dispersive X-ray microanalysis in a TEM times C/D (0.4); C/D, carbon:dry weight ratio (0.437); CHN, CHN analyser; ICA, international carbon analyser.

[a] Values calculated from authors results.
[b] Values are averages from several methods.

**TABLE V**

Some equations that have been used to predict specific growth rate ($\mu$, $h^{-1}$; $\mu_d$, $day^{-1}$) from the frequency of dividing cells (FDC) or dividing and divided cells (FDDC) for planktonic bacterial populations

| Equation[a] | Reference |
|---|---|
| $\ln\mu = 0.299\ \mathrm{FDC} - 4.961$ | Newell and Christian (1981) |
| $\mathrm{FDC} = 85.2\mu + 0.5$ | Larsson and Hagstrom (1982) |
| $\ln\mu = 0.081\ \mathrm{FDC} - 3.73$ | Hanson *et al.* (1983) |
| $\mathrm{FDC} = 36.3\mu + 3.5$ | Hagstrom (1984) |
| $\mu = 0.0357\ \mathrm{FDDC} - 0.4219$ | Davis and Sieburth (1984) |
| $\mathrm{FDC} = 1.42 + 2.71\mu_d$ | Turley and Lochte (1985) |

[a] Where authors have used different temperatures and the one closest to 20°C was used.

*al.*, 1979) by incubating natural bacteria in a suitable medium. It is usual to use 3-μm-filtered natural water to provide bacteria free from predators, to dilute with 0.2-μm-filtered water and to add yeast extract as a nutrient at several concentrations (0–100 mg $l^{-1}$). The growth rate is estimated in batch culture from increases in total count during the exponential phase. In one case the calibration has been done in dialysis bags *in situ* in the Irish Sea (Turley and Lochte, 1985).

There is much discussion in the literature about the validity of the different FDC–μ equations (Newell *et al.*, 1986). Newell and Christian (1981) suggest that the logarithmic relationship is best, but others continue to use linear equations (Hagstrom, 1984; Turley and Lochte, 1985). Some workers have found that the FDC–μ relationship is highly dependent on temperature and so suggest that different equations should be used at different temperatures (Larsson and Hagstrom, 1982; Hagstrom, 1984). In other studies temperature dependence has not been found (Hanson *et al.*, 1983).

Despite these differences the production ($\mu \times$ biomass) of planktonic bacteria, estimated from FDC-derived growth rates, is very close to those obtained by other techniques (Newell and Fallon, 1982). However, investigators should be careful because FDC is not always proportional to growth rate, as was found in deep water (60 m) from the Irish Sea ($r^2 = 0.25\%$; Turley and Lochte, 1985). Many studies have avoided converting FDC to growth rate and have used it directly as an indicator of bacterial division (Meyer-Reil, 1983; Turley *et al.*, 1986; Goulder, 1988).

Using FDC in sediment presents special problems. It seems generally accepted that when the FDC–μ equations derived for planktonic populations are used for sediment to estimate productivity the results are too high (Newell and Fallon, 1982; Fallon *et al.*, 1983). Moriarty (1984) suggested that this was because the bacteria on sediments remained together after division for longer than in water. However, other research, based on a mathematical modelling approach to analysing bacterial microcolony counts, has suggested that sediment bacteria are very mobile (Davidson and Fry, 1987). So the reason for the inapplicability of FDC for benthic production estimates is not yet understood. Therefore in this habitat counts of dividing cells should be used without conversion.

Recently it has been suggested that frequency of divided and dividing cells (FDDC) can be used to estimate growth rate (Davis and Sieburth, 1984). This approach gave a good linear regression (Table V; $r^2 = 95\%$) with a *Vibrio* sp. and makes counting easier because more bacteria are scored (FDDC = 13–38%; Davis and Sieburth, 1984). It also gave sensible planktonic production estimates that agreed with other methods (Newell and Turley, 1987).

Dividing cells are defined as described previously and divided cells are pairs of bacteria with identical morphology that are separated by a gap much smaller than the cell radius and where the cell orientation indicates recent division. The rationale for counting divided cells is as follows. Most fluorochromes used in epifluorescence microscopy stain mainly DNA and RNA, so only the centre of the cell is stained. As the cell wall is unstained bacteria can be joined after division but will still appear to have a gap between them (Fig. 4). When the FDDC is calculated it is necessary to be

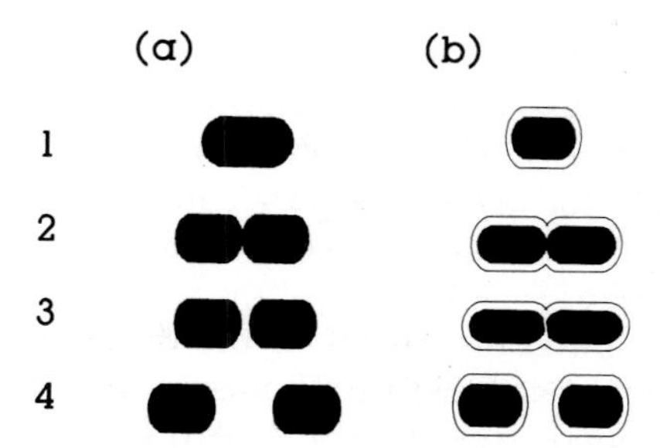


**Fig. 4.** Diagrammatic representation of rod-shaped bacteria at different stages of division as seen in fluorochrome-stained natural bacteria on membrane filters (a) and the same cells with the position of the outer part of the cell wall drawn in (b). The types of cells are: 1, single cell before division; 2, single dividing cell with a clear invagination; 3, pair of closely spaced divided cells immediately after division but before separation; 4, pair of single cells after division and complete separation.

very careful how the total count is calculated or errors and inconsistencies will result. It is best to count single and invaginated cells as one and pairs of divided cells as two, even though they count as one divided cell in the numerator of the FDDC ratio.

At present there seems no logical reason to use any particular one of the FDC–μ or FDDC–μ equations to estimate growth rate, so it is probably best to use these ratios only as growth indicators where possible. If growth rate estimates are required then it will probably be safest to perform some calibration experiments with the water being studied.

## V. Techniques requiring incubation

These techniques have mainly been applied to aquatic microbiology but many should also be useful in soil studies. They all attempt to count the number of bacteria that can actively metabolize, although they use a variety of indices of activity. Most of the methods have been developed in the last 10–15 years, as a response to the very low proportion of bacteria that appeared to be viable by plate count and the conviction by many microbial ecologists that many more bacteria in nature were really living.

Two general approaches which have been very well discussed in the literature will not be considered again here. These methods are micro-autoradiography, which can assess the uptake of radiolabelled substrates such as amino acids or DNA precursors like thymidine into individual cells, and immunofluorescence microscopy, which allows enumeration of specific strains or species of bacteria. Readers who want access to the literature should read Newell *et al.* (1986) for details of micro-autoradiography and Bohlool and Schmidt (1980) for immunofluorescence.

All the methods discussed in this section require incubation and so the samples must be transported to the laboratory quickly before processing. Most can be fixed after the incubation period before counting is undertaken. The details of counting methodology considered in detail in the last section apply equally well here.

There have been few comparisons of the different methods that have been proposed to count actively metabolizing bacteria, so a best method cannot be recommended. Newell *et al.* (1986) discusses the results of one comparison in detail using Chesapeake Bay water; their conclusions are very similar to a parallel study on the same water (Tabor and Neihof, 1984). The proportions of active bacteria (Fig. 5) varied significantly throughout the year and between the methods. No single method was best

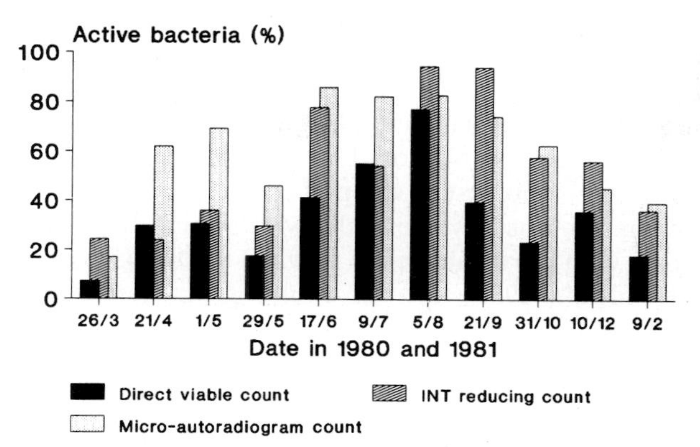


**Fig. 5.** The percentage of active bacteria in Chesapeake Bay assessed by three different methods (drawn from data presented by Newell *et al.*, 1986).

at all sampling times. The direct viable count (DVC) estimated by division inhibition (see Section V.A) never gave the highest proportion. Although the values were lowest in winter and highest in summer one method always showed at least 25% to be active. The consistently high results and the very high summer percentages of up to 94% indicate the high viability of bacteria from this habitat and the great value of these techniques for understanding aquatic microbiology. Similarly high percentages have been found in many other studies although the values for less eutrophic waters are sometimes lower (e.g. 0.7–7.9% by DVC in offshore ocean water; Kogure *et al.*, 1980).

In another study, using freshwater bacteria (Quinn, 1984), the count of respiring organisms was higher than the direct viable count. However, Maki and Remsen (1981) found no difference in other freshwater samples. Similarly Roszak and Colwell (1987) found no real difference between the direct viable count and micro-autoradiographic counts of bacteria taking up radiolabelled thymidine and glutamic acid.

The overall conclusion from these comparisons must be that no single method is best in all samples, on all occasions. However, all give similar proportions of active bacteria, so one of two alternative strategies should probably be adopted. Choose one method that works well in the habitat being studied and use this alone for comparative work. Use several methods together if it is vital to obtain the most accurate estimate of metabolically active bacteria.

### A. Division inhibition

This approach assesses the DVC and has also been called a count of substrate-responsive bacteria (Peele and Colwell, 1981) and synthetically active bacteria (Tabor and Neihof, 1984). I will use DVC in deference to Kogure *et al.* (1979) who first developed the method and coined the term. The method has been used successfully with natural bacterial populations in field studies, with pure cultures in survival studies (Roszak *et al.*, 1984; Roszak and Colwell, 1987) and for counting arsenate-resistant bacteria (Zelibor *et al.*, 1987).

The basic approach is to add an inhibitor of cell division (usually nalidixic acid) and a substrate (usually yeast extract) as sterile membrane-filtered solutions to a sample in a small glass flask. The mixture is incubated to allow the cells to grow on the substrate, but not to start dividing, and it is then fixed in formalin. The total number of bacteria and the number of elongated or enlarged cells are then counted, the DVC being the ratio of these two counts. There are several variations to the methodological details recommended by researchers who have used the technique (Table VI) and some of the key points of the method will be described below.

Nalidixic acid inhibits cell division of many Gram-negative bacteria but is ineffective against Gram-positive bacteria. Hence, Kogure *et al.* (1984) suggested using a mixture of nalidixic, piromidic and pipemidic acids to extend the range of organisms affected to include more Gram-negative and some Gram-positive bacteria. They also claimed that longer incubations were possible with this mixture, which made the elongated and enlarged cells easier to count. This mixture might well be useful if the approach is to be extended to soil where Gram-positive bacteria are abundant. However, other workers do not seem to have found it neccessary for water.

The concentration of nalidixic acid used has varied between 2.5 mg $l^{-1}$ and 100 mg $l^{-1}$. Tabor and Neihof (1984) compared three concentrations over a 12-month period and found only one significant difference when 60 mg $l^{-1}$ was best. This implies that the original and most commonly used concentration of 20 mg $l^{-1}$ is satisfactory.

The substrate used can affect the results obtained. Peele and Colwell (1981) have found that simple substrates like glucose, glutamate and vitamin-free casamino acids are not satisfactory, but yeast extract and tryptone work well. These authors also did a comprehensive test on substrate concentration and found that 500 mg $l^{-1}$ was optimal for both complex substrates. Their results agreed well with those of Tabor and Neihof (1984) who found incubation without added substrate unsatisfactory on most occasions. In one study yeast extract gave precipitation problems and so 50 mg $l^{-1}$ had to be used (Quinn, 1984).

**TABLE VI**
Methodological details used to estimate numbers of active bacteria in aquatic habitats by the division inhibition method

| Inhibitor and concentration (mg $l^{-1}$) | Yeast extract concentration (mg $l^{-1}$) | Incubation | | Reference |
|---|---|---|---|---|
| | | Time (h) | Temperature (°C) | |
| NA, 20 | 250 | 6 | 20 | Kogure *et al.* (1979)<br>Kogure *et al.* (1980)<br>Maki and Remsen (1981) |
| NA, 20 | 50–1500 (best = 500) | 6 | *In situ* | Peele and Colwell (1981) |
| NA, 20+<br>PA, 10+<br>PPA, 10 | 250 | 8 | 20 | Kogure *et al.* (1984) |
| NA, 20 or 50 | 50 | 10–13 | 20 | Quinn (1984) |
| NA, 20, 60 or 100 | 0 or 250 | 6 | *In situ* (0–28.3) | Tabor and Neihof (1984) |
| NA, 2.5 | 300 | 24 | 25 | Zelibor *et al.* (1987) |
| NA, 20 | 250 | 6–24 | 25 | Roszak and Colwell (1987) |
| NA, 20 | 250 | 18 | 25 | Al-Hadithi and Goulder (1989) |

NA, nalidixic acid, 1-ethyl-1,4-dihydro-7-methyl-4-oxo-1,8-naphthyridine-3-carboxylic acid; PA, piromidic acid, 8-ethyl-5,8-dihydro-5-oxo-2-pyrrolidinopyrido-2,3 d pyrimidine-6-carboxylic acid; PPA, pipemidic acid, 8-ethyl-5,8-dihydro-5-oxo-2-(1-piperazinyl)-pyrido 2,3 d pyrimidine-6-carboxylic acid trihydrate.

Incubation time and temperature are also important. The incubation time must be long enough to give large enough cells for easy counting. Most people have used 6–8 h, because dividing cells are apparent after 9–10 h (Kogure *et al.*, 1984). However, Quinn (1984) did not see dividing cells until >14 h and so used 10–13 h as the incubation time. Some workers have used very long incubation times, up to 24 h (Zelibor *et al.*, 1987; Roszak and Colwell, 1987; Al-Hadithi and Goulder, 1989). The importance of incubation temperature cannot be assessed properly as the correct research has not yet been done. However, the use of both *in situ* (0–28.3 °C; Tabor and Neihof, 1984) and fixed (20–25 °C; Table VI) seem satisfactory.

Most users of this technique count cells that are judged by eye to be longer or more swollen than before incubation. This might lead to two types of problem. Cells originally large before incubation and that do not enlarge during incubation would be scored wrongly positive. Tabor and Neihof (1984) did not think this was a real difficulty because in their comparative study all such cells also score as active by other methods. The work of Roszak and Colwell (1987) supported this conclusion with experiments that used a combination of division inhibition and micro-autoradiography. The other problem is that small cells might enlarge but still not score as active because they would not be big enough. Al-Hadithi and Goulder (1989) have tried to overcome these problems by measuring length distributions of bacteria before and after incubation. They recommended that by choosing a critical length (3.3–5.7 μm in their experiments) and subtracting the number before incubation from the number after the best possible DVC would be obtained. They showed that this alternative procedure could give up to 67% higher counts than their idea of the conventional method in which cells >16 μm were scored after incubation only. However, careful inspection of their data shows that by counting cells >6–8 μm after incubation only would give comparable results in most cases with much less effort.

The studies outlined earlier show clearly that division inhibition worked well for planktonic bacteria in fresh and salt water and for epiphytic bacteria. There is one report of its success in freshwater sediment (Quinn, 1984). However in my research group the method has not worked in marine sediment samples (Fry and Griffiths, unpublished) despite extensive trials with different inhibitors, concentrations and incubation times. The main reasons for this appeared to be adsorption onto or inhibition by the sediment, because neither freshwater bacteria nor a pure culture of *P. putida* elongated properly with sediment present, although they did so in control incubations without the sediment.

In principle it should be possible to automate the counting of membrane filters prepared for DVC estimation by using TV image analysis to measure lengths of cells. Singh *et al.* (1989) have described such a technique but have only used it with pure cultures in the laboratory. They used short incubation times (4 h) which gave relatively little cell enlargement from 1.8 μm to 5.7 μm for *E. coli*, which is remarkably similar to the optimum critical length suggested by the work of Al-Hadithi and Goulder (1989). Longer incubations gave areas of entangled filaments which were hard to count without extensive editing. It seems most likely that this automated approach should work well with planktonic samples that are relatively particle free.

## B. Labelling with tetrazolium salts

Active electron transport systems are an almost universal indication of bacterial metabolic activity. The tetrazolium salt 2-(*p*-iodophenyl)-3-(*p*-nitrophenyl)-5-phenyl tetrazolium chloride (INT) is preferentially reduced by the electron transport chain of respiring bacteria with the deposition of opaque, red formazan crystals. These crystals are sufficiently large and dense to be clearly visible within actively respiring cells by bright-field and epifluorescence microscopy. They are also formed in cyanobacteria, some algae and yeasts (Trevors, 1984). So actively respiring bacteria can be enumerated by incubating with INT and then counting the proportion of bacteria containing one or more dark formazan particles. This approach works in all types of habitat including sediment (Quinn, 1984; Swannell and Williamson, 1988) and soil (Macdonald, 1980).

The method was first thoroughly described by Zimmerman *et al.* (1978), and most studies use this basic technique, sometimes with modifications. Firstly this basic method will be described and then the modifications discussed. A 10-ml portion of sample suspended in water is mixed with 1 ml of 0.2% (w/v; aqueous solution) INT, to give a 200 mg $l^{-1}$ final concentration, and incubated for 20 min at the *in situ* temperature. The reaction is stopped with 0.1 ml of 37% (v/v) formaldehyde and storage for a month was possible at this stage. Zimmerman *et al.* (1978) then filtered the reaction mixture through a 0.2-μm Sudan Black-stained, polycarbonate membrane filter, and stained the bacteria on the filter with 100 mg $l^{-1}$ acridine orange for 2 min. After removal of the fluorochrome by filtration the filter was mounted in immersion oil and viewed for counting by epifluorescence and bright-field microscopy, a combination which was found to be useful to spot small, weakly fluorescing bacteria containing formazan particles.

Almost everyone who has used this method has used a final concentration of 200 mg $l^{-1}$ INT, although Swannell and Williamson (1988) used 400 mg $l^{-1}$. Only very short incubation times are needed with INT because the maximum number of respiring cells is reached after only 2 min (Zimmerman *et al.*, 1978). Despite this incubation for 30 min (Dutton *et al.*, 1983; Swannell and Williamson, 1988) and 45 min (Tabor and Neihof, 1984) have also been used. Sometimes extra substrate has been added during the INT incubation to encourage greater activity and hence stronger INT reduction. Two studies (Maki and Remsen, 1981; Dutton *et al.*, 1983) have used yeast extract and nalidixic acid, to stop cell division, incubated for 6 h, but this did not increase the count of respiring bacteria. Conversely, the studies that Macdonald (1980) carried out with soil films showed that a mixture of NADH and NADPH (both at 3 g $l^{-1}$) increased the proportion of INT-reducing bacteria by at least four times. In support, Quinn (1984) found that adding 0.08 g $l^{-1}$ NADPH, 0.5 g $l^{-1}$ NADH and 4.0 g $l^{-1}$ sodium succinate as nutrients increased counts by 37%. Although originally *in situ* temperatures were used some investigations have used room temperature which was presumably about 20 °C (Bitton and Koopman, 1982; Quinn, 1984).

It is not necessary to use the precise fluorochrome staining conditions used by Zimmerman *et al.* (1978). Quinn (1984) successfully used 10 mg $l^{-1}$ acridine orange and stained before filtration as discussed earlier (Section IV.B). Swannell and Williamson (1988) compared the effectiveness of several fluorochromes and found that ethidium bromide worked well but that DAPI and auramine fluorescence was inhibited by the formazan deposits.

One of the major problems with the Zimmerman *et al.* (1978) method is that the formazan deposits dissolve in the immersion oil used as mountant (Tabor and Neihof, 1982). Dutton *et al.* (1983) report that 25% of the formazan grains were lost after 60 min. This problem has been avoided with a very complex procedure using a gelatin matrix (Tabor and Neihof, 1982; 1984). This involves embedding the bacteria in a gelatin film on a coverslip, drying the film, staining with acridine orange, removing the filter and spraying the gelatin film with a gelatin mist. The coverslip is then put face down on a microscope slide and the bacteria, which are impregnated in the dried gelatin film, are counted. This method makes counting very easy using a mixture of epifluorescence and bright-field microscopy, because the granular filter background seen under bright-field illumination is not visible because the filter is not present in the gelatin matrix. This approach has certainly given much higher proportions of respiring bacteria than when immersion oil was used (Zimmerman *et al.*, 1978, 5–36% INT positive; Tabor and Neihof, 1984, 25–94%). Newell (1984) has improved the gelatin matrix method further by using an antioxidant layer instead of the gelatin mist. This modification improved the brightness of the fluorescence and worked particularly well for highly particulate water samples, but was not any simpler than the original technique.

However, there are other approaches that also work well. Quinn (1984) obtained very high counts (11–66%) with extra substrates, without using bright field to check small cells, but with immersion oil as mountant. Also Dutton *et al.* (1983) successfully used a saturated solution of fructose as mountant, which was gently boiled until 10% of its weight was lost. In my laboratory (Fry and Griffiths, unpublished) we have used warm 0.5% agar as mountant which does not dissolve INT formazan and found by comparison that paraffin oil gives very similar counts if done quickly (< *ca* 2 h) after preparation.

Sewage bacteria have been studied using another modification of the basic INT method (Bitton and Koopman, 1982; Dutton *et al.*, 1983). In this method, which they call MINT, malachite green is used to stain INT-treated cells. The treated bacteria are filtered through a cellulosic membrane filter, the filter is dried, cleared in immersion oil, stained with 1.0% (w/v) malachite green and counted under saturated fructose. They used this method primarily for looking at filamentous bacteria but it seemed to work well for all bacteria because the epifluorescent total count was not different from that obtained with MINT.

In conclusion it would seem that INT reduction can be used in most habitats. The best method would appear to use substrates (NADPH, NADH and succinate) and to mount using the gelatin matrix method. However, if the highest possible counts are not essential simple mounting in saturated fructose, warm agar or paraffin oil, followed by counting under epifluorescence alone should suffice.

### C. Vital stains

Several fluorogenic esters have been used as vital stains in ecological studies. The one which has been used most is fluorescein diacetate (FDA), but others like fluorescein dibutyrate and 3-*O*-methyl fluorescein phosphate (Pomeroy and Johannes, 1968) have also been used. These are polar non-fluorescent esters which enter the cell and are then hydrolysed by esterases to the non-polar fluorescein which is retained within the cell. Thus unlike conventional fluorescent stains like acridine orange and DAPI only active organisms with esterase activity will fluoresce under suitable illumination. Thus they are attractive for counting active bacteria.

They can be used successfully with different types of microorganisms such as yeasts (Paton and Jones, 1975), bacteria and fungi (Chet, 1986). However, they gave very variable results in some of the earlier studies with aquatic samples and so were not considered reliable (Jones and Simon, 1975). Chrzanowski *et al.* (1984) tested the dye with 11 Gram-positive and Gram-negative bacteria and found that all the former but only one of the latter fluoresced with acceptable brightness. These results were explained by transport difficulties with the Gram-negative bacteria. As most aquatic bacteria are Gram negative it seems unlikely that FDA will be of great use in this habitat, although 6–24% of the bacteria from a pond were found to be active by the method (Chrzanowski *et al.*, 1984).

Soil studies have been much more successful with both bacteria (Lundgren, 1981) and fungi (Chet, 1986). Lundgren (1981) showed that 78% of

111 soil bacterial isolates fluoresced after FDA staining and that 2–83% of the bacteria counted with acridine orange in 10 different soils were active. Thus it is basically this technique which will be described here.

An FDA stock solution is made by dissolving 0.25 g in 25 ml of acetone; this will keep for at least a month at 4 °C. Homogenized soil is allowed to settle for 1 min in a 500-ml measuring cylinder to remove large particles. Subsamples are appropriately diluted in membrane-filtered 0.5 M phosphate buffer (pH 8.5) and stained with 10 mg $l^{-1}$ FDA for 3 min. This is done by adding 10 μl of stock solution to 10 ml of diluted sample. The stained suspension is then filtered onto a 0.2-μm membrane filter as described previously for total counting and enumerated under blue excitation with a normal acridine orange filter set.

### D. Microcolony formation

It is possible to count the number of bacteria able to grow into microcolonies by microscopic observation after short periods of incubation. This approach has the advantage that it counts organisms that can both grow and divide directly without the use of inhibitors. It has mainly been used for aquatic habitats but also works well with soil bacteria in suspensions relatively free of particulate matter (Bakken and Olsen, 1987). Microcolony counts were first done by growing and concentrating the natural bacteria on membrane filters (Straskrabova, 1973; Kunicka-Goldfinger and Stronkowska, 1977; Meyer-Reil, 1977). This approach was particularly difficult because of the problems of viewing living bacteria on this fibrous or granular support. For this reason a slide culture technique was developed (Fry and Zia, 1982b), which was much easier to use and this method will be described below.

Firstly the bacteria must be concentrated. The density gradient method of Bakken (1985) is best for soil bacteria. Water samples (10 ml) can be centrifuged at 2000 *g* for 20 min at room temperature in sterile, tapered glass centrifuge tubes. The supernatant must be carefully removed by gentle vacuum to leave 0.3 ml and the bacteria resuspended by vigorous vortex mixing.

Thin (0.8–1 mm thick), glass microscope slides are etched with a 1 × 1 cm square and stored in 95% ethanol. They are sterilized by flaming, which also removes grease, and spread with 0.5 ml of molten, agar at 65 °C on the side that was not etched. Casein–peptone–starch agar proved the best medium and should be filtered through a coarse filter paper in a drying oven at about 65 °C before use. Stocks of sterile agar can be stored at 4 °C in boiling tubes and melted before use carefully in a boiling water bath. The agar film is then allowed to set at room temperature on a cool, levelled surface. Immediately after setting 10 μl of bacterial concentrate is spread on the agar over the etched square with a sterile, platinum wire loop. Absorption takes about 10 min after which the agar is covered with glass coverslips. The slides are then incubated, on supports, in a high-humidity incubation chamber made from Petri dishes containing dampened filter paper at 10 °C for 24 h and 72 h.

After incubation the slides are observed under phase contrast microscopy using a flat-field, high numerical aperture (1.3–1.4) 100× or 63× oil immersion objective. It is necessary to use a long working distance (2.0–2.2 mm) condenser to allow the illuminating light to be focused in the same focal plane as the bacteria. This ensures that the microcolonies are seen clearly. Single cells and microcolonies are then counted to give the total number of bacterial units. There are various ways to calculate the viability or percentage of active bacteria. Fry and Zia (1982a) recommend the use of the ratio of the number of microcolonies and the total number of single cells and microcolonies after incubation. This avoids counting before incubation and worked well on samples from 10 fresh waters. Incubation for two times is required because the bacteria in the various waters grow differently; when working at one site the incubation time giving highest viabilities should be used.

## References

Agar, A. W. (1988). *Microscope Anal.* **July,** 19–22.
Al-Hadithi, S. A. and Goulder, R. (1989). *Lett. Appl. Microbiol.* **8,** 87–90.
Anderson, J. R. and Slinger, J. M. (1975a). *Soil Biol. Biochem.* **7,** 205–209.
Anderson, J. R. and Slinger, J. M. (1975b). *Soil Biol. Biochem.* **7,** 211–215.
Babiuk, L. A. and Paul, E. A. (1970). *Can. J. Microbiol.* **16,** 57–62.
Baker, J. H. (1988). *In* "Methods in Aquatic Bacteriology" (B. Austin, Ed.), pp. 171–191, John Wiley, Chichester.
Bakken, L. R. (1985) *Appl. Environ. Microbiol.* **49,** 1482–1487.
Bakken, L. R. and Olsen, R. A. (1983). *Appl. Environ. Microbiol.* **45,** 1188–1195.
Bakken, L. R. and Olsen, R. A. (1987). *Microb. Ecol.* **13,** 103–114.
Bitton, G. and Koopman, B. (1982). *Appl. Environ. Microbiol.* **43,** 964–966.
Bjørnsen, P. K. (1986). *Appl. Environ. Microbiol.* **51,** 1199–1204.
Bjørnsen, P. K., Rieman, B., Pock-Steen, J., Nielsen, T. G. and Horstead, S. J. (1989). *Appl. Environ. Microbiol.* **55,** 1512–1518.
Bohlool, B. B. and Schmidt, E. L. (1980). *Adv. Microb. Ecol.* **4,** 203–241.
Bradbury, S. (1977). *In* "Analytical and Quantitative Microscopy" (G. A. Meek and H. Y. Elder, Eds), pp. 91–116, Cambridge University Press, Cambridge.
Bradbury, S. (1979). *J. Microsc.* **115,** 137–150.
Bradbury, S. (1981). *In* "Eleventh Annual Congress in Anatomy, Part B: Advances in the Morphology of Cells and Tissues" (M. A. Galina, Ed.), pp. 129–150, Alan R. Liss, New York.
Bratbak, G. (1985). *Appl. Environ. Microbiol.* **49,** 1488–1493.
Bratbak, G. and Dundas, I. (1984). *Appl. Environ. Microbiol.* **48,** 755–757.
Burns, R. G. and Slater, J. H. (1982). "Experimental Microbial Ecology", Blackwell Scientific Publishers, Oxford.

Cairns, J. (1982). *In* "Microbial Interactions and Communities, Volume 1" (A. T. Bull and J. H. Slater, Eds), pp. 249–285, Academic Press, London.
Chet, I. (1986). *In* "Perspectives in Microbial Ecology" (F. Mergusar and M. Gantar, Eds), pp. 489–491, Slovene Society for Microbiology, Ljubljana, Yugoslavia.
Christian, R. R., Hanson, R. B. and Newell, S. Y. (1982). *Appl. Environ. Microbiol.* **43,** 1160–1165.
Chrzanowski, T. H., Crotty, R. D., Hubbard, J. G. and Welch, R. P. (1984). *Microb. Ecol.* **10,** 179–185.
Clarke, K. R. and Joint, I. R. (1986). *Appl. Environ. Microbiol.* **51,** 1110–1120.
Coleman, A. (1980). *Limnol. Oceanogr.* **25,** 948–951.
Costerton, J. W., Nickel, J. C. and Ladd, T. I. (1986). *In* "Bacteria in Nature. Volume 2. Methods and Special Applications in Bacterial Ecology" (J. S. Poindexter and E. R. Leadbetter, Eds), pp. 49–84, Plenum Press, New York.
Dale, D. G. (1974). *Limnol. Oceanogr.* **19,** 509–518.
David, A. W. and Paul, J. H. (1989). *J. Microbiol. Meth.* **9,** 257–266.
Davidson, A. M. and Fry, J. C. (1987). *Microb. Ecol.* **13,** 31–45.
Davis, P. G. and Sieburth, J. M. (1984). *Mar. Ecol. Prog. Ser.* **19,** 237–246.
Doddema, H. J. and Vogels, G. D. (1978). *Appl. Environ. Microbiol.* **36,** 752–754.
Dodson, A. N. and Thomas, W. H. (1964). *Limnol. Oceanogr.* **9,** 455–456.
Dutton, R. J., Bitton, G. and Koopman, B. (1983). *Appl. Environ. Microbiol.* **46,** 1263–1267.
Fallon, R. D., Newell, S. Y. and Hopkinson, C. S. (1983). *Mar. Ecol. Prog. Ser.* **11,** 119–127.
Ferguson, R. L., Buckley, E. N. and Palumbo, A. V. (1984). *Appl. Environ. Microbiol.* **47,** 49–55.
Fliermans, C. B. and Schmidt, E. L. (1975). *Arch. Hydrobiol.* **76,** 33–42.
Francisco, D. E., Mah, R. A. and Rabin, A. C. (1973). *Trans. Am. Microsc. Soc.* **92,** 416–421.
Fry, J. C. (1988). *In* "Methods in Aquatic Bacteriology" (B. Austin, Ed.), pp. 27–72, John Wiley, Chichester.
Fry, J. C. and Davies, A. R. (1985). *J. Appl. Bacteriol.* **58,** 105–112.
Fry, J. C. and Humphrey, N. C. B. (1978). *In* "Techniques for the Study of Mixed Populations" (D. W. Lovelock and R. Davies, Eds), pp. 1–29, Academic Press, London.
Fry, J. C. and Zia, T. (1982a). *J. Gen. Microbiol.* **128,** 2841–2850.
Fry, J. C. and Zia, T. (1982b). *J. Appl. Bacteriol.* **53,** 189–198.
Fry, J. C., Goulder, R. and Rimes, C. (1985). *J. Appl. Bacteriol.* **58,** 113–115.
Fuhrman, J. A. (1981). *Mar. Ecol. Prog. Ser.* **5,** 103–106.
Gayford, C. G. and Richards, J. P. (1970). *J. Appl. Bacteriol.* **33,** 342–350.
Getliff, J. M. and Fry, J. C. (1989). *Bin. Comput. Microbiol.* **1,** 93–100.
Getliff, J. M. and Fry, J. C. (1990). *Bin. Comput. Microbiol.* **2,** 55–57.
Giovannetti, M. and Mosse, B. (1980). *New Phytologist* **84,** 489–500.
Godinho-Orlandi, M. J. L. and Jones, J. G. (1981). *J. Gen. Microbiol.* **123,** 81–90.
Goulder, R. (1977). *J. Appl. Bacteriol.* **43,** 399–405.
Goulder, R. (1987). *Lett. Appl. Microbiol.* **4,** 29–32.
Goulder, R. (1988). *Freshwater Biol.* **19,** 405–416.
Hagstrom, A. (1984). *In* "Current Perspectives in Microbial Ecology" (M. J. Klugg

and C. A. Reddy, Eds), pp. 495–501, American Society for Microbiology, Washington, DC.

Hagstrom, A., Larsson, U., Horstedt, P. and Normark, S. (1979). *Appl. Environ. Microbiol.* **37,** 805–812.

Hanson, R. B., Shafer, D., Ryan, T., Pope, D. H. and Lowery, H. K. (1983). *Appl. Environ. Microbiol.* **45,** 1622–1632.

Harper, R. M., Fry, J. C. and Learner, M. A. (1981). *Freshwater Biol.* **11,** 227–236.

Heldal, M., Norland, S. and Tumyr, O. (1985). *Appl. Environ. Microbiol.* **50,** 1251–1257.

Hobbie, J. E., Daley, R. J. and Jasper, S. (1977). *Appl. Environ. Microbiol.* **33,** 1225–1228.

Jackman, P. J. H. (1989). *In* "Computers in Microbiology, a Practical Approach" (T. N. Bryant and J. W. T. Wimpenny, Eds), pp. 25–40, IRL Press, Oxford.

Jones, J. G. (1974). *Limnol. Oceanogr.* **19,** 540–543.

Jones, J. G. (1979). "A Guide to Methods for Estimating Microbial Numbers and Biomass in Fresh Water", Freshwater Biological Association, Ambleside.

Jones, J. G. and Simon, B. M. (1975). *J. Appl. Bacteriol.* **39,** 317–329.

Kirchman, D. L. and Ducklow, H. W. (1987). *In* "Detritus and Microbial Ecology in Aquaculture" (D. J. W. Moriarty and R. S. V. Pullin, Eds), pp. 54–82, International Centre for Living Aquatic Resources Management, Manila, Philippines.

Kirchman, D. L. and Mitchell, R. (1982). *Appl. Environ. Microbiol.* **43,** 200–209.

Kirchman, D. L., Sidga, J., Kapulscinski, R. and Mitchell, R. (1982). *Appl. Environ. Microbiol.* **44,** 376–382.

Kogure, K. and Koike, I. (1987). *Appl. Environ. Microbiol.* **53,** 274–277.

Kogure, K., Simidu, U. and Taga, N. (1979). *Can. J. Microbiol.* **25,** 415–420.

Kogure, K., Simidu, U. and Taga, N. (1980). *Can. J. Microbiol.* **26,** 318–323.

Kogure, K., Simidu, U. and Taga, N. (1984). *Arch. Hydrobiol.* **102,** 117–122.

Krambeck, C., Krambeck, H. and Overbeck, J. (1981). *Appl. Environ. Microbiol.* **42,** 142–149.

Kunicka-Goldfinger, W. and Stronkowska, E. (1977). *Acta Microbiol. Pol.* **26,** 199–205.

Larsson, U. and Hagstrom, A. (1982). *Mar. Biol.* **67,** 57–70.

Lee, S. and Fuhrman, J. A. (1987). *Appl. Environ. Microbiol.* **53,** 1298–1303.

Lundgren, B. (1981). *Oikos* **36,** 17–22.

Lundgren, B. (1982). *Soil Biol. Biochem.* **14,** 537–542.

Lundgren, B. (1984). *Soil Biol. Biochem.* **16,** 283–284.

Macdonald, R. M. (1980). *Soil Biol. Biochem.* **12,** 419–423.

Maeda, M. and Taga, N. (1983). *La Mer* **21,** 202–204.

Maki, J. S. and Remsen, C. C. (1981). *Appl. Environ. Microbiol.* **41,** 1132–1138.

Meyer-Reil, L. A. (1977). *In* "Microbial Ecology of a Brackish Water Environment" (G. Rheinheimer, Ed.), pp. 223–236, Springer-Verlag, New York.

Meyer-Reil, L. A. (1983). *Mar. Biol.* **77,** 247–256.

Meyer-Reil, L. A., Dawson, R., Liebezeit, G. and Tiedge, H. (1978). *Mar. Biol.* **48,** 161–171.

Milthorpe, B. K. and Taylor, I. W. (1980). *J. Histochem. Cytochem.* **28,** 1224–1232.

Montagna, P. A. (1982). *Appl. Environ. Microbiol.* **43,** 1366–1372.

Monteseinos, E., Esteve, I. and Guerrero, R. (1983). *Appl. Environ. Microbiol.* **45,** 1651–1658.

Moriarty, D. J. W. (1984). *In* "Heterotrophic Activity in the Sea" (J. E. Hobbie and P. J. Le B. Williams, Eds), pp. 217–231, Plenum Press, New York.
Nagata, T. (1986). *Appl. Environ. Microbiol.* **52,** 28–32.
Newell, R. C. and Turley, C. M. (1987). *S. Afr. J. Mar. Sci.* **5,** 717–734.
Newell, S. Y. (1984). *Appl. Environ. Microbiol.* **47,** 873–875.
Newell, S. Y. and Christian, R. R. (1981). *Appl. Environ. Microbiol.* **42,** 23–31.
Newell, S. Y. and Fallon, R. D. (1982). *Microb. Ecol.* **8,** 33–46.
Newell, S. Y., Fallon, R. D. and Tabor, P. S. (1986). *In* "Bacteria in Nature. Volume 2. Methods and Special Applications in Bacterial Ecology" (J. S. Poindexter and E. R. Leadbetter, Eds), pp. 1–48, Plenum Press, New York.
Norland, S., Heldal, M. and Tumyr, O. (1987). *Microb. Ecol.* **13,** 95–101.
Olsen, R. A. and Bakken, L. R. (1987). *Microb. Ecol.* **13,** 59–74.
Palumbo, A. V., Ferguson, R. L. and Rublee, P. A. (1984). *Appl. Environ. Microbiol.* **48,** 157–164.
Parsons, T. R., Maita, Y. and Lalli, C. M. (1984). "A Manual of Chemical and Biological Methods for Seawater Analysis", Pergamon Press, Oxford.
Paton, A. M. and Jones, S. M. (1975). *J. Appl. Bacteriol.* **38,** 199–200.
Peele, E. R. and Colwell, R. R. (1981). *Can. J. Microbiol.* **27,** 1071–1075.
Pettipher, G. L. and Rodrigues, U. M. (1982). *J. Appl. Bacteriol.* **53,** 323–329.
Pomeroy, L. R. and Johannes, R. E. (1968). *Deep-Sea Res.* **15,** 381–391.
Porter, K. G. and Feig, Y. S. (1980). *Limnol. Oceanogr.* **25,** 943–948.
Quinn, J. P. (1984). *J. Appl. Bacteriol.* **57,** 51–57.
Ramsay, A. J. (1978). *N. Z. J. Mar. Freshwater Res.* **12,** 265–269.
Rice, A. L., Billett, D. S., Fry, J., John, A. W. G., Lampitt, R. S., Mantoura, R. F. C. and Morris, R. J. (1986). *Proc. R. Soc. Edin.* **88B,** 265–279.
Rimes, C. A. and Goulder, R. (1986). *Freshwater Biol.* **16,** 633–651.
Robarts, R. D. and Sephton, L. M. (1981). *J. Limnol. Soc. S. Afr.* **7,** 72–74.
Rodina, A. G. (1972). "Methods in Aquatic Microbiology", University Park Press, Baltimore.
Roser, D., Nedwell, D. B. and Gordon, A. (1984). *J. Appl. Bacteriol.* **56,** 343–347.
Roser, D. (1980). *Soil Biol. Biochem.* **12**, 329–336.
Roser, D. J. Nedwell, D. B. and Gordon, A. (1984). *J. Appl. Bacteriol.* **56,** 343–347.
Roser, D. J., Bavor, H. J. and McKersie, S. A. (1987). *Appl. Environ. Microbiol.* **53,** 1327–1332.
Roszak, D. B. and Colwell, R. R. (1987). *Appl. Environ. Microbiol.* **53,** 2889–2983.
Roszak, D. B., Grimes, D. J. and Colwell, R. R. (1984). *Can. J. Microbiol.* **30,** 334–338.
Russ, J. C. (1986) "Practical stereology", Plenum Press, New York.
Scavia, D. and Laird, G. A. (1987). *Limnol. Oceanogr.* **32,** 1017–1033.
Schallenberg, M., Klaff, J. and Rasmussen, J. B. (1989). *Appl. Environ. Microbiol.* **55,** 1214–1219.
Sieracki, M. E. and Webb, K. L. (1986). *EOS Trans. Am. Geophys. Union* **49,** 1298.
Sieracki, M. E., Johnson, P. W. and Sieburth, J. M. (1985). *Appl. Environ. Microbiol.* **49,** 799–810.
Sieracki, M. E., Reichenbach, S. E. and Webb, K. L. (1989). *Appl. Environ. Microbiol.* **55,** 2762–2772.

Singh, A., Pyle, B. H. and McFeters, G. A. (1989). *J. Microbiol. Meth.* **10,** 91–101.

Straskrabova, V. (1973). *Acta Hydrochim. Hydrobiol.* **1,** 433–454.

Swannell, R. P. J. and Williamson, F. A. (1988). *FEMS Microbiol. Ecol.* **53,** 315–324.

Tabor, P. S. and Neihof, R. A. (1982). *Appl. Environ. Microbiol.* **43,** 1249–1255.

Tabor, P. S. and Neihof, R. A. (1984). *Appl. Environ. Microbiol.* **48,** 1012–1019.

Trevors, J. T. (1984). *CRC Crit. Rev. Microbiol.* **11,** 83–100.

Trolldenier, G. (1973). *Bull. Ecol. Res. Comm. (Stockholm)* **17,** 53–59.

Turley, C. M. and Lochte, K. (1985). *Mar. Ecol. Prog. Ser.* **23,** 209–219.

Turley, C. M., Newell, R. C. and Robins, D. B. (1986). *Mar. Ecol. Prog. Ser.* **33,** 59–70.

Van Veen, J. A. and Paul, E. A. (1979). *Appl. Environ. Microbiol.* **37,** 686–692.

Velji, M. I. and Albright, L. J. (1986). *Can. J. Microbiol.* **32,** 121–126.

Vollenweider, R. A. (1974). "A Manual on Methods for Measuring Primary Production in Aquatic Environments", I.B.P. Handbook No. 12, 2nd Edn, Blackwell Scientific Publications, Oxford.

Watson, S. W., Novitsky, T. J., Quinby, H. L. and Valois, F. W. (1977). *Appl. Environ. Microbiol.* **33,** 940–946.

Weise, W. and Rheinheimer, G. (1978). *Microb. Ecol.* **4,** 175–188.

Wynn-Williams, D. D. (1985). *Soil Biol. Biochem.* **17,** 739–746.

Wynn-Williams, D. D. (1986). *In* "Perspectives in Microbial Ecology" (F. Mergusar and M. Gantar, Eds), pp. 191–200, Slovene Society for Microbiology, Ljubljana, Yugoslavia.

Wynn-Williams, D. D. (1988). *Polarforschung* **58,** 239–249.

Yamamoto, N. and Lopez, G. (1985). *J. Exp. Mar. Biol. Ecol.* **90,** 209–220.

Zelibor, J. L., Doughten, M. W., Grimes, D. J. and Colwell, R. R. (1987). *Appl. Environ. Microbiol.* **53,** 2929–2934.

Zimmerman, R. (1977). *In* "Microbial Ecology of a Brackish Water Environment" (G. Rheinheimer, Ed.), pp. 103–120, Springer-Verlag, New York.

Zimmerman, R., Iturriaga, R. and Becker-Birck, J. (1978). *Appl. Environ. Microbiol.* **36,** 926–935.

# 3
# Different Types of Continuous Culture in Ecological Studies

J. C. GOTTSCHAL

*Department of Microbiology, Biological Centre, University of Groningen, Kerklaan 30, 9751 NN Haren, The Netherlands*



## I. Introduction

The chemostat has become the most widely used apparatus for studying microorganisms under constant environmental conditions. It is designed to create a self-adjusting steady-state condition ideally suited for physiological studies since it provides a most convenient way of keeping microorganisms in a well-defined physiological condition over long periods of time. The overwhelming number of original papers and reviews on the use of the chemostat in microbial physiology adequately demonstrates its enormous importance for these types of microbiological studies. The following list of references serves to provide access to both the more recent and the older literature (Tempest, 1970; Harder and Dijkhuizen, 1976, 1983; van der Hoeven and de Jong, 1984; Matin, 1981; Gottschal and Dijkhuizen, 1988).

METHODS IN MICROBIOLOGY
VOLUME 22 ISBN 0-12-521522-3

The question of whether the chemostat would also be useful in ecological studies came to the fore somewhat later, after the apparatus had already become popular with many microbial physiologists. This hesitancy in using the chemostat for ecological research is perhaps quite understandable in view of the scepticism and warnings also expressed by one of the pioneers in marine microbial ecology. In discussing the steady state in a chemostat it was noted that "... the chemostat is an almost ideal and a legitimate tool in microbial physiology [but] its induced time independence of steady state represents an utterly unnatural situation" (Jannasch, 1974). This prudent attitude was perhaps even more explicitly expressed in a review by Jannasch and Mateles (1974) who stated with respect to natural environments: "... the absence of exponential growth and anything but complete mixing, the formation of niches, complex interactions, inconstancy of external conditions, and so on, produce heterogeneity of a degree that renders the continuous culture concept (especially the chemostat concept) irrelevant for a useful description or analysis". Yet in spite of these firm warnings the chemostat has eventually become a widely used tool in ecological studies. In fact it has been this very complexity of natural ecosystems, as referred to in the above citations, which has prompted ecologists to look for an experimental model system which could be used to study isolated basic phenomena in the ecology of microbes. It is this invaluable property of the chemostat which has made it the pre-eminent tool for studying the influence of environmental factors on the growth of pure and mixed cultures of bacteria under precisely controlled conditions. Extensive descriptions of the numerous applications of the chemostat in ecological research have appeared in many reviews (Gottschal, 1986; Gottschal and Dijkhuizen, 1988; van der Hoeven and de Jong, 1984; Veldkamp and Jannasch, 1972; Veldkamp, 1977; Kuenen and Harder, 1982; Jannasch and Mateles, 1974; Rhee, 1980; Brown *et al.*, 1978; Tempest and Neyssel, 1978). As long as the danger is recognized of equating the chemostat with certain natural systems the chemostat will continue to play an essential role in ecological studies; even more so if this technique is used in combination with other types of open culture systems and with studies directly in the field.

Both the theory and the practice of chemostat operation have been covered extensively elsewhere (Evans *et al.*, 1970; Bazin, 1981; Pirt, 1975; Novick and Szilard, 1950; Herbert *et al.*, 1956). Therefore the emphasis of the present contribution is on a few non-standard deviations or extensions of the conventional chemostat in ecological studies. Since nowadays high-quality chemostat equipment can be obtained readily from many different commercial firms only those technical aspects of the equipment which relate to the modifications of the standard chemostat will be discussed.

## II. The chemostat

### A. Theory

The following summary of the basic theory of the chemostat serves as a brief reminder of the most important factors governing the growth of bacteria under nutrient-limited conditions. For more detailed information the reader is especially advised to consult the treatments of this subject by Tempest (1970), Pirt (1975) and Bazin (1981).

In a chemostat the fresh medium is supplied at a constant rate ($F$). Since the culture fluid is removed at the same rate a constant volume ($V$) is maintained and the dilution rate ($D$) of the chemostat is defined as $F/V$. The composition of the medium is such that, in the most simple case, growth in the chemostat results in (almost) complete consumption of only one of the nutrients, the growth-limiting substrate, with a concentration of $S_r$ in the feed. The specific rate at which growth proceeds will at first be maximal but upon depletion of the limiting nutrient this rate will rapidly decline according to the well-known empirical relationship proposed by Monod (1942):

$$\mu = \mu_{max} s/(K_s + s) \tag{1}$$

in which $\mu$ (= specific rate of growth) equals $(1/x)(dx/dt)$, with $x$ representing biomass, $s$ the actual concentration of the growth-limiting nutrient and $K_s$ the half saturation constant for growth. Due to the supply of fresh medium growth does not stop completely but continues at a rate determined by the feed rate. The combined effect of growth and dilution eventually leads to a steady state in which $\mu = D$:

$$dx/dt = \mu x - Dx = (\mu - D)x = 0 \tag{2}$$

Such a steady state can be obtained for values of $D$ below $D_c$ (= critical dilution rate) which in most cases is close to $\mu_{max}$:

$$D_c = \mu_{max}[S_r/(K_s + S_r)] \tag{3}$$

Under steady-state conditions the change in substrate concentration is given by the following balance:

$$ds/dt = D(S_r - s) - \mu x/Y = 0 \tag{4}$$

in which $s$ represents the limiting substrate concentration in the culture and $Y$ a field coefficient defined as the quantity of cells produced per substrate consumed. From equation 4, solved for $x$, it follows that

$$\bar{x} = Y(S_r - \bar{s}) \tag{5}$$

and from equation 1, solved for $s$, it follows that

$$\bar{s} = K_s D/(\mu_{max} - D) \tag{6}$$

In these formulations $\bar{x}$ and $\bar{s}$ denote the steady-state values for biomass and limiting substrate, respectively.

Although the theory is based on the assumption of only one growth-limiting nutrient, multiple substrate-limited growth in chemostats has been studied frequently [reviewed by Harder and Dijkhuizen (1976, 1982)]. Some confusion as to whether two or more substrates can simultaneously become truly growth-limiting has existed for some time (Gottschal, 1986), but this confusion is resolved by making a distinction between nutrients used for distinctly different metabolic purposes (e.g. nitrogen + phosphate, carbon source + electron acceptor, etc.) and those used in physiologically similar processes (e.g. different carbon sources, different electron acceptors, etc.) In general only in the latter case may two or more nutrients contribute to one and the same intracellular requirement and thus simultaneously limit growth. The possibility of growing bacteria under conditions of multiple substrate limitation doubtless is one of the very important features of the chemostat, making it an extremely useful tool in ecological studies. This applies to growth of pure cultures but perhaps even more so to the study of mixed cultures (Gottschal, 1986; Gottschal and Dijkhuizen, 1988).

## B. Equipment

Ever since the original description of a chemostat appeared in print (Novick and Szilard, 1950) this culture system has continued to kindle the creativeness of those who work with it. This has led to the development of an enormous number of more or less useful concepts. The choice of the most appropriate type of chemostat should be based primarily on the degree of sophistication required. Descriptions of a wide range of different chemostat designs can be found in the literature (Evans *et al.*, 1970; Koch, 1971; Veldkamp and Kuenen, 1973; Laanbroek, 1978; Kuenen and Harder, 1982; Radjai *et al.*, 1984; Titus *et al.*, 1984). In addition to fermentors, specifically designed and used by scientists to accommodate flexible, bench-scale research activities, a considerable choice of chemostats is offered by commercial firms. Some of the most experienced in this field are New Brunswick Scientific Co., New Brunswick, USA; Applikon B.V., Schiedam, The Netherlands; L.H. Fermentation, Slough, UK; Braun-Melsungen, Melsungen A.G., W. Germany; Gallenkamp, Loughborough, UK.

What follows is a brief description of a bench-scale laboratory-type chemostat built according to a design which has proven its effectiveness over many years of laboratory practice. Its usefulness is based especially on a combination of the following properties: a very high degree of flexibility; low cost: the apparatus is constructed mainly of readily available standard materials and measuring equipment; both strictly anaerobic and aerobic cultivation are possible without major changes; long-term reliability; well suited for analysis of the composition of the in- and outgoing gases. A detailed diagram of this chemostat is presented in Fig. 1. This diagram is to a large extent self-explanatory but several features of the equipment will be discussed.

The culture vessel is made of Pyrex glass with a side arm positioned such that a working volume is obtained of approximately 500 ml. The lid is made of black neoprene rubber (10 mm thickness) which is clamped on the flange of the vessel with four screws. Holes are drilled in the lid to fit tubing and electrodes. Mixing and aeration are carried out using a Teflon-coated magnetic bar (with a length of approx. 5 cm). This type of aeration is quite sufficient for cultures with a relatively low density (up to approx. 400 mg protein per litre), but at higher densities in combination with high dilution rates more sophisticated aeration techniques are required.

For most experiments it is of vital importance that the feed rate remains exactly the same and is known precisely. For this purpose peristaltic tube pumps are best suited. Those made by Verder, Düsseldorf, W. Germany, have proved extremely reliable and are available in several versions covering a wide range of possible flow rates. A most critical component of the medium-supply system is the type of tubing used in the feed line and especially in the peristaltic pump. For aerobic cultivation the use of silicon tubing is definitely most convenient because it is flexible, autoclavable, transparent, inert and durable. However, since oxygen diffuses extremely rapidly through this material, it cannot be used for growth of oxygen-sensitive anaerobes and may even pose problems when oxygen consumption balances are made in aerobic cultures. A very good alternative for anaerobic cultures is black butyl rubber tubing, which is flexible, autoclavable, relatively inert and durable. It is advisable (and cheap!) to use ordinary PVC tubing for those parts of the oxygen-free nitrogen gassing system which need not be autoclaved, since PVC is highly resistant to oxygen diffusion. However, as tubing for the peristaltic pump, both of these alternatives are quite unsuitable. In our experience by far the best material available today

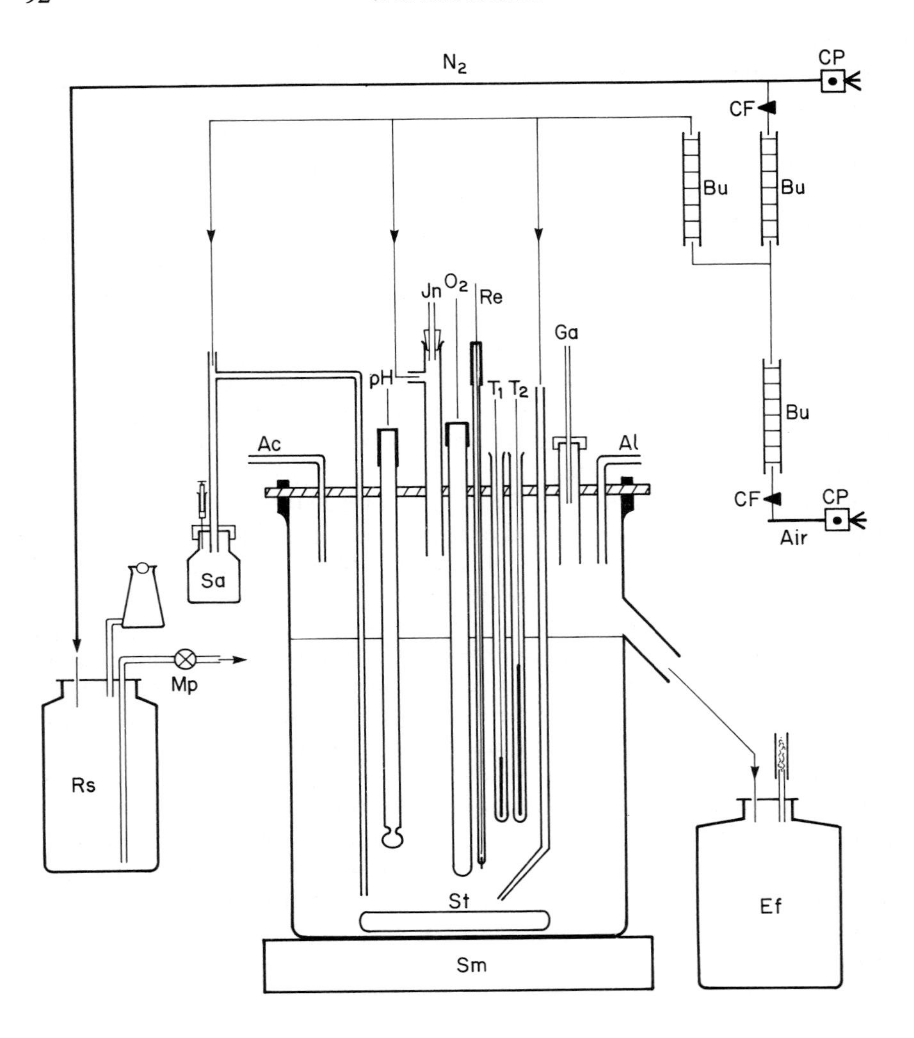


**Fig. 1.** Schematic drawing of a small-scale (approx. 500 ml working volume) low-cost glass chemostat. Rs, medium reservoir; Sa, sampling bottle; Sm, stirring motor; St, magnetic stirrer bar; Ef, effluent vessel; Ac and Al, titration-inlet for acid and alkali; pH, autoclavable pH electrode; In, medium inlet; Mp, peristaltic medium pump; $O_2$, oxygen electrode; Re, redox electrode; T1 and T2, temperature sensor and heating element, respectively; Ga, possible gas outlet for analysis; Bu, burette with soap solution for measuring the gas flow; Cf, constant flow regulator; Cp, constant pressure regulator. All gases pass cotton wool filters (not shown) before entering the fermentor. The $N_2$ is freed of traces of oxygen by passage over heated copper turnings.

for this purpose is Marprene rubber tubing, available from Smith and Nephew Watson-Marlow, Falmouth, UK. This material combines the flexibility and durability of silicon rubber with the low oxygen diffusivity of PVC and butyl rubber; however, it is not transparent.

Temperature control can be achieved in several ways. Placing the entire chemostat vessel in an incubator room or in a temperature-controlled water bath is one solution. A more flexible approach is the use of a water-jacketed fermentor vessel or the application of a heating element in combination with a temperature sensor and an electronic temperature controller. For this latter solution we use a Pt100 temperature sensor and a heating element of 125 W with an alternating voltage supply of 36 V. Both elements are fitted in Pyrex glass tubes (inner diameter 9 mm) immersed in the culture liquid. To improve heat transfer to the culture, the glass tubes contain a few ml of silicone oil. This kind of heating device has proved extremely reliable and accurate (within 0.5 °C) in our standard 500-ml chemostats over a temperature range 25–75 °C.

In dense cultures the pH will change due to metabolic activities of the bacteria, but also in dilute cultures with relatively low buffering capacity this will happen. Since this latter situation is very common particularly in ecological studies, pH control is usually required. Currently there is an enormous choice of combination pH electrodes (which include a reference cell), most of which will function satisfactorily. A most important property for their use in continuous cultures is that they must be heat sterilizable. In our experience Ingold double junction electrodes (Ingold AG, Urdorf, Switzerland) have proved extremely reliable both with respect to sterilization and stability in long chemostat runs. For using these electrodes in a pH-control system any high-quality pH meter can be used which offers a facility to activate a small (peristaltic) tube pump when a chosen set point is reached. Perhaps the most critical aspect of such a control system is the right type of tubing in combination with a given pump. Again Marprene tubing has appeared a very good choice (see above) with small peristaltic pumps for example from Smith and Nephew Watson-Marlow, Falmouth, UK.

Measurement of the oxygen partial pressure as a measure of the dissolved oxygen concentration in aerobic cultures is strongly advised since the metabolism of most microorganisms is strongly affected by this parameter. The most convenient and very sensitive oxygen probes presently available are those based on the polarographic measuring principle. Those produced by Ingold AG are at this moment the best choice. They can be used to detect partial pressures down to at least 0.1% of air saturation. To fully exploit the capabilities of these electrodes at low oxygen tensions we have used full-scale calibration gases containing only 1% or 2% oxygen in nitrogen. The electrodes are fully autoclavable and their Teflon/silicone membranes last for many months of continuous operation and, if necessary, can easily be replaced. Since their working principle demands the constant application of a polarizing voltage, they require dedicated measuring equipment. However, Ingold can supply the electrodes with a small preamplifier that generates the required voltage and provides an output current which can be measured directly with any simple meter used for the more conventional galvanic-type oxygen probes. The Ingold electrodes can also be used very effectively as the sensor in control loops designed to maintain constant oxygen tension. Technically the most simple solution for this is feedback control of the stirring speed. In our laboratory fermentors this type of regulation allows very accurate control, typically to within 0.2% air saturation at the extended 0–2% scale.

In spite of the quality of these modern oxygen probes their sensitivity is as yet insufficient to be of any use when cultures are grown under conditions of oxygen limitation. For such cultures we have successfully used redox probes. These probes consist of a 1-cm platinum rod ($\phi = 1$ mm) which extends approximately 5 mm from a 20-cm glass tube in which a thin platinum wire connects the electrode tip with a flexible lead which can be plugged into any sensitive, high-impedance mV meter. The only further requirement is that the reference side of the pH electrode also serves as the reference for the redox (Eh) measurement. The usual Ag/AgCl in 3 M KCl reference cell has a potential of approximately +210 mV relative to the standard hydrogen electrode which can be checked with any standard redox buffer system (for example, obtainable from Ingold AG). Although in this way the redox value of the culture can be measured accurately, more importantly these probes can be used to control oxygen tension at a level much below the detection limit of the oxygen probes. Through feedback regulation based on stirring speed the redox reading in our cultures can be maintained constant to within approx. 3 mV. As long as the oxygen concentration is really low (typically below 0.5% air saturation) and the culture liquid composition does not change significantly the redox reading responds sharply to any change in aeration. More information on the use of this type of regulation can be found in Ohta and Gottschal (1988a, b), Radjai *et al.* (1984) and Gottschal and Szewzyk (1985).

## C. Examples

### 1. *Enrichments*

For many ecological studies performed in the laboratory selective enrichment represents a crucial step. It is either used to obtain the desired physiological type of bacteria or bacterial community, or it serves to support hypotheses concerning the competitive strength of a given bacterium or community. Although enrichment in batch culture is still by far the most common procedure, very often the use of a chemostat would be more appropriate. Over the last 20 years the use of chemostats for selective enrichment has been reviewed quite adequately (Jannasch, 1967; Schlegel and Jannasch, 1967; Veldkamp, 1976; Brown *et al.*, 1978; Parkes, 1982; Gottschal, 1986; Gottschal and Dijkhuizen, 1988). Therefore I will restrict myself to a brief summing up of the major characteristics of chemostat enrichments and give a few illustrative examples.

Chemostat selection occurs under conditions of substrate limitation of some sort and thus will favour those species which are most successful in scavenging the limiting nutrients while growing at suboptimal growth rates. Since this situation will be more common in nature than prolonged periods of nutrient excess it is likely that the organisms obtained in this way are more representative of those prevailing in the natural environment.

Unlike batch cultures chemostats offer the possibility of maintaining growth conditions and thus selection pressure constant for prolonged periods of time. This may encompass conditions of programmed change resulting in, for example, light/dark cycles, pH, temperature or aerobic/anaerobic transitions, thus making the chemostat a very precise selection tool.

Since under conditions of nutrient limitation several nutrients can limit growth (rate) simultaneously, selection can be directed towards versatile microorganisms, specialized in using (many) different substrates at the same time. This type of bacteria could easily be missed by only employing batch culture techniques as they might not possess the highest $\mu_{max}$ values on single substrates.

Chemostats can be used very successfully to select for and isolate microbial communities. This contrasts with batch culture selection in which simple microbial food chains will largely appear as successional changes in populations, whereas a continuous flow system allows simultaneous growth of the entire community.

The classical demonstration of the existence of bacteria which are specialized in growing at very low nutrient concentrations has been given by Jannasch (1967). Using different reservoir concentrations and various dilution rates (0.05–0.5 $h^{-1}$) different species became dominant. Isolation of the various species and subsequent studies in pure and defined mixed cultures revealed that a *Spirillum* sp. (isolated at a low dilution rate) was most competitive at $D < 0.55\ h^{-1}$ whereas at higher values (and in batch culture) a *Pseudomonas* sp. (isolated at relatively high $D$ values) always grew faster. This represents the first well-documented example of 'crossing $\mu$–$s$ relationships' which in later years have been found for many other types of bacteria

(Veldkamp, 1977; Kuenen and Harder, 1982; Kuenen *et al.*, 1977; Matin and Veldkamp, 1978; Laanbroek *et al.*, 1983; Gottschal, 1985).

In the course of our work on the ecology of 'mixotrophic' thiobacilli (capable of both heterotrophic and autotrophic growth) we have noted that these 'generalists' never gain dominance in batch cultures. They are always outgrown by much faster growing 'specialized' heterotrophs and autotrophs. However, competition experiments between such a 'mixotroph' (*Thiobacillus versutus*), a heterotroph and a specialized 'autotroph' in mixed substrate-limited chemostats revealed that the 'generalist' gained a strong competitive advantage under such conditions (Gottschal *et al.*, 1979; Gottschal and Kuenen, 1980b). The ability of *T. versutus* to use acetate (a 'heterotrophic' substrate) and thiosulfate (an 'autotrophic' substrate) simultaneously made it possible to coexist with both its competitors and (depending on the relative amounts of acetate and thiosulfate in the feed) in some cases to dominate the culture almost completely. To find out whether this ability would be a general property of facultatively autotrophic thiobacilli chemostat enrichments were set up using mixtures of acetate or glycolate and thiosulphate in the feed. This type of selective enrichment resulted in the isolation of 'mixotrophic' thiobacilli from a number of freshwater environments (Gottschal and Kuenen, 1980a). It should be noted that although dominance of 'mixotrophs' was obtained after 5–20 volume changes, the cultures were by no means pure (see Table I). A considerable background of secondary species was evident and will probably remain whatever the duration of the enrichment. This is certainly not characteristic of growth on mixtures of substrates as in our experience single nutrient-limited enrichments never yield really pure cultures. This is most probably explained by some leakage or excretion of metabolic products from both viable and lysing cells of the primary population.

In a relatively new field of research, the microbial degradation of xenobiotic compounds, the benefits of selective chemostat enrichments have also become quite evident. A good example of such enrichments is the isolation of several very complex microbial communities capable of degrading various man-made herbicides such as Dalapon, Dicamba, Lontrel, Fenuron and others (Daughton and Hsieh, 1977; Senior *et al.*, 1976; Senior, 1977; Slater, 1981). A most striking feature of these stable communities is the presence of more than one primary herbicide-consuming species. A fascinating web of mutual dependencies in combination with the complexity of the primary substrate is responsible for the observed stability of these mixed cultures. A somewhat similar situation may exist in mixed cultures grown on halogenated aromatic compounds (Hartmann *et al.*, 1979; Reineke and Knackmuss, 1984; Schmidt *et al.*, 1985; Schmidt, 1987; Taeger *et al.*, 1988; Oltmanns *et al.*, 1988), although many of these compounds

**TABLE I**

Outcome of chemostat enrichments after 15–20 volume changes at a dilution rate of $0.05\,h^{-1}$ with various mixtures of thiosulphate and acetate as growth-limiting substrates (after Gottschal and Kuenen, 1980a)

| Inoculum | Substrate concentrations in the reservoir (mM) | Dominant population | Secondary population |
|---|---|---|---|
| Canal | Thiosulphate (30) + acetate (5) | Mixotroph | Heterotroph |
| Canal | Thiosulphate (10) + acetate (15) | Mixotroph | Heterotroph |
| Ditch | Thiosulphate (30) + acetate (5) | Mixotroph | Autotroph + heterotroph |
| Ditch | Thiosulphate (20) + acetate (10) | Mixotroph | Heterotroph |
| Ditch | Thiosulphate (10) + acetate (15) | Mixotroph | Heterotroph |
| Marine sediment | Thiosulphate (30) + acetate (5) | Autotroph + heterotroph | Heterotroph |

can be degraded completely by pure cultures. An interesting chemostat enrichment of bacteria able to grow on 3-chlorobenzoate (3CB) has been described by Dorn *et al.* (1974). The fermentor was inoculated with a mixed sample of sewage sludge and allowed to adapt to growth on benzoate. After 1 week the benzoate was gradually, over a period of 2 weeks, replaced by 3CB. Five weeks later a steady-state culture had been obtained from which agar plates were inoculated, resulting in the isolation of *Pseudomonas* sp. strain B13. This species, which possesses efficient ortho-cleaving chlorocatechol-degrading enzymes, has subsequently been used in elegant enrichment studies using more recalcitrant aromatics. In one of these studies a phenol-degrading mixed culture isolated from soil was inoculated together with *Pseudomonas* sp. strain B13 into a chemostat in which phenol together with an increasing amount of 2-, 3- and 4-chlorophenol were used as the major carbon and energy sources. Eventually complete degradation of the chlorophenols was observed. It was further shown that the ortho-cleavage pathway was acquired by hybrids of the phenol-degrading species from strain B13 (Schmidt *et al.*, 1983). Interestingly a control culture without strain B13 did not degrade the chlorophenols and washout of this culture took place. It was concluded that, instead of the strain B13-type

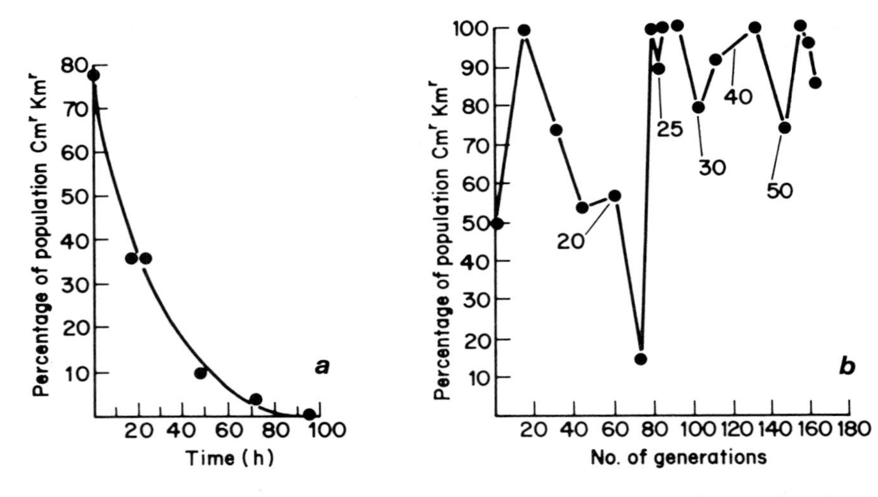


**Fig. 2.** (a) Loss of a plasmid by *B. subtilis* BC1 in non-selective batch culture. Zero time refers to inoculation in the batch culture medium. (b) Proportion of the Cm- and Km-resistent (=$Cm^rKm^r$) organisms in a Cm-containing chemostat culture of *B. subtilis* BC1. The arrows denote the increases in the Cm concentration in the medium reservoir with the figures representing the new concentrations ($mg\,l^{-1}$). (From Fleming *et al.*, 1988.)

ortho cleavage, meta-cleavage pathways had been induced resulting in the formation of toxic ('suicide') metabolites (Bartels *et al.*, 1984).

In an overview by Dykhuizen and Hartl (1983) the enormous selective pressure exerted on phenotypically pure cultures by prolonged chemostat cultivation has once more been stressed. Although this phenomenon does not represent enrichment in its usual meaning it does show how cells with only very small genetic differences within a 'pure' culture can be selectively enriched in a chemostat. A recent illustration of this type of enrichment was provided by the work of Fleming *et al.* (1988) on the isolation of *Bacillus subtilis* strains with improved plasmid stability. Cultivation of *B.subtilis* strain BC1, carrying a plasmid encoding for chloramphenicol (Cm) and kanamycin (Km) resistance and for the production of heat-stable $\alpha$-amylase, in batch culture resulted in the loss of the plasmid from most of the cells within 96 h (Fig. 2a). However, by imposing a selective pressure on Cm resistance in carbon-limited chemostat cultures by including increasing amounts of Cm in the feed (Fig. 2b), *B.subtilis* strains could be isolated that had retained their plasmid in 96-h batch cultures. The authors further showed that the increased stability did not result from incorporation in the host DNA or from altered properties of the plasmid itself. Ecologically the most interesting observation made by these authors is that the strains with the higher plasmid stability isolated at the end of the chemostat run (after 160 generations) had almost three times higher $\mu_{max}$ values than the original plasmid-containing strain. Apparently the 'hindrance' caused by the plasmid has been overcome. The strains improved in this way are now able to outcompete the parent plasmid-carrying strain both in batch and in chemostat culture.

### 2. *Mixed cultures*

A major difficulty in the study of bacteria in their natural habitats is our poor knowledge of the exact physicochemical conditions, the precise identity and number of available growth substrates and the influence of the numerous other (micro)organisms present in the same habitat. In this field the chemostat is particularly useful as it enables the ecologist to create well-defined two-, three- (or even more) membered mixed cultures and to study them under precisely defined growth conditions. It is hoped that with this approach basic principles underlying the natural ecosystem can be understood. Against this background it is understandable that the ability of various species to compete for one or more substrates has been studied so extensively (Veldkamp, 1976, 1977; Veldkamp and Jannasch, 1972; Matin and Veldkamp, 1978; Jannasch and Mateles, 1974; Fredrickson, 1977; Kuenen *et al.*, 1977; Laanbroek *et al.*, 1983, 1984; Veldkamp *et al.*, 1984; Gottschal *et al.*, 1979). For example the results from competition experiments between versatile and specialized thiobacilli, mentioned above, have revealed the fundamental importance of metabolic flexibility for species with a low $u_{max}$. This has now been confirmed for (facultatively) anaerobic species (Nanninga *et al.*, 1986; Laanbroek *et al.*, 1979; Dykhuizen and Davies, 1980; Kuenen and Robertson, 1984). A further important conclusion from these results and from mathematical description of growth under mixed substrate limitation is the importance of cell yield for the competitiveness of the versatile species (Gottschal, 1986). Briefly, the higher the yield of the versatile species on the non-shared substrate, the more effectively it competes with the specialized species for the shared substrate. Of course, for the more general case of several species competing for several substrates, yield may thus also be decisive for a species' competitiveness.

Another very interesting field of research in which the use of a chemostat seems quite indispensible is that of the interactions between the various trophic levels of the pelagic microbial ecosystem (Azam *et al.*, 1983). From competition experiments between algae and bacteria for limiting phosphate the conclusion could be drawn that in many cases bacteria outcompete the algae. In mixed cultures in which the algae are the sole source of organic carbon for the bacteria (through excretion and lysis) this commensalistic relationship prevents complete washout of the algae: stable mixed cultures can be obtained (Bratbak and Thingstad, 1985; Currie and Kalff, 1984; Pengerud *et al.*, 1987; Mayfield and Inniss, 1978). However, additional supply of organic carbon will inevitably lead to the washout of the algae, a situation not likely to be encountered in natural environments. Further experimentation along these lines by Pengerud *et al.* (1987) elegantly showed that the introduction of a third trophic level by inoculating a bacterivorous nanoflagellate allowed establishment of stable three-membered communities over a very wide range of glucose/phosphate ratios. This approach, particularly in combination with mathematical modelling, seems especially well suited for investigating the extremely complex relationships between (in)organic pollution, algal proliferation and predation in marine and freshwater ecosystems.

Interspecies hydrogen transfer, a mutualistic type of interaction, is believed to play a crucial role in most, if not all, anaerobic ecosystems (Wolin, 1982; Conrad *et al.*, 1986; Thiele and Zeikus, 1988). Again chemostat cultivation may be the best way to investigate in detail the consequences of this interaction for the microorganisms involved. To my knowlege Ianotti *et al.* (1973) were the first to use this approach in studying $H_2$ transfer between *Ruminococcus albus* (a cellulolytic species from the rumen) and *Vibrio succinogenes*. In coculture under glucose-limited conditions *R. albus* was shown to obtain considerably more energy from the fermentation of glucose than in pure culture, mainly as a result of a shift in its fermentation pattern towards more acetate production. This phenomenon has since been reported for many other fermentative species, both in batch and in chemostat cultures. More recently Traore *et al.* (1983) demonstrated that due to $H_2$ transfer a mixed culture of *Desulfovibrio vulgaris* and *Methanosarcina barkeri* could grow with lactate or pyruvate in the absence of sulphate. The methanogen successfully replaced sulphate as the electron acceptor. In the chemostat under lactate (or pyruvate) limiting conditions the concentration remained extremely low, indicating a very tight coupling between both production and consumption, whereas in batch culture transient accumulation of $H_2$ occurred. Clearly, the first situation with electron donor limitation represents a more ecologically relevant condition.

An important aspect of many ecosystems is the non-homogeneity (Wimpenny, 1981; Wimpenny *et al.*, 1983). This property is, almost by definition, not reproduced in a chemostat which requires perfect mixing for proper operation. An interesting consequence of this is that in many ecosystems aerobic and anaerobic conditions will prevail virtually next to one another. Fluctuations in the rate of supply and/or consumption of oxygen will result in considerable variations in the actual oxygen concentrations 'sensed' by both aerobic and anaerobic bacteria. Thus although the natural inhomogeneity is hard to reproduce in a chemostat some of its consequences for certain microorganisms can be studied most successfully in well-controlled chemostats. In some of our studies on the influence of very low oxygen concentrations on (facultative) anaerobes we have used redox-controlled chemostats (see above). A facultatively anaerobic *Vibrio* sp., isolated from the top few centimeters of estuarine tidal mud flats (Gottschal and Szewzyk, 1985) was grown under dual limitation of glucose and oxygen. Varying the redox value at which the culture was poised resulted in a gradual shift from purely fermentative growth to fully aerobic growth. In between these extremes the organism smoothly adapted to the limiting amounts of oxygen by fermenting the glucose for which not enough oxygen was available for respiration (Fig. 3). In subsequent experiments it was shown that if under these oxygen-limiting conditions the sulphate-reducing bacterium *Desulfovibrio* sp. strain HL21 was introduced into the culture, stable coexistence was obtained. Clearly, the facultative anaerobe was able to consume the oxygen, which was sparged continuously in the culture, effectively enough to prevent intoxification of the sulphate-reducing species. Recently we have obtained similar results with cocultures of an obligate aerobe (a *Pseudomonas* sp.) and an obligate anaerobe (a *Veillonella* sp.) (*in prep.*). In this case lactate was used as the shared limiting organic substrate with oxygen limiting the growth of *Pseudomonas* sp. Furthermore, in recently performed chemostat enrichments under oxygen limiting conditions, methanogens were shown to coexist with strictly aerobic species (Gerritse *et al.* 1990). It seems likely that this type of protection will contribute considerably to the occurrence of strict anaerobes in aerobic environments.

So far I have quite deliberately restricted the choice of examples to a few cases of relatively well-defined mixed cultures. Yet, several investigations have been reported in which attempts were made to reproduce in the chemostat entire communities as they occur in their natural habitats (Veilleux and Rowland, 1983; Slyter *et al.*, 1966; Freter *et al.*, 1983; Marsh *et al.*, 1983). So far this type of investigation has not really brought us much further in understanding the respective ecosystems. A chemostat, it should be stressed again, is not meant to simulate *in vivo* systems. For that purpose one needs far more sophisticated experimental equipment ['microcosms' (Pritchard and Bourquin, 1984)] and measurements directly 'in the field'.

### 3. *Controlled non-continuous cultures*

One of the major characteristics of a chemostat culture, the continuity of growth conditions, also yields the greatest conceptual problem for using it in ecological research. It is very hard to find natural habitats which show

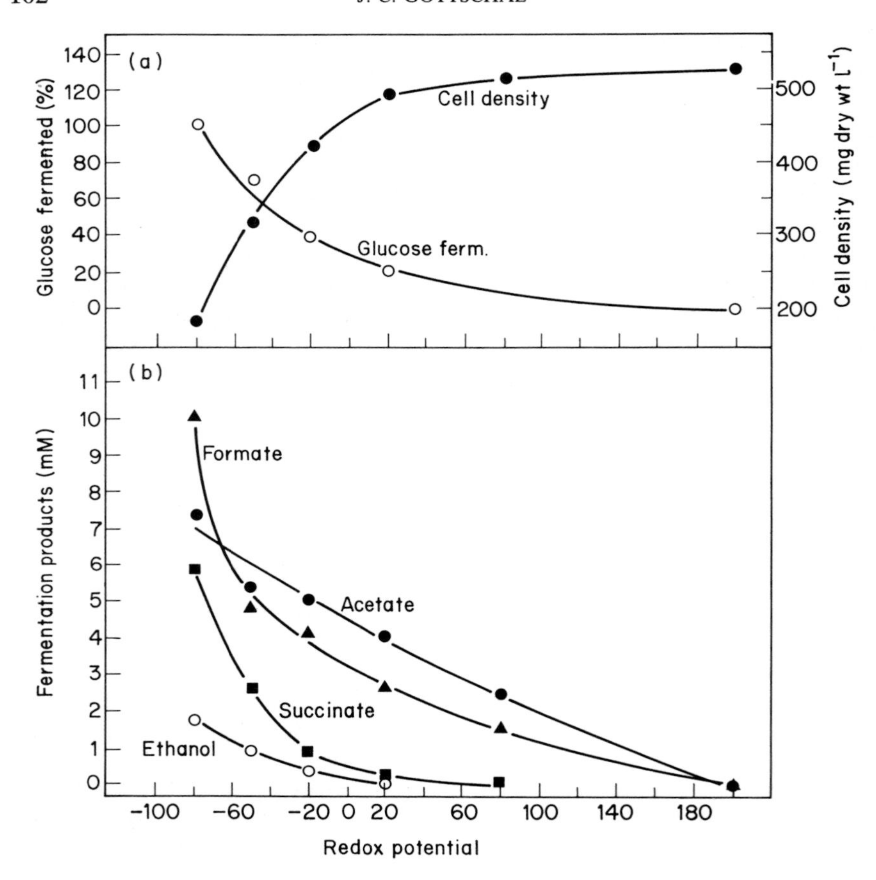


**Fig. 3.** Growth of *Vibrio* HP1 in continuous culture at a dilution rate of 0.08 $h^{-1}$ limited by both oxygen and glucose at various levels of oxygen supply. (a) Percentage of glucose fermented, relative to total consumption (○—○) and cell density (●—●) at different aeration rates. (b) Fermentation products detected in steady-state cultures as a function of the oxygen supply: succinate (■—■), acetate (●—●), formate (▲—▲), and ethanol (○—○). (From Gottschal and Szewzyk, 1985.)

a degree of constancy with respect to nutrient availability and physico-chemical conditions which comes anywhere near that established in a continuous culture at steady state (Wimpenny, 1981). But due to its other very important property, precise control of the growth environment, the chemostat offers excellent opportunities for studying microbial growth under finely regulated alternating conditions. The equipment required for this type of experimentation is essentially the same as discussed in Section II.B. The only additional requirement is a timed switching device allowing a switch between two (or more) different controlled conditions. For simple switching between different states of light intensity, type or degree of growth limitation, pH, temperature or dilution rate most electrical switching clocks are quite adequate. However, far more sophisticated time-dependent transitions can be made using (micro) computer-controlled chemostats. Instead of the simple switching from the one to the other state almost any gradual shift between two (or more) environmental conditions could be programmed, which would indeed add enormously to the versatility of the chemostat in ecological studies. In fact computer-controlled fermentors have become quite common in industry and the equipment required for laboratory use can also be obtained from most firms. Yet, to my knowledge no studies have been reported in which these new facilities have been exploited to any significant degree in chemostat studies in microbial ecology.

An interesting example of the influence of short-term environmental changes on the competitiveness of physiologically rather similar bacterial species was given by Leegwater (1983). In this study mixed cultures were subjected to either a continuous limiting supply of glucose or to pulse-wise addition of this carbon source. In a mixed culture of *Bacillus* sp. and *Klebsiella aerogenes* with a continuous supply of glucose, the *Bacillus* sp. always outcompeted the former species whereas with a pulsed addition of glucose this result was reversed. A similar result was obtained with mixed cultures of *K. aerogenes* and *Escherichia coli*. Under continuous glucose-limiting conditions, at dilution rates above $0.2\,h^{-1}$, *K. aerogenes* always outgrew *E. coli* whereas a pulsed glucose supply resulted in washout of *K. aerogenes*. Finally, when *Bacillus subtilis* Marburg was grown together with *E. coli* under continuous limitation the latter was outcompeted but a pulsed supply of glucose resulted in *E. coli* outcompeting the *Bacillus*. The physiological basis for these results has not yet been resolved but of course it is tempting to speculate these observations reflect an adaptation of both enteric bacteria to a life of 'feast and famine', so typical for gastrointestinal tracts. Surely a lot more laboratory and field work will have to be done before we can explain why the bacilli, typical inhabitants of soil, seem so much better adapted to a more continuous supply of nutrients. However, one could rightly be sceptical with respect to using chemostats in studying survival strategies of spore-forming soil inhabitants.

An environment for which a fluctuating or pulsed supply of nutrient is very characteristic is the mouth ecosystem. The complex community of the dental plaque is faced with intermittent periods of substrate supply, pH change and severe limitation. For this reason van der Hoeven *et al.* (1985) studied the competition between various oral *Streptococcus* spp. in chemostats under alternating conditions of glucose excess and limitation. All *Streptococcus mutans* strains tested were outcompeted by a *S. sanguis* strain or a *S.milleri* strain at a dilution rate of $0.2\,h^{-1}$ and a pH of 7.0 under continuous limitation of glucose (10 mM in reservoir). These experiments were then repeated with hourly pulses of concentrated glucose (final concentration in the culture = 10 mM) superimposed on the continuous supply of glucose. At the moment of the glucose additions the pH regulation was interrupted to allow a rapid drop in pH down to 5.5 in 9 min followed by automatic KOH titration to restore a culture pH of 7.0 20 min after the glucose pulse was given. The most significant result was that under such discontinuous growth conditions the *S. mutans* strains either outcompeted the other *Streptococcus* spp. or at least coexisted over long periods of time.

As part of our study on the ecology of versatile, flexible 'mixotrophic' thiobacilli (see also Section II.C.1) the competitiveness of these organisms was also studied under conditions of alternating supply of organic (acetate) and inorganic (thiosulphate) growth substrates (Gottschal *et al.*, 1981a, b; Beudeker *et al.*, 1982; Kuenen *et al.*, 1985). It was found that in spite of their superior capacity to grow with continuous addition of mixtures of these two different substrates, alternating supply resulted in their washout if both a heterotrophic and autotrophic 'specialist' were present in the culture. Apparently the metabolic flexibility of the 'mixotroph' is incompatible with a continuously changing substrate supply since such a regime results in the rapid loss of oxidative/uptake capacity for a given substrate during its absence. Under such conditions the physiological rigidity of more specialized species makes sure that after a period in which a certain substrate is absent immediate uptake is guaranteed upon its reappearance.

Clearly phototrophic (micro)organisms are most evidently subjected to alternating conditions of growth due to their dependence on light for energy generation. Therefore it is not surprising that for studying their ecology these organisms are grown frequently under various light–dark regimes (van Liere, 1979; van Gemerden, 1974; Rhee, 1980; Rhee *et al.*, 1981; van Gemerden and Beeftink, 1983; Loogman, 1982; van Gemerden and de Wit, 1989). The term 'cyclostat' was introduced by Chisholm *et al.* (1975) for cultures that were subjected to a 24-h cycle of an identical light–dark pattern. For a very useful review on the use of cyclostats and some theoretical considerations of this type of algal cultivation the reader is referred to Rhee (1980) and Rhee *et al.* (1981). Some interesting studies of this type of cultivation can also be found in the work on mixed cultures of anaerobic phototrophic bacteria (van Gemerden, 1974; Kuenen *et al.*,

1985; van Gemerden and de Wit, 1989). Among the earliest examples are competition experiments in sulphide-limited chemostats between the purple sulphur bacteria *Chromatium weissii* and *C. vinosum* (van Gemerden, 1974). In the presence of continuous light *C. vinosum* always outcompeted *C. weissii* completely, but an intermittent light–dark regime established stable coexistence of the two species. This result was explained by the finding that in the dark period sulphide accumulated, which was most rapidly taken up and oxidized by *C. weissii* upon reillumination, whereas *C. vinosum* appeared capable of most rapid growth in the presence of very low, rate-limiting sulphide concentrations. Another, very recent example concerns the ecology of the purple sulphur bacterium *Thiocapsa roseopersicina* which very often represents the dominant species immediately underneath the top layer of oxygenic phototrophic organisms in laminated microbial mats (Chisholm *et al.*, 1975). In this ecosystem *T. roseopersicina* is exposed to rather drastic environmental fluctuations such as changes in the concentration of oxygen (20–200 μM in some cases) and sulphide in combination with the daily changes in illumination. *T. roseopersicina* is physiologically extremely versatile as it is not only capable of anaerobic phototrophic growth but also grows chemolithotrophically with oxygen and thiosulphate (as electron donor) in the dark. However, in the field this bacterium will be confronted with alternating (semi)-aerobic + light and anaerobic + dark conditions and the influence of these particularly adverse conditions needs attention when studying its ecology. When sulphide-limited continuous cultures ($D = 0.03\ h^{-1}$) in continuous light received various alternating regimes of (semi)-aerobic (= 50 μm $O_2$) and fully anaerobic conditions enhanced rates of photopigment synthesis during the anaerobic period compensated for the lack of synthesis during the aerobic period. During the semi-aerobic conditions photosynthesis continued, glycogen was stored intracellularly and only pigment synthesis was repressed. Applying an additional light–dark regime coinciding with the aerobic–anaerobic regime showed that in the dark the glycogen served as an energy source for the synthesis of photopigment which in the subsequent semi-aerobic + light period resulted in sufficient growth to prevent washout.

### 4. *Multistage chemostats*

With the appearance of the chemostat (Monod, 1950; Novick and Szilard, 1950) a model system became available that eliminated heterogeneity with respect to time. In such a culture system the cells, once in steady state, no longer possess a history, that is, their physiological state has become independent of any previous conditions of growth. This property was by no means considered desirable for all applications. Especially for biosynthesis of certain products (in industrial processes) and for obtaining maximal cell and product yields a cultivation system was required that allowed, on a continuous basis, the cells to pass the various developmental stages of growth (Ricica, 1970; Ricica and Dobersky, 1981; Herbert, 1961, 1964). Yet however important these considerations may be for industrial fermentations they are of much less concern to the microbial ecologist who wants to exploit the chemostat as a practical model system for features of microbes in the natural environment. The problem there is that the chemostat's homogeneity is in overt contrast with the enormous temporal and spatial heterogeneity 'in the field' (Wimpenny, 1981; Billen, 1982; Wimpenny *et al.*, 1983). It is mainly for this reason that in recent years several attempts have been made to include some degree of heterogeneity in continuous culture systems. This approach may be particularly useful in studying natural processes which are based on the activity of complex communities in which the participating microbial populations are spatially separated. In such cases the populations depend on the exchange of solutes (and cells). This aspect of community structure may be approached using spatially separated chemostat vessels. However, this does not mean that 'simple' chemostats are no longer of much use in studying natural ecosystems. In my opinion too much progress has been made using simple chemostats and still too little has been learned from coupled chemostat systems to state that "rules applying to simpler levels . . . may be irrelevant to an understanding of behaviour at a higher level of complexity" (Wimpenny *et al.*, 1983). It is very tempting indeed to increase the complexity of chemostat-based systems, but it should not be attempted to simulate the complexity of the ecosystem under study. Their very strength is their simplicity, providing the opportunity to study phenomena which 'in the field' are too complex to be approached experimentally.

What follows are a few interesting examples of the use of some different types of multistage chemostats. It is hoped that these applications provide the best illustration of what may or may not be accomplished by using coupled chemostats.Theoretical analyses of the various types of multistage systems have appeared elsewhere (Herbert, 1964; Ricica, 1970; Ricica and Dobersky, 1981; Pirt, 1975; Lovitt and Wimpenny, 1981; Powell and Lowe, 1964).

The most simple type of multistage chemostat, the two-stage cascade, has been most extensively used for studying predator–prey relationships such as predation of bacteria by ciliates (Sambanis and Fredrickson, 1987; Jost *et al.*, 1973; Swift *et al.*, 1982; Glaser, 1988; Curds and Cockburn, 1971). The prey bacteria are grown in a standard chemostat under carbon- and energy-limiting conditions and the outflow of this culture serves as the feed for the second stage which contains the predator. Changing the

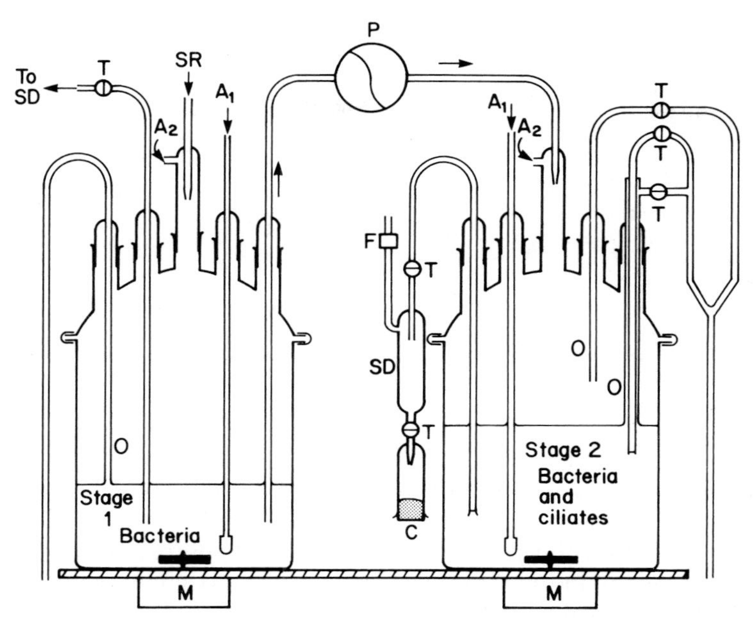


**Fig. 4.** Two-stage continuous culture apparatus, $A_1$, sterile air inlet for aeration; $A_2$, air inlet to prevent contamination of the feed line; C, cotton wool plug; F, air filter; M, magnetic stirrer; O, constant-level overflow device; P, peristaltic pump; T, tap; SD, sampling device; SR, sterile medium input. (From Curds and Cockburn, 1971.)

working volume of either stage and/or the setting of the pumps used to feed the first and/or the second stage allows precise control of the growth rate of both cultures. The main reason for using this type of equipment is the opportunity it provides to study the growth kinetics of the predator under steady-state conditions, without growth of the bacteria. However, in recent years this latter assumption has been shown to be wrong in several cases (Sambanis and Fredrickson, 1987). Bacteria have been demonstrated to grow at the expense of lysis and excretion products from the ciliates. Yet, this finding does not invalidate the use of a two-stage cascade system, as it remains the most convenient way to manipulate predator and prey independently.

In spite of the fact that the first reported experiments using this technique for predator–prey studies were done some 18 years ago (Curds and Cockburn, 1971) the basic design of the equipment has not changed (Fig. 4). The use of a pump between the two stages is to be preferred over siphoning

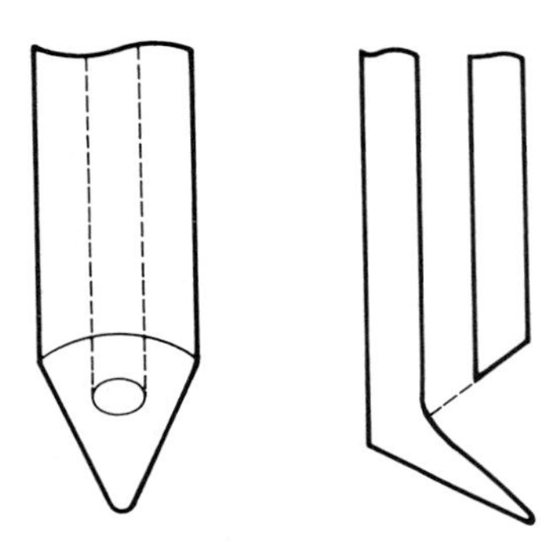

**Fig. 5.** Tip of the overflow line of the chemostat vessel. The inside diameter of the tubes is 2 mm. (From Sambanis and Fredrickson, 1987.)

systems as it is a more precise way of flow control. Siphoning is a much cheaper solution but it does induce considerable irregularities in flow rate, which, especially with small working volumes (<approx. 100 ml), affects the dilution rate of the second stage. To improve this situation the aeration gas flow may be used to force the culture liquid through a tube with its tip at the surface of the liquid. According to Sambanis and Fredrickson (1987) the actual shape of the tip can have considerable influence on the 'smoothness' of the flow (Fig. 5).

A very elegant recent study by Glaser (1988) nicely illustrates the use of such two-stage cultures. In his study a possible explanation was sought for the well-recognized but still insufficiently explained observation that the addition of protozoa to acquatic microcosms enhances rates of decomposition (Fenchel and Jørgensen, 1977; Sherr *et al.*, 1982). It was hypothesized that protozoa might contribute to overall decomposition rates by directly consuming dissolved organic matter. This idea was born out from the fact that a widely distributed protozoan, *Tetrahymena pyriformis*, is quite capable of taking up amino acids and sugars and can even grow axenically in defined medium. When *T. pyriformis*, grown in the second stage on *E. coli* cells supplied from a first stage, was exposed to a pulse or continuous addition (from a second feed directly into the second stage) of glycine or histidine rapid uptake by the protozoan was shown with starting amino acid concentrations of 20–40 $\mu$g C ml$^{-1}$. This uptake did affect the ciliate or bacterial populations slightly and it was reasoned that, given the high amino acid concentrations used, the effect would be much less in the field. The potential ciliate amino acid uptake rates and field data of the entire natural population were of the same order of magnitude. However, the uptake affinity constants for the protozoa and values measured in the field almost certainly rule out a significant contribution of these protozoa to overall decomposition in many natural habitats. In those cases, particularly in benthic ecosystems, in which high nutrient and ciliate concentrations could be present (at surfaces and immediately around organic particles), ciliates might play a direct role in organic matter decomposition. It must be stressed that many alternative explanations have been suggested (see Glaser, 1988) which could all contribute to the enhancement of mineralization by protozoa. Most of these possibilities have not yet been sufficiently verified and some of them could be tested quite adequately by using simple two-stage chemostat cultures.

Although extensions of the very simple two-stage sequential chemostat arrangement have been used in some effluent treatment systems it appears that its use in the study of natural ecosystems has hardly been considered. Perhaps the only well-documented case is a study of the anaerobic breakdown of glucose and benzoate by the anaerobic community of a marine sediment (Thompson *et al.*, 1983). The aim of the investigation was to examine the flow of carbon in the anaerobic community with respect to the type of intermediates involved. The system consisted of a five-stage chemostat series which was successfully operated for a period of approximately 2 years. Indeed, it was shown that by using this technique all five vessels were different with respect to the relative contribution of the major steps in the anaerobic mineralization of organic carbon to methane and carbon dioxide. The method permits sampling and experimental manipulation of the intermediate stages which might, in some cases, be difficult to achieve with single-stage cultures. However, too little experimental work has yet been done to allow a balanced judgement of the actual importance of such a fairly complicated cultivation technique for ecological studies.

The same may be said of the even more complicated bidirectional multistage chemostat system: the gradostat (Lovitt and Wimpenny, 1981; Wimpenny *et al.*, 1983). The system consists of any number of chemostats, connected in series, with a medium reservoir and a sink for spent culture at both ends (Fig. 6). The purpose of this set-up is not only to create separate habitats, as in the above unidirectional multistage system, but in addition to create opposing gradients of solutes. It is claimed that the gradostat can be used in many different areas of microbial ecology, but especially when studies are made of ecosystems in which adjacent habitats are linked by the transfer of solutes (and cells). As an example of the use of this system *Paracoccus denitrificans* was grown in the presence of two opposing gradients of succinate and nitrate (Lovitt and Wimpenny, 1981). It was shown that most extensive growth occurred in vessel 3, less in vessels 2 and 4 and relatively little growth in vessels 1 and 5. This result confirms the expectations, as at both ends one of the two essential nutrients will

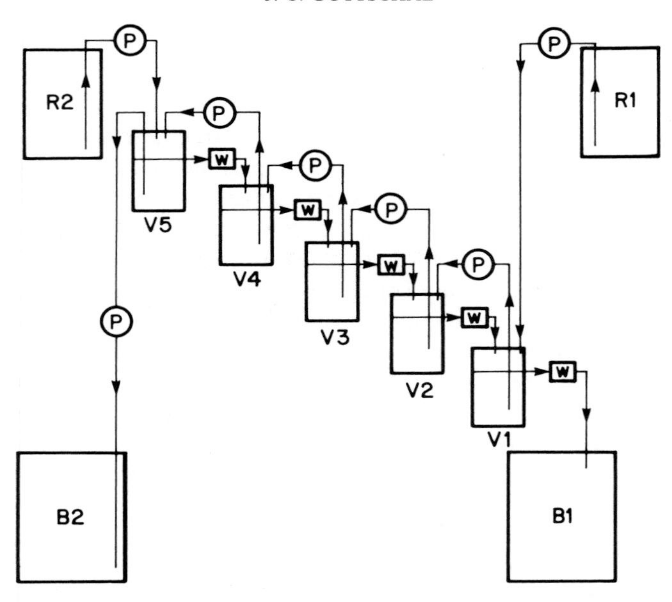


**Fig. 6.** Schematic drawing of a gradostat. The vessels are fed with medium from reservoir R1 in an upward direction, with transfer between the vessels by a series of pumps (P). Medium from reservoir R2 is fed down the sequence of vessels over a series of weirs (w). B1 and B2 are the effluent vessels. (From Lovitt and Wimpenny, 1981.)

largely have been consumed in the preceding vessels. In another experiment opposing gradients of oxygen and glucose were created in a gradostat inoculated with a mixed culture of a facultative anaerobe, *Bacillus* sp., and *Clostridium butyricum.* This time total cell count was roughly equal in all five vessels with the facultative anaerobe dominating at the aerated side and the anaerobe at the glucose (anaerobic) side.

The gradostat does indeed provide a sophisticated means of creating distinct physicochemical 'niches' which remain linked (as in natural habitats) through exchange of solutes and cells. This may in some cases be a very important feature for the study of a given ecosystem. However, up to now no examples of such studies seem to have appeared in the literature, although attempts to use the gradostat in an ecophysiological study of the rhizosphere are forthcoming (E. G. Niemann, personal communication). It is not unlikely that this reluctance is at least partly due to the considerable complexity and the resulting high vulnerability to mechanical failure. It should also be noted that, in spite of the claims made by those advocating

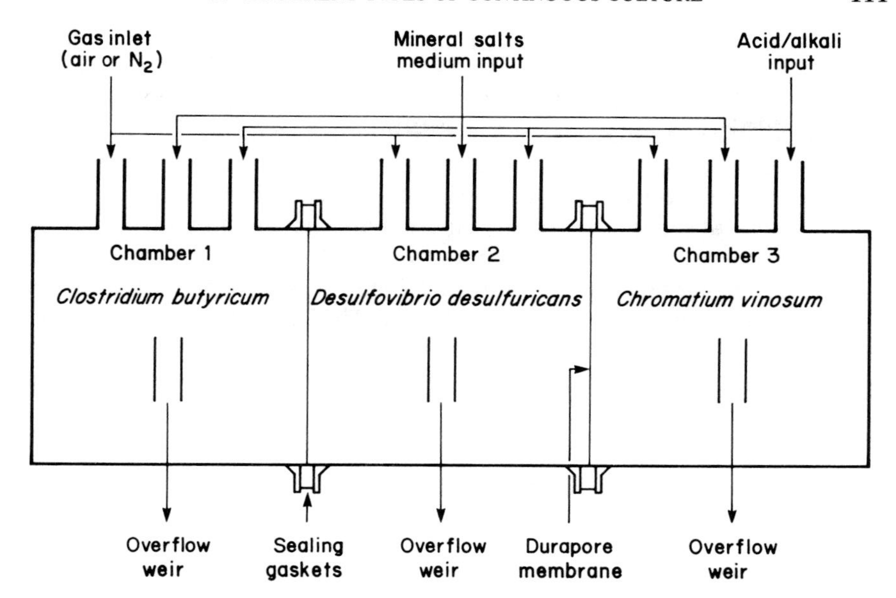


**Fig. 7.** A simplified schematic of a three-stage diffusion chemostat. (From Keith and Herbert, 1985.)

the gradostat (Wimpenny *et al.*, 1983; Lovitt and Wimpenny, 1981), growth of bacteria in 'opposing gradients' should not be taken too literally. Firstly, each one of the chemostats represents one set of physicochemical conditions and in most practical cases the difference with the neighbouring vessels will be considerable, and far from gradual. Secondly, in the above example with an opposing glucose–oxygen gradient it is very difficult to see that much (if any) glucose will ever reach the vessel next to the first one receiving the glucose-containing feed, in which *C. butyricum* will consume all glucose supplied to this vessel.

Yet another type of multistage chemostat system, developed by Keith and Herbert (1985) and named a "bidirectional continuous flow compound diffusion chemostat", has been used recently to study microbial interactions. Schematically the equipment is shown in Fig. 7. It is characterized by three chemostats of approx. 1 l working volume, each with their own external feed and overflow (which is not fed into a neighbouring vessel), physically separated from each other by Durapore semipermeable filters (Millipore UK Ltd, Harrow, UK) with a pore size of 0.2 μm and a surface of approx. 150 $cm^2$. Clearly this arrangement permits exchange of metabolic products whilst cells are retained in the separate compartments. To demonstrate its feasibility for the study of microbial interactions the equipment was first used to study the carbon flow in a simplified anaerobic community of a glucose-fermenting *C.butyricum*, a sulphate-reducing *Desulfovibrio desulfuricans* consuming the fermentation products and a phototrophic *Chromatium vinosum* recycling the sulphide produced in the second chemostat. In a subsequent set of experiments an ammonium-oxidizing *Nitrosomonas* sp., a nitrite-consuming *Nitrobacter* sp. and a nitrate-respiring *Vibrio* sp. were used to study the highly coupled processes in the microbial nitrogen cycle (MacFarlane and Herbert, 1985). The system proved sufficiently reliable to operate successfully for at least 30 days and, due to its apparent versatility and relative simplicity, may prove very useful for a variety of ecological studies.

## III. The phauxostat

### A. Theory

In the chemostat the growth rate of bacteria is limited by the *external* control of the rate at which fresh medium is fed to the culture. This rate can be varied over a large range of values but at the very high end of this range the population density gradually drops until at a given value a stable population density can no longer be maintained. At this ‘critical dilution rate’ (see Section II.A) the culture is washed out from the chemostat. The only way to maintain cells growing at or very near this critical dilution rate at an appreciable density is to switch over to some kind of internal control of the supply rate of fresh medium. Obviously this control should be based on a growth-dependent parameter. The most direct parameter, turbidity as a measure of cell density, was used in the first continuous culture based on such an internal control (Bryson and Szybalski, 1952). However, reliable construction of such a ‘turbidostat’ has always been difficult, especially because of problems associated with the accurate measurement of turbidity either directly inside the fermentor or in an external flow-through cuvet. Over the years various ingenious devices have been constructed to overcome the inevitable interfering effect of wall growth on the photometric measurements. But because the only requirement for the control loop of a turbidostatic culture is that a growth-dependent parameter of the growing culture needs to be measured continuously, turbidity is by no means the only one. In 1969 Watson was the first to report the use of the $CO_2$ content of the output gas of a fermentor, measured ‘on line’ with an infrared analyser, as the controlling factor for the medium pump (Watson, 1969).

Although a very elegant method, it is not particularly practical for routine use at laboratory scale. Some years later the use of a new type of feedback system was published (Martin and Hempfling, 1976), based on growth-dependent pH change, which in most cultures is sufficient to allow a very sensitive and accurate feedback control simply by triggering the medium pump with a signal from the pH meter. This principle appeared very useful, simple and reliable. Since its first use in pure cultures of *E. coli* B it was used successfully for a number of different types of aerobic and anaerobic cultures. The following examples and a brief description of the theory, as worked out by Martin and Hempfling (1976) and Rice and Hempfling (1985), illustrate its applicability both for physiological and ecological studies.

Assuming that during growth protons are excreted into the culture liquid the change in concentration can be expressed as

$$dH^+/dt = \mu xh + D[H_R^+] - D[H_C^+] - DB_R \tag{7}$$

in which $\mu$ is the specific growth rate ($h^{-1}$), $x$ the culture density (mg dry weight $l^{-1}$), $h$ the stoichiometry of proton formation per mg dry weight of cells, $D$ the dilution rate ($h^{-1}$), $[H_R^+]$ the proton concentration in the reservoir medium, $[H_C^+]$ the proton concentration in the culture and $B_R$ the buffer capacity of the reservoir medium. The latter may be defined as the quantity of acid or alkali required to change the pH of 1 l of medium to the pH of the culture liquid. For the simplest situation in which the medium pH is only slightly different from that in the culture, equation 7 may be written as

$$dH^+/dt = \mu xh - DB_R \tag{8}$$

which in case of a steady state reduces to

$$\mu xh = DB_R \tag{9}$$

and with $\mu = D$ reduces further to

$$x = B_R/h \tag{10}$$

This result demonstrates that the culture density is directly proportional to the buffering capacity of the medium, assuming that $h$ is independent of $B_R$. Combining equation 10 with the general nutrient balance of continuous cultures (equation 4, Section II.A) results in

$$s = S_r - B_R/hY_s \tag{11}$$

Finally, combining this result with the Monod equation for the description of $\mu$ (equation 1, Section II.A) yields

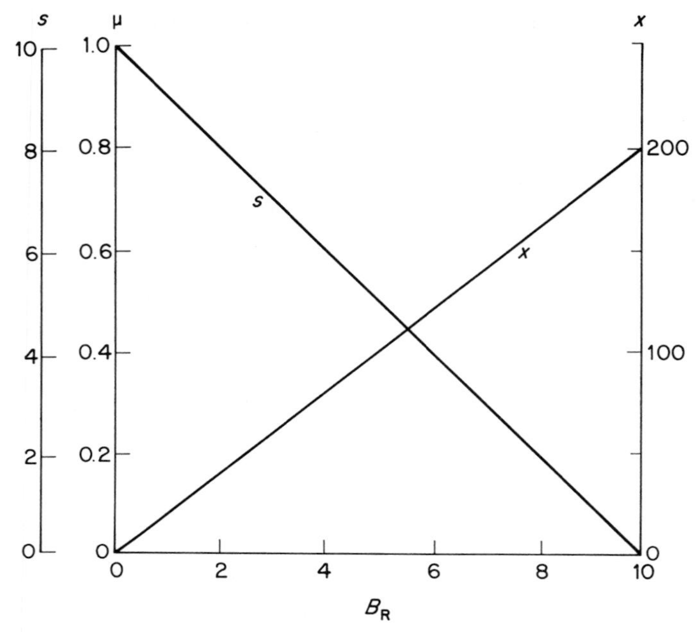


**Fig. 8.** Major parameters for growth in a phauxostat. $s$ (substrate concentration in the culture), $\mu$ (specific growth rate) and $x$ (cell density) as a function of the buffering capacity in the reservoir medium. Arbitrarily chosen values: $Y_s = 20$, $h = 0.05$, $K_s = 0.1$, $S_r = 10$ and $\mu_{max} = 1$.

$$\mu = \frac{\mu_{max} S_r - B_R/hY_s}{K_s + S_r - B_R/hY_s} \tag{12}$$

These theoretical considerations may be visualized as shown in Fig. 8, in which the major operating parameters have been plotted against the buffering capacity of the reservoir medium.

Although in the above formulations the assumption was made that growth-dependent net proton excretion occurs, the phauxostat will operate equally well if net proton consumption takes place, as for example during growth on organic acids. Another assumption, namely that the specific growth rate in the phauxostat will always be very near $\mu_{max}$, does not necessarily hold in all cases. In fact from equation 12 it can be seen that in theory the buffering capacity could be manipulated in such a way that the culture substrate concentration becomes rate limiting thus providing an elegant overlap with the conventional chemostat mode of growth. A further particularly useful property is the opportunity to regulate the rate of growth by varying the availability of a gaseous nutrient, such as oxygen, without a need to monitor and control its dissolved concentration and without the danger of culture washout due to too high a setting of the dilution rate, as may readily occur in chemostat mode. Finally, the phauxostat offers the hitherto little-explored possibility of studying the growth of bacteria as a function of cell density (by adjusting the buffering capacity), which could be of particular interest when inhibitory metabolic (end)products are formed. Again due to its self-regulatory properties washout of the culture is precluded.

## B. Equipment

One of the attractive features of phauxostat cultivation is that it requires only minor adaptations to the conventional chemostat equipment. The kind of chemostat apparatus described in Section II.B is suited perfectly to be used both in chemostat and phauxostat mode provided the following two technical considerations are taken care of.

(i) The principle of pH-dependent control of the medium supply demands that the medium pump is connected to the switched output of the pH-control unit. Depending on whether during growth net proton formation or consumption occurs a medium is used which is more alkaline or acidic, respectively, than the pH maintained in the culture. Since it will often be desirable to manipulate the culture density the buffering capacity of the medium should be easy to change. Although changing the concentration of the buffer itself or the acidity/alkalinity of the medium would seem the obvious way to do this, a far more flexible way is by using two pumps: one for the buffered medium, the other for a separate flow of acid or alkali. By changing the concentration of the acid or alkaline solution alone or in combination with changes of the setting of the pump for the medium flow the relative contribution of acid or alkali to the total flow into the culture can be changed and thus the buffering capacity of the medium. This method was used by Oltmann *et al.* (1978) for the cultivation of *Proteus mirabilis, Klebsiella aerogenes* and *Bacillus licheniformis.*

(ii) In order to measure the specific growth rate of the culture the dilution rate must be determined accurately as it will equal the specific rate of growth when the culture has reached a steady state. To accomplish this, two different techniques may be adopted: metering the input of medium or the output of culture liquid directly or indirectly from recording the time period the medium (and titration) pump(s) are turned on. In both cases it is of great importance to measure over a considerable length of time to obtain a dependable mean flow rate, as the supply occurs intermittently.

It will be evident that accuracy of the measurement much depends on the frequency of the pump switching. Most reliable values in relatively short times are obtained if the sensitivity of the pH meter is high and the hysteresis of the set point control is small (e.g. 0.01–0.05 pH units), resulting in nearly continuous pumping. This also ensures the best results with respect to the constancy of the substrate concentration in the culture. Whilst the inflowing medium (and titrant) may be metered using standard burettes in the feed line, in our experience the outflowing culture may be measured adequately by incorporating a calibrated metering tube in the overflow tubing (Fig. 9). The advantage of using this type of metering tube is that it automatically flows through if filled up. Starting the measurement only requires emptying the tube by opening the clamp at the bottom. A more sophisticated, though usually no more accurate way of measuring the flow of medium, is through recording the total time the medium and titrant pumps have been working at an accurately known rate over a given period of time. This may be done by tracing the pH signal of the electrode at a sensitive scale on a mV recorder and subsequently measuring the total length of time during which the medium flow caused a detectable declining (or increasing) tracing on the chart paper. Of course it is also possible to design a more sophisticated electronic device which detects whether the pumps are on or off and for how long. Obviously this kind of information sent to a computer-based control unit could constitute the basis of a fully automated continuous culture system.

## C. Examples

This section will be relatively short due to the paucity of ecologically relevant examples. Given the fact that the first well-documented description of theory and use of the phauxostat appeared approximately 13 years ago (Martin and Hempfling, 1976), it is perhaps surprising that there are only a very limited number of studies illustrating its use in microbiology. Although this may be partly because its applicability is less universal than that of the chemostat, it is also probably due to its conceptual ties to the turbidostat with its reputation for unreliability. Even the theory of phauxostat cultivation is at some points still in need of firm experimental support. In their pioneering study on the growth of *E. coli* B on glucose in a phauxostat (Martin and Hempfling, 1976) reasonable agreement between theory and practice was observed (Fig. 10). Yet, these authors had to conclude that the assumption of $h$ (the stoichiometry of proton excretion) being independent of buffer capacities did not hold, as they found an almost threefold increase calculated from the two different slopes of the lines in

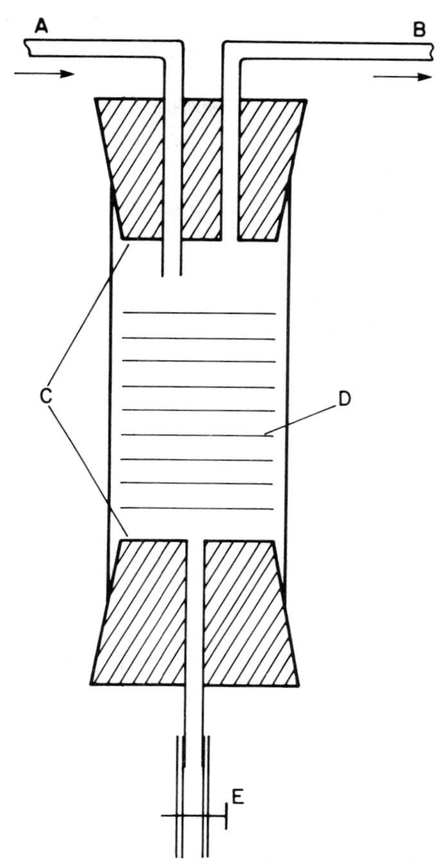


**Fig. 9.** Schematic drawing of a calibrated tube for measuring the liquid outflow from the phauxostat. A, culture liquid from the phauxostat; B, overflow into an effluent reservoir; C, rubber bung; D, calibration marks; E, clamped tubing.

Fig. 10. No explanation for this change could be provided. Several years later this observation was also reported for growth of the same bacterium on a different substrate (succinate) under conditions of supposed succinate limitation (Rice and Hempfling, 1985). This time a gradual increase of $h$ with increasing buffering capacity was observed up to five times the initial value. It was suggested that this resulted from a strong decrease in cell yield as a result of high maintenance requirements at low growth rates. Since the increase of $h$ is thus largely offset by the decrease in yield (see

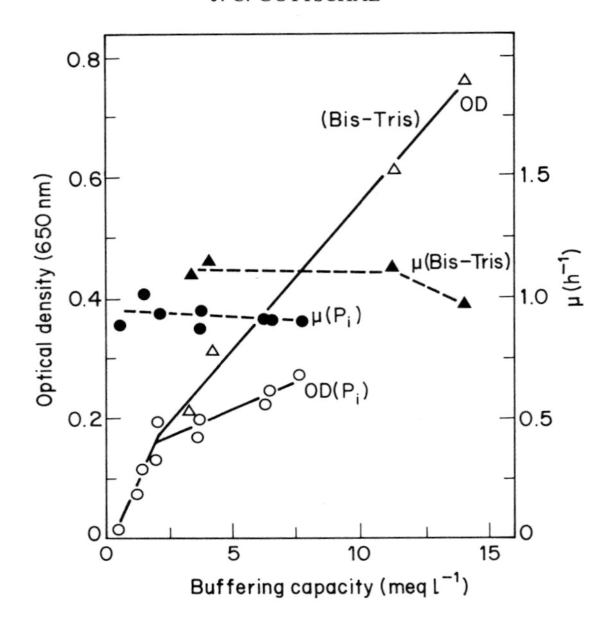


**Fig. 10.** Population density and specific growth rate of *E. coli* B as a function of the buffering capacity with Tris (Bis-Tris)- or phosphate (Pi) buffer. (From Martin and Hempfling, 1976.)

equation 11) the theory of phauxostat operation might nevertheless still hold. However, this point clearly needs further experimentation.

Although the phauxostat was designed for continuous growth at maximum specific rates it may also be used for nutrient-limited growth under certain conditions. A most significant example of this is growth under oxygen limitation (Rice and Hempfling, 1978). These authors clearly showed that by decreasing the rate of aeration the specific growth rate could be depressed to very low values leaving the cell density relatively unchanged. The self-adjusting principle of the phauxostat ensures that with decreasing availability of oxygen the time required to change the culture pH sufficiently to start the medium pump again becomes longer. In other words the availability of oxygen per unit biomass becomes less and consequently the specific growth rate drops and the cells grow oxygen limited. Especially for gaseous nutrients this method is very useful as it only requires a constant gas supply and renders precise control unnecessary.

Substrate-limited growth was also studied in mixed cultures of lactic acid-producing bacteria in milk-based media for continuous yogurt production (MacBean *et al.*, 1979). In the complex milk medium used the sugar concentration is much in excess if compared with the N source. The consequence of this for growth in a phauxostat is that two modes of growth can be distinguished: at relatively low cell densities growth rate will be maximal but at higher densities the nitrogen source will eventually limit the cell density which will no longer increase proportionally with the buffer capacity. Since sugar is still in excess acid production continues but the more slowly increasing biomass now requires more time to produce enough acid to trigger the medium pump. The culture grows at a specific growth rate lower than $\mu_{max}$: growth has become nitrogen limited. In the study by MacBean *et al.* (1979) stable coexistence of *Lactobacillus bulgaricus* and *Streptococcus thermophilus* was explained by showing that *L. bulgaricus* grew at its maximum rate whereas *S. thermophilus* (having a threefold higher $\mu_{max}$) grew at that same rate but nitrogen limited. This situation of nutrient-limited growth in the phauxostat will in general be possible if a nutrient other than that responsible for acid or alkali formation, such as phosphate, sulphate, etc. (Rice and Hempfling, 1985) becomes limiting.

Finally, the phauxostat has been used successfully in studies on the anaerobic hydrolysis of sugar polymers (e.g. cellulose, chitin, etc.). This is a particularly useful application as it has always been rather difficult to cultivate polymer-hydrolysing anaerobes on a continuous basis in a precisely controlled manner. In principle chemostat cultivation is possible but a particular problem is the accurately metered supply of suspensions of such insoluble substrates, although to some extent this difficulty may be overcome by growing the cultures under a limitation other than that of the carbon and energy source. A special difficulty with cultures of polymer-degrading (an)aerobes is the estimation of the specific growth rate due to their strong tendency to grow attached to the polymeric substrates. Kistner *et al.* (1983) have used the phauxostat to measure specific growth rates and cellulose solubilization rates of several cellulolytic anaerobes from the rumen. The cellulose suspension was supplied from a reservoir immediately above the culture by triggering an electropneumatic valve. In this way stable continuous cultures were obtained which grew at specific rates varying from 0.20 to 0.60 $h^{-1}$ for various cellulolytic species.

Very recently we have studied the growth of two different anaerobic chitin-hydrolysing bacteria in mixed culture in the phauxostat (Pel *et al.*, 1990). The phauxostat proved an excellent tool for studying the competition for chitin (supplied as a concentrated suspension by a peristaltic pump) between both anaerobes in the presence of a commensalistic non-chitinolytic species.

## IV. Concluding remarks

The level of complexity of most natural ecosystems is such that relatively few microbial ecologists are courageous enough to endeavour direct experimenting in the field. Instead, they choose to study microbial ecology in the laboratory. However illogical such an approach may seem it has nevertheless resulted in the gathering of an enormous amount of information on how microorganisms thrive in their natural habitats. And it is certainly true that the use of continuous cultures has contributed immensely to the success of this approach. It is indeed through the principle of continuous cultivation under precisely defined conditions that particular environmental influences can be singled out and studied in great detail. The use of chemostats thus demonstrates that in spite of its 'utterly unnatural constancy and homogeneity' it does make it possible to obtain some understanding of microbial life in nature. Yet, this is no reason for microbial ecologists to revel in complacency. On the contrary, the only way to further augment our knowledge of natural ecosystems will be through a constant struggle to find the proper balance between laboratory work and field observations. And it will be equally important to continue using the chemostat and its various modifications or extensions in a creative way so as to help answer the questions which relate directly to the complicated problems facing those of us who study the microbial world outside the laboratory. It is hoped that the present chapter may somehow contribute to the achievement of this goal.

## References


Azam, F., Fenchel, T., Field, J. G., Gray, J. S., Meyer-Reil, L. A. and Thingstad, T. F. (1983). *Mar. Ecol. Prog. Ser.* **10,** 257–263.

Bartels, I., Knackmuss, H. J. and Reincke, W. (1984). *Appl. Environ. Microbiol.* **47,** 500–505.

Bazin, M. J. (1981). *In* "Continuous Cultures of Cells. Vol. I" (P. H. Calcott, Ed.), pp. 27–62, CRC Press Inc., Boca Raton, Florida.

Beudeker, R. F., Gottschal, J. C. and Kuenen, J. G. (1982). *Ant. Leeuwenhoek J. Microbiol.* **48,** 39–51.

Billen, G. (1982). *In* "Sediment Microbiology" (D. B. Nedwell and C. M. Brown, Eds), pp. 15–52, Academic Press, London.

Bratbak, G. and Thingstad, T. F. (1985). *Mar. Ecol. Prog. Ser.* **25,** 23–30.

Brown, C. M., Ellwood, D. C. and Hunter, J. R. (1978). *In* "Techniques for the Study of Mixed Populations" (D. W. Lovelock and Davies, M. Eds), pp. 213–222, Academic Press, London.

Bryson, V. and Szybalski, W. (1952). *Science* **116,** 45–51.

Chisholm, S. W., Stross, R. G. and Nobbs, P. A. (1975). *J. Phycol.* **11,** 367–373.

Conrad, R., Schink, B. and Phelps, T. J. (1986). *FEMS Microbiol. Ecol.* **38,** 353–360.
Curds, C. R. and Cockburn, A. (1971). *J. Gen. Microbiol.* **66,** 95–108.
Currie, D. J. and Kalff, J. (1984). *Microb. Ecol.* **10,** 205–216.
Daughton, C. G. and Hsieh, D. P. H. (1977). *Appl. Environ. Microbiol.* **34,** 175–184.
Dorn, E., Hellwig, M., Reineke, W. and Knackmuss, H. J. (1974). *Arch. Microbiol.* **99,** 61–70.
Dykhuizen, D. and Davies, M. (1980). *Ecology* **61,** 1213–1227.
Dykhuizen, D. E. and Hartl, D. L. (1983). *Microbiol. Rev.* **47,** 150–168.
Evans, C. G. T., Herbert, D. and Tempest, D. W. (1970). *In* "Methods in Microbiology, Vol. 2." (J. R. Norris and D. W. Ribbons, Eds), pp. 277–327, Academic Press, London.
Fenchel, T. M. and Jørgensen, B. B. (1977). *Adv. Microb. Ecol.* **1,** 1–58.
Fleming, G. F., Dawson, M. T. and Patching, J. W. (1988). *J. Gen. Microbiol.* **134,** 2095–2101.
Fredrickson, A. G. (1977). *Ann. Rev. Microbiol.* **31,** 63–87.
Freter, R., Stauffer, E., Cleven, D., Holdeman, L. V. and Moore, W. E. C. (1983). *Infect. Immun.* **39,** 666–675.
Gemerden, H. van (1974). *Microb. Ecol.* **1,** 104–110.
Gemerden, H. van and Beeftink, H. H. (1983). *In* "The Phototrophic Bacteria: Anaerobic Life in the Light. Studies in Microbiology, Vol. 4" (J. G. Ormerod, Ed.), pp. 146–185, Blackwell Scientific Publications, Oxford.
Gemerden, H. van and Wit, R. de (1989). *In* "Microbial Mats. Physiological Ecology of Benthic Microbial Communities" (Y. Cohen and E. Rosenberg, Eds), pp. 313–319, ASM, Washington, DC, USA.
Gerritse, J., Schut, F. and Gottschel, J. C. (1990). FEMS Microbiol: Lett., **66,** 87–93.
Glaser, D. (1988). *Microb. Ecol.* **15,** 189–201.
Gottschal, J. C. (1985). *Ant. Leeuwenhoek* **51,** 473–494.
Gottschal, J. C. (1986). *In* "Bacteria in Nature, Vol. 2" (E. R. Leadbetter and J. S. Poindexter, Eds), pp. 261–292, Plenum Press, New York.
Gottschal, J. C. and Dijkhuizen, L. (1988). *In* "Handbook of Laboratory Model Systems for Microbial Ecosystems", pp. 19–49, CRC Press Inc., Boca Raton.
Gottschal, J. C. and Kuenen, J. G. (1980a). *FEMS Microbiol. Lett.* **7,** 241–247.
Gottschal, J. C. and Kuenen, J. G. (1980b). *Arch. Microbiol.* **126,** 33–42.
Gottschal, J. C. and Szewzyk, R. (1985). *FEMS Microbiol. Ecol.* **31,** 159–170.
Gottschal, J. C., Vries, S. de and Kuenen, J. G. (1979). *Arch. Microbiol.* **121,** 241–249.
Gottschal, J. C., Pol, A. and Kuenen, J. G. (1981a). *Arch. Microbiol.* **129,** 23–28.
Gottschal, J. C., Nanninga, H. J. and Kuenen, J. G. (1981b). *J. Gen. Microbiol.* **126,** 85–96.
Harder, W. and Dijkhuizen, L. (1976). *In* "Continuous Culture 6. Applications and New Fields" (A. C. R. Dean, D. C. Ellwood, C. G. T. Evans and I. Mellinge, Eds), pp. 297–314, Ellis Horwood, Chichester.
Harder, W. and Dijkhuizen, L. (1982). *Phil. Trans. R. Soc. Lond. B.* **297,** 459–480.
Harder, W. and Dijkhuizen, L. (1983). *Ann. Rev. Microbiol.* **37,** 1–24.
Hartmann, J., Reineke, W. and Knackmuss, H. J. (1979). *Appl. Environ. Microbiol.* **37,** 421–428.

Herbert, D. (1961). *In* "Continuous Culture of Microorganisms", pp. 21–53, Soc. Chem. Ind. Monograph No. 12, London.
Herbert, D. (1964). *In* "Continuous Cultivation of Microorganisms, 2nd Symp." (I. Malek, K. Beran, and J. Hospodka, Eds), pp. 23–44, Publ. House Czechoslovak Acad. Sci., Prague.
Herbert, D., Elsworth, R. and Telling, R. C. (1956). *J. Gen. Microbiol.* **14,** 601–622.
Hoeven, J. S. van der and Jong, M. H. de (1984). *In* "Continuous Culture 8: Biotechnology, Medicine and Environment" (A. C. R. Dean, D. C. Ellwood and C. G. T. Evans, Eds), pp. 89–110, Ellis Horwood Ltd, Chichester.
Hoeven, J. S. van der, Jong, M. H. de, Camp, P. J. M. and Kieboom, C. W. A. van den (1985). *FEMS Microbiol. Ecol.* **31,** 373–379.
Ianotti, E. L., Kafkewitz, D., Wolin, M. J. and Bryant, M. P. (1973). *J. Bacteriol.* **114,** 1231–1240.
Jannasch, H. W. (1967). *Arch. Mikrobiol.* **59,** 165–173.
Jannasch, H. W. (1974). *Limnol. Oceanogr.* **19,** 716–720.
Jannasch, H. W. and Mateles, R. I. (1974). *Adv. Microb. Physiol.* **11,** 165–212.
Jost, J. L., Drake, J. F., Fredrickson, A. G. and Tsuchiya, H. M. (1973). *J. Bacteriol.* **113,** 834–840.
Keith, S. M. and Herbert, R. A. (1985). *FEMS Microbiol. Ecol.* **31,** 239–248.
Kistner, A., Kornelius, J. H. and Miller, G. S. (1983). *S. Afr. J. Anim. Sci.* **13,** 217–220.
Koch, A. L. (1971). *Adv. Microbial. Physiol.* **6,** 147–217.
Kuenen, J. G. and Harder, W. (1982). *In* "Experimental Microbial Ecology" (R. G. Burns and J. H. Slater, Eds), pp. 342–367, Blackwell Scientific Publications, Oxford.
Kuenen, J. G. and Robertson, L. A. (1984). *In* "Current Perspectives in Microbial Ecology" (M. J. Klug and C. A. Reddy, Eds), pp. 306–313, ASM, Washington.
Kuenen, J. G., Boonstra, J., Schroder, H. G. J. and Veldkamp, H. (1977). *Microbial. Ecol.* **3,** 119–130.
Kuenen, J. G., Robertson, L. A. and Gemerden, H. van (1985). *Adv. Microb. Ecol.* **8,** 1–59.
Laanbroek, H. J. (1978). PhD Thesis, University of Groningen, Groningen.
Laanbroek, H. J., Smit, A. J., Klein Nulend, G. and Veldkamp, H. (1979). *Arch. Microbiol.* **120,** 61–66.
Laanbroek, H. J., Geerligs, H. J., Peynenburg, A. A. C. M. and Siesling, J. (1983). *Microb. Ecol.* **9,** 341–354.
Laanbroek, H. J., Geerligs, H. J., Sijtsma, L. and Veldkamp, H. (1984). *Appl. Environ. Microbiol.* **47,** 329–334.
Leegwater, M. P. M. (1983). PhD Thesis, University of Amsterdam, Amsterdam.
Liere, L. van (1979). PhD Thesis, University of Amsterdam, Amsterdam.
Loogman, J. G. (1982). PhD Thesis, University of Amsterdam, Amsterdam.
Lovitt, R. W. and Wimpenny, J. W. T. (1981). *J. Gen. Microbiol.* **127,** 261–268.
MacBean, R. D., Hall, R. J. and Linklater, P. M. (1979). *Biotechnol. Bioeng.* **21,** 1517–1541.
MacFarlane, G. T. and Herbert, R. A. (1985). *FEMS Microbiol. Ecol.* **31,** 249–254.
Marsh, P. D., Hunter, J. R., Bowden, G. H., Hamilton, I. R., McKee, A. S., Hardie, J. M. and Ellwood, D. C. (1983). *J. Gen. Microbiol.* **129,** 755–770.
Martin, G. A. and Hempfling, W. P. (1976). *Arch. Microbiol.* **107,** 41–47.

Matin, A. (1981). *In*. "Continuous Culture of Cells, Vol. 2" (P. H. Calcott, Ed.), pp. 69–97, CRC Press Inc., Boca Raton, Florida.
Matin, A. and Veldkamp, H. (1978). *J. Gen. Microbiol.* **105,** 187–197.
Mayfield, C. I. and Inniss, W. E. (1978). *Microb. Ecol.* **4,** 331–344.
Monod, J. (1942). "Recherches sur la Croissance des Culture Bacteriennes" Hermann and Cie, Paris.
Monod, J. (1950). *Ann. Inst. Pasteur, Paris* **79,** 390–410.
Nanninga, H. J., Drent, W. J. and Gottschal, J. C. (1986). *FEMS Microbiol. Ecol.* **38,** 321–329.
Novick, A. and Szilard, L. (1950). *Science* **112,** 715–716.
Ohta, H. and Gottschal, J. C. (1988a). *FEMS Microbiol. Ecol.* **53,** 79–86.
Ohta, H. and Gottschal, J. C. (1988b). *FEMS Microbiol. Lett.* **50,** 163–168.
Oltmann, L. F., Schoenmaker, G. S., Reijnders, W. N. M. and Stouthamer, A. H. (1978). *Biotechnol. Bioeng.* **20,** 921–925.
Oltmanns, R. H., Rast, H. G. and Reineke, W. (1988). *Appl. Microbiol. Biotechnol.* **28,** 609–616.
Parkes, R. J. (1982). *In* "Microbial Interactions and Communities, Vol. 1" (A. T. Bull and J. H. Slater, Eds), pp. 45–102, Academic Press, London.
Pel, R., Wÿngaard, A. J. van den, Epping, E. and Gottschel, J. C. (1990). *J. Gen. Microbiol.* **136,** 695–704.
Pengerud, B., Skjoldal, E. F. and Thingstad, T. F. (1987). *Mar. Ecol. Prog. Ser.* **35,** 111–117.
Pirt, S. J. (1975). "Principles of Microbe and Cell cultivation" Blackwell, Oxford.
Powell, E. O. and Lowe, J. R. (1964). *In* "Continuous Cultivation of Microorganisms, 2nd Symp." (I. Malik, K. Beran and J. Nospodka, Eds), pp. 45–57, Publ. House of the Czechoslovak Acad. Sci., Prague.
Pritchard, P. H. and Bourquin, A. W. (1984). *Adv. Microb. Ecol.* **7,** 133–215.
Radjai, M. K., Hatch, R. T. and Cadman, T. W. (1984). *Biotechnol. Bioneng.* **S14,** 657–680.
Reineke, W. and Knackmuss, H. J. (1984). *Appl. Environ. Microbiol.* **47,** 395–402.
Rhee, G. Y. (1980). *Adv. Aquat. Microbiol.* **2,** 151–203.
Rhee, G. Y., Gotham, I. J. and Chisholm, S. W. (1981). *In* "Continuous Culture of Cells" (P. Calcott, Ed.), pp. 159–186, Cleveland, Ohio.
Rice, C. W. and Hempfling, W. P. (1978). *J. Bacteriol.* **134,** 115–124.
Rice, C. W. and Hempfling, W. P. (1985). *Biotechnol. Bioeng.* **27,** 187–191.
Ricica, J. (1970). *In* "Methods in Microbiology, Vol 2" (J. R. Norris and D. W. Ribbons, Eds), pp. 329–348, Academic Press, London.
Ricica, J. and Dobersky, P. (1981). *In* "Continuous Cultures of Cells, Vol. 1" (P. H. Calcott, Ed.), pp. 63–96, CRC Press Inc., Boca Raton, Florida.
Sambanis, A. and Fredrickson, A. G. (1987). *J. Gen. Microbiol.* **133,** 1619–1630.
Schlegel, H. G. and Jannasch, H. W. (1967). *Ann. Rev. Microbiol.* **21,** 49–70.
Schmidt, E. (1987), *Appl. Microbiol. Biotechnol.* **27,** 94–99.
Schmidt, E., Hellwig, M. and Knackmuss, H. J. (1983). *Appl. Environ. Microbiol.* **46,** 1038–1044.
Schmidt, E., Bartels, I. and Knackmuss, H. J. (1985). *FEMS Microb. Ecol.* **31,** 381–389.
Senior, E. (1977). PhD Thesis, University of Kent.
Senior, E., Bull, A. T. and Slater, J. H. (1976). *Nature* **263,** 476–479.
Sherr, B. F., Sherr, E. B. and Berman, T. (1982). *Limnol. Oceanogr.* **27,** 765–769.

Slater, J. H. (1981). *In* "Mixed Culture Fermentations" (M. E. Bushell and J. H. Slater, Eds), pp. 1–24, Academic Press, London.
Slyter, L. L., Bryant, M. P. and Wolin, M. J. (1966). *Appl. Microbiol.* **14,** 573–578.
Swift, S. T., Najita, I. Y., Ohtaguchi, K. and Fredrickson, A. G. (1982). *Biotechnol. Bioeng.* **24,** 1953–1964.
Taeger, K., Knackmuss, H. J. and Schmidt, E. (1988). *Appl. Microbiol. Biotechnol.* **28,** 603–608.
Tempest, D. W. (1970). *Adv. Microb. Physiol.* **4,** 223–250.
Tempest, D. W. and Neyssel, O. M. (1978). *Adv. Microb. Ecol.* **2,** 105–153.
Thiele, J. H. and Zeikus, J. G. (1988). *Appl. Environ. Microbiol.* **54,** 20–29.
Thompson, L. A., Nedwell, D. B., Balba, M. T., Banat, I. M. and Senior, E. (1983). *Microb. Ecol.* **9,** 189–199.
Titus, J. A., Luli, G. W., Dekleva, M. L. and Strohl, W. R. (1984). *Appl. Environ. Microbiol.* **47,** 239–244.
Traore, A. S., Fardeau, M. L., Hatchikian, C. E., LeGall, J. and Belaich, J. P. (1983). *Appl. Environ. Microbiol.* **46,** 1152–1156.
Veilleux, B. G. and Rowland, I. (1983). *J. Gen. Microbiol.* **123,** 103–115.
Veldkamp, H. (1976). *In* "Continuous Culture 6: Applications and New Fields" (A. C. R. Dean, D. C. Ellwood, C. G. T. Evans and J. Melling, Eds), pp. 315–328, Ellis Horwood Ltd., Chichester.
Veldkamp, H. (1977). *Adv. Microb. Ecol.* **1,** 59–94.
Veldkamp, H. and Jannasch, H. W. (1972). *J. Appl. Chem. Biotechnol.* **22,** 105–123.
Veldkamp, H. and Kuenen, J. G. (1973). *Bull. Ecol. Res. Comm. (Stockholm)* **17,** 347–355.
Veldkamp, H., Gemerden, H. van, Harder, W. and Laanbroek, H. J. (1984). *In* "Current Perspectives in Microbial Ecology" (M. J. Klug and C. A. Reddy, Eds), pp. 279–290, ASM, Washington.
Watson, T. G. (1969). *J. Gen. Microbiol.* **59,** 83–89.
Wimpenny, J. W. T. (1981). *Biol. Rev.* **56,** 295–342.
Wimpenny, J. W. T., Lovitt, R. W. and Coombs, J. P. (1983). *In* "Microbes in Their Natural Environments, 34th Symp. Soc. Gen. Microbiol." (J. H. Slater, R. Whittenbury and J. W. T. Wimpenny, Eds), pp. 67–117, Cambridge University Press, Cambridge.
Wolin, M. J. (1982). *In* "Microbial Interactions and Communities, Vol. I, (A. T. Bull and J. H. Slater, Eds), pp. 323–356, Academic Press, London.

# 4
# Mathematical Methods in Microbial Ecology

M. BAZIN AND A. MENELL

*Department of Microbiology, King's College London, University of London, London, UK*



METHODS IN MICROBIOLOGY
VOLUME 22 ISBN 0-12-521522-3

## I. Introduction

As in other branches of science, in ecology, mathematics is used to describe phenomena, make predictions about them and aid in understanding them. In a single chapter it is not feasible to cover all the mathematical methods that have been used to investigate problems in microbial ecology. We will describe some applications, mostly concerned with microbial population dynamics, by giving a series of examples. The central theme of the chapter is the construction and analysis of mathematical models of microbial ecosystems. Mathematical models have a multiplicity of functions. They can be used to predict events in time and space with the potential of having quite practical applications. In agriculture, the timing and amount of fertilizer required for a particular situation is useful for farmers to know and mathematical models have been derived for this purpose. Models for variation in sewage strength, eutrophication, nitrification and many other ecological processes have been devised. An advantage of describing such phenomena in mathematical terms is that sometimes formal procedures can be applied to optimize them. In chemical engineering and biotechnology, optimization, often by some form of on-line computer control, is becoming increasingly important in economic terms.

Purely predictive models are quite often arbitrary. That is to say, the mathematical functions employed need not represent the underlying ways that the system works. On the other hand, mechanistic models incorporate attempts to explain phenomena on the basis of how it is thought they operate. It is upon such models that we concentrate in this chapter. If the scientific method is regarded, rather simplistically, as proposing a hypothesis based on prior observations, testing the predictions arising from the hypothesis by experimentation and then accepting or rejecting the hypothesis on the basis of the results, mechanistic models can serve a very useful function as vehicles for hypotheses. Mathematical analysis can be applied to generate predictions and appropriate experiments performed. We will not be concerned with experimentation here but the approach to scientific investigation outlined above does require experiments to test the predictions generated. Often these are most appropriately carried out in a laboratory environment. The question as to whether or not such an approach can be used justifiably to study 'natural' systems then arises. We will not argue the case. Rather, we take ecology to be the study of the interactions between organisms and their environment. Whether or not these interactions are studied under 'natural' or 'unnatural' conditions (whatever these terms mean) seems to us to be irrelevant provided that the methods employed are scientifically acceptable.

We begin our discussion of how mathematics has been applied to microbial ecology with a description of how the physical environment of microbial ecosystems may be represented mathematically. Mostly, this involves the derivation of mass balance equations. We then introduce some of the ways that the kinetics of microbial growth and interaction have been described. Using these two sets of information we derive a few exemplary models which are really idealized descriptions of particular microbial ecosystems. We emphasize that these mathematical models do not describe 'real' systems but are intended to serve as ways in which well-posed hypotheses can be tested. We then go on to outline some of the mathematical techniques that have been used to generate predictions from the models. This involves mainly the ways in which solutions of sets of differential equations can be obtained or approximated. We begin this section with quite simple mathematical methods and then progress. However, in no case do the techniques involve mathematics beyond that taught in college-level intermediate calculus courses. We conclude with a section that contains some rather specialized mathematical applications.

## II. The physical environment of microbial ecosystems

By the term 'physical environment' we mean the inanimate surroundings of microorganisms in a microbial ecosystem. The two most important aspects of this topic are whether or not the microbial population is well mixed and whether or not there is an input and output of material and/or energy to and from the system.

### A. Homogeneous ecosystems

In such systems we assume that the microorganisms are randomly distributed in the liquid phase and do not adhere to surfaces. In effect, microbial populations are treated as if they were dissolved in the solution which surrounds them.

#### *1. Closed systems*

A system in which there is no input or output of energy or material is said to be thermodynamically closed. Such is the case for an idealized batch culture. Closed microcosms containing several species have been studied experimentally and apparently can survive for several years. However, the main advantage of a theoretical study of closed ecosystems is conceptual as it leads naturally to a consideration of open ecosystems.

For a population at concentration $X$, the simplest useful equation for microbial growth is

$$\frac{\mathrm{d}X}{\mathrm{d}t} = \mu X \tag{1}$$

where $\mu$ is the per capita rate of increase or specific growth rate of the microbial population. As it stands, this equation represents a lack of any type of interaction between the organisms and the abiotic part of the environment.

### 2. *Open systems*

Such systems have an input and output of mass and/or energy. Practically all microbial ecosystems are of this type. In practice, only mass balances are usually considered but for some large-scale biotechnological processes heat production is often a problem and should be taken into account.

For a well-mixed ecosystem of constant volume, $V$, and rate of flow through the system, $f$, the balance with respect to biomass is

change in biomass = input of biomass + change due to growth or death − output of biomass

For a system to which there is no biomass input and no death, this may be written as

$$\frac{\mathrm{d}X}{\mathrm{d}t} = \mu X - \frac{f}{V}X \tag{2}$$

The experimental realization of such a system is the chemostat where the dilution rate, $D$, is equated to $f/V$.

## B. Heterogeneous environments

Many microbial environments cannot be regarded as being well mixed even by applying the simplifying assumptions inherent in the above discussion. Such is the case when a significant part of the microbial flora of an ecosystem adheres to a solid substratum. We will take as an example of this sort of ecosystem an idealized column of soil. The basic characteristics of such a system are the downward flow of water and nutrients and the attachment of the microbial population to the soil particles. Assuming that these particles are inert with respect to biological activity, we have to consider equations for mass balance which describe change with respect to time and also with respect to distance down the column. For a population of microorganisms in such a system, as in other ecosystems, growth is often dependent on the concentration, $S$, of some limiting nutrient. The nutrient will be used by the biomass and give rise to gradients down the column. In the absence of microbial activity, Fick's first and second laws can be used to derive the following continuity equation (Nye and Tinker, 1977):

$$\frac{\partial S}{\partial t} = D^* \frac{\partial^2 S}{\partial z^2} - q \frac{\partial S}{\partial z} \tag{3}$$

Here $z$ is the distance down the column and $q$ is the average net rate of flow of fluid across a unit cross section of the column which is called the specific discharge. The constant $D^*$ is known as the dispersion coefficient and together with the second order term in equation 3 is supposed to account for diffusion of substrate within the solute (water) and movement by mass flow of the solute. In the presence of microorganisms an extra term on the right hand side of equation 3 must be added to describe the way in which the substrate is being utilized. In addition, further equations must be written to describe substrate-dependent microbial growth. Examples of such equations are given in Section III.

Equation 3 is a partial differential equation because it has two independent variables, $z$ and $t$. The preceding differential equations we have mentioned are ordinary because they have only one independent variable, $t$. As we shall see later, partial differential equations are considerably more difficult to analyse than their ordinary counterparts.

## III. Microbial growth kinetics

We describe here some of the mathematical functions that have been used to represent the specific growth rate of microbial populations. More extensive lists of such functions are given by Pirt (1975), Bailey and Ollis (1986), Howell (1983) and Roels (1983) as well as many other workers. We will be concerned with relatively few of these kinetic equations.

### A. Single-species populations

If the generation time of a population reproducing by binary fission is a constant, $\nu$, then it can be shown (Menell and Bazin, 1988) that the specific growth rate, $\mu$, is also a constant such that

$$\mu = \frac{\log 2}{\tau} * \tag{4}$$

However, only for quite specialized purposes (such as the so-called exponential phase of growth in batch culture) can $\mu$ be regarded as constant. Usually it decreases as population density increases. If this decrease is linear, then

$$\mu = k_1 - k_2 X \tag{5}$$

where $k_1$ and $k_2$ are constants.

As indicated previously, there are situations where growth is considered to be dependent on a single growth-limiting substrate. The most commonly employed kinetic equation reflecting this idea is that of Monod (1942) which for substrate at concentration $S$ takes the form

$$\mu = \frac{\mu_m S}{K_s + S} \tag{6}$$

where $K_s$ is known as the saturation constant and is numerically equal to the concentration of substrate required to support growth at half the maximum specific growth rate, $\mu_m$.

## B. Microbial interactions

Detailed accounts of the many ways in which microorganisms interact with each other have been given by Fredrickson (1977, 1983), Fredrickson and Stephanopoulos (1981), Kuenen (1983), Slater and Bull (1978), Bazin (1981) and other workers. For the models of microbial interaction we develop later, we will require only one additional kinetic equation in addition to those we have described already. In a predator–prey interaction, the predator specific growth rate, $\lambda$, is sometimes considered to be directly proportional to prey concentration ($H$). Thus

$$\lambda = \beta_2 H \tag{7}$$

where $\beta_2$ is a constant.

Another kinetic equation for $\lambda$ is based on the Monod function. This takes the form

$$\lambda = \frac{\lambda_m H}{L_s + H} \tag{8}$$

*We use the term log throughout the chapter to denote logarithm to the base e.

where $\lambda_m$ is the maximum specific growth rate of the predator and $L_s$ is the saturation constant.

## IV. Mathematical models of microbial ecosystems

Here we will combine the equations of mass balance introduced in Section II with the kinetic expressions of Section III to produce several models which will serve as examples of the ways in which mathematical techniques of analysis have been applied. We have numbered each model and prefixed each of them with the letter M.

### A. Unrestricted growth in a closed environment

For a population with a constant generation time, the specific growth rate, $\mu$, is also constant. Implicit in this statement is the proposition that the physical nature of the environment does not affect the specific growth rate. In a closed environment then, growth may be described by our first and simplest model

$$\frac{dX}{dt} = \mu X \qquad \text{M1}$$

### B. Density-dependent growth in a closed environment

Using the density-dependent expression of equation 5, letting $k_1 = \mu_m$ and $k_2 = \mu_m/K$ and recalling that, by definition, $\mu = dX/X dt$ we obtain

$$\frac{dX}{dt} = \left(\mu_m - \frac{\mu_m}{K} X\right) X \qquad \text{M2}$$

This equation is known as the logistic equation. In ecological terms, $\mu_m$ is known as the intrinsic rate of increase and $K$ is the carrying capacity of the environment. Microbiologically, $\mu_m$ is the maximum specific growth rate and $K$ is related to the growth yield. Inherent in the logistic equation is the idea that some interaction with the environment takes place through the parameter $K$ although no mechanisms are indicated.

### C. Single-species Monod-type growth in an open system

Combining the mass balance equation 2 and assuming Monod growth kinetics defined in equation 1, the equations for change in biomass and growth-limiting substrate concentrations are

$$\frac{\mathrm{d}X}{\mathrm{d}t} = \frac{\mu_m S X}{K_s + S} - DX$$

$$\frac{\mathrm{d}S}{\mathrm{d}t} = D(S_r - S) - \frac{\mu_m S X}{Y(K_s + S)} \qquad \text{M3}$$

where $S_r$ is the input concentration of nutrient and $Y$ is the growth yield. Notice that a second equation for change in substrate has to be included in order to specify the way in which this dependent variable changes with time. This model can be used to represent a closed system simply by equating the dilution rate, $D = f/V$, to zero.

### D. Lotka–Volterra predation dynamics in a closed system

Consider the situation in which predator ($P$) growth is defined by equation 7 and prey ($H$) grows exponentially according to equation 1. In addition, let us assume that the predator dies exponentially in the absence of prey and that the decrease of prey as a result of predation is directly proportional to the increase in predator. Then it follows (Menell and Bazin, 1988) that

$$\frac{\mathrm{d}H}{\mathrm{d}t} = \alpha_1 H - \beta_1 HP$$

$$\frac{\mathrm{d}P}{\mathrm{d}t} = -\alpha_2 P + \beta_2 HP \qquad \text{M4}$$

The above constitute the simplest form of the so-called Lotka–Volterra equations.

### E. Monod-type predation in an open ecosystem

Models in which the specific growth rates of both prey and predator are represented by Monod functions have been investigated by Fredrickson *et al.* (1973), Curds and Bazin (1977), Dent *et al.* (1976), Canale (1970) and other workers. From equations 6 and 8 the specific growth rates of prey and predator can be written as

$$\mu = \frac{1}{H}\frac{\mathrm{d}H}{\mathrm{d}t} = \frac{\mu_m S}{K_s + S}$$

$$\lambda = \frac{1}{P}\frac{\mathrm{d}P}{\mathrm{d}t} = \frac{\lambda_m H}{L_s + H}$$

Combining these expressions with mass balance relationships of the form of equation 2, rearranging and adding a third equation for prey substrate balance yields

$$\frac{dP}{dt} = \frac{\lambda_m HP}{L_s + H} - DP$$

$$\frac{dH}{dt} = \frac{\mu_m SH}{K_s + S} - \frac{\lambda_m HP}{W(L_s + H)} \qquad \text{M5}$$

$$\frac{dS}{dt} = D(S_r - S) - \frac{\mu_m SH}{Y(K_s + S)}$$

where $W$ is the yield of predator per unit of prey consumed and the other terms have already been defined.

### F. Competition in an open system based on Monod kinetics

Assuming Monod kinetics, in a well-mixed ecosystem in which two microbial populations at concentrations $X_1$ and $X_2$ compete for a single growth-limiting substrate at concentration $S$ the equations of balance are

$$\frac{dX_1}{dt} = \frac{\mu_{m1} SX_1}{K_{s1} + S} - DX_1$$

$$\frac{dX_2}{dt} = \frac{\mu_{m2} SX_2}{K_{s2} + S} - DX_2 \qquad \text{M6}$$

$$\frac{dS}{dt} = D(S_r - S) - \frac{\mu_{m1} SX_1}{Y_1(K_{s1} + S)} - \frac{\mu_{m2} SX_2}{Y_2(K_{s2} + S)}$$

In these equations the subscripts refer to the relevant species and the symbols are as defined previously.

### G. Substrate utilization in a soil column

As an example of a model of a heterogeneous ecosystem we will consider the conversion of a single dissolved substrate by a fixed (constant) amount of microbial biomass in an idealized soil column. Equation 3 then becomes

$$\frac{\partial S}{\partial t} = D^* \frac{\partial^2 S}{\partial z^2} - q \frac{\partial S}{\partial z} - kX_c \qquad \text{M7}$$

where $X_c$ is biomass concentration and $k$ is a first-order reaction constant.

As the reader will appreciate later in this chapter, this model requires rather lengthy analysis. Fortunately, references to extensions of the model to two-species systems and systems in which the biomass grows are available. These will be given in the next section.

## V. Methods of analysis

We now use the models we have derived to illustrate some of the mathematical methods that have been used to determine how they behave and to generate experimentally testable predictions. We will find that in some cases these predictions need not actually be tested—either the results are already known or they are of a form that cannot be physically realized. Such outcomes are, of course, regarded very warmly by mathematical biologists. They mean that bench or field work need not yet be embarked upon and theoretical investigations with equations and computers can continue!

In what follows we have attempted to progress from relative simplicity to relative difficulty in mathematical terms. The models we have derived do not follow the same progression and we use some of them repeatedly and not in the same order that they were introduced. Finally before we begin on what is really the heart of the chapter, we should mention the many references to our text (Menell and Bazin, 1988). These are made because we think it can be of some value to readers with limited mathematical expertise and partly because we cannot bring ourselves to repeat yet again the same material that we have laboured over for quite a considerable amount of time.

### A. Analytical methods I

The simple model M1 may be rearranged to give

$$\frac{\mathrm{d}X}{X} = \mu\,\mathrm{d}t \qquad (9)$$

Integrating both sides of this equation gives the solution

$$\log X = \mu t + c \qquad (10)$$

Here $c$ is the constant of integration and represents the initial conditions of the system, i.e. the value of $\log X$ at $t = 0$. If the initial value of $X$ is $X_0$ then its identity to $c$ becomes evident and equation 10 can be rewritten as

$$\log X = \mu t + \log X_0 \tag{11}$$

which, by exponentiating both sides, gives

$$X = X_0 \, e^{\mu t} \tag{12}$$

Thus the prediction from model M1 is that the microbial population will increase exponentially without limit. The hypothesis on which this model was based, that the time between generations was constant, could reasonably be rejected on the basis of this prediction.

For the logistic model M2, rearrangement to separate the variables gives

$$\frac{dX}{(K - X)X} = \frac{\mu_m}{K} \, dt \tag{13}$$

Integrating the right hand side of this equation is straightforward

$$\int \frac{\mu_m}{K} \, dt = \frac{\mu_m}{K} t + c \tag{14}$$

The left hand side is usually solved using partial fractions and is detailed by Menell and Bazin (1988, p. 181). They express the solution in terms of the way $t$ changes with respect to population density. Rearranging this result yields

$$X = \frac{K X_0 \, e^{\mu_m t}}{X_0(e^{\mu_m t} - 1) + \mathrm{K}} \tag{15}$$

(We should mention here that most workers would resort to a table of integrals in order to obtain the solution rather than starting from first principles.) This results in a prediction, shown in Fig. 1, whereby the population density changes sigmoidally with respect to time tending towards a maximum value of $K$. Such behaviour is approximated reasonably well by several microbial species growing in closed systems and the underlying hypothesis, that the specific growth rate is a decreasing linear function of population density, cannot be rejected.

## B. Steady-state analysis

There is little in the way of a mechanistic basis for the hypothesis underlying the logistic equation. On the other hand, the Monod kinetics associated with model M3 relate the concentration of a growth-limiting nutrient to the specific growth rate of a microbial population. This relationship is similar to the Michaelis–Menten expression for enzyme kinetics and

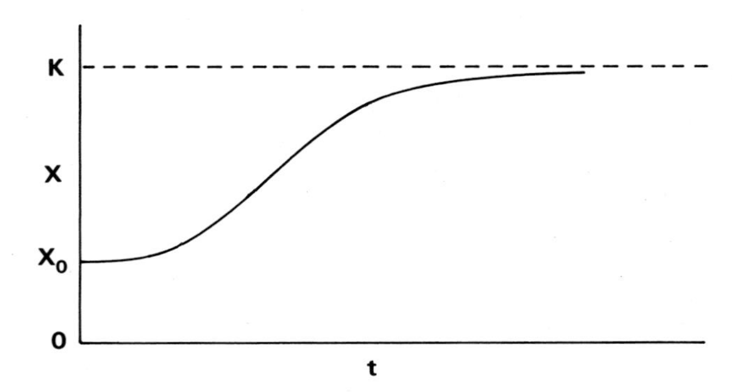


**Fig. 1.** Relationship between population density and time according to the logistic equation.

attempts have been made to derive the Monod function using the same sort of steady-state assumptions used by enzyme kineticists. At least to some extent then, model M3 is mechanistic in nature.

Analysis of model M3 is considerably more difficult than that for M2. Model M3 consists of a set of nonlinear differential equations for which an explicit solution does not exist. By this we mean that the time-dependent behaviour of $X$ and $S$ cannot be obtained from standard formulae in tables of integrals or by the numerous methods covered in elementary calculus courses. Unfortunately, this is the case for many biological systems which seem to be characterized by their nonlinearity. Therefore, other methods of obtaining predictions must be found. One of the simplest of these is based on studying systems at steady state. Such analyses apply only to thermodynamically open systems in which steady state is defined as the condition obtained when the dependent variables cease to change with respect to time. At steady state then, the derivatives may be equated to zero. For model M3,

$$\frac{\mathrm{d}X}{\mathrm{d}t} = \frac{\mathrm{d}S}{\mathrm{d}t} = 0 \tag{16}$$

By letting the steady-state values of $X$ and $S$ be $X_s$ and $S_s$ respectively, we can write

$$0 = \frac{\mu_m S_s X_s}{K_s + S_s} - DX_s \tag{17}$$

$$0 = D(S_r - S_s) - \frac{\mu_m S_s X_s}{Y(K_s + S_s)} \tag{18}$$

From equation 17 we can see immediately that

$$\frac{\mu_m S_s}{K_s + S_s} = D \tag{19}$$

i.e. at steady state the dilution rate of the system is numerically equal to the specific growth rate of the population. Rearrangement of equations 17 and 18 gives

$$S_s = \frac{DK_s}{\mu_m - D} \tag{20}$$

$$X_s = Y(S_r - S_s) \tag{21}$$

These relationships are depicted diagrammatically in Fig. 2. Note that they apply only for $\mu_m > D$. Otherwise, negative concentrations result which are a physical impossibility. The predictions conform reasonably well with steady-state experimental data but at low concentrations of substrate maintenance and endogenous metabolism effects must be taken into account (Pirt, 1975).

When model M6, representing competition for a single growth-limiting nutrient in an open ecosystem, is subjected to steady-state analysis we find that for positive (i.e. realistic) values of $X_{1s}$, $X_{2s}$ and $S_s$, the following relationships must hold:

$$D = \frac{\mu_{m1} S_s}{K_{s1} + S_s} = \frac{\mu_{m2} S_s}{K_{s2} + S_s} \tag{22}$$

$$S_s = \frac{DK_{s1}}{\mu_{m1} - D} = \frac{DK_{s2}}{\mu_{m2} - D} \tag{23}$$

These imply that in order for both species to survive, their specific growth rates must be identical. This is only possible if

$$K_{s2}\mu_{m1} > K_{s1}\mu_{m2}$$

and

$$K_{s2} > K_{s1}$$

or

$$K_{s2}\mu_{m1} < K_{s1}\mu_{m2}$$

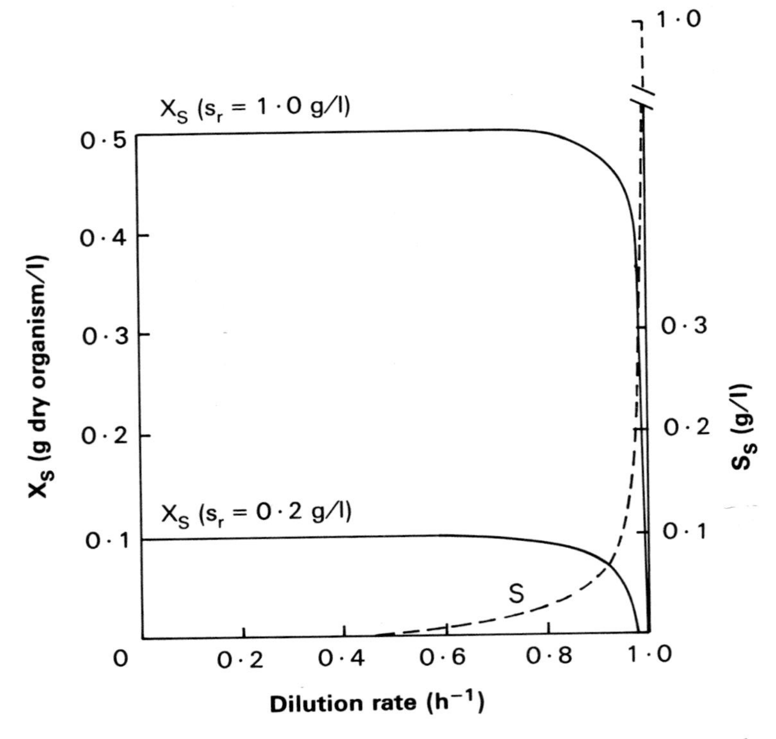


**Fig. 2.** Steady-state biomass and substrate concentrations in chemostat culture assuming Monod kinetics. Adapted from Pirt (1975).

and

$$K_{s2} < K_{s1}$$

If either of these conditions are met then there is a unique dilution rate at which survival of both species is theoretically possible:

$$D = \frac{K_{s2}\mu_{m1} - K_{s1}\mu_{m2}}{K_{s2} - K_{s1}}$$

or (24)

$$D = \frac{K_{s1}\mu_{m2} - K_{s2}\mu_{m1}}{K_{s1} - K_{s2}}$$

In practice, however, even with well-controlled chemostat cultures, the dilution rate cannot be held with sufficient precision. Outside the laboratory, the maintenance of such constancy is even more unlikely.

This analysis of competition in an open ecosystem constitutes a formal argument in support of Gause's principle of competitive exclusion which states that no two species can occupy the same niche.

### C. Phase plane analysis

The graphical representation of the behaviour of dependent variables with respect to each other constitutes a phase plane plot. We will use the Lotka–Volterra equations of model M4 to illustrate such an analysis. These equations are nonlinear and do not admit to an analytical solution. Dividing $\mathrm{d}H/\mathrm{d}t$ by $\mathrm{d}P/\mathrm{d}t$ gives

$$\frac{\mathrm{d}H}{\mathrm{d}P} = \frac{\alpha_1 H - \beta_1 PH}{-\alpha_2 P + \beta_2 PH} \tag{25}$$

Rearranging this equation results in

$$\beta_2\,\mathrm{d}H - \frac{\alpha_2}{H}\,\mathrm{d}H = \frac{\alpha_1}{P}\,\mathrm{d}P - \beta_1\,\mathrm{d}P \tag{26}$$

and integrating each term leads to

$$\beta_2 H - \alpha_2 \log H = \alpha_1 \log P - \beta_1 P + c \tag{27}$$

where $c$ is the constant of integration and can be evaluated from the initial conditions $H_0$, $P_0$:

$$c = \beta_2 H_0 - \alpha_2 \log H_0 - \alpha_1 \log P_0 + \beta_1 P_0 \tag{28}$$

Equation 27 represents a family of closed curves, each dependent on $c$, the initial conditions. The point about which these trajectories move is called the equilibrium or stationary point of the system. The equilibrium point is found by evaluating the independent variables, $H$ and $P$, when their derivatives are set equal to zero. If these values are designated $H_e$ and $P_e$, respectively then,

$$H_e = \frac{\alpha_2}{\beta_2}$$

and

$$P_e = \frac{\alpha_1}{\beta_1} \tag{29}$$

In Fig. 3 we show a phase plane plot of the Lotka–Volterra system. This has been derived numerically as described in the next section.

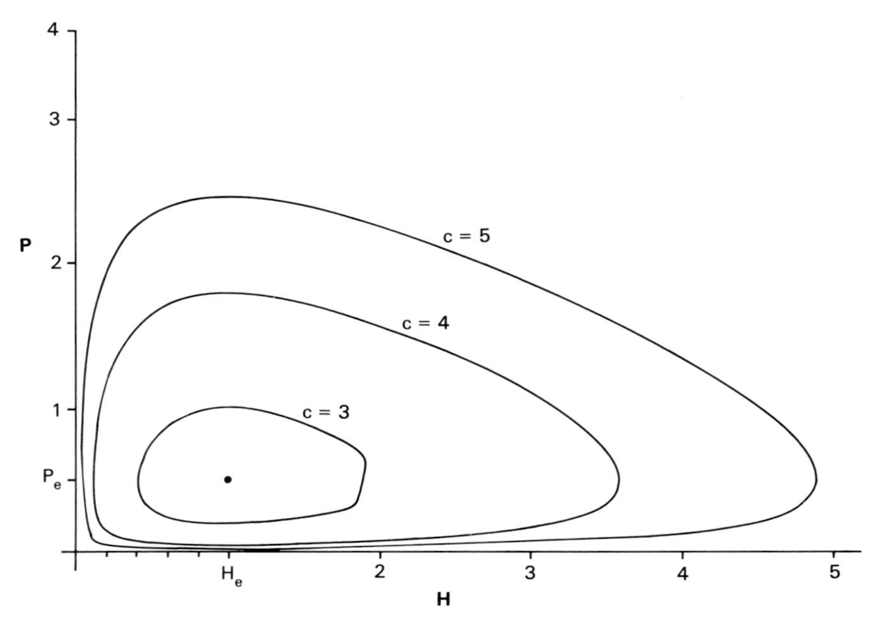


**Fig. 3.** Phase plane plot of predator, *P*, against prey, *H*, resulting from the Lotka–Volterra equations.

## D. Numerical approximation

Although analytical solutions for the equations for most of our models are not known, techniques exist for approximating them. These are known as numerical methods. They involve much arithmetic calculation and are usually undertaken on a digital computer. Fortunately, extremely efficient packages for performing such calculations are available on most mini- and mainframe-computers. The disadvantage of such methods is that the values of all the coefficients in the equations to be analysed need to be known. If they are not known then it is difficult to assess the global behaviour of the system.

Often differential equations of the sort we are dealing with are critically dependent on the constants and parameters associated with the variables. Relatively small changes in their values or the ratios between them may result in qualitatively different responses. For example, under some circumstances the variables may oscillate while under others they may change smoothly and monotonically. As we shall see later this is the case with some predator–prey models. Using numerical methods it is often difficult to determine all the types of responses that a system may produce.

Newton's method of finding a solution (root) to an equation is based on the idea that if $x_0$ is a first approximation to the root of the equation $f(x) = 0$, then

$$x_1 = x_0 - \frac{f(x_0)}{f'(x_0)} \tag{30}$$

(where the prime notation represents a derivative) is a better approximation. In general, then

$$x_{n+1} = x_n - \frac{f(x_n)}{f'(x_n)}$$

Letting $\alpha_1 = \alpha_2 = \beta_2 = 1$ and $\beta_1 = 2$, equation 27 can be rewritten as

$$f(P) = 2P + H - \log P - \log H - c$$

so that

$$P_{n+1} = P_n - \frac{2P_n + H - \log P_n - \log H - c}{2 - \dfrac{1}{P_n}}$$

$$= \frac{P_n\left(2 - \dfrac{1}{P_n}\right) - (2P_n - H - \log P_n - \log H - c)}{2 - \dfrac{1}{P_n}}$$

Therefore,

$$P_{n+1} = \frac{\log P_n + \log H + c - 1 - H}{2 - \dfrac{1}{P_n}} \tag{31}$$

In order to plot the phase-plane curve represented by equation 27, we follow the following progression:

(i) Assign a value to $c$.
(ii) Give successive values to $H$, say from 0.1 to 4 in steps of 0.1.
(iii) Solve equation 31 for values of $P$, given one value of $H$. Repeat for all values of $H$.

The last step involves the following:

Locate the solution roughly using the principle that if $f(a) > 0$ and $f(b) < 0$ or if $f(a) < 0$ and $f(b) > 0$ then there is a zero of $f(x)$ between $x = a$ and $x = b$.

Use Newton's method, with $0.5(a + b)$ as a first approximation, to find the root more accurately.

```
>LIST
    5 VDU2
   10 C=5;P=0.01
   15 INPUT H
   16 PRINT "H= ";H
   20 Q=2*P+H-LN(P)-LN(H)-C
   30 P=P+0.01
   40 R=2*P+H-LN(P)-LN(H)-C
   50 IF SGN(Q)+SGN(R)=0 THEN PRINT "ROOT BETWEEN ";P-0.01;" AND
      ";P
   60 IF P>4 GOTO 100
   70 GOTO 20
  100 INPUT P
  110 S=(LN(P)+LN(H)+C-1-H)/(2-1/P)
  120 1=ABS:P*S)
  130 IF T>=0.0001 THEN PRINT S ELSE PRINT "ROOT= ":S:STOP
  150 P=S
  160 GOTO 110
  170 VDU3
```

**Fig. 4.** Basic program to illustrate Newton's method.

A simple Basic program for performing these calculations is presented in Fig. 4 and example results are shown in Fig. 3.

Numerical methods for approximating the solutions of ordinary differential equations are described well by Morris (1974) and we refer the reader to this text for further details. Based on such methods, specialized computer packages exist for approximating the solutions of ordinary differential equations. These simulation languages, as they are sometimes called, include CSMP, MIMIC and DARE IIB. They usually include a simple plotting routine for line printer output and are simple to use. An example of output from a CSMP program written by Curds (Curds and Bazin, 1977) is reproduced in Fig. 5.

### E. Analog simulation

Analog computers, or differential analysers as they are also called, have been used widely in electrical and aeronautical engineering for solving sets of differential equations. An analog computer functions by representing sets of equations as electrical circuits. Details of how they may be used in ecological applications are given by Patten (1971). Essentially, they consist of four main elements with the properties and symbolic representation given in Table I.

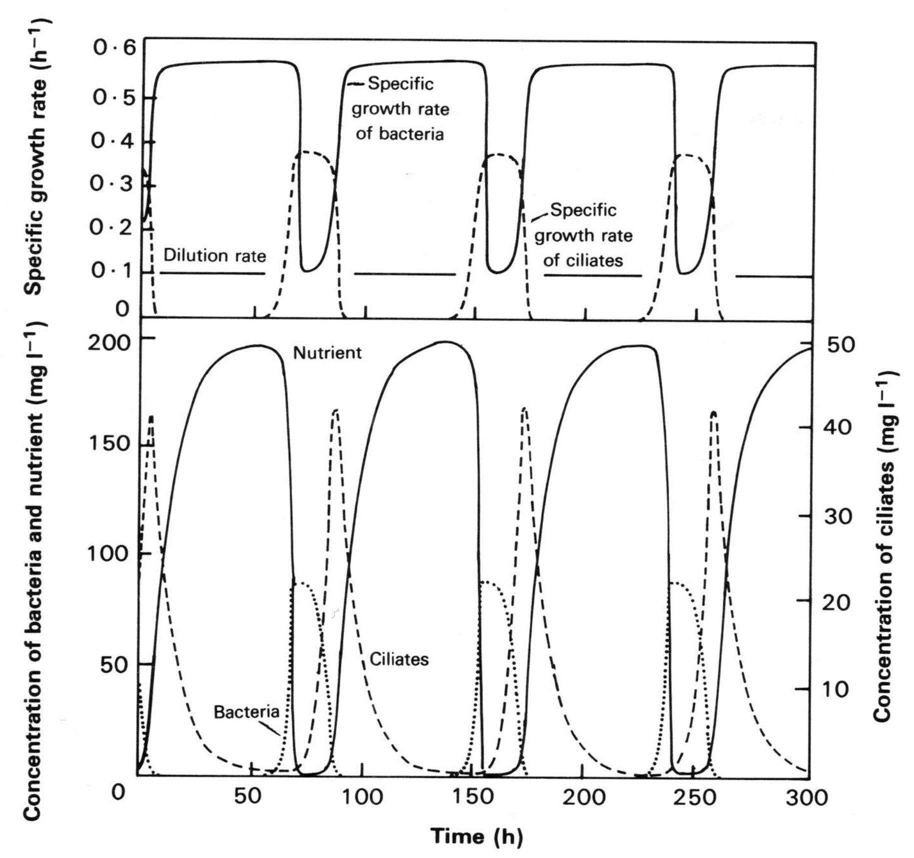


**Fig. 5.** A CSMP simulation of a nutrient–bacteria–ciliate chemostat system as described in Curds and Bazin (1977).

In order to solve differential equations on an analog computer it is first necessary to design a circuit diagram. As a simple example, we will use model M2, the logistic equation. This can be rewritten as

$$\frac{dX}{dt} = \mu_m X - \frac{\mu_m}{K} X^2 \tag{32}$$

Let the initial value of $X$ be $X_0$. The first step in designing a circuit is to write down the input into integrators. In this case we have a single differential and therefore need only one integrator:

**TABLE I**
Basic operations of an analog computer

| Operation | Component | Symbol<br>Input voltage/Output voltage |
|---|---|---|
| Multiplication by a constant | Potentiometer | $X$ → (R) → $k\,X$ |
| Addition (and subtraction) | Adder | $X$, $Y$ → $-(X+Y)$ |
| Integration | Integrator | $X$, $Y$ → $-\int(X+Y)\mathrm{d}t$ |
| Multiplication of variables | Multiplier | $X$, $Y$ → $XY$ |

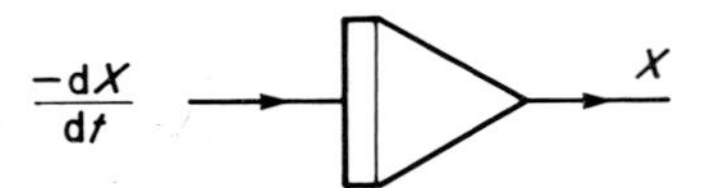


Initial conditions are set on a potentiometer which connects directly to the integrator. The potentiometer is supplied by a current from a voltage source in the computer with a reference value of 1. Thus we have:

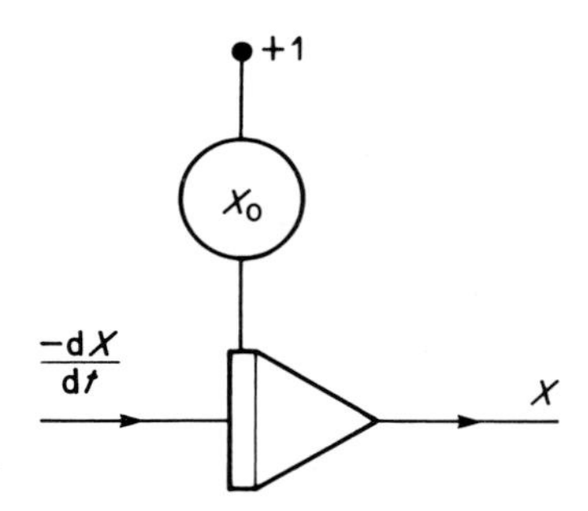


We now need to construct the components of $\mathrm{d}X/\mathrm{d}t$ using equation 32 and the symbols from Table I.

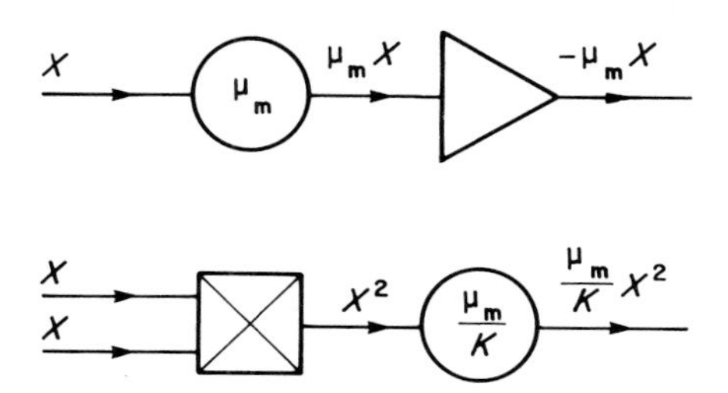


Notice we have used an adder to reverse the sign of $\mu_m X$. The sum of the outputs is $-\mu_m X + \mu_m X^2/K = -dX/dt$. If we feed these values into an integrator it will sum and integrate them to give $X$. This output can be used to supply the multiplier and the potentiometer set at a value of $\mu_m$. The completed circuit thus becomes:

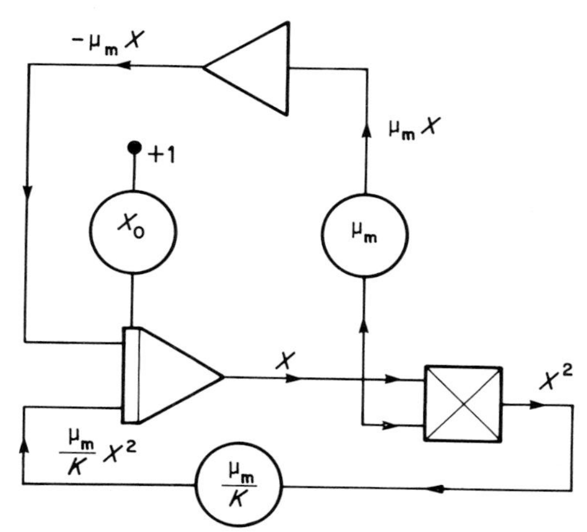


The circuit diagram is then used to connect appropriate sockets on the computer with wire connectors. This is called 'patching in' the circuit. The output from the integrator, $X$, can then be recorded with a voltmeter or chart recorder to determine its time-dependent behaviour. An example of analog computer output is given in Fig. 6.

Using an analog computer to determine the behaviour of nonlinear models suffers from the same disadvantage as that described for numerical methods. In addition, analog computation is limited by the input and output voltages that the machines can handle. Thus even moderately wide ranges of values for the variables are difficult to handle. This necessitates the scaling of the equations so that the inputs and outputs fall within the voltage range of the computer. Scaling is often mathematically difficult and mainly for this reason the use of analog computers in ecology is diminishing.

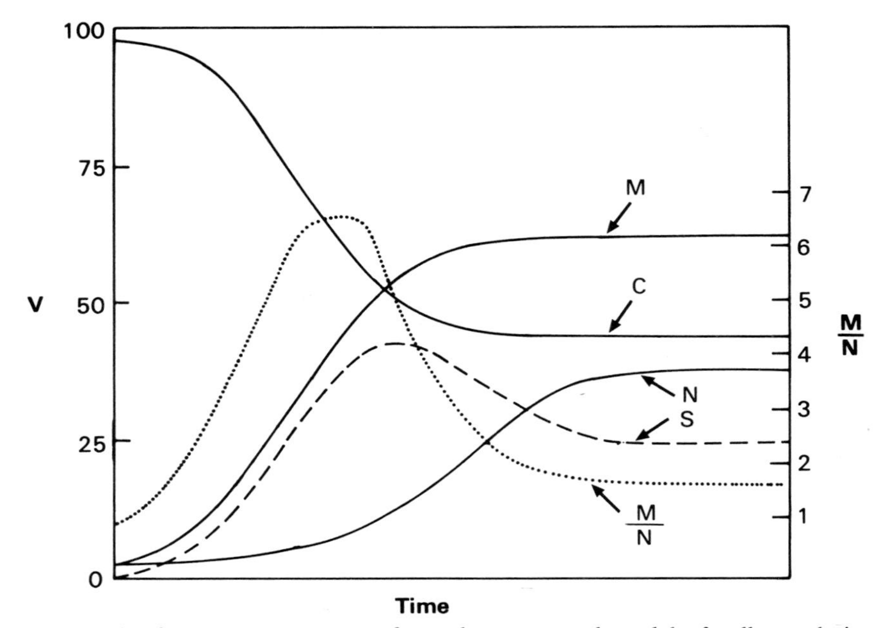


**Fig. 6.** Analog computer output from the structured model of cell population growth of Williams (1967). The symbols represent the biomass concentration ($M$), nutrient ($C$), the synthetic portion of the biomass ($S$) and the structural-genetic portion of the biomass ($N$).

## F. Stability analysis

This method, which is widely used in chemical engineering, determines the behaviour of systems near equilibrium (or steady state in the case of open systems). As indicated in Section V.B and V.C equilibria and steady states are determined by equating the differential equations to zero and solving for the values of the dependent variables. As we shall see later, some systems have more than one of these stationary points. The goal of stability analysis is to determine the system behaviour near these points by eliminating second and higher order terms in order to obtain a set of linear differential equations which can be solved explicitly. The steps in this procedure may be summarized as follows:

(i) Transform the variables of the system by subtracting from each of them their stationary point value.

(ii) Consider the behaviour of these new variables close to the new equilibrium value, i.e. near to zero. Thus they will be of numerically small value.

(iii) Eliminate products and higher order terms on the basis that their numerical values will be very small and can be ignored.

(iv) Solve the resulting linear differential equations to determine the behaviour of the system near equilibrium.

In this section we will analyse three models using this Liapounov stability analysis as it is sometimes called.

First, let us consider the Lotka–Volterra equations of model M4. We define new variables for $H$ and $P$ such that

$$x = H - H_e$$

and

$$y = P - P_e$$

where $H_e = \alpha_2/\beta_2$ and $P_e = \alpha_1/\beta_1$ are the equilibrium values of the system. Substituting into the first equation of M4, we get

$$\begin{aligned}\frac{dx}{dt} = \frac{dH}{dt} &= \alpha_1(x + H_e) - \beta_1(x + H_e)(y + P_e)\\ &= \alpha_1 x + \alpha_1 H_e - \beta_1 xy - \beta_1 P_e x - \beta_1 H_e P_e - \beta_1 H_e y\\ &= (\alpha_1 H_e - \beta_1 H_e P_e) + x(\alpha_1 - \beta_1 P_e) - \beta_1 H_e y - \beta_1 xy \qquad (33)\end{aligned}$$

By definition of equilibrium, the first two terms of equation 33 sum to zero. Thus

$$\frac{dx}{dt} = -\beta_1 H_e y - \beta_1 xy \qquad (34)$$

Similarly, for the second equation of model M4,

$$\begin{aligned}\frac{dy}{dt} = \frac{dP}{dt} &= -\alpha_2(y + P_e) + \beta_2(y + P_e)(x + H_e)\\ &= -\alpha_2 y - \alpha_2 P_e + \beta_2 xy + \beta_2 H_e y + \beta_2 P_e x + \beta_2 H_e P_e\\ &= (-\alpha_2 P_e + \beta_2 H_e P_e) + y(-\alpha_2 + \beta_2 H_e) + \beta_2 P_e x + \beta_2 xy\\ &= \beta_2 P_e x + \beta_2 xy \qquad (35)\end{aligned}$$

As we are looking for values of $x$ and $y$ near the equilibrium point, their values will be small. Their products will be very small so we can ignore them. Thus we have

$$\frac{dx}{dt} = -\beta_1 H_e y \tag{36}$$

and

$$\frac{dy}{dt} = \beta_2 P_e x \tag{37}$$

Differentiating equation 36 gives

$$\frac{d^2x}{dt^2} = -\beta_1 H_e \frac{dy}{dt} \tag{38}$$

and substituting in equation 37 we get

$$\frac{d^2x}{dt^2} = -\beta_1 \beta_2 H_e P_e x$$

$$= -\beta_1 \beta_2 \frac{\alpha_1}{\beta_1} \frac{\alpha_2}{\beta_2} x$$

$$= -\alpha_1 \alpha_2 x$$

i.e.

$$\frac{d^2x}{dt^2} + \alpha_1 \alpha_2 x = 0 \tag{39}$$

In a similar manner we can obtain

$$\frac{d^2y}{dt^2} + \alpha_1 \alpha_2 y = 0 \tag{40}$$

The solution of equation 39 is

$$x = A_h \cos(\sqrt{\alpha_1 \alpha_2} t + p_h) \tag{41}$$

where $A_h$ is the amplitude of the oscillations that result and $p_h$ is the phase constant. Their values depend on the initial conditions. If we assume that at time $t = 0$, x is at its maximum then $p_h = 0$. Similarly, the solution of equation 40 is

$$y = A_p \sin(\sqrt{\alpha_1 \alpha_2} t + p_p) \tag{42}$$

where again $A_p$ and $p_p$ depend on the initial conditions. Figure 7 shows an example of simple Lotka–Volterra behaviour when $p_h$ and $p_p$ are both zero.

The second system we will analyse by the Liapounov method is model M3, growth of a single species in an open ecosystem assuming Monod kinetics. These equations may be rewritten as

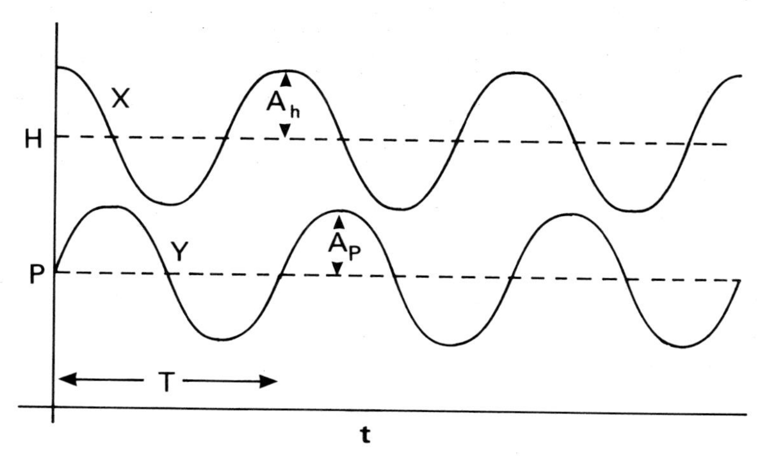


**Fig. 7.** Results of analysing the Lotka–Volterra equations near equilibrium. The transformed prey and predator variables are $x$ and $y$ and the broken lines labelled $H$ and $P$ represent the equilibrium values of the populations. The amplitudes of the prey and predator populations are $A_h$ and $A_P$, respectively and the period of the oscillations is $T$.

$$\dot{X} = \mu(S)X - DX \tag{43}$$

$$\dot{S} = DS_r - \frac{\mu(S)X}{Y} - DS \tag{44}$$

Here we have written a dot over a symbol to indicate its derivative with respect to time. This will make it easier to represent equations and their analysis. In equations 43 and 44 $\mu(S)$ represents a function with argument $S$. In the case we are considering it is the Monod function, so

$$\mu(S) = \frac{\mu_m S}{K_s + S} \tag{45}$$

As we are considering an open ecosystem, it has (potentially) at least one steady state and we designate $X$ and $S$ at steady state as $X_s$ and $S_s$. Our new variables are therefore

$$x = X - X_s$$

and

$$y = S - S_s$$

In terms of the new variables the Monod function becomes

$$\mu(y + S_s) = \frac{\mu_m(y + S_s)}{K_s + (y + S_s)} \tag{46}$$

This function can be expanded in a Taylor series (Menell and Bazin, 1988, p. 113) as follows:

$$\mu(y + S_s) = \mu(S_s) + y\frac{d\mu(S)}{dS_s} + \begin{matrix}\text{second and higher}\\ \text{order terms}\end{matrix}$$

$$= \frac{\mu_m S_s}{K_s + S_s} + \frac{(K_s + S_s)\mu_m - \mu_m S_s}{(K_s + S_s)^2} y + \cdots \tag{47}$$

In the method we are employing here, we assume that second order terms can be eliminated. Therefore by substitution into equations 43 and 44, the system becomes

$$\dot{X} = \frac{d(x + X_s)}{dt} = \frac{dx}{dt} = \dot{x}$$

$$= \left[\frac{\mu_m S_s}{K_s + S_s} + \frac{\mu_m K_s}{(K_s + S_s)^2} y\right](x + X_s) - D(x + X_s) \tag{48}$$

$$\dot{S} = \frac{d(y + S_s)}{dt} = \frac{dy}{dt} = \dot{y}$$

$$= DS_r - \left[\frac{\mu_m S_s}{K_s + S_s} + \frac{\mu_m K_s}{(K_s + S_s)^2} y\right]\frac{x + X_s}{Y} - D(y + S_s) \tag{49}$$

Multiplying out equation 48 results in

$$\dot{x} = \frac{\mu_m S_s x}{K_s + S_s} + \frac{\mu_m S_s X_s}{K_s + S_s} + \frac{\mu_m K_s xy}{(K_s + S_s)^2}$$

$$+ \frac{\mu_m K_s}{(K_s + S_s)^2} X_s y - Dx - DX_s \tag{50}$$

By definition,

$$\frac{\mu_m S_s X_s}{K_s + S_s} - DX_s = 0 = \frac{\mu_m S_s x}{K_s + S_s} - Dx$$

Therefore, the first two and last two terms of equation 50 sum to zero. The third term is second order ($yx$) and so it too can be eliminated. Therefore, we are left with

$$\dot{x} = \frac{\mu_m K_s}{(K_s + S_s)^2} X_s y = \frac{\mu_m K_s - \mu_m S_s + \mu_m S_s}{(K_s + S_s)^2} X_s y$$

$$= \frac{\dfrac{\mu_m K_s}{K_s + S_s} - \dfrac{\mu_m S_s}{K_s + S_s} + \dfrac{\mu_m S_s}{K_s + S_s}}{K_s + S_s} X_s y$$

$$= \frac{\dfrac{\mu_m (K_s + S_s)}{K_s + S_s} - D}{K_s + S_s} X_s y$$

$$= \frac{\mu_m - D}{K_s + S_s} X_s y \quad (51)$$

This can be expressed as

$$\dot{x} = ay \quad (52)$$

where

$$a = \frac{\mu_m - D}{K_s + S_s} X_s$$

Similar treatment of equation 49 yields

$$\dot{y} = -\frac{D}{Y} x - \left(D + \frac{a}{Y}\right) y \quad (53)$$

In equations 52 and 53 we have two linear equations but they cannot be solved directly because they both contain two dependent variables. We deal with this by first differentiating equation 53 to get

$$\ddot{y} = -\frac{D}{Y} \dot{x} - \left(D + \frac{a}{Y}\right) \dot{y} \quad (54)$$

By substituting in equation 52 and rearranging we obtain

$$\ddot{y} + \left(D + \frac{a}{Y}\right) \dot{y} + \frac{D}{Y} ay = 0 \quad (55)$$

This is a second order differential equation with a solution of the form (Menell and Bazin, 1988, p. 211)

$$y = A\,e^{\omega_1 t} + B\,e^{\omega_2 t} \quad (56)$$

where $A$ and $B$ are constants dependent on the initial conditions of the system and $\omega_1$ and $\omega_2$ are known as eigenvalues. The eigenvalues are obtained from what is called the auxiliary equation derived from the differential equation. For the case at hand:

$$\omega^2 + \left(D + \frac{a}{Y}\right)\omega + \frac{D}{Y}a = 0 \tag{57}$$

Applying the quadratic formula*

$$\begin{aligned}
\omega &= \frac{-\left(D + \frac{a}{Y}\right) \pm \sqrt{\left(D + \frac{a}{Y}\right)^2 - \frac{4Da}{Y}}}{2} \\
&= \frac{-\left(D + \frac{a}{Y}\right)^2 \pm \sqrt{\left(D - \frac{a}{Y}\right)^2}}{2} \\
&= \frac{-\left(D + \frac{a}{Y}\right) \pm \left(D - \frac{a}{Y}\right)}{2}
\end{aligned} \tag{58}$$

Therefore,

$$\omega_1 = -D$$

and

$$\omega_2 = -a/Y$$

For $\mu_m > D$, $S_s > 0$ and $X_s > 0$, all of which inequalities must be true for the system to be non-trivially realizable, $a$ will always be positive so $\omega_2$ will always be negative. Similarly, $\omega_1$ will always be negative. Therefore, both exponents in equation 56 will have negative values and as $t$ increases so $y$ will get closer and closer to zero. That is to say, the system will move towards steady state smoothly and monotonically. Thus $S = y + S_s$ will tend towards $S_s$, its steady-state value. A similar analysis can be performed on $x$ to show that $X$ behaves in a similar manner.

A system that moves towards its equilibrium or steady state in this way is said to be stable. Stability is dependent on the sign of the eigenvalues. If all the eigenvalues of a system are negative, it is stable. If one or more are positive it is unstable—the system moves away from the stationary point. As a result of the analysis we have performed here we can say that model M3 admits to a stable solution. This sort of behaviour is commonly observed in chemostat cultures and is depicted diagrammatically in Fig. 8.

*For the quadratic equation $ax^2 + bx + c = 0$, $x = (-b \pm \sqrt{b^2 - 4ac})/2a$.

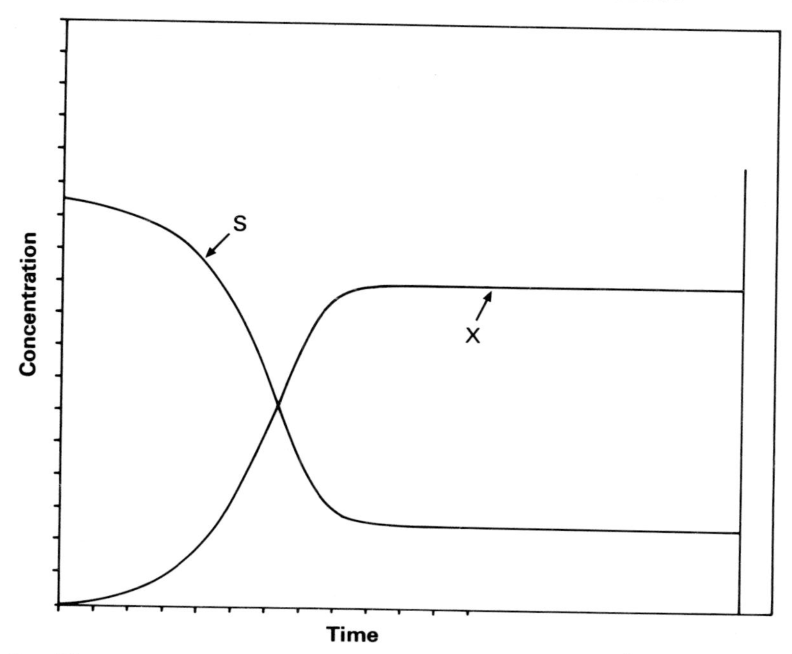


**Fig. 8.** Chemostat dynamics assuming Monod kinetics. Biomass concentration, $X$, and substrate concentration, $S$.

Our third example of stability analysis uses model M5, a predator–prey interaction in an open ecosystem based on Monod kinetics. Before we linearize this model we will simplify the original equations by making them dimensionless, a practice commonly employed by chemical engineers. We define new variables as follows:

$$\theta = Dt, x = S/S_r, y = H/YS_r, z = P/YWS_r$$

and let

$$a = K_s/S_r, b = L_s/YS_r, A = \mu_m/D, B = \lambda_m/D$$

By substituting these relationships into the equations of the model and cancelling we obtain

$$\frac{dy}{d\theta} = \frac{Axy}{a + x} - \frac{Byz}{b + y} - y \tag{59}$$

$$\frac{dz}{d\theta} = \frac{Byz}{b + y} - z \tag{60}$$

$$\frac{dx}{d\theta} = 1 - \frac{Axy}{a + x} - x \tag{61}$$

We can reduce this system of three equations to two by assuming, reasonably, that

$$H/Y + P/YW + S = S_r$$

for all time so that (Canale, 1970)

$$y + x + z = 1$$

Substituting this relationship into equations 59 and 60 gives

$$\frac{dy}{d\theta} = \frac{A(1 - y - z)y}{a + 1 - y - z} - \frac{Byz}{b + y} - y \tag{62}$$

$$\frac{dz}{d\theta} = \frac{Byz}{b + y} - z \tag{63}$$

We now change variables such that

$$u = y - y_s$$

and

$$v = z - z_s$$

Letting

$$g(y, z) = \frac{Bz}{b + y}$$

$$h(y) = \frac{By}{b + y}$$

and

$$f(y, z) = \frac{A(1 - y - z)}{a + 1 - y - z}$$

we obtain

$$\frac{du}{d\theta} = f(u + y_s, v + z_s)(u + y_s) - g(u + y_s, v + z_s)(v + z_s) - (v + z_s) \tag{64}$$

$$\frac{dv}{d\theta} = h(u + y_s)(v + z_s) - (v + z_s) \tag{65}$$

Expanding the functions in equation 64 by Taylor's series gives us

$$\begin{aligned}\frac{du}{d\theta} &= \left[f(y_s, z_s) + \frac{\partial f}{\partial y_s} u + \frac{\partial f}{\partial z_s} v + O^2\right](u + y_s) \\ &\quad - \left[g(y_s, z_s) + \frac{\partial g}{\partial y_s} u + \frac{\partial g}{\partial z_s} v + O^2\right](v + y_s) \\ &\quad - (v + y_s)\end{aligned} \tag{66}$$

where $O^2$ represents second and higher order terms. Eliminating these and substituting in for the derivatives in equation 66 produces

$$\begin{aligned}\frac{du}{d\theta} &= \left[\frac{A(1-y_s-z_s)}{a+1-y_s-z_s} + \frac{(a+1-y_s-z_s)(-A)+A(1-y_s-z_s)}{(a+1-y_s-z_s)^2} u \right. \\ &\quad \left. + \frac{(a+1-y_s-z_s)(-A)-A(1-y_s-z_s)}{(a+1-y_s-z_s)^2} v\right](u+y_s) \\ &\quad - \left[\frac{Bz_s}{b+y_s} + \frac{(b+y_s)(0)-Bz_s(1)}{(b+y_s)^2} u \right. \\ &\quad \left. + \frac{(b+y_s)B-Bz_s(0)}{(b+y_s)^2} v\right](u+y_s)-(u+y_s)\end{aligned} \tag{67}$$

Multiplying out and making the appropriate cancellations results in

$$\begin{aligned}\frac{du}{d\theta} &= \left[\frac{A(1 - y_s - z_s)}{a + 1 - y_s - z_s} - \frac{Aa}{(a + 1 - y_s - z_s)^2} u \right. \\ &\quad \left. - \frac{Aa}{(a + 1 - y_s - z_s)^2} v\right](u + y_s) \\ &\quad - \left[\frac{Bz_s}{b + y_s} - \frac{Bz_s}{(b + y_s)^2} u + \frac{B}{b + y_s} v\right](u + y_s) - u - y_s\end{aligned} \tag{68}$$

At steady state, by definition,

$$0 = \frac{A(1 - y_s - z_s)y_s}{a + (1 - y_s - z_s)} - \frac{By_s z_s}{b + y_s} - y_s$$

so that when second order terms are eliminated, equation 68 reduces to

$$\frac{du}{d\theta} = \left(\frac{z_s}{By_s} - \frac{Aay_s}{(a + x_s)^2}\right)u - \left(1 + \frac{Aay_s}{(a + x_s)^2}\right)v \tag{69}$$

Similar treatment of equation 63 gives

$$\frac{dv}{d\theta} = \frac{bz_s}{By_s}u \tag{70}$$

In matrix notation the system can be written as

$$\begin{bmatrix} \frac{du}{d\theta} \\ \frac{dv}{d\theta} \end{bmatrix} = \begin{bmatrix} c_1 & c_2 \\ c_3 & c_4 \end{bmatrix} \begin{bmatrix} u \\ v \end{bmatrix} \tag{71}$$

where

$$c_1 = \frac{z_s}{By_s} - \frac{Aay_s}{(a + x_s)^2}$$

$$c_2 = -\left(1 + \frac{Aay_s}{(a + x_s)^2}\right)$$

$$c_3 = \frac{bz_s}{By_s}$$

$$c_4 = 0$$

The central matrix of equation 71 is called the stability matrix. Using the elements it contains, the auxiliary equation can be obtained directly. In determinant form this is

$$\begin{vmatrix} c_1 - \omega & c_2 \\ c_3 & c_4 - \omega \end{vmatrix} = 0 \tag{72}$$

Multiplying out this equation we get

$$(c_1 - \omega)(c_4 - \omega) - c_2c_3 = 0$$

Therefore,

$$\omega^2 - (c_1 + c_4)\omega - (c_2c_3 - c_1c_4) = 0$$

Using the quadratic formula and substituting in for $c_1$, $c_2$, $c_3$ and $c_4$, we obtain

$$\omega = \left\{\left[\frac{z_s}{By_s} - \frac{Aay_s}{(a+x_s)^2}\right] \pm \left[\left(\frac{z_s}{By_s} - \frac{Aay_s}{(a+x_s)^2}\right)^2 - 4\left(1 + \frac{Aay_s}{(a+x_s)^2}\right)\frac{bz_s}{By_s}\right]^{1/2}\right\} \Big/ 2 \qquad (73)$$

Canale (1970) investigated this system in detail. He found that the solutions depended critically on the values of $A$, $B$, $a$ and $b$, that is the governing coefficients, $L_s$, $Y$, $S_r$, $K_s$, $\mu_m$, $\lambda_m$ and $D$, of the original equations. In particular, for some ranges of values the type of response depends only on the value of the dilution rate, $D$. Recalling that the general solution of the system takes the form

$$y = A_1 \, e^{\omega_1 t} + B_1 \, e^{\omega_2 t}$$

$$z = A_2 \, e^{\omega_1 t} + B_2 \, e^{\omega_2 t}$$

with the $A$s and $B$s constants, the form of the solution depends on the eigenvalues $\omega_1$ and $\omega_2$. For stability these must be negative. If in equation 73

$$c_1^2 > -4c_2c_3$$

and

$$c_1 + \sqrt{c_1^2 + 4c_2c_3} < 0$$

both eigenvalues will be negative and all variables will move smoothly and monotonically towards their steady-state values. Such behaviour occurs at relatively high dilution rates ($D < \mu_m$ and $\lambda_m$) and constitutes what is called a stable node. It is depicted diagrammatically in $H$–$P$ phase space in Fig. 9.

For smaller values of $D$ it is possible for the term in the square root sign of equation 73 to become negative, i.e. $c_1^2 < -4c_2c_3$. In this case the eigenvalues are expressed as complex conjugate numbers

$$\omega = n \pm pi$$

where $n$ and $p$ are real numbers and $i^2$ is defined as being equal to $-1$ (Menell and Bazin, 1988, p. 200). Thus the solutions take the form

$$A \, e^{(n+pi)t} + B \, e^{(n-pi)t}$$

Now it can be shown (Menell and Bazin, 1988, p. 204) that

$$e^{i\theta} = \cos\theta + i\sin\theta$$

That is to say the solution is sinusoidal. Thus $x$, $y$ and $z$ (and therefore $S$, $H$ and $P$) become periodic with time. For $n < 0$, the solution is said to be

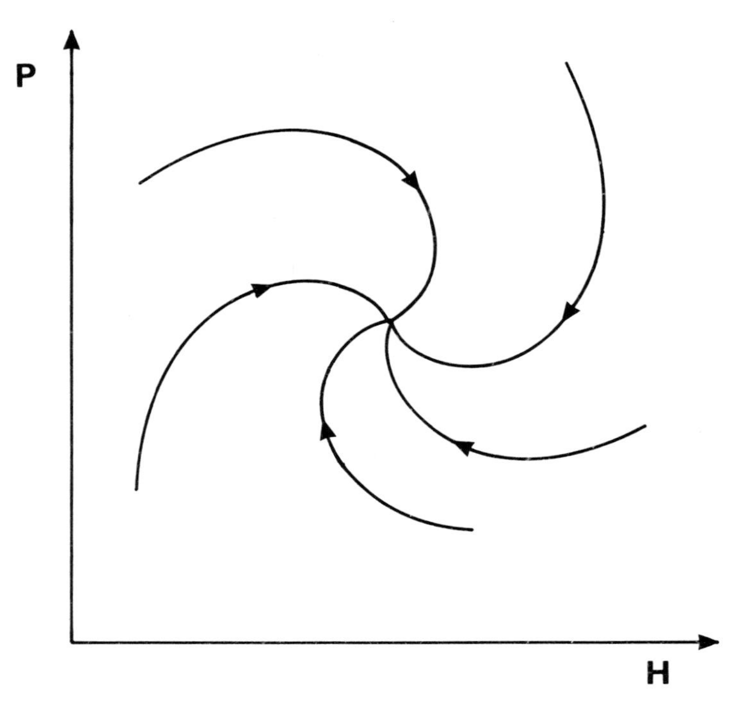


**Fig. 9.** Predator ($P$)–prey ($H$) phase plane plot showing stable node behaviour.

a stable focus and the three variables exhibit damped oscillations. The $H$–$P$ phase plane plot in this case is shown in Fig. 10.

For $n > 0$ an unstable focus results. The system is periodic with oscillations of increasing amplitude. The original variables ($S$, $H$ and $P$) can never become negative so there is a physical limitation of behaviour of this sort. It is possible for $H$ and $P$ to become zero in which case the system collapses. However, what is likely to happen is that the system becomes constrained, essentially by the amount of nutrient being received, $S_r$. What results then is what is called a limit cycle which settles down into regular oscillations. The $H$–$P$ phase plane plot of such behaviour is shown in Fig. 11. Notice that there is only a single closed trajectory as compared to the family of curves obtained from the Lotka–Volterra equations and shown in Fig. 3. In the latter case the trajectories, and therefore the amplitudes of the oscillations, are dependent on the initial conditions. In a limit cycle

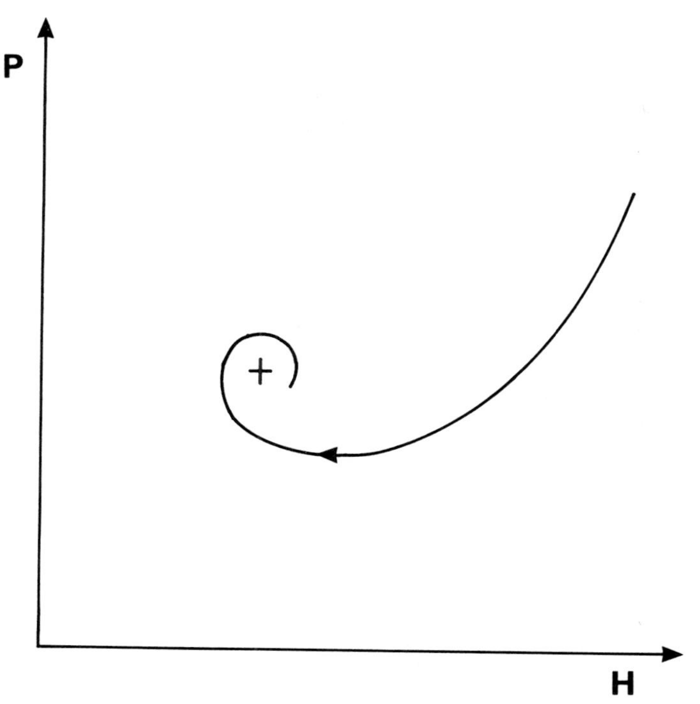


**Fig. 10.** Predator ($P$)–prey ($H$) phase plane plot showing stable focus behaviour.

this is not the case; the amplitudes of the oscillations are independent of the initial conditions.

Finally, we will mention what happens when $c_1 = 0$ so that the eigenvalues are pure imaginary numbers. This results in what is called a vortex and is in fact the solution obtained when the Lotka–Volterra equations are analysed by the method outlined above.

## G. Analytical methods II

In this section we analyse model M7 which represents a column of soil containing a fixed amount of microbial biomass, $X_c$, which utilizes nutrient at concentration $S$ at rate $kX_c$. The following analysis is *relatively* simple because $X_c$ is constant. For more complex situations analytical methods are usually not available and numerical methods must be employed (see

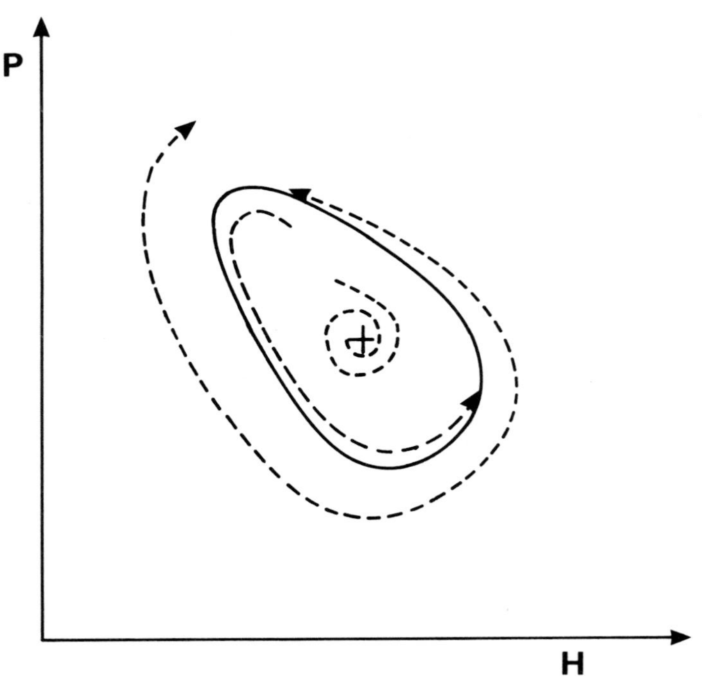


**Fig. 11.** Predator ($P$)–prey ($H$) phase plane plot showing limit cycle behaviour.

for example, Smith, 1985). Unfortunately, these are not as easy to apply as those for ordinary differential equations although this situation is changing rapidly. For further examples that have been treated analytically we refer the reader to Cho (1971) and Saunders and Bazin (1973) who analysed nitrification in soil ecosystems.

In what follows we shall use Laplace transforms, denoted by a bar over a variable, to solve the equation of model M7. A useful reference is Churchill (1944). The Laplace transform of a function is defined as

$$\mathcal{L}f(x) = \int_0^\infty e^{-px} f(x)\, dx$$

Also included is the convolution rule, defined for two functions $f(t)$ and $g(t)$ as

$$\mathcal{L}^{-1}\{F(p)G(p)\} = f(t) * g(t) = \int_0^t f(t-u)g(u)\, du$$

the error function,

$$\mathrm{erf}(x) = \frac{2}{\sqrt{\pi}} \int_0^x \mathrm{e}^{-\eta^2}\, \mathrm{d}\eta$$

and the complementary error function,

$$\mathrm{erfc}(x) = 1 - \mathrm{erf}(x)$$

Taking Laplace transforms of the terms in the equation of model M7, we obtain

$$p\bar{S} - S(z, 0) = D^* \frac{\mathrm{d}^2 \bar{S}}{\mathrm{d}z^2} - q \frac{\mathrm{d}\bar{S}}{\mathrm{d}z} - \frac{kX_\mathrm{c}}{p} \tag{74}$$

and by rearrangement, we get

$$D^* \frac{\mathrm{d}^2 \bar{S}}{\mathrm{d}z^2} - q \frac{\mathrm{d}\bar{S}}{\mathrm{d}z} - p\bar{S} = -S(z, 0) + \frac{kX_\mathrm{c}}{p} \tag{75}$$

The auxiliary equation is

$$D^* \omega^2 - q\omega - p = 0 \tag{76}$$

so that

$$\omega = \frac{q \pm \sqrt{q^2 + 4D^* p}}{2D^*} \tag{77}$$

Thus

$$\bar{S} = \mathrm{e}^{qz/2D^*} (A\, \mathrm{e}^{\sqrt{q^2 + 4D^* p}\, z/2D^*} + B\, \mathrm{e}^{-\sqrt{q^2 + 4D^*}\, z/2D^*}) + \frac{S(z, 0)}{p} - \frac{kX_\mathrm{c}}{p^2} \tag{78}$$

where $A$ and $B$ are constants. $\bar{S}$ must tend to 0 as $p$ tends to $\infty$ so $A = 0$. Therefore,

$$\bar{S} = B\, \mathrm{e}^{qz/2D^*}\, \mathrm{e}^{-\sqrt{q^2 + 4D^* p}\, z/2D^*} + \frac{S(z, 0)}{p} + \frac{kX_\mathrm{c}}{p^2} \tag{79}$$

With $S(x, 0) = 0$, that is to say there is no nutrient present at $t = 0$,

$$\bar{S} = B\, \mathrm{e}^{qz/2D^*}\, \mathrm{e}^{-\sqrt{q^2 + 4D^* p}\, z/2D^*} - \frac{kX_\mathrm{c}}{p^2} \tag{80}$$

and with constant input of nutrient at concentration $S_0$, $S(0, t) = S_0$,

$$\bar{S} = \frac{S_0}{p} \quad \text{at } z = 0$$

Therefore,

$$\frac{S_0}{p} = B - \frac{kX_c}{p^2}$$

and

$$B = \frac{S_0}{p} + \frac{kX_c}{p^2}$$

Equation 80 then becomes

$$\bar{S} = \left(\frac{S_0}{p} + \frac{kX_c}{p^2}\right) e^{qz/2D^*} e^{-\sqrt{q^2+4D^*p}\, z/2D^*} - \frac{kX_c}{p^2}$$

$$= S_0 e^{qz/2D^*} \frac{1}{p} e^{-\sqrt{q^2+4D^*p}\, z/2D^*}$$

$$+ kX_c e^{qz/2D^*} \frac{1}{p^2} e^{-\sqrt{q^2+4D^*p}\, z/2D^*} - \frac{kX_c}{p^2} \tag{81}$$

Let

$$f_1(p) = S_0 e^{qz/2D^*} \frac{1}{p} e^{-\sqrt{q^2+4D^*p}\, z/2D^*}$$

$$f_2(p) = kX_c e^{qz/2D^*} \frac{1}{p^2} e^{-\sqrt{q^2+4D^*p}\, z/2D^*}$$

and

$$f_3(p) = -\frac{kX_c}{p^2}$$

Then

$$S(z,t) = \mathcal{L}^{-1} f_1(p) + \mathcal{L}^{-1} f_2(p) + \mathcal{L}^{-1} f_3(p) \tag{82}$$

From standard tables of Laplace transforms we can find

$$\mathcal{L}^{-1} f_1(p) = \frac{S_0}{2}\left[\operatorname{erfc}\left(\frac{z - qt}{2\sqrt{D^*t}}\right) + e^{qz/D^*} \operatorname{erfc}\left(\frac{z + qt}{2\sqrt{D^*t}}\right)\right]$$

$$\mathcal{L}^{-1} f_3(p) = -kX_c t \tag{84}$$

We now need to evaluate $\mathcal{L}^{-1} f_2(p)$. We start by considering

$$\mathcal{L}^{-1}(e^{-a\sqrt{p}}) = \frac{a}{2\sqrt{\pi}} t^{-3/2} e^{-a^2/4t}$$

Therefore,

$$\mathcal{L}^{-1}(e^{-a\sqrt{p+h}}) = \frac{a}{2\sqrt{\pi}} e^{-ht} t^{-3/2} e^{-a^2/4t} \tag{85}$$

where

$$a = \frac{z}{\sqrt{D^*}} \quad \text{and} \quad h = \frac{q^2}{4D^*}$$

Then

$$\mathcal{L}^{-1}[\exp(-z/\sqrt{D^*}\sqrt{p + q^2/4D^*})] = \mathcal{L}^{-1}(e^{-\sqrt{q^2+4D^*p}\, z/2D^*})$$

$$= \frac{z}{\sqrt{D^*}2\sqrt{\pi}} e^{-tq^2/4D^*} t^{-3/2} e^{-z^2/4D^*t}$$

$$= \frac{z}{2\sqrt{\pi D^*}} t^{-3/2} e^{-(q^2t/4D^* + z^2/4D^*t)} \tag{86}$$

Using the convolution rule,

$$\mathcal{L}^{-1}\left(\frac{1}{p^2} e^{-\sqrt{q^2+4D^*p}z/2D^*}\right) = t \divideontimes \frac{z}{2\sqrt{\pi D^*}} t^{-3/2} e^{-(q^2t/4D^* + z^2/4D^*t)}$$

Therefore,

$$\mathcal{L}^{-1}f_2(p) = kX_c\, e^{qz/2D^*} \frac{z}{2\sqrt{\pi D^*}} [t \divideontimes t^{-3/2} e^{-(q^2t/4D^* + z^2/4D^*t)}]$$

$$= \frac{kX_c z}{2\sqrt{\pi D^*}} e^{qz/2D^*} \int_0^t (t-u)u^{-3/2} e^{-(q^2u/4D^* + z^2/4D^*u)}\, du$$

$$= \frac{kX_c zt}{2\sqrt{\pi D^*}} e^{qz/2D^*} \int_0^t u^{-3/2} e^{-(q^2u/4D^* + z^2/4D^*u)}\, du$$

$$- \frac{kX_c z}{2\sqrt{\pi D^*}} e^{qz/2D^*} \int_0^t u^{-1/2} e^{-(q^2u/4D^* + z^2/4D^*u)}\, du$$

$$= I_1 - I_2 \tag{87}$$

where $I_1$ and $I_2$ are the first and second terms of equation 87.
We now evaluate $I_1$ and $I_2$.

$$I_1 = \frac{kX_c zt}{2\sqrt{\pi D^*}} e^{qz/2D^*} \int_0^t u^{-3/2} e^{-(q^2u/4D^* + z^2/4D^*u)} du$$

$$= \frac{kX_c zt}{2\sqrt{\pi D^*}} e^{qz/2D^*} \int_{\frac{z}{2\sqrt{D^*t}}}^{\infty} \frac{4\sqrt{D^*}}{z} e^{-\left(\eta^2 + \frac{[qz/4D^*]^2}{\eta^2}\right)} d\eta \tag{88}$$

where

$$\eta = \frac{z}{2\sqrt{D^*u}} \quad d\eta = \frac{-z}{4\sqrt{D^*}} u^{-3/2} du$$

$$\frac{q^2u}{4D^*} = \frac{q^2}{4D^*}\frac{z^2}{4D^*\eta^2} = (q/4D^*)^2/\eta^2$$

Therefore,

$$I_1 = kX_c t\, e^{qz/2D^*} \frac{2}{\sqrt{\pi}} \int_{\frac{z}{2\sqrt{D^*t}}}^{\infty} e^{-\left(\eta^2 + \frac{[qz/4D^*]^2}{\eta^2}\right)} d\eta \tag{89}$$

Let

$$a = \frac{qz}{4D^*}$$

and

$$b = \frac{q}{2}\sqrt{\frac{t}{D^*}}$$

Then

$$I_1 = kX_c t\, e^{qz/2D^*} \frac{2}{\sqrt{\pi}} \int_{a/b}^{\infty} e^{-(\eta^2 + a^2/\eta^2)} d\eta$$

$$= \frac{kX_c t}{2}\left\{\mathrm{erfc}\left(\frac{z - qt}{2\sqrt{D^*t}}\right) + e^{qz/D^*}\, \mathrm{erfc}\left(\frac{z + qt}{2\sqrt{D^*t}}\right)\right\} \tag{90}$$

We now consider $I_2$. Defining

$$\nu = \frac{q}{2}\sqrt{\frac{u}{D^*}}, \quad d\nu = \frac{q}{4\sqrt{D^*}} u^{-1/2} du$$

$$\frac{z^2}{4D^*u} = \frac{z^2}{4D^*}\frac{q^2}{4D^*\nu^2} = \frac{(qz/4D^*)^2}{\nu^2}$$

Then,

$$I_2 = \frac{kX_c z}{2\sqrt{\pi D^*}} e^{qz/2D^*} \int_0^{q\sqrt{t}/2\sqrt{D^*}} \frac{4\sqrt{D^*}}{q} e^{-(\nu^2 + [qz/4D^*]^2/\nu^2)} d\nu \tag{91}$$

$$= \frac{kX_c z}{q} e^{qz/2D^*} \frac{2}{\pi} \int_0^{q\sqrt{t}/2\sqrt{D^*}} e^{-(\nu^2 + [qz/4D^*]^2/\nu^2)} d\nu$$

Consider

$$I(K) = \frac{2}{\sqrt{\pi}} \int_0^k e^{-(\nu^2 + a^2/\nu^2)} d\nu \tag{92}$$

$$= \frac{2}{\sqrt{\pi}} e^{-2a} \int_0^k e^{-\left(\nu - \frac{a}{\nu}\right)^2} d\nu$$

Let

$$t = v - \frac{a}{v}, \quad v^2 - vt - a = 0$$

$$v = \frac{1}{2}\left(t + \sqrt{t^2 + 4a}\right), \quad \frac{dv}{dt} = \frac{1}{2}\left(1 + \frac{t}{\sqrt{t^2 + 4a}}\right)$$

Then

$$I(K) = \frac{2}{\sqrt{\pi}} e^{-2a} \frac{1}{2} \int_{-\infty}^{T} e^{-t^2} \left(1 + \frac{t}{\sqrt{t^2 + 4a}}\right) dt \tag{93}$$

where

$$T = K - \frac{a}{K}$$

So,

$$I(K) = \tfrac{1}{2} e^{-2a} \frac{2}{\sqrt{\pi}} \int_{-\infty}^{T} e^{-t^2} dt + \tfrac{1}{2} e^{-2a} \frac{2}{\sqrt{\pi}} \int_{-\infty}^{T} \frac{t\, e^{-t^2}}{\sqrt{t^2 + 4a}} dt$$

$$= \tfrac{1}{2} e^{-2a} \left\{ \frac{2}{\sqrt{\pi}} \int_{-\infty}^{0} e^{-t^2} dt + \frac{2}{\sqrt{\pi}} \int_0^T e^{-t^2} dt \right\}$$

$$+ \tfrac{1}{2} e^{-2a} \frac{2}{\sqrt{\pi}} \int_{-\infty}^{T} \frac{t\, e^{-(t^2 + 4a) + 4a}}{\sqrt{t^2 + 4a}} dt$$

$$= \tfrac{1}{2}\,\mathrm{e}^{-2a}\,\{1 + \mathrm{erf}(T)\} + \tfrac{1}{2}\,\mathrm{e}^{2a}\,\frac{2}{\sqrt{\pi}}\int_{-\infty}^{T} \frac{t\,\mathrm{e}^{-(t^2+4a)}}{\sqrt{t^2+4a}}\,\mathrm{d}t$$

$$= \tfrac{1}{2}\,\mathrm{e}^{-2a}\,\{1 + \mathrm{erf}(T)] + \tfrac{1}{2}\,\mathrm{e}^{2a}\,\frac{2}{\sqrt{\pi}}\int_{\infty}^{\sqrt{T^2+4a}} \mathrm{e}^{-u^2}\,\mathrm{d}u$$

$$= \tfrac{1}{2}\,\mathrm{e}^{-2a}\{1 + \mathrm{erf}(T)\} - \tfrac{1}{2}\,\mathrm{e}^{2a}\,\frac{2}{\sqrt{\pi}}\int_{K+\frac{a}{K}}^{\infty} \mathrm{e}^{-u^2}\,\mathrm{d}u$$

$$= \tfrac{1}{2}\,\mathrm{e}^{-2a}\left\{1 + \mathrm{erf}\left(K - \frac{a}{K}\right)\right\} - \tfrac{1}{2}\,\mathrm{e}^{2a}\,\mathrm{erfc}\left(K + \frac{a}{K}\right) \tag{94}$$

Now,

$$K = \frac{q\sqrt{t}}{2\sqrt{D^*}}$$

and

$$a = \frac{qz}{4D^*}$$

Therefore,

$$\frac{a}{K} = \frac{qz}{4D^*}\,\frac{2\sqrt{D^*}}{q\sqrt{t}} = \frac{z}{2\sqrt{D^*t}}$$

so that

$$I_2 = \frac{kX_c z}{2q}\,\mathrm{e}^{qz/2D^*}\left\{\mathrm{e}^{-2a}\left[1 + \mathrm{erf}\left(\frac{q\sqrt{t}}{2\sqrt{D^*}} - \frac{z}{2\sqrt{D^*t}}\right)\right]\right\}$$

$$- \frac{kX_c z}{2q}\,\mathrm{e}^{qz/2D^*}\left\{\mathrm{e}^{2a}\,\mathrm{erfc}\left(\frac{q\sqrt{t}}{2\sqrt{D^*}} + \frac{z}{2\sqrt{D^*t}}\right)\right\}$$

$$= \frac{kX_c z}{2q}\,\mathrm{e}^{qz/2D^*}\,\mathrm{e}^{-qz/2D^*}\left\{1 + \mathrm{erf}\left[-\left(\frac{z-qt}{2\sqrt{D^*t}}\right)\right]\right\}$$

$$- \frac{kX_c z}{2q}\,\mathrm{e}^{qz/2D^*}\,\mathrm{e}^{qz/2D^*}\,\mathrm{erfc}\left(\frac{z+qt}{2\sqrt{D^*t}}\right)$$

$$= \frac{kX_c z}{2q}\left\{1 - \mathrm{erf}\left(\frac{z-qt}{2\sqrt{D^*t}}\right)\right\} - \frac{kX_c z}{2q}\,\mathrm{e}^{qz/D^*}\,\mathrm{erfc}\left(\frac{z+qt}{2\sqrt{D^*t}}\right)$$

$$= \frac{kX_c z}{2q} \operatorname{erfc}\left(\frac{z - qt}{2\sqrt{D^* t}}\right) - \frac{kX_c z}{2q} e^{qz/D^*} \operatorname{erfc}\left(\frac{z + qt}{2\sqrt{D^* t}}\right) \tag{95}$$

Thus,

$$\begin{aligned}
\mathscr{L}^{-1} f_2(p) &= I_1 - I_2 \\
&= \frac{kX_c t}{2}\left\{\operatorname{erfc}\left(\frac{z - qt}{2\sqrt{D^* t}}\right) + e^{qz/D^*} \operatorname{erfc}\left(\frac{z + qt}{2\sqrt{D^* t}}\right)\right\} \\
&\quad - \frac{kX_c z}{2q} \operatorname{erfc}\left(\frac{z - qt}{2\sqrt{D^* t}}\right) + \frac{kX_c z}{2q} e^{qz/D^*} \operatorname{erfc}\left(\frac{z - qt}{2\sqrt{D^* t}}\right) \\
&= \frac{kX_c}{2q}\left\{e^{qz/D^*} (z + qt) \operatorname{erfc}\left(\frac{z + qt}{2\sqrt{D^* t}}\right)\right. \\
&\quad \left. - (z - qt) \operatorname{erfc}\left(\frac{z - qt}{2\sqrt{d^* t}}\right)\right\}
\end{aligned} \tag{96}$$

and finally,

$$\begin{aligned}
S(z, t) &= \frac{S_0}{2}\left\{\operatorname{erfc}\left(\frac{z - qt}{2\sqrt{D^* t}}\right) + e^{qz/D^*} \operatorname{erfc}\left(\frac{z + qt}{2\sqrt{D^* t}}\right)\right\} \\
&\quad + \frac{kX_c}{2}\left\{e^{-qz/D^*} (z + qt) \operatorname{erfc}\left(\frac{z + qt}{2\sqrt{D^* t}}\right)\right. \\
&\quad \left. - (z - qt) \operatorname{erfc}\left(\frac{z - qt}{2\sqrt{D^* t}}\right)\right\} - kX_c t \\
&= \left\{\frac{S_0}{2} - \frac{kX_c}{2q}(z - qt)\right\} \operatorname{erfc}\left(\frac{z - qt}{2\sqrt{D^* t}}\right) \\
&\quad + \left\{\frac{S_0}{2} + \frac{kX_c}{2q}(z + qt)\right\} e^{qz/D^*} \operatorname{erfc}\left(\frac{z + qt}{2\sqrt{D^* t}}\right) - kX_c t \quad (S \geqslant 0)
\end{aligned} \tag{97}$$

This function is plotted in Fig. 12. Despite the intricacies of the mathematics, the visual representation is not spectacular!

## VI. Special applications

In this section we consider some applications of mathematics to microbial ecology which do not readily fit into the scheme we have so far employed in this chapter.

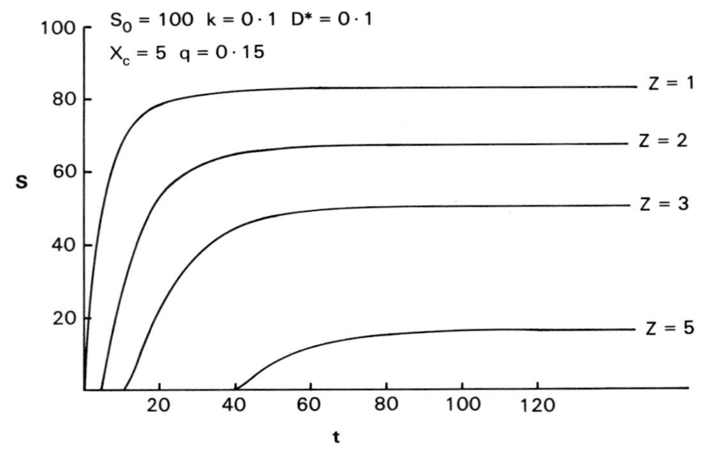


**Fig. 12.** Concentration of nutrient ($S$) in an idealized column of soil containing a constant and homogeneously distributed concentration of biomass as a function of time at various depths down the column. These results were obtained using equation 97.

## A. Stability of multispecies ecosystems

It has long been held in ecology that complexity leads to stability of ecosystems (Elton, 1958). By complexity we mean both the species diversity and the number of interactions between species. Gardner and Ashby (1970) showed on theoretical grounds that, at least for linearly connected systems, this was not the case. This was accomplished by performing stability analyses on sets of Lotka–Volterra equations representing multispecies systems. In model M4, a predator–prey interaction was represented. By adding a third term to each of the equations of this model, logistic growth of the two interacting populations can be represented. In addition, the equations can be generalized to model other types of association by adjusting the signs in front of the coefficients in the equations. For a two-species system assuming logistic growth the equations can be written as

$$\dot{X}_1 = \alpha_1 X_1 + \beta_{12} X_1 X_2 + \beta_{11} X_1^2 \tag{98}$$

$$\dot{X}_2 = \alpha_2 X_2 + \beta_{21} X_1 X_2 + \beta_{22} X_2^2 \tag{99}$$

where $X_1$ and $X_2$ are the population densities. For $\beta_{12} < 0$ and $\alpha_2 < 0$, $\beta_{11} = \beta_{22} = 0$, $\alpha_1 > 0$ and $\alpha_2 < 0$, the Lotka–Volterra predator–prey model we have already quoted results. For $\beta_{11} < 0$, the prey is assumed to grow logistically. For $\beta_{21} < 0$ and $\beta_{12} < 0$, competition is represented. When both of these coefficients are positive a mutualistic association results. For an $n$-

species system the equations can be conveniently written as

$$\dot{X}_i = \alpha_i X_i + \beta_{ij} \sum_{j=1}^{n} X_i X_j \tag{100}$$

When such equations are linearized close to equilibrium, we get

$$\dot{\mathbf{x}} = \mathbf{A}\mathbf{x}$$

where **A** is the stability matrix. A minimum necessity for stability is that all the eigenvalues of the system must be negative. For this to be the case all the diagonal elements of the stability matrix must also be negative. The other elements of the matrix represent the interactions. The more non-zero off-diagonal elements there are, the greater the number of interactions. Gardner and Ashby (1970) called the fraction of non-zero elements the connectivity, $C$. Thus $C$ represents the number of interactions and $n$ the number of species present. Gardner and Ashby investigated such a system in order to relate complexity and stability. They did this by assigning randomly chosen numbers between $-1$ and $-0.1$ to the diagonal elements of an $n \times n$ matrix in order to ensure that *a priori* the system was not unstable. Then they assigned random numbers between $-1$ and $+1$ to $C\%$ of the off-diagonal elements. They then determined whether or not the system was stable using the appropriate auxiliary equation. They repeated this procedure many times. The fraction of stable solutions that resulted they equated to the probability of stability. This work was repeated by Daniels and Mackay (1974) and the results of these workers are reproduced in Fig. 13. It is immediately apparent that the likelihood of stability decreases as both $C$ and $n$ increase. Thus at least for linearly connected systems, complexity appears to reduce stability, contrary to what might have been expected.

## B. An application of catastrophe theory

Catastrophe theory is concerned with phenomena that change suddenly. It was developed by the French mathematician, Rene Thom, and is centred on what he calls the seven elementary catastrophes. These are equations which are supposed to be mathematical generalizations of systems that move rapidly from one state to another. Catastrophe theory has been applied to a wide variety of phenomena ranging from the sudden collapse of box bridges and light patterns on the bottom of swimming pools (caustics) to the occurrence of prison riots and the onset of *anorexia nervosa*. Saunders (1980) gives an elementary introduction to catastrophe theory but a rigorous approach requires a sound understanding of modern mathematics.

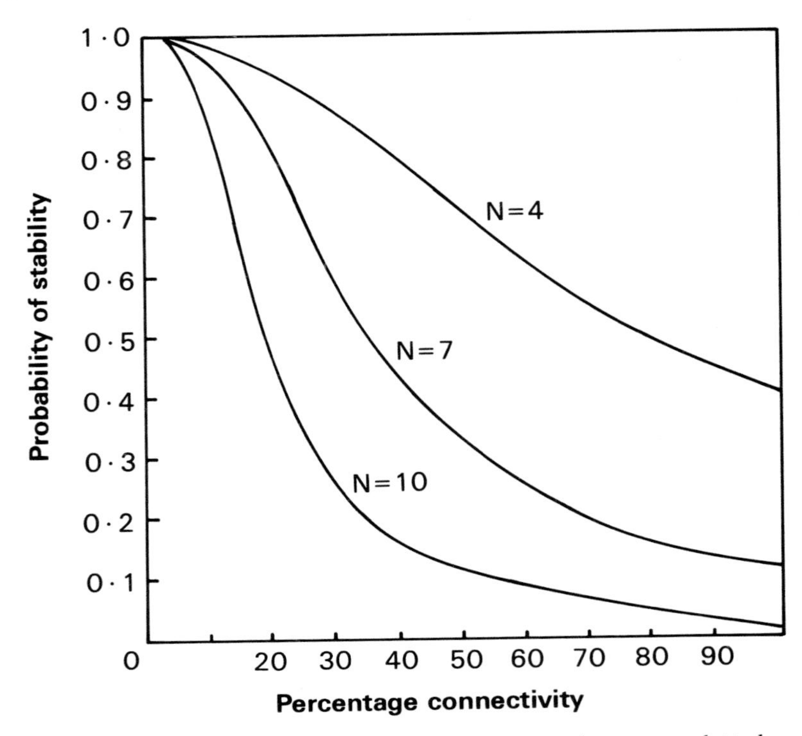


**Fig. 13.** Probability of stability of linearly connected systems plotted against connectivity, $C$, for systems of $N$ equations. Adapted from Daniels and Mackay (1974).

In this section we describe how catastrophe theory has been used in the study of a microbial predator–prey system. In this study slime mould amoebae and *Escherichia coli* were grown together in chemostat culture (Dent *et al.*, 1976). The results obtained were compared to those of model M5 and found to be both qualitatively and quantitatively different. This model is based on the hypothesis that the predator specific growth rate is a Monod function. When the predator density was plotted semi-logarithmically against time (Fig. 14) the results could be approximated by a series of straight lines with alternating positive and negative slopes. These slopes represent the specific rate of change of the predator population and when added to the dilution rate of the system, equate to the predator specific growth rate, $\lambda$. As can be seen by inspection of Fig. 14, this value changes quite suddenly at each relative maximum and minimum. This characteristic led Bazin and Saunders (1978)

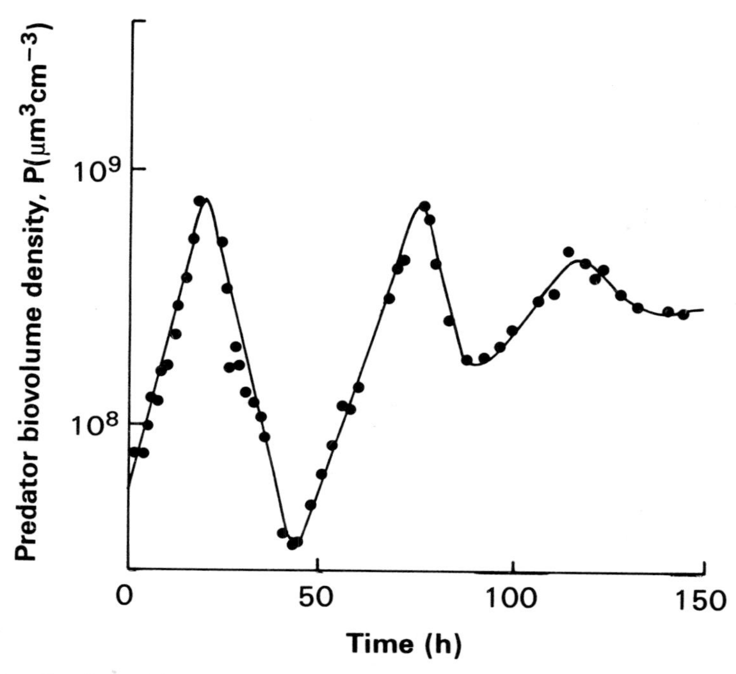


**Fig. 14.** Semilogarithmic plot of predator population density against time from a chemostat culture. Data from Dent *et al.* (1976). After Bazin and Saunders (1978).

to interpret the data in terms of the second elementary catastrophe, the cusp. This required the identification of one state variable and two control variables. The former is that part of the system which exhibits sudden changes. For the case at hand, it is the predator specific growth rate. The other two variables control the state variable.

When Bazin and Saunders (1978) used prey density, $H$, and time, $t$, as control variables the results shown in Fig. 15 were obtained. In this figure the times at which the predator density changes suddenly are marked on the curve representing prey density. When these points are connected, they map out a triangular area equivalent to the catastrophe set. (They can be transformed by arbitrary diffeomorphisms to trace the actual cusp-shaped area of the catastrophe set but for this general outline we need not go into such details.) Although at first glance the behaviour might appear to be of a form predicted by Thom's theory, the sudden changes in the state variable, $\lambda$, occur as the control trajectory, $H$, enters the catastrophe set rather than when it

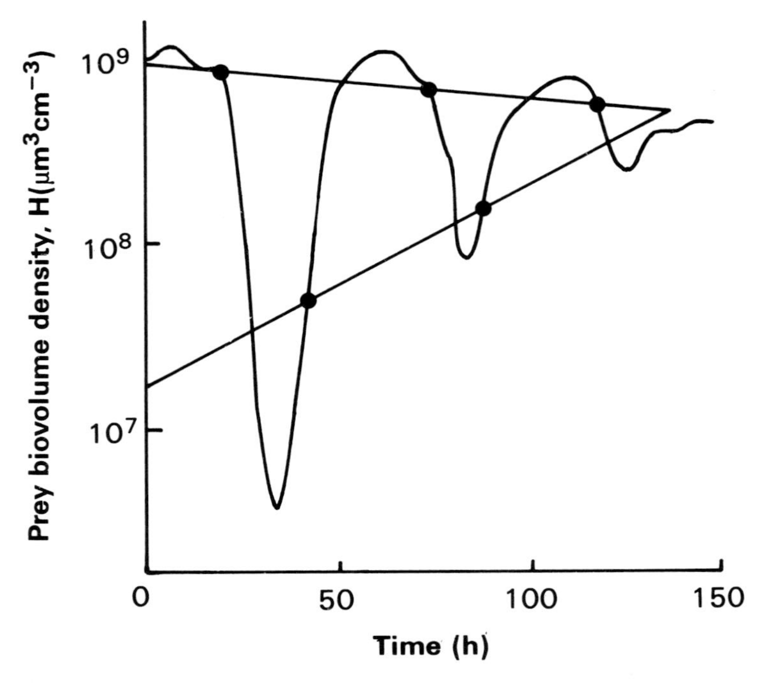


**Fig. 15.** Prey concentration plotted against time. The times at which the predator specific growth rate changed suddenly are marked by solid circles. These are connected up to produce a rough estimation of the catastrophe set of a cusp catastrophe. Data from Dent *et al.* (1976). After Bazin and Saunders (1978).

leaves it. This is contrary to what is expected in catastrophe theory. The sudden changes should occur as the control variable leaves the catastrophe set, not as it enters it as indicated in Fig. 15. Therefore, the authors chose another control variable to replace $H$. This was $H/P$, the ratio of the prey and predator densities. When the procedure outlined above was repeated with this new control variable, the results shown in Fig. 16 were obtained. These conform to those expected. In this case the sudden changes in $\lambda$ occur as the control variable, $H/P$, leaves the catastrophe set.

What the above study illustrates is that catastrophe theory can be used to identify the critical variables of a system. In the example we quote $H/P$, rather than just $H$, seems to be controlling predator specific growth rate. These ideas have been confirmed by experiments with two-stage chemostat cultures (Bazin *et al.*, 1983). They imply that an appropriate function for prey specific growth rate might take the form of that suggested

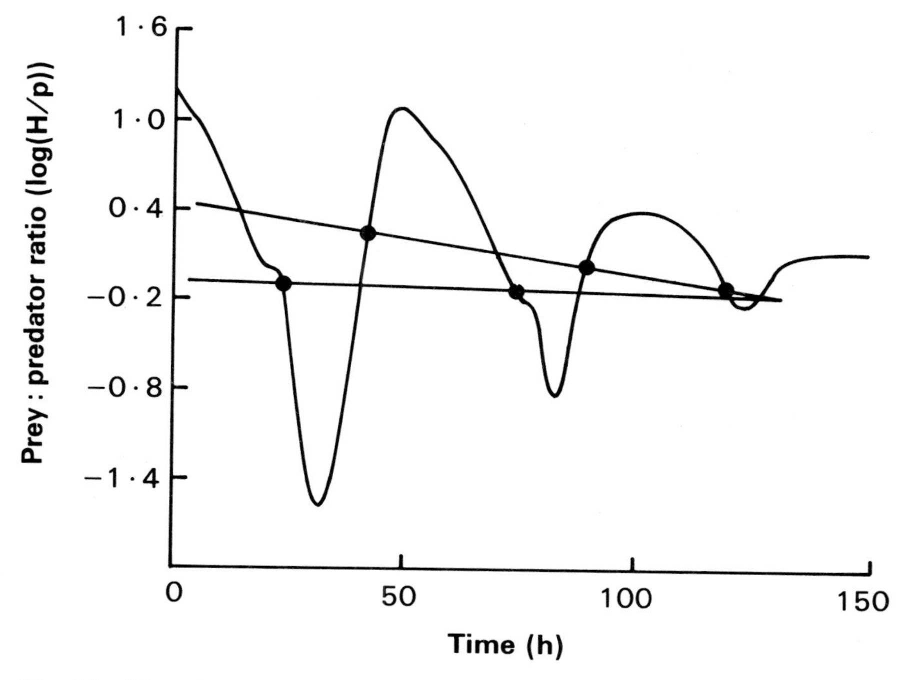


**Fig. 16.** Here the ratio of prey to predator has been used as a control variable and is plotted against time. As in Fig. 15, the sudden changes in predator specific growth rate are marked as solid circles and then connected up. In this case they occur as the control variable leaves the triangular region representing the catastrophe set. This conforms to expectation from catastrophe theory. After Saunders (1980).

by Contois (1959) for bacterial growth rather than that of Monod. This function takes the form

$$\lambda = \frac{\lambda_m H/P}{L_s + H/P} = \frac{\lambda_m H}{L_s P + H} \tag{101}$$

## C. Chaos

A dynamic system which can be described deterministically but behaves in a manner that cannot be distinguished from the sample function of a random process is said to be chaotic (May, 1976). Chaotic dynamics have been investigated by mathematicians for several decades but only recently

**TABLE II**

```
 20  SET MODE 80
 30  FOR R := 2.8 TO 4 STEP 1e-02
 40    FOR I := 1 TO 100
 50      X := R * X * (1 - X)
 60    NEXT I
 70    FOR J := 1 TO 300
 80      X := R * X * (1 - X)
 90      POINTS X * 250, R * 70 - 140
100    NEXT J
110  NEXT R
```

has the subject found application in science and engineering (Grebogi *et al.*, 1987). Within ecology there has been much speculation about chaos but to our knowledge no experimental investigations have been undertaken. A simple example of chaotic behavior results when logistic growth is represented as a difference equation:

$$N_{t+1} = \left(\mu_m - \frac{\mu_m}{K} N_t\right) N_t \tag{102}$$

where $N_t$ is the number of individuals at time $t$ and the other symbols are as defined for equation model M2. This equation differs from the differential form of the logistic equation in that the number of organisms is expressed in terms of discrete time steps rather than continuously. By letting $X = N/K$, equation 102 may be reduced to a single-parameter expression:

$$X_{t+1} = \mu_m(1 - X_t)X_t \tag{103}$$

A simple Basic program for analysing this equation for values of $\mu_m$ between 0.3 and 4.0 is given in Table II and the results are shown in Fig. 17. In this figure the value of $X$ reached after several iterations is plotted as a function of $\mu_m$, the values of which range between 2.8 and 4.0. At lower values of $\mu_m$, $X_t$ has unique values. As $\mu_m$ increases, however, the curve branches indicating that two final values may be attained. Further branches occur and then chaotic behaviour occurs when a multitude of final values are reached.

Although chaos is often associated only with difference equations (May, 1976), it can occur in continuous systems particularly if they are subjected to periodic or randomly changing environmental conditions (Markus *et al.*, 1987). An example of such a system we have investigated (Trollope,

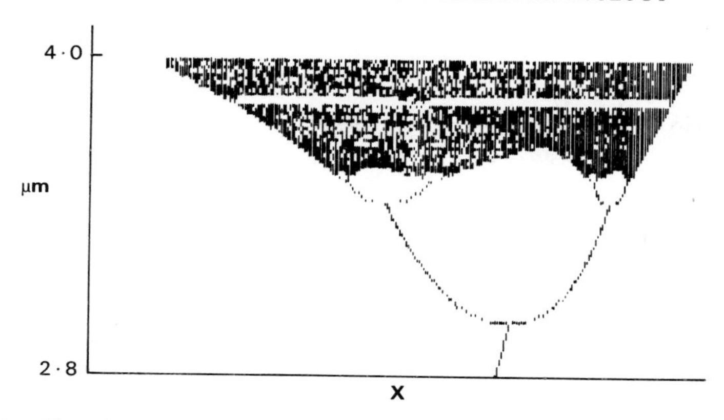


**Fig. 17.** Chaotic behaviour resulting as $\mu_m$ is reduced in value and the equilibrium value of $X_t$ from equation 103 is plotted.

unpublished) is a model of predator–prey dynamics. The system can be represented by the equations of model M5 except that it does not have a constant volume. Specifically, $D = f/V$ where

$$V = V_0 + ft \qquad \text{for } V < V_m$$

$$V = V_0 \qquad \text{otherwise}$$

and $V_0$ is the minimal volume of the system and $V_m$ is the maximal volume. The model represents an ecosystem the volume of which increases from $V_0$ to $V_m$ at rate $f$. It then empties and the process repeats itself. This might be regarded as an idealization of what happens in small pools of water or in the soil. The laboratory realization of such a system is a repeated fed-batch culture.

As the system increases in volume so the effective dilution rate decreases. Values for the parameters and coefficients of the model can be chosen such that the system exhibits the three types of behaviour discussed in Section V.F, i.e. stable node, stable focus and limit cycle. Effectively, at the beginning of each new cycle the system will have different initial conditions so that different cycle-dependent responses are expected. Such a result was achieved by solving the equations numerically. Figure 18 shows a typical simulation. For clarity, only the behaviour of the predator variable is illustrated. It is very difficult to discern any regularity in the way $P$ changes and provisionally, we have regarded this response as an example of chaotic behaviour. Perhaps of more significance, however, are the results shown in the following two figures. In Fig. 19 a set of values for the parameters and coefficients were chosen which resulted in non-chaotic dynamics.

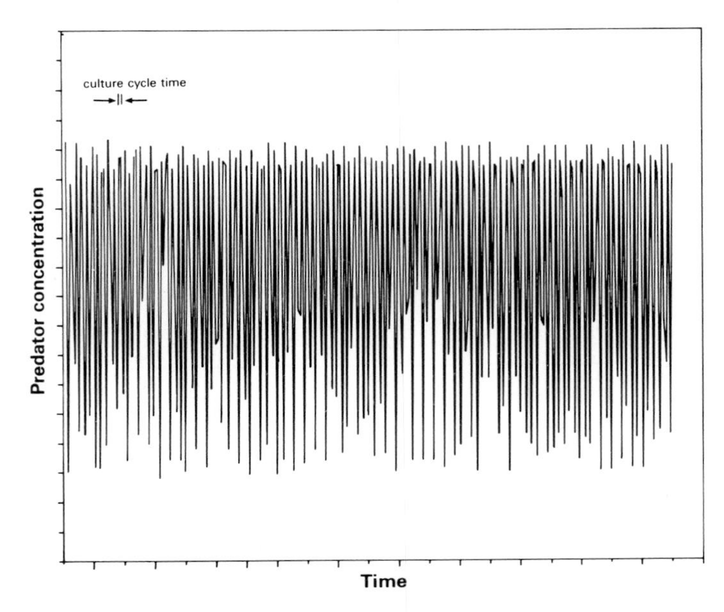


**Fig. 18.** Simulation of a repeated fed-batch culture of a predator–prey system. The predator population is plotted against time and appears to show chaotic behaviour. After Trollope (unpublished).

Imposing 5% random error on the values of the variables after each step in the integration routine induced what appears to be chaos, as illustrated in Fig. 20.

As we have indicated previously, the mathematical models we have introduced represent ideal systems. Real ecosystems are subject to environmental variation that far exceeds the 5% random error we used in the fed-batch simulation illustrated in Fig. 20. Therefore if this model is even a rough approximation of predator–prey activity in the 'natural' environment, then chaotic behaviour is likely to occur in the field. Indeed, Schaffer (1987) provides evidence for such occurrences. It follows then that at least in some cases attempting to relate field data to the predictions of well-posed hypotheses could be a task verging on the impossible. On the other hand, laboratory experiments in which parameters can be controlled by the investigator could use to advantage the qualitative change in behaviour from non-chaotic to chaotic dynamics. Perhaps then modellers should not venture out into the field but restrict their investigations to theory and laboratory experimentation.

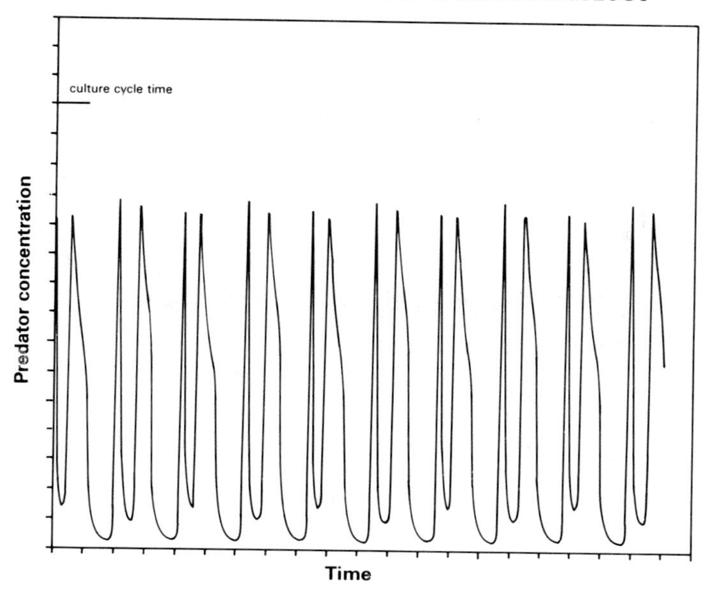


**Fig. 19.** Non-chaotic behaviour in a repeated fed-batch predator–prey system. After Trollope (unpublished).

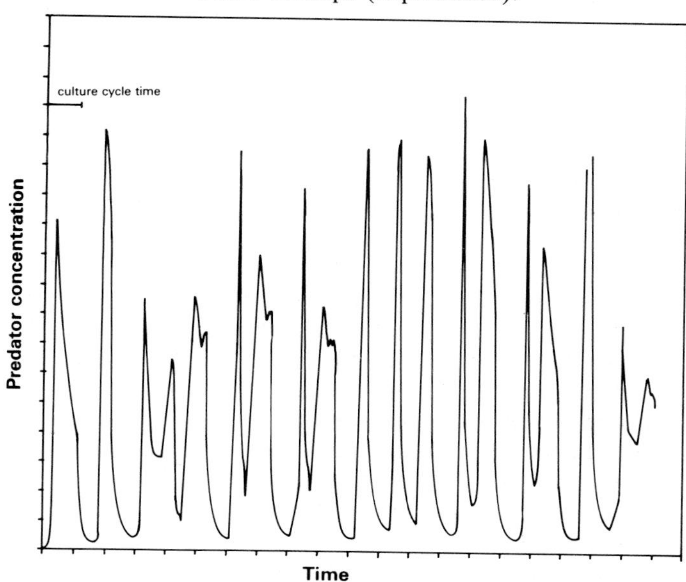


**Fig. 20.** As in Fig. 19 but with 5% random error imposed. This causes much more than 5% variability in the predator population density and the results may represent chaotic behaviour. After Trollope (unpublished).

## References

Bailey, J. E. and Ollis, D. F. (1986). "Biochemical Engineering Fundamentals", 2nd Edn, McGraw-Hill, New York.
Bazin, M. J. (1981). *In* "Mixed Culture Fermentations" (M. E. Bushell and J. H. Slater, Eds), p. 25, Academic, London.
Bazin, M. J. and Saunders, P. T. (1978). *Nature (London)* **275,** 52–54.
Bazin, M. J., Curds, C., Dauppe, A., Owen, B. and Saunders, P. T. (1983). *In* "Foundations of Chemical Engineering. Kinetics and Thermodynamics in Biological Systems" (H. W. Blanch, E. T. Papoutsakis and G. Stephanopoulos, Eds), p. 253, American Chemical Society, Washington.
Canale, R. P. (1970). *Biotechnol. Bioeng.* **12,** 353–378.
Cho, C. M. (1971). *Can. J. Soil Sci.* **51,** 339–360.
Churchill, R. V. (1944). "Modern Operational Mathematics in Engineering", McGraw-Hill, New York.
Contois, D. E. (1959). *J. Gen. Microbiol.* **21,** 40–50.
Curds, C. R. and Bazin, M. J. (1977). *In* "Advances in Aquatic Microbiology. Vol. 1", p. 115, Academic, London.
Daniels, J. and Mackay, A. L. (1974). *Nature (London)* **251,** 49–50.
Dent, V. E., Bazin, M. J. and Saunders, P. T. (1976). *Arch. Microbiol.* **109,** 187–195.
Elton, C. S. (1958). "The Ecology of Invasions by Animals and Plants", Methuen, London.
Fredrickson, A. G. (1977). *Ann. Rev. Microbiol.* **31,** 63–87.
Fredrickson, A. G. (1983). *In* "Foundations of Chemical Engineering. Kinetics and Thermodynamics in Biological Systems" (H. W. Blanch, E. T. Papoutsakis and G. Stephanopoulos, Eds), p. 201, American Chemical Society, Washington.
Fredrickson, A. G. and Stephanopoulos, G. (1981). *Science* **213,** 972–979.
Fredrickson, A. G., Jost, J. L., Tsuchiya, H. M. and Ping-Hwa Hsu (1973). *J. Theor. Biol.* **38,** 486–526.
Gardner, M. R. and Ashby, W. R. (1970). *Nature (London)* **228,** 74.
Grebogi, C., Ott, E. and Yorke, J. A. (1987). *Science* **238,** 632–638.
Howell, J. A. (1983). *In* "Mathematics in Microbiology" (M. Bazin, Ed.), p. 37, Academic Press, London.
Kuenen, J. G. (1983). *In* "Foundations of Chemical Engineering. Kinetics and Thermodynamics of Biological Systems" (H. W. Blanch, E. T. Papoutsakis and G. Stephanopoulos, Eds), p. 229, American Chemical Society, Washington.
Markus, M., Hess, B., Rossler, J. and Kiwi, M. (1987). *In* "Chaos in Biological Systems" (H. Degn, A. V. Holden and L. F. Olsen, Eds), p. 267, Plenum, New York.
May, R. M. (1976). *Nature (London)* **261,** 459–467.
Menell, A. and Bazin, M. (1988). "Mathematics for the Biosciences", Ellis Horwood, Chichester.
Monod, J. (1942). "Recherches sur la Croissance des Cultures Bacteriennes", 2nd Edn, Hermann, Paris.
Morris, W. D. (1974). "Differential Equations for Engineers and Applied Scientists", McGraw-Hill, Maidenhead.
Nye, P. H. and Tinker, P. B. (1977). "Solute Movement in the Soil–Root System", Blackwell, Oxford.

Patten, B. C. (1971). *In* "Systems Analysis and Simulation Ecology. Vol. 1" (B. C. Patten, Ed.), p. 4, Academic, New York.
Pirt, S. J. (1975). "Principles of Microbe and Cell Cultivation", Blackwell, Oxford.
Roels, J. A. (1983). "Energetics and Kinetics in Biotechnology", Elsevier, Amsterdam.
Saunders, P. T. (1980). "An Introduction to Catastrophe Theory", Cambridge University Press, Cambridge.
Saunders, P. T. and Bazin, M. J. (1973). *Soil Biol. Biochem.* **5,** 547–557.
Schaffer, W. M. (1987). *In* "Chaos in Biological Systems" (H. Degn, A. V. Holden and L. F. Olsen, Eds), p. 233, Plenum, New York.
Slater, J. H. and Bull, A. T. (1978). *In* "Companion to Microbiology" (A. T. Bull and P. M. Meadow, Eds), p. 181, Longman, London.
Smith, G. D. (1985). "Numerical Solution of Partial Differential Equations", 3rd Edn, Clarendon Press, Oxford.

# 5

# Methods to Study the Bacterial Ecology of Freshwater Environments

G. H. HALL, J. G. JONES, R. W. PICKUP AND B. M. SIMON

*Institute of Freshwater Ecology, The Ferry House, Far Sawrey, Ambleside, Cumbria LA22 0LP, UK*



METHODS IN MICROBIOLOGY
VOLUME 22 ISBN 0-12-521522-3

The aim of the microbial ecologist is to understand the interactions between microorganisms and their environment. This goal can be achieved using a variety of techniques but particularly those of enumeration of the organisms and assessment of their activity. Such techniques can be directed towards the total community or separate populations (synecology) or to individual species within a population (autecology). Selection of suitable methodology will depend on the group of microorganisms under study and the habitat in which they are found. This chapter is concerned with bacteria in lakes and is divided into assessment of numbers and activity with regard to synecological or autecological studies. It is not intended to provide detailed descriptions of techniques as virtually all methods will have to be adapted for the habitat under study. This is particularly true of sediment systems for which reagents may have to be modified due to changes in the composition of the sediment.

## I. The lake environment

Most deep lakes in temperate regions experience a seasonal cycle of thermal (or density) stratification and destratification and this is a major factor in controlling bacterial activity (Jones, 1982). Throughout winter and into spring the water column is isothermal and well oxygenated. Increased solar radiation raises the temperature of the surface waters and the related changes in density give rise to distinct zones. The surface water (the epilimnion) is warm and well mixed whilst the deep water (the hypolimnion) remains cool. These two layers are separated by a water mass (the metalimnion) which is characterized by steep temperature gradients. The temperature difference, and associated changes in density, ensure that mixing between the epilimnion and hypolimnion is minimal and the latter can be treated essentially as a closed system (Jones, 1982). The air–water interface has been recognized as an important and active environment which after application of specialized sampling procedures can be treated like other water samples. The surface sediments in contact with these zoned water masses exhibit similar temperature cycles and can be broadly defined as shallow (littoral) or deep (profundal) deposits.

A lake can be more or less productive depending on its trophic status. This has two components, the cultural (enrichment from the catchment and man's activities) and the morphometric (the volume/area relationship of the lake basin). Generally a large deep lake is less likely to be eutrophic than a shallow lake simply due to the dilution of incoming nutrients and the greater volume of hypolimnetic water which is required to be de-oxygenated. The degree to which the hypolimnion becomes deoxygenated will depend on the activity of benthic organisms and the input of suitable electron donors. This in turn affects the redox profiles and subsequent microbial activity in both the water and the sediment.

## II. Sampling

The techniques for sampling the water and the sediment have been reviewed elsewhere (Collins *et al.*, 1973; Herbert, 1988). Any sampling procedure will disturb the natural conditions but it is important to minimize such effects in order to obtain reliable measurements. Consequently, sample bottles should be flushed with at least three times their own volume of water prior to stoppering in such a way as to eliminate, or reduce, gaseous exchange or entrapment of air. Gradients within the water column may be steep and conventional sampling devices can straddle different water layers (Jones, 1975). Under such circumstances the sample should be mixed within the sampler prior to filling the sample bottle. If smaller scale resolution is required pump samplers (Davison, 1977) should be employed with the intake pipes placed horizontally rather than vertically. Moreover, water samples can be obtained at closely defined intervals from the sediment water interface using electronic devices attached to appropriate sampling gear (Cunningham and Davison 1980). It is good practice to sample deep water at defined distances from the sediment rather than from the water surface. It is difficult to obtain representative sediment samples because of the existence of steep physical and chemical gradients (Carlton and Wetzel, 1985) often associated with specific bacterial activities (Jones, 1979a). Coring devices cause less disturbance than grab samplers.

An impractical degree of replication may be required to work to a given level of statistical significance. Appropriate example calculations relating the mean and variance of a given variable to the appropriate significance level are provided by Elliott (1977). Such a test is worthwhile performing if only to demonstrate how far the practical limits of replication fall short of what is possible. The degree of variability of microbiological data, on both temporal and spatial scales, is reported by Jones and Simon (1980b) further emphasizing the caution which must be exercised in interpreting data from a single sample.

## III. Handling procedures

Transportation of samples from the site to the laboratory should avoid any temperature change although the cooling of sample bottles is acceptable.

Exposure of the samples to natural illumination is often overlooked although inhibitory effects, particularly in sediment deposits, have been observed (Hall, 1986).

If necessary samples can be preserved using formaldehyde (2%), glutaraldehyde (2%) or mercuric chloride (10 mg $l^{-1}$). Samples should be fixed immediately but care should be taken as the preservative may quench analytical procedures (Riemann and Schierup, 1978).

Bacterial populations respond rapidly to enclosure in sample containers therefore all procedures should start as soon as practicably possible after sampling. Bacterial activity will change more rapidly than the community structure although some bacteria are so sensitive (e.g. to traces of oxygen) that caution must be exercised at all stages of handling. Water samples should be shaken vigorously for about 30 s whilst benthic material (or highly particulate water samples) requires homogenization or sonication. Additions of surface active or anti-foam agents can be made but the effects on viability or activity may be significant. All treatments involve a compromise between disaggregation of clumps and cell disruption. The removal of bacteria from surfaces requires specialized handling procedures which are discussed in Fry (1988) and Goulder (1987). Suspended bacterial populations may require concentration by centrifugation or filtration. The former is cumbersome for quantitative recovery and therefore filtration is the preferred method. Although bacteria are probably the most robust of the aquatic microbial community filtration procedures should be applied with caution. Gentle concentration is provided by the method of Dodson and Thomas (1964) although for routine use gentle suction pressure can be applied across a suitable membrane.

## IV. Enumeration of bacteria in the mixed community

For many years it has been recognized that the total number of bacteria in a water sample is much greater (often more than two orders of magnitude) than that capable of producing visible growth in or on culture media. Techniques have been developed which attempt to quantify the total population (direct or total counts), the culturable population (viable count), and the non-culturable population which can be independently shown to have activity or cellular integrity (the metabolically active). The last category of organisms has become increasingly important as ecologists recognize the limitations of total and viable counting procedures. A full discussion of these bacterial states can be found in Roszak and Colwell (1987a).

## A. Direct counts

Direct bacterial counts can be made in chambers, on membrane filters after filtration of the sample, or on natural or artificial surfaces. Various designs of counting chambers have been described which have their own particular characteristics, however they all suffer from several practical limitations. The concentration of samples and subsequent transfer of material to the chamber can be subject to major errors in volume calculations. These can be minimized by accurate weighing and re-suspension procedures (Fry, 1988) and the depth of sample in the counting chamber should be determined on each occasion (Jones, 1979c). Movement of motile organisms within the sample can be discouraged by the addition of polyvinyl alcohol, methyl cellulose or agar (Jones, 1979c).

Direct counts of bacteria on black membrane filters by epifluorescence microscopy have become the most frequently used method for total population estimates as they permit rapid, easy and quantitative estimation of aquatic bacteria. Fluorochromes coupled to particular cell components fluoresce when excited with light of a suitable wavelength. The fluorochromes in use stain either the nucleic acid or protein components of the cell although specific reaction sites may still be uncertain. A full description of fluorochromes and staining procedures can be obtained from the tables presented by Jones (1979c) and Fry (1988). The quality of the image obtained depends upon the light source, optical filters used, optical characteristics of the microscope, the membrane filter, the fluorochrome and the staining procedures. A full description is beyond the scope of this discussion and reference should be made to Jones and Simon (1975), Daley (1979), Roszak and Colwell (1987a) and an excellent review by Fry (1988). The performance characteristics of different membranes should only be compared using filters of similar pore size and if automated counting procedures are used (see Fry, this volume, Chapter 2) polycarbonate membranes are preferred.

Bacteria attached to surfaces can be removed using standard procedures and direct counting methods applied to the suspension (Goulder, 1987). However, the removal procedure will destroy the community structure and valuable information could be lost. Natural surfaces are rarely flat making microscopy extremely difficult, if not impossible. Moreover, the spatial heterogeneity of populations may require a large number of observations to obtain quantitative data. Although some success has been achieved by grinding stone surfaces flat and allowing recolonization (Jones, 1977), most enumeration techniques have been developed on more regular surfaces such as the leaves of macrophytes (Fry and Humphrey, 1978; Hossell and Baker, 1979). An alternative approach is to allow colonization of introduced artificial surfaces. Many different materials have been used (Fletcher, 1979) but such artificial surfaces may possibly be selective. However, in a recent review Wardell (1988) pointed out that colonization of artificial surfaces causing biofouling or corrosion, or producing biofilms for sewage treatment, is of major economic importance. The production of polysaccharides within biofilms may interfere with the uptake of stains by cells, although specialized procedures have been developed to overcome this (Allison and Sutherland, 1984).

Direct counts of bacteria can also be made on thin layers of agar coated on glass slides. An accurately measured small volume of bacterial suspension is added to the agar. The water is absorbed leaving the cells on the agar surface which are then counted using phase-contrast microscopy. Fry and Zia (1982b) found excellent correlation between such counts and corresponding epifluorescence counts, although the latter counts were approximately twice as high.

Electron microscopes have been used to obtain total bacterial counts of filtered samples. Harris *et al.* (1972) describe the preparation of cellulose ester membranes for use with TEM but most applications are for the scanning electron microscope (Overbeck, 1974; Bowden, 1977; Schmaljohann *et al.*, 1987). Bacteria can penetrate up to 10 μm into 0.22-μm pore size cellulose ester membranes (Jones, 1979c) which is acceptable for transmitted light or epifluorescence microscopy but not for SEM techniques. The flat surface of polycarbonate filters provides a uniform optical plane which is essential for this type of work.

### B. Viable counts

The viable count of a bacterial community is obtained after growth on a suitable medium containing carbon and energy sources. As all media are selective and not all bacteria are recoverable, viable counts are very rarely quantitative. A medium based on casein, peptone and starch (Collins and Willoughby, 1962) has given the highest counts of aquatic bacteria (Jones, 1970; Fry and Zia, 1982a) although more concentrated media may be appropriate for heavily polluted waters (Pike *et al.*, 1972). Incubation temperatures between 10 and 20°C are recommended for up to 23 days (Jones, 1970; Fry and Zia, 1982a). Standard agar can be used to gel the medium although silica gel can be used for isolating chemolithotrophs (Collins, 1969). For more specialized procedures purified agars (Maiden and Jones, 1984) or gelling agents which solidify on warming (Gardener and Jones, 1984) have proved useful. Pour plates are less successful than spread plates presumably as aquatic bacteria are heat sensitive at the temperatures of molten agar (Buck, 1979). Other viable count techniques include the drop plate method, slide culture after suspension of the slide in water (Mara, 1972) or sediments (Godinho-Orlandi and Jones, 1981b) and microcolony counts after concentration by centrifugation (Fry and Zia, 1982a). Fry and Zia (1982a) found that the majority of microcolonies consisted of only two cells, representing one division cycle, but the proportion of viable cells (up to 45%) was greater than that obtained using conventional procedures. Samples have also been concentrated on membrane filters and incubated on pads soaked in medium (Straskrabova, 1972) or in complex flow-through systems (Kunicka-Goldfinger, 1973).

Counts obtained using most probable number procedures are usually higher than those from conventional plate counts (Jones, 1979c). This may be attributed to the less harsh nature of the treatment and continuation of interactions which normally occur within the sample.

## C. Metabolically active bacteria

Such organisms constitute a proportion of the population which cannot be cultured using standard procedures but can be directly shown to be metabolically active or physiologically responsive under experimental conditions. This fraction of the population has been referred to as pseudosenescent (Postgate, 1976) or somnicells (Roszak and Colwell, 1987a). Its detection is important as direct counting procedures cannot differentiate active and inactive bacteria and viable counts are necessarily selective. A discussion on the physiological state of bacteria within the environment can be found in Roszak and Colwell (1987a) as well as a full bibliography on methods used to distinguish metabolically active and inactive cells. As the proportion of active bacteria within the community is of most interest to the ecologist such methods are frequently combined with total bacteria counts using epifluorescence microscopy.

Hobbie *et al.* (1977) speculate on the possibility of using acridine orange and epifluorescence microscopy to distinguish active and inactive cells due to the differential fluorescence of DNA and RNA. However, many other factors have been shown to influence the dominant fluorescent colour (Roszak and Colwell, 1987a) which would invalidate any conclusions. Similarly, whilst other staining procedures may be satisfactory under controlled laboratory conditions, they fail to realize their full potential when applied to sample material (Roszak and Colwell, 1987a).

### *1. Microautoradiography*

Radiolabelled substrates are added to the sample and after incubation the bacteria are concentrated by filtration and unincorporated label is removed. The filter is then placed in close proximity to a photographic emulsion and the incorporated radioactive substrate within bacterial cells can be detected after development of the emulsion and counting the exposed silver grains under transmitted light or phase-contrast microscopy. For detailed descriptions of the method as applied to aquatic samples reference should be made to Stanley and Staley (1977), Simek (1986) and O'Carroll (1988). Application of single labelled compounds (Stanley and Staley, 1977) could be selective for specific populations and therefore use of a suite of substrates may provide more representative counts (Simek, 1986; Roszak and Colwell, 1987a). Counts of unlabelled bacteria on the filter can also be made permitting simple calculation of the proportion of active bacteria within the community (Simek, 1986). Tabor and Neihof (1982) compared different preparation techniques and also superimposed acridine orange staining directly on the micro-autoradiogram. O'Carroll (1988) noted that, if such techniques are used, careful selection of the sample fixative is necessary. It is important to include appropriate sample blanks for autoradiographic studies. Marcussen *et al.* (1984) found that the background counts obtained varied between lakes, and this necessitated changing the criteria used to define cell activity on the photographic emulsion. The developed silver grains may also obscure the source material which could be a single cell or a microcolony. Straskrabova and Simek (1984) checked for such effects by comparing total counts with the sum of labelled and unlabelled cell counts. If these values differed by more than 10% the results were discarded.

### 2. *Electron transport activity*

Dehydrogenases associated with electron transport activity can competitively transfer electrons to 2-(*p*-iodophenyl)-3-(*p*-nitrophenyl)-5-phenyl tetrazolium chloride (INT). This is reduced to the insoluble red-coloured INT-formazan which accumulates as granules within the cell. These deposits can be observed by bright-field microscopy on thin membrane filters (Zimmerman *et al.*, 1978). The filters can be simultaneously stained with fluorochrome dyes for total bacterial counts although the fluorescence of very active cells can be obscured by the large deposits of formazan (Maki and Remsen, 1981). Improvements to the technique have been made for marine samples (Taber and Neihof, 1982) and could be applied to freshwater systems. Roszak and Colwell (1987a) state that it is difficult to detect intracellular deposits of the many small bacteria present in environmental samples, and Zimmerman *et al.* (1978) propose a detection limit of 0.4 μm (cell diameter). Bacteria of this size and less form a significant proportion of the natural aquatic community and therefore the usefulness of the method may be limited. The INT can be reduced by both aerobic and anaerobic respiratory chains and therefore it is important that the conditions of the assay are the same as those in the original sample (Graf and Bengtsson, 1984).

### 3. *Nalidixic acid*

Nalidixic acid inhibits DNA synthesis in susceptible prokaryotes and when added to a water sample, which has also been enriched with yeast extract, the bacteria which respond to the substrate grow but are unable to divide (from Roszak and Colwell, 1987a). They become enlarged and elongated and can be easily resolved using epifluorescence microscopy (Kogure *et al.*, 1979). Peele and Colwell (1981) noted that counts varied with the substrate added and that the highest were obtained with yeast extract. Low concentrations of substrate should be used to prevent inhibition of oligotrophic bacteria or substrate-accelerated death. Nalidixic acid is only active against Gram-negative bacteria (the majority of those found in fresh water) and attempts to broaden the antibiotic spectrum did not affect the counts obtained (Kogure *et al.*, 1984). Application of the technique to some lake sediments may be difficult as many filamentous organisms are already present (Godinho-Orlandi and Jones, 1981a) and bacteria from such a relatively rich nutrient environment may respond differently (Maki and Remsen, 1981). In addition, conventional rod-shaped bacteria, such as those belonging to the genus *Pseudomonas*, may produce filamentous forms (several hundred μm in length) in dilute media (Jensen and Wollfolk, 1985).

### 4. *Fluorescein diacetate*

Fluorescein diacetate (FDA) may be taken up by some bacteria and hydrolysed by non-specific esterase enzymes to yield fluorescein. This does not pass readily across the cell membrane, accumulates, and can be observed by fluorescence microscopy after excitation with blue light. This process requires an intact membrane and active enzymes and therefore should be a function of metabolically active bacteria. However, Chrzanowski *et al.* (1984) observed differences between the counts of active bacteria in a water sample using this and ETS methodology. Further investigation indicated that FDA does not enter all bacteria and fluorescence is often weak and fades rapidly. The method is therefore of limited usefulness.

## V. Enumeration of specific populations of bacteria

### A. Direct counts

The application of fluorescent antibodies can be used to detect specific populations (see Stanley *et al.*, 1979). Such antibodies are prepared against individual isolates and therefore a number of these must be prepared, mixed and applied as a suite of probes. The success of such an approach initially depends on the ability to culture all the individual serotypes likely to be present within the population (Stanley *et al.*, 1979). Given the limitations of culture techniques such success would appear to be unlikely.

The presence of methanogenic bacteria in enrichment cultures may be detected by the fluorescence of coenzyme $F_{420}$ under ultraviolet light (Doddema and Vogels, 1978; Jones *et al.*, 1982). However, many other organisms may exhibit fluorescence at this excitation wavelength, and the genuine fluorescence of methanogens may also be quenched by the presence of the reducing substances almost invariably found in environmental samples.

### B. Viable counts

The use of selective media has provided some information on the population ecology of certain groups of bacteria. A list of the media most commonly used is provided by Schneider and Rheinheimer (1988). The limitations of viable counts, particularly with regard to viable but non-culturable cells, apply equally, if not more so, to the use of selective media.

## VI. Morphologically distinct bacteria

### A. Direct counts

Bacteria generally have a low morphological diversity but it is possible to recognize a number of organisms based on cell morphology. Planktonic bacteria have been recognized on cleared membrane filters (Jones, 1975, 1978; Clark and Walsby, 1978). The observation of similar organisms in sediment deposits is not always possible but relative distribution patterns can be observed by adhesion of the cells to agar-coated slides inserted into the sediment cores (Jones, 1981). Filamentous organisms can make up to 50% of the bacterial biomass in sediment deposits (Godinho-Orlandi and

Jones, 1981b) but their morphological characteristics may be unstable and therefore a 'pragmatic' key to their identification was devised (Jones and Jones, 1986). Many of these morphologically distinct organisms have never been isolated and investigations of distribution patterns may provide clues to their metabolic capability and cultural requirements.

The majority of aquatic bacteria, however, do not possess recognizable morphological characteristics and therefore autecological studies require additional biochemical techniques. Fluorescent antibody (FA) techniques are proving increasingly promising in this respect. Controls to test the specificity of the labelled antibody should be performed and the degree of cross-reaction with other organisms ascertained (Belser and Schmidt, 1978). Cross-reactivity can be decreased by adsorbtion of the antibody preparation with pure strains of the undesirable species or careful selection of a monoclonal antibody specific for a single antigenic determinant not present in the non-target species. The production of antibody requires the initial isolation of the target species. This isolate is best obtained from the highest positive dilution of an MPN series which will provide the dominant culturable form which may, or may not, be the ecologically dominant form. A disadvantage of immunological techniques is that only positive results can be interpreted. Negative results may be due to absence of target in the sample or to an alteration of the antigenic determinant. Non-specific adsorption of the FA to detrital material can be overcome by pretreatment of the sample with a different fluorochrome, usually rhodamine tagged to gelatin.

### B. Viable counts

Using micromanipulation procedures it is possible to move individual bacteria across an agar surface and, if growth occurs obtain a microcolony which is then known to have developed from a single cell (Skerman *et al.*, 1983). The technique deserves wider application as it provides a promising first step in the isolation of 'new' bacteria.

## VII. Bacterial activity

The natural environment results in the expression of the full range of bacterial metabolic activity. The heterogeneity of habitats, the presence of generally low substrate concentrations and interactions between different groups of organisms contribute to the problems of measuring bacterial activity in natural material. This review is not a catalogue of all possible metabolic activities which might be measured. It is an attempt to categorize appropriate methodologies and to provide examples based primarily on experience in this laboratory.

Bacterial activity in fresh waters is controlled by the supply of electron donors (predominantly organic) and electron acceptors from both autochthonous and allochthonous sources. Under aerobic conditions oxygen will be the preferred electron acceptor, then as it is depleted it will be replaced, in order, by nitrate, sulphate and carbon dioxide. This sequence of utilization may be observed, for example, in the hypolimnion of eutrophic lakes. In sediments similar events occur on a much smaller depth scale, the redox boundary being maintained by the input of electron donors and acceptors. As oxygen becomes depleted in the overlying water the redox boundary moves from the sediment into the water column (Jones, 1979a). Bacterial activity in sediment is largely controlled by the supply of electron acceptors (Cappenberg, 1988) and the redox boundary is a zone of considerable activity due to their depletion and regeneration.

The major geochemical cycles are tightly linked and driven by the process of primary production. For example, the decomposition of organic matter will result in the regeneration of inorganic nitrogen species and associated organotrophic and lithotrophic processes. To understand such a complex system it is necessary to quantify each component process. The bacterial activity associated with the processes may be determined in terms of metabolic capability and/or growth. As the dissolved organic material in lakes is largely refractory (Geller, 1986) and concentrations tend to remain relatively constant (in Straskrabova and Simek, 1988) changes in metabolic activity have been related to either increasing activity per cell or an increase in the proportion of active cells (Straskrabova and Simek, 1988) and therefore one should not always associate increased activity with increased biomass.

Recent reviews on the measurement of bacterial activity (Findlay and White, 1984; O'Carroll, 1988) have each described four approaches. These, however, were not wholly consistent and therefore, for simplicity, they have been reduced in this chapter to two distinct categories: (a) the measurement of the net flux of metabolic substrates or products from which activities can be calculated by the application of assumptions or conversion factors: (b) measurement of assimilation, mineralization, uptake kinetics or transformation of substrates which can be readily identified. The application of mathematical models based on estimates obtained using the above approaches may be used for predictive purposes but discussion of this aspect is beyond the scope of this chapter.

## A. Element fluxes

### *1. Whole lake approach*

The accuracy of such an approach depends on the mobility of the element in question and therefore is of most use in the relatively closed system of the hypolimnion. Although exchange can occur through turbulence, sedimentation and evolution of gases, the necessary corrections required to approximate a closed system have been made (Jones, 1982). In such systems measurement of the net changes of electron acceptors or accumulation of products should provide an estimate of the relative importance of decomposition processes. Aerobic respiration rapidly consumes oxygen within the hypolimnion and using such measurements the hypolimnetic areal oxygen deficit (HAOD) may be calculated. Although the relationship between HAOD and lake productivity is not simple (Charlton, 1980) limits can be defined for oligotrophic and eutrophic conditions (Mortimer, 1941). Despite the close links between HAOD and primary production, the measurement of oxygen consumption must be combined with changes in concentration of all electron acceptors and the determination of lithotrophic processes to obtain a realistic estimate of carbon flux. Jones and Simon (1980a) measured the net change of sulphate, sulphide, nitrate, ammonia, oxygen, methane and nitrogen in the deeper waters of a eutrophic lake. Using the equations of Richards (1965) they calculated the importance of individual processes in carbon mineralization by relating theoretical carbon dioxide production to the measured flux. Gaseous end products such as methane and nitrogen may be collected in gas traps, but corrections must be applied for biological reoxidation and gas stripping as the bubbles travel (Jones, 1982). The diversity of fermentation reactions and low concentrations of end products makes quantification difficult but an estimate of the role of such reactions in decomposition can be obtained as the difference of measured $CO_2$ flux and that calculated from respiratory activity. Similar methods have been used by Kelly *et al.* (1988) to describe the role of all electron acceptors in anaerobic decomposition and a further mathematical treatment of mass balance data is presented by Priscu *et al.* (1986).

Using the flux, or mass balance, approach it is difficult to identify the site of activity. There is evidence that processes in the sediment dominate changes within the hypolimnion and it may be useful to determine the extent of direct influence of the sediment on the water column. A first approximation of this has been described by Edberg (1976). The logarithm of the change in concentration of any chosen determinant is related in a linear manner to the distance from the sediment. The upper limit of influence is defined when the relationship deviates from linearity.

The interpretation of data can also be affected by the morphometry of the lake basin. Davison and Finlay (1986) observed similar distribution of sulphide in the water column and ferrous sulphide formation on the surface sediments of two adjacent lakes with depths of 15 m and 4 m respectively. The cycling of sulphur therefore looked similar in both lakes. However, the shallower lake supported phototrophic oxidation of sulphide as the zone of anaerobiosis came close to the water surface. In the deeper lake the anaerobic zone of sulphide formation was below the maximum depth of penetration of physiologically active light which precluded the occurrence of photo-oxidative processes.

Analysis of the concentrations of nutrients in the inflow and outlet of lakes will indicate if the lake is a net source or sink of a particular chemical species. However, due to incomplete mixing or long retention times the interpretation of such data in terms of bacterial activity is limited. Such an approach is best applied during the circulation period for lakes with short retention times.

*2. Sample approach.*

Assuming that representative and reproducible samples can be obtained the incubation time must be short to avoid changes in bacterial activity and community structure which occur quite soon after enclosure of the sample material. This is particularly problematical for planktonic samples because bacterial activities are low and therefore sensitive and precise analytical techniques are required to detect small changes in water chemistry (Talling, 1973). The length of incubation can be extended if the bacterial population is monitored using direct counting techniques (Jones, 1977). The greater numbers and activity of bacteria in the sediment deposits have enabled the extensive use of sediment water systems in isolated samples (Findlay and White, 1984). The disturbance caused by sampling procedures can be avoided by isolating a defined area of sediment and overlying water in specially constructed chambers (Sweerts *et al.*, 1986). The modern design of such chambers allows manipulation of experimental conditions but their operation requires the use of divers. Lack of access to such facilities requires that samples must be taken and removed from the system under investigation. Manipulation within the laboratory allows experimental conditions to be readily modified. The most common modification is to section the sediment into discrete layers; this destroys established substrate gradients within the intact sediment and therefore affects the activity to be measured (Findlay and White, 1984). Conditions within the slurry are often further modified to maximize the activity being estimated and therefore such estimates should be regarded as potential activities rather than actual activities. Relative estimates of activity between samples can be made provided all the slurry conditions and procedures are the same.

Chemical analysis of the interstitial water to obtain concentration gradients of key electron acceptors or metabolic products can provide estimates of bacterial activity (Cappenberg, 1988) but caution should be exercised in the interpretation of substrate pool sizes obtained in this way. A small pool size could indicate a minor metabolic role or a rapid turnover time, moreover a large pool size is difficult to interpret as analytically available substrate may not be the same as biologically available substrate.

It is preferable to use intact core systems to study aquatic microbial processes. Although the pioneering studies on sediment oxygen uptake by Mortimer (Mortimer, 1941) used such systems, further use has been largely ignored despite the acceptance of the limitations of slurry procedures. This is reflected by the few available comparisons between activities measured in slurries and intact cores (Findlay and White, 1984). Despite the paucity of such studies, techniques are available that facilitate the use of intact core systems. The use of these would be applicable to any study using intact cores and therefore the tested procedures will be discussed. Whole sediment cores can be incubated (Jones, 1976) or, to facilitate statistical analyses of data obtained under different conditions, the original core can be subsampled into mini-cores (Hall and Jefferies, 1984). Procedures to maintain anaerobiosis whilst subsamples are taken have been attempted although these were not always successful (Jones and Simon, 1985). Large intact cores can be incubated in specially designed incubation vessels (Gardner *et al.*, 1987) or large blocks of sediment sampled which facilitate smaller subsamples being taken or allow replicate determination to be used (Carlton and Wetzel, 1985). Reconstituted cores have been used which are claimed to restore *in situ* micro zones rapidly (Jones, 1977). Additions to cores can be made as point sources, using long syringe needles, and techniques to obtain even distribution of added material have been described although up to 24 h may be required for added material to equilibrate with natural substrate pools (Jones and Simon, 1984). Sediment core tubes can be modified by drilling holes in the side which allows horizontal additions to be made at defined depths in the core profile (Jorgensen, 1978; Jones *et al.*, 1982). The overlying water can be subsampled after gentle mixing procedures to disrupt gradients formed due to lack of turbulence and modification of the sampling system can eliminate exposure of the sample to the atmosphere (Jones, 1976). The sediment layers can be subsampled through modified core tubes (Cappenberg, 1974), by extrusion (Hall and Jefferies, 1984) or by rapid freezing followed by mechanical separation (Jones and Simon, 1985). Care should be exercised during the freezing of samples as expansion of water can distort the core profile.

Despite the arguments for using intact core procedures, data from such studies are sparse. The mass balance approach using slurried sediment is very much restricted to oxygen uptake studies and nitrogen cycle reactions. Although there is little reason for such restriction the lack of application is presumable due to the availability of radioisotopes of the other major nutrients (C, S and P) which would provide rapid and sensitive techniques to study microbial processes.

A by-product of all metabolic processes, which is often overlooked, is heat and measurement of the heat production has much potential for ecological studies. However, the output of heat would estimate the activity of the whole community and useful information on specific activities would be lost. Microcalorimetric procedures have concentrated therefore on the effect of perturbations on community activity (Lock and Ford, 1985). Other restrictions on the use of the technique concern the sensitivity and cost of appropriate apparatus, however recent innovations circumvent these restrictions to some extent and we await more use of such approaches (Lock and Ford, 1983).

## B. Detection of substrates or products

A description of methods defined under this approach would not be possible and therefore we propose to present examples of each technique and its application to fresh water. This serves only to describe the fundamental principle and with such information it may be possible to develop equivalent techniques particularly as knowledge of metabolic processes and instrument technology improves. The detection methods include (1) formation of a coloured end product, (2) detection of end product by analytical procedures and (3) use of radioisotopes or stable isotopes.

### *1. Coloured end product*

Such techniques can be directed towards different populations within the bacterial community. Active dehydrogenase activity (present in respiring bacteria) can be detected by addition of the tetrazolium salt (INT) which is oxidized within the cell to the pink–red-coloured insoluble formazan. This technique has been used to distinguish metabolically active members of the bacterial community (King and Parker, 1988). An estimate of dehydrogenase activity can be obtained by measurement of the extinction (at 540 nm) of the formazan after extraction from the cells using organic solvents (Ohle, 1972). The extraction step is eliminated if Triton X-100 is included in the assay solution. Such modification of buffer solutions and enzyme extraction procedures suitable for freshwater systems have been described by Jones and Simon (1979). More specifically, assays for alkaline phosphatase activity using *p*-nitrophenyl phosphate as substrate are available. The substrate is degraded releasing nitrophenol which can be determined colorimetrically at 418 nm (Jones, 1972). A number of such substrates are commercially available or can be prepared in the laboratory (see Reichardt, 1988). Similarly less complex substrates attached to dyes are available as an aid to rapid identification procedures (APIZYM, Api-System S.A., 38390 Montalieu, Vercieu, France) and may be of use provided they have sufficient sensitivity for planktonic or benthic samples.

### 2. *Analysis of end product*

The requirements for suitable analysis of any end product are that the product must be (a) specific, (b) sensitive and (c) not liable to interference from other components of the reaction mixture. Using such criteria, assays based on colorimetry, fluorimetry (Jones, 1979) and gas chromatography (Jones and Simon, 1981) have been developed. The conditions within the assay are frequently modified to increase the sensitivity of the method or maximize the activity being determined. In an assay for nitrate reductase the sediments were treated to extract active enzymes from the bacterial cells (Jones, 1979b). This procedure would detect the activity of both assimilatory and dissimilatory nitrate reductases although addition of chlorate could suppress the activity of the assimilatory enzyme. The extraction procedure removed any constraints on enzyme activity due to transport of the electron acceptor into the cell. Moreover, the activity was maximized by providing a chemically reduced environment and non-limiting concentrations of nitrate (Jones *et al.*, 1980). Similarly the activities of malate and lactate dehydrogenases were maximized for assay by addition of their respective substrates. The preparation of enzyme extracts may not be reproducible either within or between samples due to the effects of particulate size or non-selective absorption of enzyme to non-biological material (Reichardt, 1988).

Gas chromatographic techniques are probably the most convenient analytical procedures as quenching or interference effects are minimal. The assays must be performed on closed systems to facilitate analysis of the gaseous end products. Analysis of the head space gases in vials sealed with neoprene or butyl rubber septa have proved to be most convenient. Precautions must be taken to avoid contamination from atmospheric gases for example, when estimating denitrification, by the accumulation of nitrogen gas (Jones and Simon, 1981). This assay was modified by the addition of acetylene to inhibit nitrous oxide reductase; nitrous oxide therefore accumulates at a rate equal to nitrate reduction and has low background levels in the atmosphere (Sorensen, 1978). This assay has proved to be unsuitable for use in fresh water (Jones and Simon, 1981). The use of differential metabolic inhibitors in studies of bacterial activity can be problematical. The use of such techniques has recently been reviewed with reference to commonly used inhibitors (Oremland and Capone, 1988).

## C. The use of labelled compounds

### *1. Carbon-14*

The use of carbon-14 radioisotopes for the determination of bacterial activity was probably stimulated by the success of using $^{14}C$ bicarbonate to estimate primary productivity. The sensitivity of the technique and the relative ease of quantitative detection of radioisotopes meant that $^{14}C$-labelled compounds were ideal tools with which to examine features of the carbon cycle, particularly heterotrophic activity, in aquatic systems. Parsons and Strickland (1962) showed that the uptake of low concentrations of organic molecules could be described by Michaelis–Menton kinetics. Wright and Hobbie (1966) described the differences between bacterial and algal uptake as the latter was best modelled by diffusion kinetics. The methods were modified to trap the $^{14}C$ carbon dioxide during incubation as respiratory losses could be a significant proportion of the labelled substrate uptake (Hobbie and Crawford, 1969). As the natural heterotrophic population would be utilizing a whole spectrum of organic compounds the use of one specific substrate was not representative. The estimate therefore is referred to as the heterotrophic potential which will vary with the particular substrate being used. Evidence has been presented (Findlay and White, 1984) that use of different substrates on similar samples gives different estimates of uptake and respiration rates and the use of a single substrate will be biased to the portion of the community utilizing that substrate. Generally, simple labile substrates such as sugars or amino acids have been used although attempts have been made to provide more representative labelled material. A major exudate product of phytoplankton, gycolic acid, has been used (Wright, 1975) as has a mixture of amino acids obtained from the lysis of algal proteins (Sepers, 1981). The uptake of $^{14}C$-labelled photosynthate by bacteria has been estimated using size fractionation techniques (Jones *et al.*, 1983; Jones and Salonen, 1985) or by preparing $^{14}C$-labelled exudate or plant material (Benner *et al.*, 1984; Cappenberg *et al.*, 1982; Watanabe, 1984; Wiebe and Smith, 1977). Such methods, however, result in low specific activities of material which, on addition to samples, disturb natural substrate concentrations. Much of the organic material in lakes is regarded as being recalcitrant and therefore attempts to estimate mineralization rates of such compounds require long incubation times which are associated with changes in the natural bacterial community (Ladd *et al.*, 1982).

There are many reviews of the application of the heterotrophic potential technique (Fry and Humphrey, 1978; Wright and Burnison, 1979; Van Es and Meyer-Reil, 1982) which do not require repeating here. It will suffice to say that there are three conditions which can apply after the addition of a radiolabelled substrate. The added substrate concentration can be less than, equal to or greater than the natural substrate concentration. To know which situation exists the natural substrate concentration must be known. This can pose difficult analytical problems. Robinson and Characklis (1984) describe a mathematical treatment of kinetic data when natural substrate concentrations are not known but many workers estimate the activity towards a range of substrate additions and analyse the data using the kinetic model discussed by Wright and Burnison (1979) and Van Es and Meyer-Reil (1982). Using such data the kinetic parameters of maximum substrate utilization rate ($V_{max}$), the affinity constant (substrate concentration at $V_{max}/2$), the natural substrate concentration ($K + S_n$) and the natural substrate turnover time ($T_t$) can be calculated although care must be exercised in the interpretation of such population parameters (Van Es and Meyer-Reil, 1982; Lewis *et al.*, 1988). The comparison of substrate turnover times in relation to the substrate pool sizes measured in sediment slurries and intact cores illustrated the interdependence of bacterial groups in anaerobic sediments (Jones, 1985). Caution must also be taken with the interpretation of the analysis of natural substrate concentrations as the analytically available substrate may not be the same as the biologically available substrate (Jones and Simon, 1984).

Uptake studies are usually performed in the dark or on prefiltered samples to reduce the effects of phytoplankton heterotrophy on the measurements. Spencer (1979) provides some evidence that bacterial uptake can be affected by illumination.

Specific application of $^{14}C$ to measure other bacterial processes can be found in Iverson *et al.* (1987) who used labelled methane to measure rates of methane oxidation and also in Hall (1982) and Hall and Jefferies (1984) who estimated the uptake of $CO_2$ in the presence and absence of a nitrification inhibitor to measure rates of nitrification.

*2. Carbon-11*

We are not aware of any application of this short-lived radioisotope (half life 20.4 min) to freshwater samples. The advantages are that it can be produced at high specific activities and detected with high sensitivity (McCallum *et al.*, 1981). Moreover, labelled amino acids and glucose have been produced although these are not carrier free and have low activity due to the time required for synthesis (McNaughton, 1986). One disadvantage of $^{11}C$ is that specialized production and counting facilities are required and if these are not readily available then the advantage of using this radioisotope will be negated by changes in the sample during transport to the facility.

*3. Tritium*

The use of tritium for micro-autoradiography to detect the metabolically active cells within the community has already been described. Using such techniques with different substrates may provide information on the spectrum of metabolic capability within the sample. However, attempts have been made to quantify the total uptake of radiolabel in micro-autoradiograms. This discussion also applies to studies using $^{14}C$-labelled substrates. Ward (1984) counted all the developed silver grains associated with one cell and used this as an index of the relative activity of individual cells. Stanley and Staley (1977) used a similar method but calculated the disintegrations per developed grain of emulsion after obtaining a calibration using pure cultures of organisms. Straskrabova and Simek (1984) and Simek (1986) measured the total uptake of label on the same samples used for preparation of autoradiograms. Using this value and the active cell counts a specific activity could be calculated. This estimate was considered to characterize best the physiological state of the bacterial community (Straskrabova and Simek, 1988).

*4. Phosphorus-32*

Uptake of phosphorus is associated with the synthesis of nucleic acids and is often used to estimate bacterial production. Such techniques are not included in the present discussion. The energetic $\beta^-$ emission is too penetrating for use with microautoradiography.

*5. Sulphur-35*

This isotope is also used to estimate bacterial productivity and could be used for micro-autoradiography. Labelled sulphate has been used to estimate rates of sulphate reduction in fresh waters which are usually low due to limiting sulphate concentrations (Jones and Simon, 1980a) although in extreme freshwater environments sulphate reduction can be an important process (Smith and Oremland, 1987). Interpretation of data from samples with oxidized zones could be difficult due to the rapid chemical oxidation of sulphides. Similarly, sulphide could be assimilated into the organic fraction and not recovered using conventional techniques (David and Mitchell, 1985). Addition of cold sulphate to intact sediment cores caused perturbation of the pool sizes of low molecular weight fatty acids (Jones *et al.*, 1982) and differences in sulphide concentrations in bottles incubated in light and dark conditions have provided an index of the activity of phototrophic bacteria (Guerrero *et al.*, 1985).

### 6. *Nitrogen-15*

There have been few applications of $^{15}N$-labelled tracers to study bacterial nitrogen transformations in fresh waters. Stewart *et al.* (1982) added $^{15}NO_3$ to a large experimental enclosure and determined the enrichment of different nitrogen fractions on both a temporal and a spatial basis. Such facilities are not readily available and, as previously discussed, interpretation of data on substrate pool sizes may be difficult in the absence of independently measured rates of nitrogen transformation. The lack of application to isolated sample material could be due to (a) the $^{15}N$ addition increasing the natural substrate concentration and (b) the analytical problems and time required for separating the different nitrogen fractions to obtain sufficient N for analysis. The amount of $^{15}N$ to be added is dependent upon the $^{15}N$ enrichment of the substrate (maximum 99 atom % excess), the natural substrate concentration and rate of transformation. Often practical constraints are imposed by the permissable incubation time and the size of sample required for analysis. For example, Brezonik (1968) perturbed surface water ammonia concentrations by addition of 100 μg $^{15}NH_3$-N per litre and in many cases could not obtain sufficient nitrate fraction for $^{15}N$ analysis. The amount of sample required for analysis by mass spectrometry is dependent upon the volume of the gas inlet system of the instrument as discussed by Fiedler (1984). Preston and Owens (1983) report that a small sample size of 10–100 nmol $N_2$ could be used on spectrometers coupled to nitrogen analysers with capillary flow rates. A maximum sample size of 10 μg N has been reported for use in optical emission spectrometers (Chatapaul and Robinson, 1979) but such techniques lack the precision obtained with mass spectrometry (Fiedler, 1984). The perturbation of natural substrate concentrations can be avoided if an isotope dilution approach is adopted as described for $^{15}NH_4^+$

(Blackburn, 1979) or $^{15}NO_3$ (Enoksson, 1986) in marine systems. The authors are unaware of any application of isotope dilution to freshwater systems.

Methods to separate the different nitrogen fractions have been adapted from available analytical techniques using, for example, steam distillation (Bremner and Keeney, 1965) or HPLC systems (Tiedje *et al.*, 1981). Methods are frequently modified (Selmar and Sorensson, 1986) and can be developed to suit specific requirements.

The recent description of nitrosation of soil organic matter by oxidized inorganic nitrogen could produce significant errors in $^{15}N$ studies (Azhar *et al.*, 1986). The extent of such reactions in freshwater sediments is not known and should be investigated.

The natural isotopic abundance of $^{15}N$ has limited its application in environments with low substrate concentrations (Ward, 1986) but due to discrimination between $^{14}N$ and $^{15}N$ in different processes it can be possible to determine the source of a nitrogenous product (from Domenach *et al.*, 1977). However, reference standards for such methods require careful preparation (Mariotti, 1984).

### 7. *Nitrogen-13*

The short half life of $^{13}N$ and the specialized facilities required for its production restrict the use of this isotope. However reports indicate that it has been used on freshwater sediments although no data are presented (Kaspar, 1985). The high specific activity of nitrogen compounds and the sensitivity with which the positron emission can be detected provide many practical advantages (Tiedje *et al.*, 1981) but the fractionation of nitrogen species necessarily requires the use of rapid and precise techniques.

## VIII. A future role for molecular biology in microbial ecology

The advances in molecular biology over the last 10 years are now being applied to microbial ecology although their application in many respects is in its infancy. Recent advances have been accelerated by the need to study the behaviour and ecological effects of the release of genetically engineered bacteria into the environment. As a consequence, this new technology is biased towards enumeration and species identification rather than population/community activity. DNA hybridization is one such technique that has been proven in the molecular biology laboratory and is now actively being applied to environmental studies. Ford and Olsen (1988) and Holben and Tiedje (1988) have suggested that hybridization, in particular, and molecular techniques, in general, have great potential for the study of microbial ecology. Advantages of this methodology include the abilities to (a) detect populations without prior culturing of the organisms, (b) detect specific organisms without the need for selectable markers (such as antibiotic resistance), (c) detect multiple populations in the same analysis and (d) detect genetic rearrangements or gene transfer in natural communities (Holben and Tiedje, 1988). Hybridization has been used to detect bacteria carrying novel phenotypes such as degradative ability (Sayler *et al.*, 1985; Pettigrew and Sayler, 1986) and mercury resistance (Barkay *et al.*, 1985; Barkay, 1987). Construction of specific probes which are unique to certain classes of microorganisms using either genomic DNA or RNA has been used to identify and enumerate individual species such as *Escherichia coli* (Echeverria and Haanstra, 1982), *Pseudomonas fluorescens* (Festl *et al.*, 1986) and low-diversity thermophilic and acidophilic populations (for review see Pace *et al.*, 1986). The potential of this methodology is further increased with advances in the isolation of total bacterial communities (MacDonald, 1986a, b) and total DNA from microbial populations (Ogram *et al.*, 1987). The total RNA isolated from natural populations (Pace *et al.*, 1986; Fox and Stackebrandt, 1987) has been used for phylogenetic classification of microorganisms based on the analysis of 16S RNA and 5S RNA. Frederickson *et al.* (1988) combined DNA hybridization with a microdilution most probable number procedure to detect as few as 10 isolates $g^{-1}$ soil of a uniquely marked *Pseudomonas putida* strain. Population analysis will be greatly enhanced by further developments in lipid analysis as described by Komazata and Suzuki (1987) and further application of fluorescent antibody techniques for isolation and enumeration of target organisms (Xu *et al.*, 1984; Porter, 1988).

It has been recognized that plasmids exist as ubiquitous and autonomously replicating units capable of spread and persistence in a large variety of bacteria from diverse environments (Hardman and Gowland, 1985). They can, under certain conditions, form their own 'population' within the bacterial community. Much attention has been focused on their function, distribution and transfer ability in the natural environment. Methods for the isolation of plasmid DNA often select for units of less than 50 kb in size. Gowland and Hardman (1985) have reviewed methods for the isolation of large plasmids which are known to occur in natural populations (Hardman and Gowland, 1985).

The methods of Wheatcroft and Williams (1981) and Hansen and Olsen (1978) have been used for the visualization of whole plasmid DNA by agarose gel electrophoresis. These produce DNA of sufficient purity for analysis by restriction endonuclease digestion which can be further manipulated to make DNA probes or for cloning (see Maniatis *et al.*,

1982). The method of Wheatcroft and Williams (1981) has been used to analyse changes in structure as visualized by endonuclease digestion with plasmids of a size >250 kilobases (Pickup *et al.*, 1983; Pickup and Williams, 1982; Kiel and Williams, 1985). Other methods which are available have been reviewed by Hardman and Gowland (1985). There is no particular method that is applicable to all systems and therefore various methods must be pursued when screening exotic bacteria for the presence of plasmid DNA.

## IX. Conclusion

As our knowledge of the activity and interaction of aquatic bacterial populations increases the limitations of various methods become apparent. Methodology will, and must, continue to develop as information, or technology, becomes available. Such developments will ultimately allow accurate assessment of the role of bacteria in freshwater environments.

## References


Allison, D. G. and Sutherland, I. D. (1984). *J. Microbiol. Methods* **2,** 93–99.

Azhar, E. S., Van Cleemput, O. and Verstraete, W. (1986). *Plant Soil* **94,** 401–409.

Barkay, T. (1987) *Appl. Environ. Microbiol.* **53,** 2725–2732.

Barkay, T., Fouts, D. L. and Olsen, B. H. (1985). *Appl. Environ. Microbiol.* **49,** 686–692.

Belser, L. W. and Schmidt, E. L. (1978). *Appl. Environ. Microbiol.* **36,** 584–588.

Benner, R., Maccubbin, A. E. and Hodson, R. E. (1984). *Appl. Environ. Microbiol.* **47,** 381–389.

Blackburn, T. H. (1979). *Appl. Environ. Microbiol.* **37,** 760–765.

Bowden, W. B. (1977). *Appl. Environ. Microbiol.* **33,** 1229–1232.

Bremner, T. M. and Keeney, D. R. (1965). *Anal. Chim. Acta* **32,** 485–495.

Brezonik, P. L. (1968). PhD Thesis, University of Wisconsin, Madison, WI, USA.

Buck, J. D. (1979). *In* "Native Aquatic Bacteria; Enumeration, Activity and Ecology" (J. W. Costerton and R. R. Colwell, Eds), Vol. 695, pp. 19–28. American Society for Testing and Materials, Special Technical Publication, ASTM, Philadelphia.

Cappenburg, Th. E. (1974). *Ant. Leeuwenhoek* **40,** 285–295.

Cappenberg, Th. E. (1988). *Arch. Hydrobiol. Beith. Ergebn. Limnol.* **31,** 307–317.

Cappenberg, T. E., Hordijk, K. A., Jonkheer, G. J. and Laumen, J. P. M. (1982). *Hydrobiologia* **91,** 161–168.

Carlton, R. G. and Wetzel, R. G. (1985). *Limnol. Oceanogr.* **30,** 422–426.

Charlton, M. N. (1980). *Can. J. Fish Aquat. Sci.* **37,** 1531–1539.

Chatapaul, L. and Robinson, J. B. (1979). *In* "American Society for Testing and Materials Technical Publication" (C. D. Litchfield and P. L. Seyfried, Eds), Vol. 673, pp. 118–127. ASTM, Philadelphia.

Chrzanowski, T. H., Crotty, R. D., Hubbard, J. G. and Welch, R. P. (1984). *Microb. Ecol.* **10,** 179–185.

Clark, A. E. and Walsby, A. E. (1978). *Arch. Microbiol.* **118,** 223–228.
Collins, V. G. (1969). *In* "Methods in Microbiology" (J. R. Norris and D. W. Ribbons, Eds), Vol. 3B, pp. 2–52, Academic Press, London.
Collins, V. G., Jones, J. G., Hendrie, M. S., Shewan, J. M., Wynn-Williams, D. D. and Rhode, M. E. (1973). *In* "Sampling – Microbiological Monitoring of Environments" (R. G. Board and D. W. Lovelock, Eds), Society for Applied Bacteriology Technical Series, No. 7, pp. 77–110, Academic Press, London.
Collins, V. G. and Willoughby, L. G. (1962). *Arch. Mikrobiol.* **43,** 294–307.
Cunningham, C. R. and Davison, W. (1980). *Freshwater Biol.* **10,** 413–418.
Daley, R. J. (1979) *In* "Native Aquatic Bacteria, Enumeration, Activity and Ecology" (T. W. Costerton and R. R. Colwell, Eds), Vol. 695, pp. 29–45, American Society for Testing and Materials, Special Technical Publication, ASTM, Philadelphia.
David, M. B. and Mitchell, M. J. (1985). *Limnol. Oceanogr.* **30,** 1196–1207.
Davison, W. (1977). *Freshwater Biol.* **7,** 393–402.
Davison, W. and Finlay, B. J. (1986). *J. Ecol.* **74,** 663–673.
Dodson, A. N. and Thomas, W. H. (1964). *Limnol. Oceanogr.* **9,** 455–456.
Domenach, A. P., Chalamet, A. and Pachiaudi, C. (1977). *Rev. Ecol. Biol. Soils* **14,** 279–287.
Doddema, H. J. and Vogels, G. D. (1978). *Appl. Environ. Microbiol.* **36,** 752–754.
Echeverria, P. and Haanstra, L. (1982). *J. Clin. Microbiol.* **16,** 1086–1090.
Edberg, N. (1976). *Vatten* **32,** 2–12.
Elliott, J. M. (1977). Scientific Publications No. 25, 2nd Edn, Freshwater Biological Association, Windermere, UK. Titus Wilson & Son, Kendal.
Enoksson, V. (1986). *Appl. Environ. Microbiol.* **51,** 244–250.
Festl, H., Ludwig, W. and Schleifer, K. G. (1986). *Appl. Environ. Microbiol.* **52,** 1190–1194.
Fiedler, R. (1984). *In* "Soil–Plant–Water Relationships" (M. F. L'Annunziata and J. O. Legg, Eds), Academic Press, London.
Findlay, R. H. and White, D. C. (1984). *Microbiol. Sci.* **1,** 90–93.
Fletcher, M. (1979). *In* "Microorganisms of Surfaces." (D. C. Ellwood, J. Melling and P. Rutter, Eds), Special Publications of the Society for General Microbiology No. 2, pp. 87–108, Academic Press, London.
Ford, S. and Olsen, B. H. (1988). *Adv. Microb. Ecol.* **10,** 45–79.
Fox, G. E. and Stackebrandt, E. (1987). *In* "Methods in Microbiology" (R. R. Colwell and R. Grigorova, Eds), Vol. 19, pp. 405–458, Academic Press.
Frederickson, J. K., Bezdicet, D. F., Brockman, F. J. and Li, S. W. (1988). *Appl. Environ. Microbiol.* **54,** 446–453.
Fry, J. C. (1988). *In* "Methods in Aquatic Bacteriology" (B. Austin, Ed.), Vol. 1, pp. 27–72, Modern Microbiological Methods. J. Wiley & Sons, Chichester.
Fry, J. C. and Humphrey, N. C. B. (1978). *In* "Techniques for the Study of Mixed Populations" (D. W. Lovelock and R. Davies, Eds), pp. 1–29, The Society for Applied Bacteriology, Technical Series No. 11, Academic Press, London.
Fry, J. C. and Zia, T. (1982a). *J. Appl. Bacteriol.* **53,** 189–198.
Fry, J. C. and Zia, T. (1982b). *J. Gen. Microbiol.* **128,** 2841–2850.
Gardener, S. and Jones, J. G. (1984). *J. Gen. Microbiol.* **130,** 731–733.
Gardner, W. S., Nalepa, T. F. and Malczyk, J. M. (1987). *Limnol. Oceanogr.* **32,** 1226–1238.
Geller, A. (1986). *Limnol. Oceanogr.* **31,** 755–764.
Godinho-Orlandi, M. J. L. and Jones, J. G. (1981a). *J. Gen. Microbiol.* **123,** 91–101.

Godinho-Orlandi, M. J. L. and Jones, J. G. (1981b). *J. Gen. Microbiol.* **123,** 81–90.
Goulder, R. (1987). *Let. Appl. Microbiol.* **4,** 29–32.
Gowland, P. C. and Hardman, D. J. (1985). *Microbiol. Sci.* **3,** 252–253.
Graf, G. and Bengtsson, W. (1984). *Arch. Hydrobiol. Beih. Ergebn. Limnol.* **19,** 249–256.
Guerrero, R., Montesinos, E., Pedros-Alio, C., Esteve, I., Mas, J., Van Gemerden, H., Hofman, P. A. G. and Bakker, J. F. (1985). *Limnol. Oceanogr.* **30,** 919–931.
Hall, G. H. (1982). *Appl. Environ. Microbiol.* **43,** 542–547.
Hall, G. H. (1986). *In* "Nitrification" (J. I. Prosser, Ed.), Vol. 20, pp. 127–156, Special Publications of the Society for General Microbiology, IRL Press, Oxford.
Hall, G. H. and Jefferies, C. (1984). *Microbiol. Ecol.* **10,** 37–46.
Hansen, J. B. and Olsen, R. H. (1978). *J. Bacteriol.* **135,** 227–238.
Hardman, D. J. and Gowland, P. C. (1985). *Microbiol. Sci.* **2,** 90–94.
Harris, J. E., McKee, T. R., Wilson, R. C. and Whitehouse, U. G. (1972). *Limnol. Oceanogr.* **17,** 784–787.
Herbert, R. A. (1988). *In* "Methods in Aquatic Bacteriology" (B. Austin, Ed.), Vol. 1, pp. 3–25, Modern Microbiological Methods, J. Wiley & Sons, Chichester.
Hobbie, J. E. and Crawford, C. C. (1969). *Limnol. Oceanogr.* **14,** 528–532.
Hobbie, J. E., Daley, R. J. and Jasper, S. (1977). *Appl. Environ. Microbiol.* **33,** 1225–1228.
Holben, W. E. and Tiedje, J. M. (1988). *Ecology* **69,** 561–563.
Hossell, J. C. and Baker, J. H. (1979). *J. Appl. Bacteriol.* **46,** 87–92.
Iverson, N., Oremland, R. S. and Klug, M. J. (1987) *Limnol. Oceanogr.* **32,** 804–814.
Jannasch, H. W. (1958). *J. Gen. Microbiol.* **18,** 609–620.
Jensen, R. H. and Wollfolk, C. A. (1985). *Appl. Environ. Microbiol.* **50,** 364–372.
Jones, J. G. (1970). *J. Appl. Bacteriol.* **33,** 679–687.
Jones, J. G. (1972). *J. Ecol.* **60,** 777–791.
Jones, J. G. (1975). *J. Appl. Bact.* **39,** 63–72.
Jones, J. G. (1976). *J. Ecol.* **64,** 241–278.
Jones, J. G. (1977). *In* "Aquatic Microbiology." (F. A. Skinner, and J. M. Sewan, Eds), pp. 1–34, The Society for Applied Bacteriology Symposium Series No. 6, Academic Press, London.
Jones, J. G. (1978). *Freshwater Biol.* **8,** 127–140.
Jones, J. G. (1979a). *J. Gen. Microbiol.* **115,** 19–26.
Jones, J. G. (1979b). *J. Gen. Microbiol.* **115,** 27–35.
Jones, J. G. (1979c). Scientific Publication No. 39, Freshwater Biological Association, Windermere, Cumbria, UK, Titus Wilson & Son, Kendal.
Jones, J. G. (1981). *J. Gen. Microbiol.* **125,** 85–93.
Jones, J. G. (1982). *In* "Sediment Microbiology" (D. B. Nedwell and C. M. Brown, Eds), Special Publications of the Society for General Microbiology Vol. 7, pp. 107–146, Academic Press, London.
Jones, J. G. (1985). *Phil. Trans. Roy Soc. (Series A)* **315,** 3–17.
Jones, J. G., Downes, M. T. and Talling, I. B. (1980). *Freshwater Biol.* **10,** 341–359.
Jones, J. G. and Jones, H. E. (1986). *In* "Perspectives in Microbial Ecology" (F. Megusar and M. Ganter, Eds), pp. 375–382, Proc. 4th Int. Symp. Microb. Ecol., Slovene Society for Microbiology, Ljubljana.
Jones, J. G. and Simon, B. M. (1975). *J. Appl. Bacteriol.* **39,** 317–329.
Jones, J. G. and Simon, B. M. (1979). *J. Appl. Bacteriol.* **46,** 305–315.
Jones, J. G. and Simon, B. M. (1980a). *J. Ecol.* **68,** 493–512.

Jones, J. G. and Simon, B. M. (1980b), *J. Appl. Bacteriol.* **49,** 127–135.
Jones, J. G. and Simon, B. M. (1981). *J. Gen. Microbiol.* **123,** 297–312.
Jones, J. G. and Simon, B. M. (1984). *J. Microbiol. Methods* **3,** 47–55.
Jones, J. G. and Simon, B. M. (1985). *Appl. Environ. Microbiol.* **49,** 944–948.
Jones, J. G., Simon, B. M. and Cunningham, C. R. (1983). *J. Appl. Bacteriol.* **54,** 355–365.
Jones, J. G., Simon, B. M. and Gardener, S. (1982). *J. Gen. Microbiol.* **128,** 1–11.
Jones, R. I. and Salonen, K. (1985). *Holarctic Ecol.* **8,** 133–140.
Jorgensen, B. B. (1978). *Geomicrobiol. J.* **1,** 11–27.
Kasper, H. F. (1985). *In* "Proceedings of International Workshop on Biological Research with Short-lived Isotopes. Lower Hutt, New Zealand" (P. E. H. Minchin, Ed.), pp. 101–104, DSIR Science Information Publishing Centre, Wellington, NZ.
Kelly, C. A., Rudd, J. W. M. and Schindler, D. W. (1988). *Arch. Hydrobiol. Beih. Ergebn. Limnol.* **31,** 333–344.
Kiel, H. and Williams, P. A. (1985). *J. Gen. Microbiol.* **131,** 1023–1033.
King, L. A. and Parker, B. C. (1988). *Appl. Environ. Microbiol.* **54,** 1630–1631.
Kogure, K., Simudu, U. and Taga, N. (1979). *Can. J. Microbiol.* **25,** 415–420.
Kogure, K., Simudu, U. and Taga, N. (1984). *Arch. Hydrobiol.* **102,** 117–122.
Komozata, K. and Suzuki, K-I. (1987). *In* "Methods in Microbiology" (R. R. Colwell and R. Grigorova, Eds), Vol. 19, pp. 161–208, Academic Press, London.
Kunicka-Goldfinger, W. (1973). *In* "Modern Methods in the Study of Microbial Ecology." (T. Rosswall, Ed), pp. 311–316, Bull. Ecol. Res. Comm. 17, NFR, Stockholm.
Ladd, T. I., Ventullo, R. M., Wallis, P. M. and Costerton, J. W. (1982). *Appl. Environ. Microbiol.* **44,** 321–329.
Lewis, D. L., Hodson, R. E. and Hwang, H. (1988). *Appl. Environ. Microbiol.* **54,** 2054–2057.
Lock, M. A. and Ford, T. E. (1983). *Appl. Environ. Microbiol.* **46,** 463–467.
Lock, M. A. and Ford, T. E. (1985). *Appl. Environ. Microbiol.* **49,** 408–412.
Maiden, M. F. J. and Jones, J. G. (1984). *J. Gen. Microbiol.* **130,** 2943–2959.
Maki, J. S. and Remsen, C. C. (1981). *Appl. Environ. Microbiol.* **41,** 1132–1138.
Maniatis, T., Fritsch, E. F. and Sambrook, J. (1982). "Molecular Cloning: A Laboratory Manual", Cold Spring Harbor Laboratory.
Mara, D. D. (1972). *Water* **6,** 1605–1607.
Marcussen, B., Nielsen, P. and Jeppesen, M. (1984). *Arch. Hydrobiol. Beih. Ergelm. Limnol.* **19,** 141–149.
Mariotti, A. (1984). *Nature* **311,** 251–252.
Meynell, G. G. and Meynell, E. (1970). "Theory and Practice in Experimental Bacteriology", Cambridge University Press, Cambridge.
MacDonald, R. M. (1986a). *Soil Biol. Biochem.* **18,** 399–406.
MacDonald, R. M. (1986b). *Soil Biol. Biochem.* **18,** 407–410.
McCallum, G. J., McNaughton, G. S., Minchin, P. E. H., More, R. D., Presland, M. R. and Stout, J. D. (1981). *Nucl. Sci. Applic.* **1,** 163–190.
McNaughton, G. S. (1986). *In* "Short-lived Isotopes in Biology. Proceedings of International Workshop on Biological Research with Short-lived Isotopes. Lower Hutt, New Zealand 1985" (P. F. H. Minchin, Ed), pp. 37–40, DSIR Science Information Publishing Centre, Wellington, NZ.
Mortimer, C. H. (1941). *J. Ecol.* **29,** 280–329.

O'Carroll, K. (1988). *In* "Methods in Aquatic Bacteriology" (B. Austin, Ed.), Modern Microbiological Methods Vol. 1, pp. 347–366, J. Wiley & Sons, Chichester.
Ogram, A., Sayler, G. S. and Barkay, T. (1987). *J. Microbiol. Methods* **7,** 57–66.
Ohle, W. (1972). *In* "Techniques for the Assessment of Microbial Production and Decomposition in Freshwaters" (Y. I. Sorokin and H. Kadota, Eds), pp. 27–28, IBP Handbook No. 23, Blackwell Scientific, London.
Oremland, R. S. and Capone, D. G. (1988). *Adv. Microbiol. Ecol.* **10,** 285–383.
Overbeck, J. (1974). *Int. Verein. Theor. Ang. Limnol.* **20,** 198–228.
Pace, N. R., Stahl, D. A., Lane, D. J. and Olsen, G. J. (1986). *Microbial Ecol.* **9,** 1–56.
Parsons, T. R. and Strickland, J. D. H. (1962). *Deep Sea Res.* **8,** 211–222.
Peele, E. R. and Colwell, R. R. (1981). *Can. J. Microbiol.* **27,** 1071–1075.
Pettigrew, C. A. and Sayler, G. S. (1986). *J. Microbiol. Methods* **5,** 205–213.
Pickup, R. W. and Williams, P. A. (1982). *J. Gen. Microbiol.* **128,** 1385–1390.
Pickup, R. W., Lewis, R. J. and Williams, P. A. (1983). *J. Gen. Microbiol.* **129,** 153–158.
Pike, E. B., Carrington, E. G. and Ashburner, P. A. (1972). *J. Appl. Bacteriol.* **35,** 309–321.
Porter, K. G. (1988). *Ecology* **69,** 558–560.
Postgate, J. R. (1976). *In* "The Survival of Vegetative Microbes" (T. R. G. Gray and J. R. Postgate, Eds), pp. 1–19, Cambridge University Press, Cambridge.
Preston, T. and Owens, N. J. P. (1983). *Analyst* **108,** 971–977.
Priscu, J. H., Spigel, R. H., Gibbs, M. M. and Downes, M. J. (1986). *Limnol. Oceanogr.* **31,** 812–831.
Reichardt, W. (1988). *Arch. Hydrobiol. Beih. Ergebn. Limnol.* **31,** 353–363.
Riemann, B. and Schrierup, H. H. (1978). *Water Res.* **12,** 849–853.
Richards, F. A. (1965). *In* "Chemical Oceanography" (J. P. Riley and G. Skirrow, Eds), Vol. 1, pp. 611–645, Academic Press, London.
Robinson, J. A. and Characklis, W. G. (1984). *Microbial Ecol.* **10,** 165–178.
Roszak, D. B. and Colwell, R. R. (1987a). *Microbiol. Rev.* **51,** 365–379.
Roszak, D. B. and Colwell, R. R. (1987b). *Appl. Environ. Microbiol.* **53,** 2889–2893.
Sayler, G. S., Shields, M. S., Tedford, E. T., Breen, A., Hooper, S. W., Sivotkin, K. M. and Davis, J. W. (1985). *Appl. Env. Microbiol.* **49,** 1295–1503.
Schmaljohann, R., Pollingher, U. and Berman, T. (1987). *Microbiol. Ecol.* **13,** 1–12.
Schneider, J. and Rheinheimer, G. (1988). *In* "Methods in Aquatic Bacteriology" (B. Austin, Ed), Modern Microbiological Methods Vol. 1, pp. 73–94, J. Wiley & Sons, Chichester.
Selmar, J. S. and Sorensson, F. (1986). *Appl. Environ. Microbiol.* **52,** 577–579.
Sepers, A. B. J. (1981). *Archiv. Hydrobiol.* **92,** 114–129.
Simek, K. (1986). *Int. Rev. Gesamten. Hydrobiol.* **71,** 593–612.
Skerman, U. B. D., Sly, L. I. and Williamson, M. L. (1983). *Int. J. Syst. Bact.* **33,** 300–308.
Smith, R. L. and Oremland, R. S. (1987). *Limnol. Oceanogr.* **32,** 794–803.
Sorensen, J. (1978). *Appl. Environ. Microbiol.* **36,** 139–143.
Spencer, M. J. (1979). *FEMS Microbiol. Lett.* **5,** 343–347.
Stanley, P. M. and Staley, J. T. (1977). *Limnol. Oceanogr.* **22,** 26–37.

Stanley, P. M., Gage, M. A. and Schmidt, E. L. (1979). *In* "Native Aquatic Bacteria, Enumeration, Activity and Ecology" (J. W. Costerton and R. R. Colwell, Eds), pp. 46–55, American Society for Testing and Materials. Special Technical Publications 695, ASTM, Philadelphia.

Straskrabova, V. (1972). *In* "Techniques for the Assessment of Microbial Production and Decomposition in Fresh Waters" (Y. J. Sorokin and H. Kadota, Eds), IBP Handbook No. 23, p. 77, Blackwell Scientific Publications, Oxford.

Straskrabova, V. and Simek, K. (1984). *Arch. Hydrobiol. Beih, Ergeben. Limnol.* **19,** 1–6.

Straskrabova, V. and Simek, K. (1988). *Arch. Hydrobiol. Beih. Ergeben. Limnol.* **31,** 55–60.

Stewart, W. D. P., Preston, T., Peterson, G. H. and Christofi, N. (1982). *Phil. Trans. R. Soc. Ser. B* **296,** 491–509.

Sweerts, J. P., Rudd, J. W. M. and Kelly, C. A. (1986). *Limnol. Oceanogr.* **31,** 330–338.

Tabor, P. S. and Neihof, R. A. (1982). *Appl. Environ. Microbiol.* **44,** 945–953.

Talling, J. F. (1973). *Freshwater Biol.* **3,** 335–362.

Tiedje, J. M., Firestone, R. B., Firestone, M. K., Betlach, M. R., Kaspar, H. F. and Sorensen, J. (1981). *In* "Advances in Chemistry Series 197. Short-lived Radio-nuclides in Chemistry and Biology" (J. W. Root and K. A. Krohn, Eds), pp. 295–315, American Chemical Society (Pub.).

Van Es, F. B. and Meyer-Reil, L. H. (1982). *In* "Advances in Microbial Ecology" (K. C. Marshall, Ed.), Vol. 6, pp. 111–170, Plenum, New York.

Ward, B. B. (1984). *Limnol. Oceanogr.* **29,** 402–410.

Ward, B. B. (1986). *In* "Nitrification" (J. I. Prosser, Ed.), Vol. 20, pp. 157–184, Special Publications of the Society for General Microbiology, IRL Press, Oxford.

Wardell, J. N. (1988). *In* "Methods in Aquatic Bacteriology" (B. Austin, Ed.), Modern Microbiol. Methods, Wiley Interscience.

Watanabe, Y. (1984). *Jap. J. Limnol.* **45,** 116–125.

Wheatcroft, R. and Williams, P. A. (1981). *J. Gen. Microbiol.* **124,** 433–437.

Wiebe, W. J. and Smith, DlF. (1977). *Microbiol. Ecol.* **4,** 1–8.

Wright, R. T. (1975). *Limnol. Oceanogr.* **20,** 626–633.

Wright, R. T. and Burnison, K. (1979). *In* "Native Aquatic Bacteria: Enumeration, Activity and Ecology" (J. W. Costerton and R. R. Colwell, Eds), pp. 140–155, American Society for Testing and Materials, Special Technical Publications 695, ASTM, Philadelphia.

Wright, R. T. and Hobbie, J. E. (1966). *Ecology* **47,** 447–464.

Xu, H. S., Roberts, N. C., Adam, L. B., West, P. A. Siebelin, R. J., Huq, A., Huq, M. I., Rahman, R. and Colwell, R. R. (1984). *J. Microbiol. Methods* **2,** 221–231.

Zimmerman, R., Itturriaga, R. and Becker-Birck, J. (1978). *Appl. Environ. Microbiol.* **36,** 926–935.

# 6

# Techniques for Estimating Bacterial Growth Rates and Production of Biomass in Aquatic Environments

D. J. W. MORIARTY

*CSIRO Marine Laboratories, P.O. Box 120, Cleveland, Queensland 4163, Australia*



## I. Introduction

For quantitative studies on the role of heterotrophic bacteria in the biogeochemical cycling of major elements, especially C and N, we need to determine values for the growth rates or productivity of the bacteria. Such values are also necessary for analysis of food webs involving decomposition of detritus. The methods described here are for the synecologist; they give average or composite values for all the heterotrophic bacteria. Methods have not yet been developed to determine the growth rates of all or most

METHODS IN MICROBIOLOGY
VOLUME 22 ISBN 0-12-521522-3

genera or species of bacteria, individually, in a particular environment. Some advances have been made at this autecological level, but will not be discussed here (e.g. see Brock, 1971).

At present, the most useful method for estimating growth rates of heterotrophic bacteria in aquatic environments is the measurement of rates of DNA synthesis with tritiated thymidine. Adenine has also been proposed as a suitable precursor, but because it and ATP are involved in so many different biochemical processes and in all organisms, results are too difficult to interpret or may be two or three orders of magnitude too high (Moriarty, 1986).

Bacterial productivity may also be estimated from rates of phosphate incorporation into phospholipid and leucine incorporation into protein; these methods will be outlined below.

## II. DNA synthesis

The rationale, advantages and sources of error in the use of thymidine for ecological studies have been reviewed by Moriarty (1986). The main points are covered briefly here. Thymidine has been selected as a precursor because it is used almost exclusively for DNA synthesis, it is not used for the biosynthesis of other macromolecules without first being degraded. Labelling of DNA by tritiated thymidine is specific to heterotrophic bacteria.

Once replication of DNA has been initiated, the synthesis usually proceeds at a fixed rate until it is complete. Thus it is possible to carry out short-term experiments (usually 10–15 minutes in warm conditions) with sediment or water without affecting rates of DNA synthesis. But changing the environment of bacteria will bring about changes in the rates of initiation of DNA synthesis and this will alter rates of thymidine incorporation after a period of time. It is necessary, therefore, to carry out experiments in the field as soon as samples are collected.

Criticisms of the thymidine method have been made, based on the premise that because compounds other than DNA are labelled the method is not sound. The criticisms are not valid, however, if a pulse-label technique is used, because only DNA is labelled significantly (Moriarty, 1986). Reports by some authors that RNA is labelled by [methyl-$^{3}$H]thymidine can be explained by inappropriate extraction techniques (Munro and Fleck, 1966; Roodyn and Mandel, 1960; Moriarty, 1986). It may also be due to binding of thymidine to lipid, which is extracted in the same fraction as RNA (Robarts *et al.*, 1986). The most likely sources of error, and therefore criticism in the literature, are the incubation period, extraction technique and isotope dilution. These topics are discussed below.

## A. Biochemistry of thymidine incorporation into DNA

Thymidine is readily incorporated into DNA via a salvage pathway, but in some bacteria the incorporation stops after a short time due to breakdown of thymidine (O'Donovan and Neuhard, 1970). *De novo* synthesis proceeds via dUMP directly to dTMP (Fig. 1). Catabolism of thymidine starts with conversion to thymine and ribose-1-phosphate by the action of an inducible phosphorylase. The best radioactive label is [methyl-$^3$H] because subsequent conversion to uracil removes the label. The tritiated methyl group can be transferred to a wide variety of compounds, but is considerably diluted. Longer incubation times than are needed for DNA labelling would be needed for significant labelling of other macromolecules.

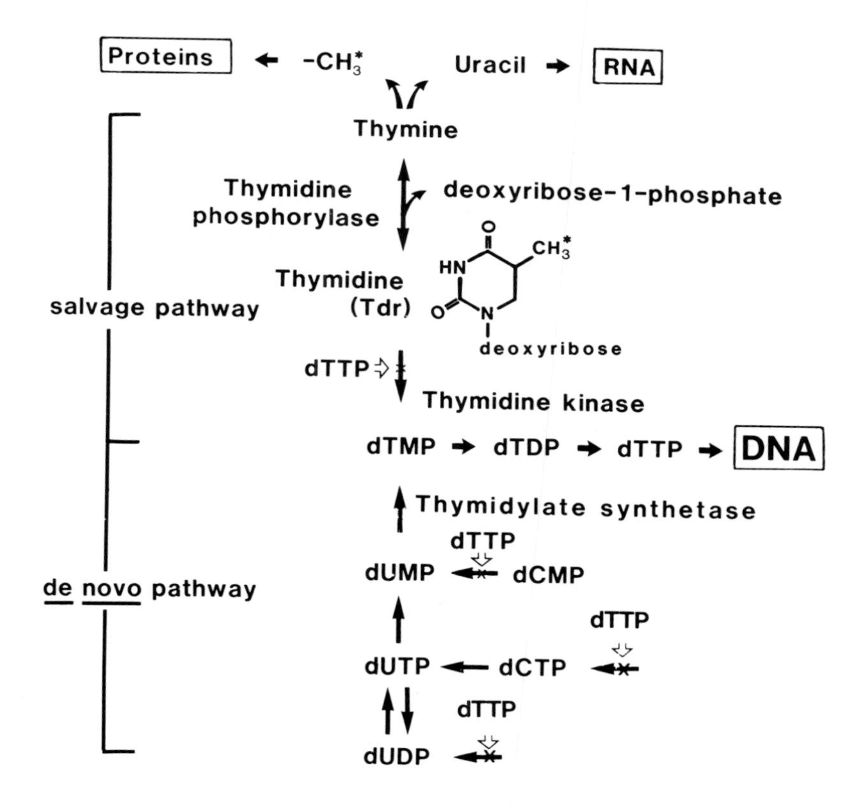


**Fig. 1.** Some pathways of thymidine metabolism. The asterisk shows the position of tritium labelling in thymidine. Sites of feedback inhibition of dTTP are indicated. (Adapted from Kornberg, 1980.)

Thymidine is converted to dTMP by thymidine kinase (Fig. 1). This enzyme must be present for labelling of DNA to occur to a significant extent. Some groups of microorganisms are now known not to contain thymidine kinase. These include many fungi, microalgae and cyanobacteria (Glaser *et al.*, 1973; Grivell and Jackson, 1968; Pollard and Moriarty, 1984). It is absent from the nuclei of various eukaryotic algae, but may be present in chloroplasts; the amount of label incorporated from tritiated thymidine into DNA in algae was slight and required much higher concentrations and much longer incubation periods than was required for bacteria, and thus does not interfere with the technique described here (Rivkin, 1986; Sagan 1965).

The lack of thymidine kinase in blue-green algae and many eukaryotic microorganisms is a considerable advantage for studies on heterotrophic bacterial production in the marine environment. Protozoa probably do contain the enzyme (Plant and Sagan, 1958; Stone and Prescott, 1964), but as explained below their contribution to labelled DNA in short-term experiments is probably small.

Although most heterotrophic bacteria are able to take up thymidine, not all can do so, probably due to a deficient cell membrane transport system (Pollard and Moriarty, 1984; Ramsay, 1974). For very detailed or critical studies it would be necessary to obtain an estimate of the proportion of bacteria that do take up thymidine in the particular environment being studied. Fuhrman and Azam (1980) have found good agreement between bacterial growth rates in sea water measured by the incorporation of thymidine and by counting the increase in cell numbers. The results of an autoradiographic study on bacteria in sea water support the view that most marine heterotrophic bacteria can utilize thymidine (Fuhrman and Azam, 1982). Furthermore, Riemann *et al.* (1987), in an extensive study with many water samples, found that the average factor for converting rates of thymidine incorporation to rates of bacterial division was within the theoretical range (see below). In other words the thymidine method gave a good estimate of bacterial growth and thus the growth of most or all bacteria was being measured.

Autotrophic and anaerobic bacteria with limited nutrient requirements may not be able to utilize thymidine, particularly if they can transport only a limited range of metabolites. Sulphate-reducing bacteria, for example, do not appear to be able to utilize exogenous thymidine. As the productivity of sulphate-reducing bacteria is inefficient, and therefore low compared to fermenters, the overall productivity of anaerobic bacteria would not be seriously underestimated in sediment (Moriarty, 1986). The growth of most fermentative anaerobes is measured with the thymidine technique (Pollard and Moriarty, 1984).

## B. Kinetics and incubation period

Bacteria, with their active transport systems, take up organic molecules much more rapidly than do algae or protozoans, and can utilize nanomolar concentrations of organic molecules in their environment more effectively (Wright and Hobbie, 1966; Fuhrman and Azam, 1980). Thus in a short time period (e.g. 10 min at 25°C, 20 min at 15°C) tritiated thymidine is taken up mostly by bacteria in a mixed community.

A long incubation time is the primary reason for radioactivity appearing in other macromolecules. Initially, DNA is the only macromolecule that is labelled at high specific activity, because thymidine is not degraded before being incorporated (Moriarty, 1986). As the incubation progresses, other macromolecules become labelled when the tritium, released from degraded thymidine, is cycled within the cell. The actual time period over which this occurs varies with conditions such as temperature, growth rates of bacteria and species composition.

The tritiated thymidine method is a pulse-labelling method; results and conclusions will be incorrect if the time period is too long. Unfortunately, a number of papers have been published by research workers who have been unaware of this.

Uptake of labelled thymidine by organisms should not be confused with incorporation into DNA (Moriarty, 1986). The thymidine incorporated into DNA might be only a small proportion of the total taken up. The period of time over which incorporation of [methyl-$^{3}$H]thymidine into DNA remains linear will depend on the concentration of thymidine, activity of thymidine phosphorylase, adsorption to sediment and bacterial growth rates. Experimental studies to determine growth rates must be carried out in the initial linear period of incorporation of label into DNA. Uptake of thymidine by cells and incorporation into other TCA-insoluble fractions are different processes, probably with different kinetics, and may be uninterpretable in a mixed population.

It is necessary to select an incubation period long enough to give an incorporation of tritium into DNA that is well above background adsorption; preferably at least ten-fold higher. Errors due to variable backgrounds are thus minimized.

A time course should be the first experiment carried out as it is a good check on technique. If values are widely scattered in the first 15 min or so, problems in sampling, pipetting or washing may be the cause. The incubation period should be short enough to ensure that DNA is the main macromolecule that is labelled. Degradation of tritiated thymidine in the cell will eventually lead to labelling of all cellular components, but this occurs more slowly than DNA synthesis in growing bacteria. Where bacteria are not growing, degradation could be very marked. Thus a time course experiment to determine the rate of labelling of DNA itself is needed. For tropical water bodies, 15 min is generally a suitable time for incubation of water samples, and 10 min for sediment. For very cold environments, where generation times are long, 2 h or more may be needed.

### C. Extraction technique

A simple procedure is used in ecological work to distinguish labelled DNA from other macromolecules (especially RNA and protein), but the results are not necessarily accurate. Material that is soluble in, but not hydrolysed by, dilute NaOH and is hydrolysed by hot dilute acid is considered to be DNA. Most proteins are not hydrolysed by hot dilute acid; RNA and some proteins are hydrolysed by dilute alkali (Roodyn and Mandel, 1960).

In order to be rigorous, however, in deciding what proportion of labelled macromolecules are DNA, more refined methods are needed. Some DNA is bound to membranes and protein in the cell, and may appear as protein in the simple extraction procedure. If the extraction in NaOH is too harsh (e.g. high temperature or concentration), protein and DNA will be degraded to some extent and may appear as RNA (Munro and Fleck, 1966; Roodyn and Mandel, 1960). Furthermore, DNA is slowly hydrolysed or fragmented in cold acid, and can be lost in supernatants after centrifuging.

It is not possible (at least not without throwing biochemists into confusion by inventing a new pathway for pyrimidine nucleotide biosynthesis) for RNA to be labelled rapidly at high specific activity by [methyl-$^3$H]thymidine. There are reports that RNA is labelled, but in fact it is likely that 'RNA' was protein and perhaps also DNA that was degraded by NaOH. To prove that RNA was labelled, degradation by pure RNase or chromatography of hydrolysed products showing label in ribonucleosides would be needed. Robarts *et al.* (1986) found that the labelled 'RNA' fraction in their samples was in fact thymidine bound to a lipid.

With the simple extraction procedures, Moriarty (1986), Moriarty *et al.* (1985b) and Pollard and Moriarty (1984) showed that all the label was found in DNA for the first 10–20 min of incubation with [methyl-$^3$H]thymidine. This conclusion has since been confirmed for seagrass sediments by studies with catabolic enzymes specific for DNA. Over 86% of the labelled macromolecules was DNA, after a 15 minute incubation (Moriarty and Pollard, 1990).

## D. Isotope dilution

The specific radioactivity of exogenous thymidine is diluted during incorporation into DNA, primarily by *de novo* synthesis of dTMP (Fig. 1). A technique for measuring the dilution of label from an exogenous precursor during synthesis of a macromolecule is to add different quantities of unlabelled precursor as well, and to measure the effect on the amount of label actually incorporated into the macromolecule (Forsdyke, 1968). This technique worked well with bacterial populations in sediments (Moriarty and Pollard, 1981). A plot of the reciprocal of isotope incorporated into DNA against total amount of thymidine added is extrapolated to give the amount of dilution of isotope in DNA itself (Fig. 2). The negative intercept on the abscissa is used to determine the specific activity of tritiated thymidine actually incorporated into DNA.

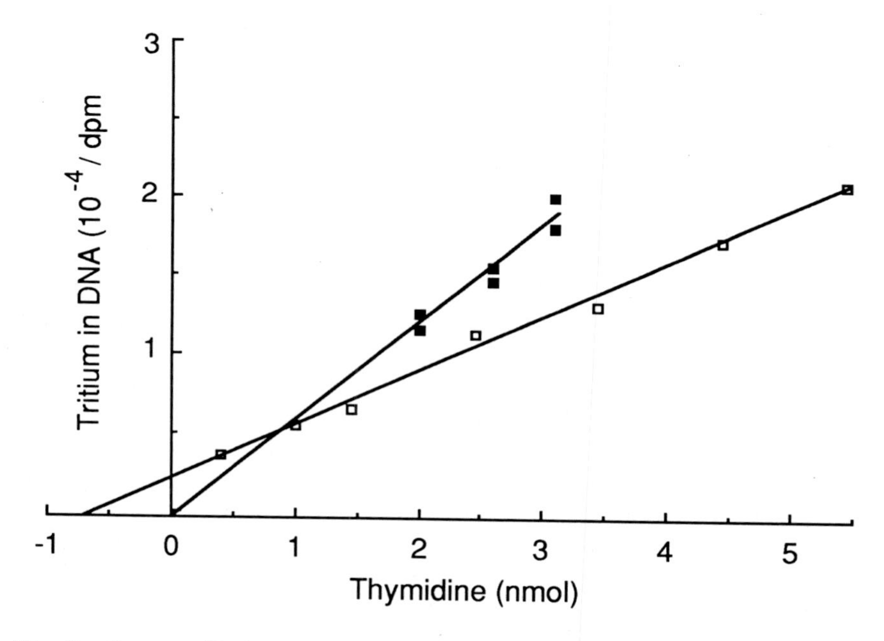


**Fig. 2.** Isotope dilution plots for incorporation of tritiated thymidine into DNA showing dilution (□) and no dilution (■). (Redrawn from Moriarty and Pollard, 1981.)

This technique measures the dilution of labelled thymidine in dTTP (the final precursor for DNA synthesis) by all other precursors of dTTP, including *de novo* synthesis. The dilution of isotope in thymidine pools that are actually being used for DNA synthesis in growing cells is all that is measured; this pool may be only a small proportion of the thymidine that is taken up by the bacteria. Thus this method is not subject to errors inherent in trying to extract and quantify nucleotides from cells.

The need to do isotope dilution experiments on every sample can generally be avoided by using a high enough concentration of thymidine to supply all the TTP required for DNA synthesis (usually 20 nM in water or 1 nmol in 100 μl of sediment). Isotope dilution is then eliminated or at least minimized because *de novo* synthesis is switched off by feedback inhibition (Moriarty, 1986; Pollard and Moriarty, 1984).

If the rate-limiting step for incorporation of tritiated thymidine into DNA is the activity of DNA polymerase, then isotope dilution can be prevented. If, however, the rate-limiting step is prior to TMP synthesis, e.g. uptake into the cell, isotope dilution will occur, but will not be detectable. At present, it seems that there are not many heterotrophic bacteria in which uptake is limiting. Experiments similar to those of Riemann *et al.* (1987) may be used to check the conversion factor, and thus indirectly to establish whether isotope dilution is likely to be a problem. High conversion factors found in earlier studies could be due to experimental technique, particularly the use of too low a thymidine concentration.

## E. Disturbance effects

It is necessary to mix radioactive compounds into sediment in order to ensure that all bacteria receive a high enough concentration to minimize isotope dilution. The incorporation of thymidine into DNA is not affected by disturbance as quickly as phospholipid or protein synthesis or uptake of organic nutrients or oxygen. Because DNA synthesis depends on many interacting processes, e.g. energy generation and protein synthesis, its regulation is complex, and not influenced by the concentration of added precursors such as thymidine (Moriarty, 1986). Thus one can disturb sediments quite considerably without immediately changing rates of DNA synthesis (Fig. 3, Moriarty *et al.*, 1985a).

At least, surface sediments can be disturbed, but there may be problems with deep, old sediments perhaps because many bacteria are dormant. In this case the long incubation time needed to detect thymidine incorporation into DNA may be long enough to stimulate resting cells to grow.

The usual effect of disturbance is a decrease in rates of DNA synthesis, especially with anoxic sediments (Fig. 3). With the methods described below, artifacts due to disturbance are absent or minimized.

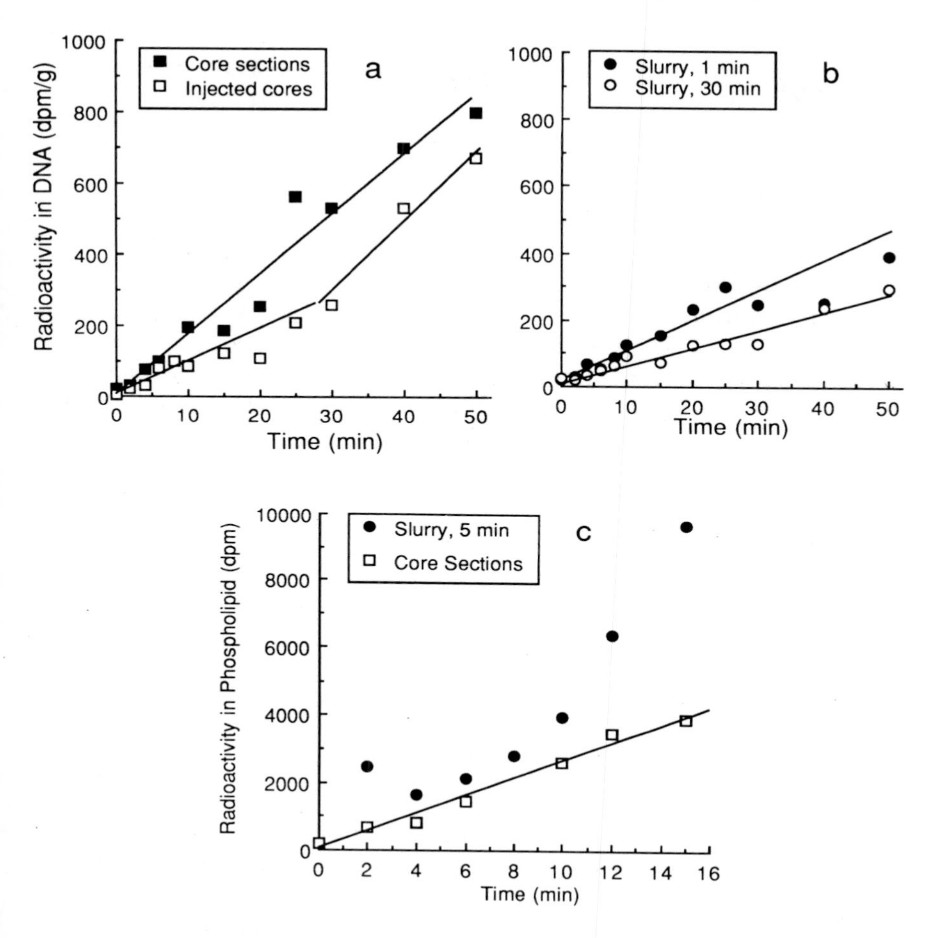


**Fig. 3.** Time course of incorporation of tritiated thymidine into DNA (a, b) and $^{32}$P-phosphate into phospholipid (c) of bacteria in seagrass bed sediments. Different techniques for dispensing sediment were used. The core sections with tritiated thymidine and both treatments with $^{32}$P-phosphate were incubated aerobically; the other treatments with thymidine were incubated anaerobically. (From Moriarty and Pollard, 1990.)

## III. Phospholipid synthesis

Phospholipids are the major lipids in bacteria and their rate of synthesis is correlated with the growth rate of bacteria (Moriarty *et al.*, 1985a; White *et al.*, 1979). The principle of the isotope dilution technique described above for thymidine may be applied to the determination of phosphate pool sizes. As with the thymidine method, it is assumed that the rate-limiting step for phosphate incorporation into phospholipid is the final step and not an intermediate one such as uptake into the cell. Rates of equilibration of phosphate pools inside and outside the cells should not be slower than the rates of incorporation. No evidence for slow (i.e. more than 1 minute) rates of equilibration have been observed for sediment samples (e.g. see Moriarty *et al.*, 1985a).

Time course experiments are needed to check that the selected incubation period is within the period over which rates of phosphate incorporation are linear. This period is shorter than that for thymidine incorporation for a given population, because rates of phospholipid incorporation are influenced more quickly by disturbance caused by the sampling procedure. See Moriarty *et al.* (1985a) for details.

## IV. Protein synthesis

In studies of detrital decomposition, the production of microbial biomass can be estimated from protein production (Simon and Azam, 1989; Kirchman *et al.*, 1986). If growth is not balanced, rates of protein synthesis will not be equivalent to rates of cell division. For example, under starvation conditions bacteria may still synthesize DNA and divide, but they become smaller and thus protein content declines (Kjelleberg *et al.*, 1982). Alternatively, where nutrients are plentiful, the bacteria increase in size and thus rates of protein synthesis will be faster than DNA synthesis. In general, however, there does seem to be a good correlation between rates of bacterial growth and protein synthesis in the sea (Simon and Azam, 1989).

Direct measurement of rates of bacterial protein synthesis would be preferable for studies of trophic dynamics in detrital systems. Protein is a major component of bacterial cell biomass and is well correlated to cell volume (Simon and Azam, 1989). Protein synthesis may be determined in the laboratory from rates of radioactive leucine incorporation. The laboratory technique has been adapted for determining rates of protein synthesis of bacteria in aquatic systems (Kirchman *et al.*, 1985; Simon and Azam, 1989).

There are problems in determining the specific radioactivity of tritiated leucine within the bacteria, and thus obtaining accurate rates of synthesis. Kirchman *et al.* (1986) used the isotope dilution procedure discussed above for thymidine. This technique gives results for productivity that agree well with those obtained with the thymidine and phospholipid methods (Moriarty, unpublished data). Simon and Azam (1989) determined dilution directly (by high-performance liquid chromatography) on extracted pools of leucine from bacteria; there is an assumption that there is only one pool of leucine in bacteria or that all pools reach isotopic equilibrium rapidly and that all extracted leucine is from growing bacteria. The latter method will not give accurate results where bacteria cannot be separated from algae and animals. Simon and Azam (1989) filtered sea water through a 0.6-μm pore size filter to remove large organisms.

Kirchman and Hodson (1984) have discussed the relationships between protein synthesis and the uptake and utilization of dissolved organic carbon compounds. The turnover of organic carbon in aquatic systems is not necessarily directly correlated with rates of bacterial growth. For a fuller understanding of the carbon cycle and trophic dynamics, studies need to be carried out on processes such as protein and lipid synthesis, bacterial respiration and bacterial growth rates.

## V. Detailed procedures

### A. Sampling and dispensing sediments

The simplest, practical procedure that works for most sediments is to drop small sections of cores (1–2 mm depth, 8 mm diameter, made from disposable polypropylene 2-ml syringes) with a spatula into tubes containing about 50 μl of filtered water, then add the radioactive precursor [see note (iv) re tubes]. The drop of water aids in removing sediment from the tip of the spatula. For studies of depth profiles, 2 mm depth samples, taken every 1 or 2 cm down the sediment profile, may be incubated with the radioactive compounds and the resulting values for activity integrated over the whole profile. A device for sampling sediments is described in Fig. 4. Where replicate samples are needed from one core, a larger core, e.g. 30 ml, may be used and 2-mm sections pooled into 2-ml corers for dispensing into tubes.

Where sediments are not consolidated enough for cores to work, such as in aquaculture ponds with much particulate organic matter mixed with clay, a slurry is satisfactory. Make a slurry of the top 2–10 mm with 10 ml of pond water. Take 100 μl (0.1 ml) for each sample. Repeat for deeper sediment depths if desired.

Subsamples should be kept with preservative (0.5% formaldehyde) for direct counts by epifluorescence microscopy. Where the sediment is entirely carbonate, such as coral reef sand, the same sample that is used for the radioactive incorporation can be readily subsampled for direct counts.

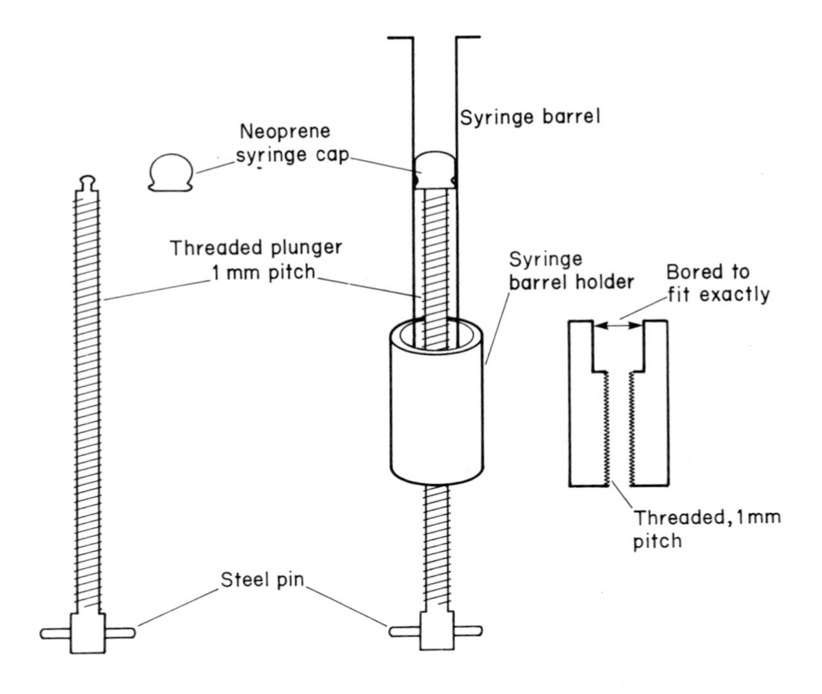


**Fig. 4.** A device for extruding measured sections of sediment cores. It is made from PVC. The threaded screw is shaped similarly to the plunger of a syringe and is fitted with a neoprene cap from a syringe plunger.

Thus in coral sands the inherent variability in results for specific growth rates from sediment is considerably lessened, because heterogeneity due to different bacterial populations on different sand grains is removed.

## B. Soil

The thymidine method could be applicable to soils. I have not tried to work with soils, but from the discussion above on disturbance effects, I see no reason why a soil slurry should not work. Sufficient distilled water should be added to make a slurry, which may be dispensed with a pipette.

## C. DNA synthesis

Tritiated thymidine should not be kept for longer than about 2 months; store as aqueous solution (sterile) with 2% ethanol (decomposition rate: 4% per month). Ethanol helps prevent the decomposition of thymidine that is caused by free radicals generated during radioactive decay. The ethanol does not need to be removed prior to the experiment, except perhaps in oligotrophic environments.

Shortly before experiments are undertaken, withdraw the total amount needed with a sterile syringe.

Two methods may be used to determine radioactivity in DNA in sediment. In the filter method, radioactivity in small molecules is removed by washing with alcohol and dilute, cold acid and then the DNA is hydrolysed and separated from the sediment and other macromolecules with hot acid. In the dialysis method, DNA is removed from the sediment as an alkaline solution and small molecules are removed by dialysis. Blank values are usually lower and more consistent with the latter method. For large numbers of samples, operator contact time is less with the dialysis method.

### *1. Procedures for water column*

(a) Collect water samples; dispense into bottles [see note (iv)]. Suggested volumes: 5 ml for aquaculture ponds, 20 ml for lake or coastal sea water, 50 ml for oligotrophic sea water.

(b) Add 1 $\mu$Ci ml$^{-1}$ sample volume of [methyl-$^{3}$H]thymidine (about 25 Ci mmol$^{-1}$) at known time intervals [see note (i)].

(c) Incubate at *in situ* temperature [see note (ii)]. Include blanks [see note (iii)].

(d) Stop reaction by adding 0.2 ml 37% formaldehyde containing 10 mg thymidine per 100 ml and buffered at about pH 8 with sodium tetraborate or bicarbonate. Store cold until water can be filtered.

(e) Filter through polycarbonate membrane filters (0.2–0.45 $\mu$m pore size). Wet the filters first to lessen capillary transport of radioactivity under the edge seals.

(f) As soon as filtration is complete, add 5 ml of 80% ethanol and draw it through. Repeat once, then wash with 1–2 ml of ice-cold 5% trichloroacetic acid (TCA) through the filter. Repeat cold TCA wash four times.

(g) Remove filter, place in tube [see note (iv)] and store if necessary.

(h) Add 2 ml 5% TCA (or 0.5 M HCl), cap the tubes, heat at 100°C for 30 min, and then cool and centrifuge if particulate matter is noticeable.

(i) Take 0.5 ml, place it in a mini-vial and add 4 ml of water-miscible scintillation fluid and count in a liquid scintillation counter.

### *2. Procedures for sediment: filter method*

(a) Collect and dispense sediment (see page 221 above). Cores should not be stored; the experiments should be carried out as soon as possible after collection. Prepare blanks [see note (iii)].

(b) Add tritiated thymidine [see note (v)]. Use [methyl-$^3$H]thymidine with a specific activity of about 10–25 Ci mmol$^{-1}$.

(c) Incubate at *in situ* temperature for predetermined time [see note (vi)]. Stop reaction by adding 10 ml ethanol : water : thymidine (80 ml : 20 ml : 10 mg); methylated spirits are a cheap and satisfactory alternative to ethanol. Store in a refrigerator or on ice for a few days if necessary. Storing in alcohol for at least one night improves extraction of excess tritium.

(d) After storage, remove the ethanol by centrifuging and aspirating off the fluid. Add 10 ml alcohol mixture and repeat centrifuging. Add 2 ml alcohol mixture, mix and transfer to a polycarbonate filter of 0.4–0.6 μm pore size using a syringe and needle to wash down remaining sediment with alcohol.

(e) If sediments contain carbonate proceed as follows: remove alcohol. Add 10 ml 20% v/v acetic acid and leave to stand in a refrigerator until effervescence stops. If carbonates remain, centrifuge, discard supernatant and repeat acid treatment. Subsample for direct counts. Pour sediment slurry onto filter.

(f) Wash filter with 2 ml 5% ice-cold TCA; repeat four times.

(g) Transfer filter to centrifuge tube; add 2 ml 5% TCA (or 0.5 M HCl) containing thymidine, about 1 mM; cap tubes; heat at 100°C for 30 min. Cool and centrifuge. Variable values can result if water is lost by evaporation, or particles of filter or sediment are transferred to the scintillation vial.

(h) Transfer 0.5 ml to mini-scintillation vials (or 1–2 ml to large vials). Add water-miscible scintillation fluid and count radioactivity. Be careful not to transfer any sediment.

### 3. *Procedures for sediment: dialysis method*

(a)–(c) Follow steps (a) to (c) above.

(d) After storage, remove the ethanol by centrifuging and aspirating off the fluid. Add 2 ml 0.3 M NaOH containing 10 mM thymidine. Heat at 100°C for 30 min.

(e) Centrifuge (2000–5000 × *g*); collect supernatant; wash pellet with 2 ml distilled water. Combine supernatants.

(f) Gather about 1 m of wet dialysis tubing (25 mm flat width, molecular weight cut-off 6000–8000) around a thin plastic tube with a small funnel attached (Fig. 5). Transfer supernatants to dialysis tubing; wash down with distilled water and tie knots between each sample. Dialyse overnight against running water. A convenient apparatus is shown in Fig. 5.

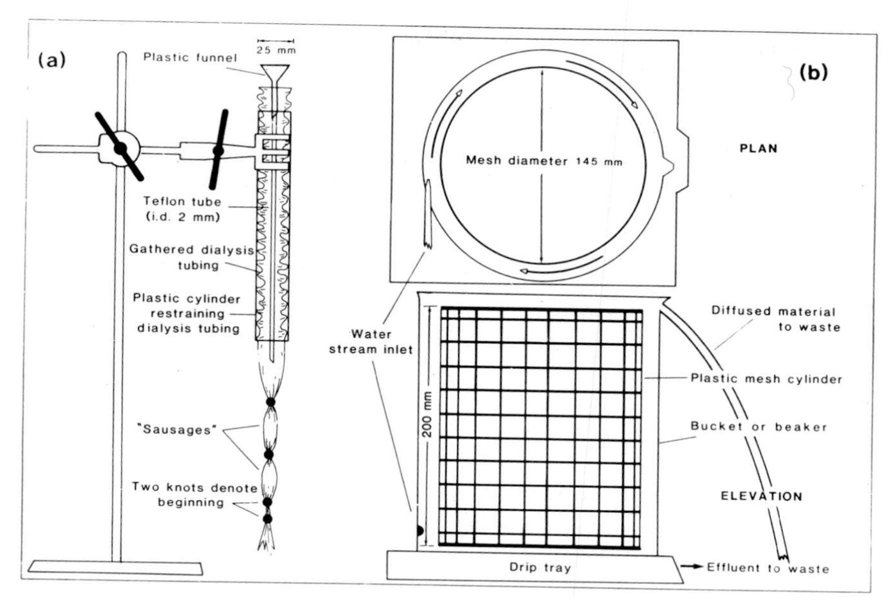


**Fig. 5.** Dialysis membrane and diffusion apparatus. (a) Filling the dialysis membrane. Apparatus for holding gathered dialysis tubing and directly delivering alkali extract to each dialysis bag. (b) The diffusion apparatus. Plan and elevation of the mesh cylinder around which the tubing containing the alkali extracts is wrapped. The use of the mesh cylinder allows efficient diffusion of small molecular weight compounds into running water without cross-contamination of the samples. (From Pollard, 1987, with permission.)

(g) Transfer all or a portion of the solution to scintillation vials and count radioactivity. (In short incubations, most of the radioactivity remaining in the dialysis bag is in DNA.)

*4. Procedures for determining isotope dilution in sediments*

(a) Shortly before use, prepare isotope dilution series with three or more solutions with different specific radioactivities and sufficient volume for at least five replicates each and a total of four blanks; keep cold, but not frozen. An example is given in Table I.

(b) Collect sediment, mix and dispense into tubes. Add thymidine solution then mix and incubate at *in situ* temperature for the selected time period [the time is temperature dependent; see note (ii)].

(c) Stop reaction with 10 ml 80% ethanol (ethanol 80 ml, water 20 ml, NaCl 2 g, thymidine 0.01 g).

(d) Proceed as in 3(d) above.

**TABLE I**

Example of solution preparation for isotope dilution experiment. This example is for 100 μl per sample with five replicates per dilution step, using [methyl-$^3$H]thymidine, specific activity 25 Ci mmol$^{-1}$, and 1 μCi μl$^{-1}$ (i.e. 25 μl = 1 nmol). A stock solution of normal thymidine at a concentration of 50 nmol ml$^{-1}$ (1 nmol/20 μl) should be made. The water used to make all solutions to the same final volume should be collected from the environment being studied and filtered before use

| Dilution step | μl $^3$H-Tdr | μl Tdr | μl water |
|---|---|---|---|
| 1 | 125 | 0 | 375 |
| 2 | 125 | 100 | 275 |
| 3 | 125 | 200 | 175 |

## 5. *Notes*

(i) The amount of isotope should be adequate to overcome isotope dilution. For most water bodies, 20 nM thymidine is usually adequate. If detritus is present, 50 nM or more may be needed. A preliminary isotope dilution experiment should be carried out for any new environment that is investigated.

(ii) The incubation time should be determined by a preliminary time course experiment. It must not exceed the time period during which the rate of DNA synthesis is linear. For tropical waters, 15 min is usually enough; colder temperate waters may require 30 min to 1 h; very cold water may require 2–12 h. For sediments with a surface temperature of about 30°C, 5 min may be enough. For temperatures around 15–25°C, 15–20 min should be adequate. Longer times will be needed for colder temperatures.

(iii) Blanks should be prepared by adding formalin to water samples and alcohol to sediment samples before radioactive material.

(iv) Disposable polypropylene vials (about 10 ml capacity) with screw caps, as used by pathology laboratories for transport of blood, are suitable for sediments and the final heating step. For water column samples polycarbonate or Teflon bottles should be used. These should be cleaned before use where traces of metals or plasticizers may influence bacteria, e.g. in the open ocean (see Fitzwater *et al.*, 1982).

(v) To prevent isotope dilution, about 1–2 nmol thymidine is needed per 50 mg dry weight of sediment. An isotope dilution experiment should be carried out initially, whenever a new environment is being studied, or where conditions change.

(vi) The recovery of DNA may be measured by adding a known amount of labelled bacteria. The bacteria are prepared by labelling a culture with

$^3$H-Tdr during log phase. Centrifuge the labelled bacteria (6000 × $g$, 20 min); resuspend in water and formaldehyde [1% v/v and sodium tetraborate pH 8–9 (25 mg/100 ml)]. Centrifuge again. Resuspend in same solution, use for recovery experiments. Determine total dpm by filtering and washing with cold TCA as for seawater (Part 1).

In order to determine the efficiency of DNA recovery, dispense sediment samples onto filters and pipette the labelled bacterial suspension directly onto the sediment. Do not wash with alcohol because false values for recovery can result from adsorption of the bacteria to walls of containers; the bacteria that were labelled in culture are free in suspension and not enveloped in slime and inorganic materials in sediment particles.

(vii) For short-term incubations, there is no significant difference in results between anoxic sediments that are mixed briefly as described here and anoxic sediments that are retained in cores and injected.

### 6. *Calculation of results*

(a) Calculate mol Tdr incorporated into DNA

(i) Convert cpm to dpm. Automatic scintillation counters can usually be programmed to convert counts per minute to disintegrations per minute. A set of quench standards is necessary.

(ii) Subtract values for blanks.

(iii) Correct for isotope dilution, if concentration of thymidine was too low to prevent dilution. If the intercept on the $x$ axis is less than zero, its absolute value $[x]$ must be added to the amount of thymidine actually added. Thus new specific activity (SA) is:

$$\mathrm{SA} = \frac{\mu\mathrm{Ci}}{(\mathrm{nmol} + [x])}$$

This is equivalent to Ci mmol$^{-1}$.

(iv) Convert dpm to mol:

$$T = \frac{\mathrm{dpm}}{\mathrm{SA}}$$

$$T = \frac{\mathrm{dpm} \times \mathrm{mmol} \times 10^{-3}}{\mathrm{Ci} \times 2.22 \times 10^{12}}$$

where:

$T$ = mol thymidine incorporated into DNA;

Ci = Curies; when tritiated thymidine is supplied, its specific activity is usually quoted as Ci mmol$^{-1}$, so it must be converted to Ci mol$^{-1}$.

The SI unit is bequerel (Bq), which equals one disintegration per second. The equivalent specific activity would be quoted as gigabequerels $\text{mmol}^{-1}$. Thus the above formula becomes:

$$T = \frac{\text{dpm} \times \text{mmol} \times 10^{-3}}{\text{GBq} \times 60}$$

where GBq = gigabequerels.

(b) Calculate rate of bacterial division from rate of DNA synthesis or rate of tritiated thymidine incorporation using the following formula:

$$N = \frac{T \times 318 \times 1/p}{w}$$

where:
$N$ = number of bacteria dividing;
$p$ = proportion thymine comprises of the four bases (A, G, C, T); an average value for $p$ is 0.25;
$w$ = amount of DNA per cell (range 1.7–5 fg); average $w$ = 2.5 fg;
318 = molecular weight of the thymidine nucleotide.

Thus, the total weight of DNA synthesized is divided by the weight of the genome. The formula condenses to:

$$N = T \times 5 \times 10^{17}$$

where $5 \times 10^{17}$ is an average theoretical factor. Some measured values are higher, which would be expected if thymidine incorporation were underestimated, due perhaps to isotope dilution or poor uptake of thymidine by some bacteria (Moriarty, 1988).

(c) Complete calculation

$$\text{No. of bacteria dividing per hour } (N\ \text{h}^{-1}) = \frac{\text{dpm} \times 10^{-3} \times 60 \times 5 \times 10^{17}}{\text{SA} \times 2.22 \times 10^{12} \times t}$$

$$= \frac{\text{dpm} \times 1.35 \times 10^{4}}{\text{SA} \times t}$$

where $t$ = incubation time in minutes and SA = Ci $\text{mmol}^{-1}$.

(d) Specific growth rate ($\mu$) is obtained by dividing the values for number of cells produced in a given time ($N$) by the total number present ($N_t$).

$$\mu = \frac{N}{N_t}$$

Generation time, or doubling time, ($g$) is the reciprocal of $\mu$ times the natural log of 2:

$$g = \frac{\ln 2}{\mu}$$

(e) To obtain productivity in terms of carbon, the average cell volume ($V$) should be calculated from sizes measured under the microscope. Assume specific gravity is 1.1 and carbon content is 22% of wet weight (see Bratbak and Dundas, 1984; Simon and Azam, 1989).

Thus carbon content ($C$) $= V \times 1.1 \times 0.22$

e.g. if $V = 0.5\ \mu m^3$

$C = 0.12 \times 10^{-12}$ g

$= 1.2 \times 10^{-13}$ g C per cell

$N \times 1.2 \times 10^{-13}$ = g C produced per unit time.

## D. Phospholipid synthesis

Purchase carrier-free $H_3{}^{32}PO_4$ immediately before experimental work. Withdraw sufficient for each experiment from the stock solution with a syringe (20 $\mu$Ci sample$^{-1}$) and dilute with distilled water to give a solution of 1 $\mu$Ci $\mu l^{-1}$. Work with high-activity materials behind clear acrylic sheets about 1 cm thick (Perspex or Plexiglas are common trade names). Sediment samples in syringes may be stored in boxes made from acrylic sheets, during extraction of labelled lipids.

The incubation with $^{32}P$ phosphate and extraction of lipids are carried out in disposable polypropylene syringes with eccentric luer tips, 30 ml capacity. Plastic tubing (polyvinyl chloride), equal in length to the syringe barrel, is fastened to the tips and bent up alongside the barrel and clamped with a rubber band.

### *1. Detailed procedure for sediment*

(a) Prepare at least 11 syringes for each type of sediment to be sampled: two blanks and three replicates of three different isotope dilutions. Add 50 $\mu$l water to each. Prepare glass scintillation vials with chloroform and methanol or water mixtures [see steps (f) and (g) below].

(b) Prepare phosphate solutions [see note (i)] in three tubes: A, B and C. To tube A add prepared (diluted) stock $^{32}PO_4$, 20 $\mu$l (20 $\mu$Ci) and 30 $\mu$l filtered water from the environment being studied. To tube B add diluted $^{32}PO_4$, 10 $\mu$l (5 nmol) phosphate and 20 $\mu$l filtered water. To tube C add diluted $^{32}PO_4$, and 20 $\mu$l (10 nmol) phosphate and 10 $\mu$l filtered water.

Each sediment sample will require 50 μl total; make up sufficient volume for total number needed.

(c) Collect and dispense sediment.

(d) Add 50 μl of $^{32}P$ phosphate solution to each. Release plastic tubing, carefully replace plunger, clamp plastic tubing and mix briefly. Place in water bath at *in situ* temperature.

(e) Incubate for predetermined time [see note (ii) above for thymidine: the principle is the same for phospholipid synthesis].

(f) Stop reaction by drawing up a mixture of 4 ml $CHCl_3$, 8 ml methanol and 50 μl 5 mM $Na_2HPO_4$ from a glass scintillation vial. Clamp tubing, shake syringe well and leave to stand for about 2 h or longer at an angle so that sediment does not settle in the luer tip. Blanks are prepared by drawing up the chloroform mixture immediately after adding the $^{32}P$.

(g) Draw up a mixture of 4 ml $CHCl_3$, 4 ml $H_2O$ and 10–20 mg of $Ca(OH)_2$ scintillation vial. Mix well and leave syringes standing at an angle overnight.

(h) Fold phase-separating filter papers (Whatman 1PS) and place in filter funnels over glass scintillation counter vials.

(i) Expel chloroform layer through tubing onto filters, which retain any sediment particles or aqueous phase.

(j) Allow chloroform to evaporate, then add scintillation fluid and count radioactivity [see note (ii)].

(k) Dilute stock $^{32}P$ phosphate solution 100 times and add 20–50 μl to scintillation vials. Count these at the same time as experimental samples. (This avoids the need to correct for decay of $^{32}P$ during the experiment.)

## 2. *Notes*

(i) A preliminary isotope dilution experiment to determine a working range of phosphate additions for routine experiments. Start with a solution of 0.5 nM $Na_2HPO_4$. Use additions of 10 μl (5 nmol) and 20 μl (10 nmol) per sample. Plot reciprocal of cpm incorporated into lipid against amount of phosphate added, as shown in Fig. 6. If the $^{32}P$ incorporated into lipids is similar to blank values, decrease amounts of phosphate. Conversely, if there is little or no apparent dilution, increase the amount of phosphate.

(ii) Usually, no quenching occurs with $^{32}P$ in a liquid scintillation counter, i.e. cpm = dpm. If there is much coloured lipid, e.g. plant pigments, quenching may occur. The chloroform solution should be bleached in strong light or with hydrogen peroxide.

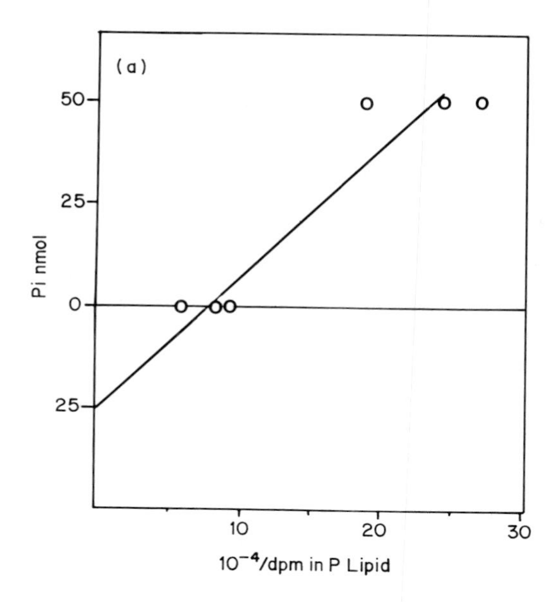


**Fig. 6.** Isotope dilution plot of $^{32}P$ incorporation into phospholipid in sediment. (From Moriarty *et al.*, 1985a, with permission.)

*3. Calculation of results*

(a) Calculate specific activity (SA) of $^{32}P$. Determine size of phosphate pool ($p$ nmol) from the intercept on the $x$ axis (Fig. 6). Calculate total cpm ($cpm_t$) added to each sample from results of step (k); do not forget the dilution factor

$$SA = \frac{cpm_t}{p \times 10^{-3}} \quad \text{dpm/}\mu\text{mol } PO_4$$

(b) Calculate rate of phosphate incorporation ($R$) into phospholipid

$$R = \frac{cpm \times 60}{SA \times t} \quad \mu\text{mol phospholipid synthesized h}^{-1}$$

where cpm is the cpm in lipid with blank values subtracted; $t$ is incubation period in min.

(c) Calculate rate of bacterial biomass production ($B$)

$$B = \frac{R}{25} \quad \text{g C h}^{-1}$$

The conversion factor of 25 μmol P $g^{-1}$ C is adapted from White *et al.* (1979) (50 μmol $g^{-1}$ dry weight).

## E. Protein synthesis

The procedure for determining rates of protein synthesis is similar to that for DNA (see above), except that [3,4,5-$^3$H]leucine is used. Rates of protein synthesis and bacterial productivity are calculated using factors reported by Kirchman *et al.* (1986), who give a full discussion of the technique. Isotope dilution and time course experiments are necessary.

Kirchman *et al.* (1986) heated water samples in TCA at 100°C for 30 min to hydrolyse nucleic acids, but I do not recommend this, because some proteins can be degraded. I have seen no evidence that nucleic acids are labelled by leucine in a short time period. For a rigorous determination of protein synthesis, analysis of labelled products with enzymes specific for various macromolecules would be needed.

### *1. Procedures for water column*

(a) As for DNA, Part C, 1, (a).

(b) Add tritiated leucine to a final concentration of 10 nM.

(c)–(g) As for DNA, Part C, 1, (c)–(g).

(h) Place filter in scintillation vial, add 1 ml ethyl acetate to dissolve the filter and then add scintillation fluid and count radioactivity.

### *2. Procedures for sediment*

(a) As for DNA synthesis, Part C, 2.

(b) Add tritiated leucine; about 0.2 nmol should be sufficient to minimize isotope dilution. The quantity added will need to be checked by an isotope dilution experiment [see note (i)].

(c)–(e) As for DNA synthesis, Part C, 2, (c)–(e).

(f) Transfer filter to screw-capped tube. Add 2 ml 2 M NaOH; heat for 2 h at 100°C. Cool and centrifuge.

(g) Transfer 0.5 ml to mini-scintillation vials (or 1–2 ml to large vials). Add water-miscible scintillation fluid and count radioactivity. Be careful not to transfer any sediment.

### *3. Notes*

(i) An isotope dilution experiment should be set up similarly to that for thymidine. Increments of 0.2 nmol leucine per sediment sample should be added.

*4. Calculation of results*

(a) Convert cpm to dpm and subtract blank values.

(b) Determine if specific activity of leucine needs to be adjusted to account for any isotope dilution. The procedure is the same as for thymidine (see above).

(c) Calculate rate of leucine incorporation ($L$) into protein.

$$L = \frac{\text{dpm} \times 60}{\text{SA} \times t}$$

$$= \frac{\text{dpm} \times (\text{mmol} \times 10^{-3}) \times 60}{\text{Ci} \times 2.22 \times 10^{12} \times t} \quad \text{mol L h}^{-1}$$

where $t$ is incubation period in minutes.

(d) Calculate rates of protein synthesis ($P$), cell ($C$) and biomass ($B$) production.

$P = L \times 1797$ g protein h$^{-1}$ (see Simon and Azam, 1989)

$B = P \times 0.86$ g C h$^{-1}$ (empirical factor, see Simon and Azam, 1989).

$C = L \times 5 \times 10^{17}$ cells produced h$^{-1}$ (see Kirchman *et al.*, 1986).

## References


Bratbak, G. and Dundas, I. (1984). *Appl. Environ. Microbiol.* **48,** 755–757.

Brock, T. D. (1971). *Bacteriol. Rev.* **35,** 39–58.

Fitzwater, S. E., Knauer, G. A. and Martin, J. H. (1982). *Limnol. Oceanogr.* **27,** 544–551.

Forsdyke, D. R. (1968). *Biochem. J.* **107,** 197–205.

Fuhrman, J. A. and Azam, F. (1980). *Appl. Environ. Microbiol.* **39,** 1085–1095.

Fuhrman, J. A. and Azam, F. (1982). *Mar. Biol.* **66,** 109–120.

Glaser, V. M., Al-Nui, M. A., Groshev, V. V. and Shestakov, S. V. (1973). *Arch. Microbiol.* **92,** 217–226.

Grivell, A. R. and Jackson, J. J. (1968). *J. Gen. Microbiol.* **54,** 307–317.

Kirchman, D. and Hodson, R. (1984). *Appl. Environ. Microbiol.* **47,** 624–631.

Kirchman, D., K'nees, E. and Hodson, R. (1985). *Appl. Environ. Microbiol.* **49,** 599–607.

Kirchman, D. L., Newell, S. Y. and Hodson, R. E. (1986). *Mar. Ecol. Prog. Ser.* **32,** 47–59.

Kjelleberg, S., Humphrey, B. A. and Marshall, K. C. (1982). *Appl. Environ. Microbiol.* **43,** 1166–1172.

Kornberg, A. (1980). "DNA Replication", W. H. Freeman, San Francisco.

Moriarty, D. J. W. (1986). *Adv. Microb. Ecol.* **9,** 245–292.

Moriarty, D. J. W. (1988). *Arch. Hydrobiol. Ergebn. Limnol.* **31,** 211–217.

Moriarty, D. J. W. and Pollard, P. C. (1981). *Mar. Ecol. Prog. Ser.* **5,** 151–156.

Moriarty, D. J. W. and Pollard, P. C. (1990). *J. Microbiol. Methods.* **11**, 127–139.
Moriarty, D. J. W., White, D. C. and Wassenberg, T. J. (1985a). *J. Microbiol. Methods* **3,** 321–330.
Moriarty, D. J. W., Pollard, P. C. and Hunt, W. G. (1985b). *Mar. Biol.* **85,** 285–292.
Munro, H. N. and Fleck, A. (1966). *In* "Methods of Biochemical Analysis" (D. Glick, Ed.), pp. 113–176, Interscience, New York.
O'Donovan, G. A. and Neuhard, J. (1970). *Bacteriol. Rev.* **34,** 278–343.
Plant, W. and Sagan, A. (1958). *J. Biophys. Biochem. Cytol.* **4,** 843–847.
Pollard, P. C. (1987). *J. Microbiol. Methods* **7,** 91–101.
Pollard, P. C. and Moriarty, D. J. W. (1984). *Appl. Environ. Microbiol.* **48,** 1076–1083.
Ramsay, A. J. (1974). *J. Gen. Microbiol.* **80,** 363–373.
Riemann, B., Bjørnsen, P. K., Newell, S. and Fallon, R. (1987). *Limnol. Oceanogr.* **32,** 471–476.
Rivkin, R. B. (1986). *J. Phycol.* **22,** 193–198.
Robarts, R. D., Wicks, R. J. and Sephton, L. M. (1986). *Appl. Environ. Microbiol.* **52,** 1368–1373.
Roodyn, D. B. and Mandel, H. G. (1960). *Biochim. Biophys. Acta* **41,** 80–88.
Sagan, L. (1965). *J. Protozool.* **12,** 105–109.
Simon, M. and Azam, F. (1989). *Mar. Ecol. Prog. Ser.* **51**, 201–213.
Stone, G. E. and Prescott, D. M. (1964). *J. Cell Biol.* **21**, 275–281.
White, D. C., Bobbie, R. J., Herron, J. S., King, J. D. and Morrison, S. J. (1979). *In* "Native Aquatic Bacteria: Enumeration, Activity, and Ecology" (J. W. Costerton and R. R. Colwell, Eds), pp. 69–81, ASTM STP 695, American Society for Testing and Materials.
Wright, R. T. and Hobbie, J. E. (1966). *Ecology* **47,** 447–464.

# 7
# Measuring Heterotrophic Activity in Plankton

JOHN E. HOBBIE
*Marine Biological Laboratory, Woods Hole, MA 02543, USA*



## I. Introduction

### A. Heterotrophic activity

In order to understand what microbes are actually doing in nature, as opposed to what microbes are capable of doing in the laboratory, ecologists must make measurements of rates of microbial processes in the real world. One obvious and important process is metabolism and growth

METHODS IN MICROBIOLOGY
VOLUME 22 ISBN 0-12-521522-3

of bacteria on organic compounds supplied from outside the cell. This heterotrophic growth decomposes most of the organic matter produced in the biosphere of land and water. In the water of lakes and oceans, measurements of bacterial heterotrophy are difficult because the great dilution of both substrates and organisms allows only low rates of activity, which laboratory techniques cannot measure, and because of the sensitivity of microbes to manipulation, which means that the rates are changed by, for example, incubations of 24 h or by filtration through a 10-μm pore size filter. For these reasons, techniques have been developed to measure low rates of heterotrophy either in undisturbed waters or in samples of lake or ocean water incubated briefly with radioisotopes.

The techniques discussed here have given a tremendous amount of information about the role, rates of activity, and factors limiting the heterotrophic bacteria in the plankton. Only some of the methods, those using radioisotopes, are presented in detail here. These are all variations on the theme of the addition of extremely low concentrations of labelled organic substrates to water samples. Anyone working in this field should understand the background and theory of this basic theme and then choose the variation of the method that best suits the needs and circumstances of the investigation.

Two kinds of measurements have been made, relative and absolute. Absolute measures give true rates of microbial metabolism and growth while relative measures give values which are positively correlated with the true rates but are not identical. For example, the incorporation of $^{14}C$ glucose into bacteria is easily measured and the rate correlates well with the rate of growth of bacteria in laboratory cultures. Glucose, however, is just one substrate of many the bacteria are using and so glucose uptake is a relative measure. Absolute rates of growth can be derived from changes in bacterial numbers (if no predators are present) or from the incorporation of thymidine or specific amino acids into DNA or protein.

## B. Oxygen measurements

The first successful measurements of heterotrophic activity were of changes in the concentration of oxygen in a water mass. Oxygen, however, constantly exchanges with the atmosphere across the water surface so the water mass must be isolated from the surface for a long enough period for the changes to build up to a measurable level. A stratified lake, where deep-lying water is isolated from the surface water by a thermocline, is one situation where oxygen changes are often measured. This deep water is also in contact with the sediment, however, so the technique is useful only for a measure of all the processes occurring in the water mass plus some occurring in the sediment. These include animal respiration and microbial oxidation of sulphide and methane.

Riley (1951) made better use of the concept in his examination of water masses at various depths of the Sargasso Sea, in the western North Atlantic Ocean. He assumed that the water mass was at one time at the surface of the ocean and that its oxygen content was at equilibrium with the atmosphere. Later, a few months to decades, the water mass has moved deeper in the ocean and is isolated from the surface waters by temperature and salinity stratification. Because the temperature of the water mass is unchanged over the months or years, the initial concentration of the oxygen may be calculated. When a water sample from the deep-lying water mass is collected and the oxygen measured, the rate of use of oxygen may be measured if the elapsed time is obtained from the physics of the ocean system. Riley measured rates of $<0.01$ to 1.0 ml $O_2$ $l^{-1}$ $year^{-1}$ with this method. Better estimates have now been made by Jenkins (1977) who used tritium ($^3H$) and its stable isotope daughter $^3He$ as a clock to calculate the period of isolation of the water mass. This method is complicated and has not been widely used.

In productive lakes and coastal oceanic waters, changes in oxygen may also be measured in short-term incubations in bottles. The measurements, which are made after a 6–24-h incubation of the water sample, will include the respiration of animals and algae unless the water is first filtered through a 1-μm mesh (e.g. Nytex). This filtration step may (Hopkinson *et al.*, 1989) or may not (Williams, 1981) increase the rate of metabolism. Hopkinson *et al.* (1989) attributed the increase to the release of control of bacteria by the removal of predatory protozoans. With great care and high-precision techniques, the oxygen method may be extended to offshore waters (Griffith, 1988).

## C. Radioisotope methods

In the late 1950s radioisotopes became available and marine biologists soon began adding them to water samples to measure first algal photosynthesis (incubations in the light with $H^{14}CO_3^-$) and then bacterial heterotrophy (incubations in the dark with $H^{14}CO_3^-$). The photosynthesis measurements worked well because it was a true tracer experiment. That is, the $H^{14}CO_3^-$ added only a tiny amount to the rather large pool of $HCO_3^-$ (2 mM in sea water) and as a result the rate of incorporation did not increase because of higher amounts of substrate. The bacterial heterotrophy measurement assumed that the dark incorporation of $H^{14}CO_3^-$ was all due to bacteria and that it represented the 6% of the total carbon in heterotrophic growth taken up as inorganic carbon (Sorokin,

1965). In reality, algae also take up some $H^{14}CO_3^-$ in the dark and this and other sources of error make the method unworkable.

Next, ecologists began to add $^{14}C$ organic compounds to water samples in experiments analogous to the $H^{14}CO_3^-$ photosynthesis measurement. Unlike the $H^{14}CO_3^-$ measurements, the concentration of substrate (organic compound) could not be measured. Also, the uptake measurement was not a tracer-level experiment because the amount of organic compound added, for example $^{14}C$-glucose, was hundreds to thousands of times greater than the natural level present; in this situation the higher amounts of substrate increased the uptake rate. Parsons and Strickland (1962) were the first to apply kinetic analysis to measurements in the sea by measuring uptake and incorporation of glucose into microbes at a series of concentrations of added glucose. They found that uptake by the entire community of microbes in a sample of sea water could be described by the same Michaelis–Menten-type equation that describes uptake of a laboratory culture. A maximum velocity of uptake ($V_{max}$) could be obtained which they called a relative heterotrophic potential. This heterotrophic potential has proven to be very useful as it correlates well with bacterial growth and activity. While it only measures one substrate of the hundreds present, the heterotrophic potential is a sensitive way of measuring when and where microbes are active, their relative rates of activity, and their response to such events as phytoplankton growth peaks and pollution.

Wright and Hobbie (1966) examined this method in detail and found that the turnover time of the organic substrate could be calculated as well as a single value for the substrate concentration ($S$) plus the half-saturation constant for uptake ($K$). This turnover time tells how fast the natural level of substrate is being cycled by microbes; the ($K + S$) gives the maximum value for the substrate concentration. The $V_{max}$ increases when microbial cells become better adapted to using a particular substrate, perhaps by an increase in the number of transport sites, and also increases when the number of cells increases. The $V_{max}$ range is four orders of magnitude from deep lakes to polluted ponds (Hobbie and Rublee, 1977).

## D. Improvements in radioisotope methods

It is obvious that some of the $^{14}C$ substrate that is taken up is respired as $^{14}CO_2$ during the experiment. This may be collected after the experiment on a piece of paper soaked with an organic base such as phenethylamine and the radioactivity measured by liquid scintillation counting (Hobbie and Crawford, 1969).

One major problem with working with a plankton sample is that many different types of organisms are present. When radiolabelled organic compounds are added to a sample, bacteria, algae, and even the larvae of many marine animals will take up the organic compound. The solution is to add extremely low amounts of substrate in the experiment. At low substrate levels, bacteria are the only group adapted for uptake and as a consequence, they are responsible for almost all of the uptake of an organic compound. One way to reduce the amount of substrate is to use tritiated compounds (Azam and Holm-Hansen, 1973). This has the disadvantage that it is difficult to correct for respiration losses of the $^{3}H$ as $^{3}H_2O$.

If the experimental question can be answered by studies of a mixed culture of bacteria, then the sample is filtered through a 1.0-μm or 0.6-μm pore size filter (such as the Nuclepore type) and the bacteria in the filtrate are allowed to grow. Their activity may be compared over time or after different treatments by measuring the uptake of a radioactively labelled compound, such as $^{14}C$-glucose, added at quite high concentrations. The uptake gives the relative heterotrophic potential or the maximum velocity of uptake. Tranvik and Höfle (1987) used 80 μg glucose $l^{-1}$ to test the ability of cultures to use easily degradable substances.

The advent of HPLC (high-performance liquid chromatography) has allowed measurement of the actual concentration of individual organic compounds in sea water that has not been modified in any way such as by desalting. These concentration data plus the turnover information allow calculation of the actual flux of different organic compounds through the bacteria (e.g. Fuhrman and Ferguson, 1986). As noted previously, the measurement of the flux of one or ten compounds does still not give the actual heterotrophic activity or growth because hundreds of compounds may be used by the natural bacteria. Such a flux measurement is an ideal relative activity measurement but does require a high level of chemical equipment and skill.

The most promising development for measuring absolute heterotrophic activity is two techniques for the direct measurement of bacterial growth. Both make use of vital components of bacteria that must be synthesized before division. The first of these techniques (Fuhrman and Azam, 1982) measures the rate of incorporation of $^{3}H$-thymidine, a precursor of DNA. Bacteria do incorporate externally supplied thymidine through a salvage pathway and the amount of thymidine incorporated may be translated into bacterial DNA and then into bacterial biomass. The rates of production measured in nature with this technique are reasonable and for the most part fit with ecological constraints such as the total amount of organic matter available. However, the technique as presently developed is not perfect and must be used with care. One problem is that sometimes there is an increase in DNA without any change in the number or biomass of the bacteria. Another problem is that it is not possible to determine the actual amount of DNA in the small bacteria that live in the plankton. Any attempt to calibrate using natural bacteria results in a rapid increase in the rate of growth and a potential switch to endogenous production of thymidine.

The second promising technique uses the incorporation of $^3$H-leucine into bacterial protein as the basic measurement. Kirchman *et al.* (1985) pointed out that most of the heterotrophic bacteria take up 10 nM leucine and incorporate it directly into protein. Recent measurements by Simon and Azam (1989) confirm that leucine is a constant fraction of the total amino acids in bacterial protein and that protein is a constant per cent of bacterial dry weight and carbon. Thus, leucine incorporation may be translated into increase in carbon (growth).

### E. Ecological importance of heterotrophic activity measurements

Heterotrophic activity measurements have proven to be a valuable ecological tool. With the use of these techniques, it is now known when during the year the bacteria are active and that bacterial activity is positively correlated with algal photosynthesis in the plankton over a range of three orders of magnitude. Up to 60% of the carbon fixed in photosynthesis is broken down by bacteria in the water column.

There are high rates of heterotrophic activity by bacteria in warm waters and low rates in cold waters. This low activity in cold waters may allow more of the algal carbon to reach the higher levels of the food chain and may in this way account for the high production of fish in colder waters (Hopkinson *et al.*, 1989; Pomeroy and Deibel, 1986).

## II. Background of methods

### A. Theory of uptake measurements

The essential elements to understand are: (1) Bacteria in natural waters live in an environment containing only a few micrograms per litre (10–100 nM) of each of the simple organic compounds used as substrates. They are well adapted to these concentrations and have transport systems with half-saturation constants that allow uptake at this low level. These transport systems are often very specific for individual substrates. The bacteria may take up many substrates simultaneously. (2) When radioactively labelled substrate is added to water samples at this concentration or lower, then bacteria take up most of the substrate while the other organisms in the water (algae, animals) take up only a little substrate. Sometimes bacteria may have multiple uptake systems for the same substrate. (3) Uptake can be described by Michaelis–Menten kinetics; it is not necessary to measure the concentration of substrate in order to determine some parameters that are ecologically useful.

These elements are explained by means of an imaginary experiment (Fig. 1) in which $^{14}C$-glucose is added to a sample of lake or ocean water, the water incubated for 1 h, the water filtered through a membrane filter, and the uptake of $^{14}C$ into the particles measured with liquid scintillation counting. Actually, the uptake was measured at 10 different concentrations of added $^{14}C$-glucose (called $A$).

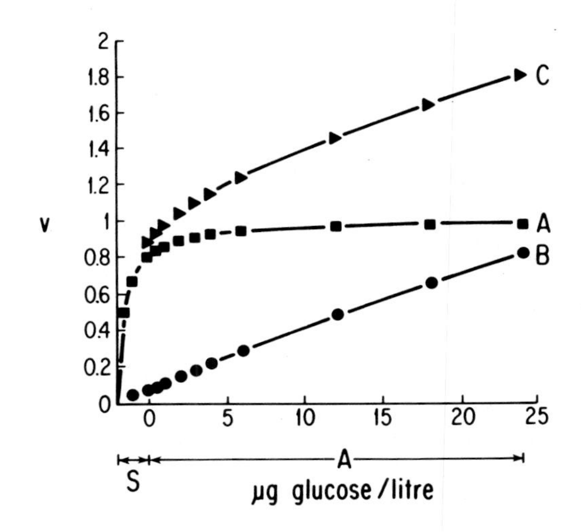


**Fig. 1.** Uptake velocity ($v$) of added $^{14}C$-glucose ($A$) in the presence of a known amount of substrate ($S$) by a bacterial population (curve A) and by a second population (curve B) (algal or bacterial population) and the uptake for the total plankton (curve C). Kinetics terms are defined in the text. The example is imaginary.

We will first assume that we have measured the natural substrate concentration ($S$) and that it is 2 µg glucose $l^{-1}$. The $S$ and the added $A$ is taken up according to curve A by a population of bacteria. There is another uptake system present in the water sample (curve B) which might belong to bacteria of another type or to algae or might even be a different transport system of the A bacteria. The uptake of the entire plankton population is the sum of bacteria plus algae/bacteria (curve C). It is obvious that uptake due to bacteria must be measured at close to the natural substrate concentration ($S$) or else there will be interference from curve B. The four additions at 1, 2, 3 and 4 μg $l^{-1}$ would give the best estimate while the measurements at 6, 12, 18, and 24 μg $l^{-1}$ would give neither information about the bacteria nor information about what was actually happening at $S$ (the imaginary vertical line where $A = 0$). The uptake of substrate by the bacteria (curve A) becomes saturated as $A$ is increased until a maximum velocity of uptake is reached ($V_{max}$). The value of the substrate concentration when the uptake rate is half the $V_{max}$ is $K$, the half-saturation constant.

## B. Mathematical description

We will now assume that we do not know the natural substrate concentration (the usual situation). As explained in Wright and Hobbie (1966), the velocity of uptake ($v$) at substrate ($S + A$) in Fig. 1 is

$$v_{(S+A)} = (f/t)(S + A) \quad (1)$$

where $f$ is the fraction of the isotope added that is taken up and $t$ is the time of incubation. Equation 1 may also be rearranged as

$$t/f = (S + A)/v_{(S+A)} \quad (2)$$

Bacterial uptake (curve A) follows a saturation curve or the Michaelis–Menten equation

$$V_{(S+A)} = V_{max}(S + A)/(K + S + A) \quad (3)$$

To better estimate $V_{max}$, a linear transformation is employed so that

$$(S + A)/v_{(S+A)} = (K + S + A)/V_{max} \quad (4)$$

Combining equations 2 and 4 gives

$$t/f = (K + S)/V_{max} + A/V_{max} \quad (5)$$

Accordingly, the data that produced curve C in Fig. 1 may be replotted according to equation 5 to produce Fig. 2. Note that this transformation is employed because it is not necessary to know the value of $S$ and in nature this is usually extremely difficult to measure. The plot employs the two values that are known, $A$ or the value of the substrate added in the experiment, and $t/f$ which is the inverse of the fractional uptake per unit of time of the radiolabelled substrate.

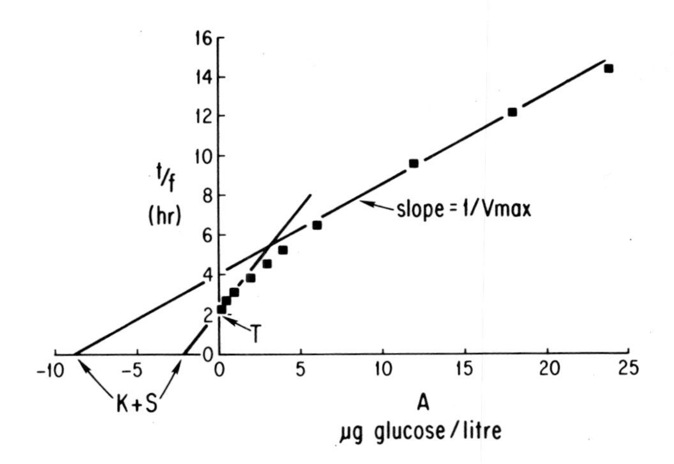


**Fig. 2.** The data of Fig. 1 (curve C) transformed by equations 4 and 5. The *Y* axis is incubation time (h) divided by the fractional uptake of the radioisotope and the *X* axis is the added $^{14}$C-glucose in μg $l^{-1}$.

## C. $V_{max}$, $T$, $-(K + S)$

It can be seen that curve C in Fig. 1 does, indeed, appear to be the result of the uptake by two populations. If we want to estimate the $V_{max}$ of the bacterial population (which is the inverse of the slope), then the portion of the curve closest to the point where $A = 0$ should be used. In this artificial example we know that $V_{max}$ is 1 (Fig. 1) so that the interference from the other population introduced an error. If the entire curve is used, the estimated $V_{max}$ is 2.3. If the four points closest to the $Y$ axis are used, the estimated $V_{max}$ is close to 1.

The example above illustrates the error possible when too high substrate concentrations are used in the measurement; but errors can also arise when too low concentrations are used. Fuhrman and Ferguson (1986) showed very well (Fig. 3) by the addition of extremely small quantities of substrate that there were two different uptake systems, one with a very low $K$ and low $V_{max}$, present in this sample of ocean water. They actually measured 4.5 nM serine in the water by HPLC which is close to the value of $K + S$ from the extrapolation of the whole curve (they ignored the two points closest to the $Y$ axis). My interpretation of the low $K$, low $V_{max}$ uptake curve is that it reveals some basic properties of the membrane transport system or of the concentrations in the space immediately next to this system.

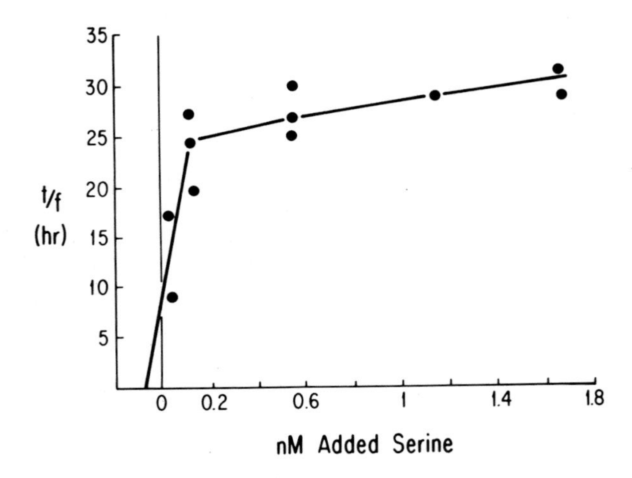


**Fig. 3.** Uptake of $^{3}$H-serine by bacteria in a water sample from New York Bight on 9 February 1984. Data plotted according to equation 5. The figure is modified from Fig. 1 of Fuhrman and Ferguson (1986).

To avoid these errors, it is usually safe to measure uptake at 10–40 nM for sugars and 1–10 nM for amino acids.

Another bit of information is the extrapolation of the curve to the $X$ axis where the intercept gives $-(K + S)$. Thus, from equation 5 when $t/f$ is 0, then $A = -(K + S)$. Neither quantity can be measured but from their sum we know a maximum value for the substrate (in Fig. 2 it was about 2 μg glucose $l^{-1}$, in Fig. 3 it is 1.4 nM if the two low points are ignored).

Finally, we may also measure the turnover time of the substrate ($T$) as the extrapolation to the $Y$ axis. From equation 5, the intercept on the $Y$ axis is $(K + S)/V_{\max}$. When $A = 0$, then from equation 4

$$(K + S)/V_{\max} = (S)/v_{(S)} \text{ or } T \tag{6}$$

because the natural concentration of substrate ($S$) divided by the velocity of uptake ($v_{(S)}$) is the turnover time, $T$. Again, neither quantity can be easily measured. In Fig. 2, $T$ is about 2 h and in Fig. 3 it is 23 h (again, ignoring two points closest to the $Y$ axis).

## D. Different types of heterotrophic activity measurements

The five types of measurements described earlier may now be described in terms of their relation to the theory given above. First, are the measurements using kinetic analysis. It is important to use as low concentrations as possible of added substrate but sometimes there must be a trade-off. The microbial activity may be so low that more isotope must be added in order to get significant uptake after an incubation of 0.5–4 h. This means that the concentrations may be higher than optimal. So the rule is to add as low concentrations as possible while still obtaining enough radioisotope in the particles on the filter for good counting statistics. At a minimum 400 cpm are necessary; several thousand counts per minute are ideal. In very clear oceanic waters, the cpm may be increased by filtering more water through the filter. When the actual $T$ and $V_{max}$ are needed, then $^{14}C$ must be used and the respired $^{14}CO_2$ collected and counted. When relative results are adequate and respiration can be ignored, then tritiated compounds may be used (these are available at much higher specific activity and so very low concentrations of added substrate are possible).

The second type of measurement is the tracer level addition. In this type of experiment, a single concentration of substrate is used but the concentration is extremely low. In Figs 2 and 3 the single addition would be very close to the $A = 0$ point. If the addition is low enough, then a good estimate of the $Y$ intercept may be obtained. For $^{14}C$-labelled compounds, different compounds are available at different specific activities and this may limit which may be used. In a recent catalog, D-U-$^{14}C$-glucose was available at 230 mCi mmol$^{-1}$ while L-3-$^{14}C$-serine was available at only 50–60 mCi mmol$^{-1}$. For $^{3}H$-labelled compounds the specific activity is much higher. For example, Azam and Holm-Hansen (1973) used glucose at 8600 mCi mmol$^{-1}$.

When the activity of bacteria alone is being measured, either in a laboratory culture or in a mixed culture of bacteria from nature grown on filtered sea or lake water, then a third type of measurement may be the easiest one to make. This is the determination of the $V_{max}$ of the culture through the addition of a single high level of substrate. In Fig. 1, curve A represents the uptake of bacteria; a high-level addition of greater than 5 μg glucose l$^{-1}$ would be adequate to estimate the $V_{max}$. The advantages are that only one quick measurement must be made and high amounts of radioactivity are incorporated. The disadvantage is that only one type of information ($V_{max}$) is obtained but for experimental manipulations this is often enough to determine an effect.

The fourth type of measurement is the use of $^{3}H$-thymidine to measure the actual growth of bacteria (Fuhrman and Azam, 1982). The method has been thoroughly reviewed by Moriarty (1986) and will not be described in detail here. The addition of thymidine is made at 5–20 nM (0.9–3.8 μg l$^{-1}$) which are very low levels but are not tracer levels. Instead, they are the concentrations at which the *de novo* synthesis of thymidine is completely inhibited. The transport mechanisms and uptake kinetics of thymidine are similar to those of other organic compounds and at these concentrations it is the bacteria which take up most of the compound. In the growth measurement, it is the incorporation into DNA, not the transport into the cell, which is measured (the radioactivity remaining in the cells is measured after an extraction with cold TCA). When this is done, then any kinetic analysis, for example to give $(K + S)$, measures both the external concentration of thymidine and the amount of thymidine internally produced which diluted the isotope after it was transported into the cell. Finally, the amount incorporated is multiplied by a factor which is the number of bacteria per mole of DNA. Once the number of bacteria produced is known, the carbon and biomass may be calculated from appropriate factors.

Overall, it appears that the results of bacterial growth analyses with thymidine give ecologically reasonable results. There are, however, enough inexplicable instances of high thymidine incorporation without bacterial growth, of problems with calibration (is there less DNA per cell in very small cells than in larger cells?), and of lack of thymidine uptake by certain types of bacteria that the method must be judged as promising but still imperfect in many ways.

The fifth measurement is of the incorporation of leucine into bacterial protein. In the technique of Kirchman *et al.* (1985) as extended by Simon and Azam (1989), $^{3}$H-leucine is added to samples at a final concentration of 10 nM. This amount overwhelms the ambient leucine pool (about 1 nM) and maximizes leucine uptake. After incubation, the bacteria are extracted with hot TCA (to hydrolyse DNA and RNA) and then filtered onto 0.45-μm pore size membrane filters for analysis of the incorporated tritium. The assumptions behind this method are that protein is a stable fraction of bacterial dry weight (63%), that leucine makes up a constant percentage of the total protein, that leucine at nM levels is taken up exclusively by marine bacteria, and that there is a consistent relationship between protein and cell volume for various sizes of bacteria. Thus, rates of production of bacterial protein may be translated directly into rates of production of bacterial carbon and biomass. With HPLC, Simon and Azam have successfully tested many of these assumptions on filtered samples that represent the bacteria between 0.2 and 0.6 μm. In most of the cases they examined, the leucine method gave comparable results to the thymidine method. This leucine method has fewer assumptions than the thymidine method for measuring bacterial heterotrophic activity and represents a very promising development.

## III. Methods

### A. Standard method

There is a standard method which is slightly modified for each of the various types of radioisotope experiments. However, once a method is chosen, tests must be made to decide upon incubation time and amount of labelled substrate to add.

Samples are collected in acid-washed bottles of glass or plastic, held at the *in situ* temperature in the dark, and the radioisotope experiment begun as soon as possible. This holding period should be no longer than 30 min. Samples may be incubated in any volume of flask but 10–20 ml of sample are convenient to filter through a 25-mm diameter filter so that large test tubes may be used. If the respired $^{14}CO_2$ is to be collected, then a 25-ml Erlenmeyer flask is suitable. Replicate or triplicate samples along with killed controls are necessary.

In clean ocean water, special techniques will often prevent contamination by trace metals (such as Cu) and increase the rate of bacterial activity. Ferguson and Sunda (1984) achieved this by using a 30-l Teflon-lined water sampler suspended on a plastic-coated hydrowire and closed with a Teflon messenger. They also carefully cleaned their glassware and rinsed with specially prepared distilled water.

Isotope is diluted in distilled water with special care to avoid contamination from organic compounds and trace metals (Ferguson and Sunda, 1984). The isotope is delivered to the sample with automatic micropipettes so that, for example, 25, 50, 75, and 100 μl will produce the desired concentrations in 10 or 20 ml of sample.

The incubation is carried out in the dark. The time should be long enough to produce good uptake by the bacteria yet short enough that there is no significant growth or adaptation during the experiment. The ideal time must be chosen in a test so that the amount of uptake increases linearly over time. Incubation is ended by the addition of buffered formalin (0.4% final concentration, pH 8). For a control, the isotope is also added to a formalin-killed sample.

When respired $^{14}CO_2$ is to be measured, the incubation is carried out in a 25-ml Erlenmeyer flask (Hobbie and Crawford, 1969). After addition of the isotope, the flask is immediately sealed with a rubber serum stopper that has a plastic cup suspended from it. The cup contains a 25 × 51 mm piece of accordion-folded chromatographic paper (Whatman No. 1). After incubation, 0.2 ml of a 2 N $H_2SO_4$ solution is injected through the septum to stop the uptake. Next, still working through the septum, 0.2 ml of an organic base (phenethylamine) is slowly added to the folded paper and the flask is then shaken for an hour at room temperature so that the $^{14}CO_2$ is absorbed. After this, the paper is placed in a scintillation vial with a scintillation cocktail and counted. Under these conditions, tests with a $^{14}CO_2$ standard showed that only 82% of the $^{14}C$ was counted even after a counting correction is made with an internal standard. This loss might be due to absorption of the light inside the filter paper. Thus, the final counts were multiplied by 1.23.

The samples are filtered through a membrane filter and rinsed with at least 5 ml of filtered sea water. The 25-mm diameter 0.45-μm pore size cellulose filters of Millipore or Sartorius work well and capture all of the bacteria. Filters are placed in standard counting vials and 1 ml ethyl acetate is added to dissolve the filter. The samples are counted with liquid scintillation and standard scintillation cocktails.

### B. Kinetic analysis

The method is that of Wright and Hobbie (1966) and Fuhrman and Ferguson (1986). The sample (10–50 ml) is incubated in flasks or in rinsed polyethylene bags (Whirlpak) with or without light. The final concentration of the added $^3H$- or $^{14}C$-labelled substrate is 10–100 nM or approximately 1–10 μg $l^{-1}$. Sugars, acetate, and amino acids may be used. The respired $^{14}CO_2$ may be collected into an organic base or the respired $^3H_2O$ calculated from the loss from solution during freeze-drying.

### C. Tracer-level additions

The method is that of Azam and Holm-Hansen (1973). For analysis of glucose uptake in sea water they added 0.01 μg glucose-6-$^3H$ to 100 ml of sample (to give 0.5 nM) and incubated for 0.5–4 h. The loss of labelled intracellular pool material was minimized by not killing the sample with formalin before washing the filter with 20 ml of iced and filtered sea water.

### D. Addition of high level of substrate

The only change used in this method is that a single concentration of labelled substrate is used. Tranvik and Höfle (1987) used 80 μg glucose $l^{-1}$ in short-term measurements of the adaptation of microbes in a lakewater culture to taking up sugar.

## E. Thymidine incorporation

The method is that of Fuhrman and Azam (1982) and Simon and Azam (1989). Methyl-1-[$^3$H]-thymidine (70 Ci mmol$^{-1}$) is added to 10-ml samples to give a final concentration of 5 nM. Formalin-treated controls are run in parallel. The incubation times are 15–45 min depending upon the temperature (incubations are carried out in the dark at the *in situ* temperature). After stopping the incubation with formalin, samples are filtered onto membrane filters, extracted with ice-cold 5% TCA and radioassayed with liquid scintillation. The thymidine incorporation rates are converted to cell multiplication rates by multiplying by the factor of $1.18 \times 10^{18}$ cells produced per mol of thymidine incorporated. Cell multiplication rates are converted to carbon production by using the cell carbon values for the given cell size (Table I).

**TABLE I**
Cell volumes, protein composition (in femtograms or $10^{-15}$ g), dry weight, and carbon content of average marine bacteria in the size range 0.026–0.4 μm$^3$ (adapted from Simon and Azam, 1989)

| Volume μm$^3$ | Protein fg | Protein %vol. | Protein %dw | Dry weight fg | Carbon fg | Carbon %dw |
|---|---|---|---|---|---|---|
| 0.026 | 12.1 | 46.5 | 61.4 | 19.7 | 10.4 | 52.0 |
| 0.036 | 14.7 | 40.8 | 62.5 | 23.5 | 12.6 | 53.6 |
| 0.050 | 17.7 | 35.4 | 63.0 | 28.1 | 15.2 | 54.2 |
| 0.070 | 21.6 | 30.9 | 62.8 | 34.4 | 18.7 | 54.3 |
| 0.100 | 26.7 | 26.7 | 62.8 | 42.5 | 23.3 | 54.7 |
| 0.200 | 40.3 | 20.2 | 63.5 | 63.5 | 35.0 | 55.1 |
| 0.400 | 60.6 | 15.2 | 63.3 | 95.8 | 53.3 | 55.7 |

## F. Leucine incorporation

In this method (Simon and Azam, 1989), [3,4,5-$^3$H]-l-leucine (140 Ci mmol$^{-1}$) is added to triplicate 10 ml samples to produce a concentration of 10 nM. Formalin-treated controls are run in parallel. Sample incubations (20–40 min) are ended by the addition of formalin and then samples are extracted with 5% TCA at 95–100°C for 30 min to hydrolyse RNA and DNA. Longer extraction times lead to significant hydrolysis of protein. The extracted samples are cooled and filtered.

One necessary correction is for the isotope dilution within the cell. Simon and Azam (1989) suggest that a factor of two is appropriate in many situations but the pool specific activity may also be measured with HPLC.

The conversion of $^3$H-leucine incorporated to carbon or biomass produced by the bacteria is made by the following formula (Simon and Azam, 1989):

$$\text{BPP} = (\text{mol } ^3\text{H-leucine incorporated})(100/7.3)(131.2)(2)(0.86) \quad (7)$$

where BPP is bacterial protein production (in g C), 100/7.3 is 100 divided by the mol % of leucine in protein, 131.2 is the formula weight of leucine, 2 is the intracellular isotope dilution of labelled leucine, and 0.86 converts protein to carbon. If the carbon per cell is known (from Table I), then BPP can also be converted into bacterial cell production. If the isotope dilution is not known, then using 1 instead of 2 in formula 7 gives a minimum estimate of BPP.

Azam (personal communication) states that even if the isotope dilution is known from HPLC measurements, the absolute value for growth is still elusive because the dilution gives the maximum estimate of BPP. We still do not know how much of the isotope dilution is due to *de novo* synthesis by the active cells and how much is due to unlabelled leucine pools of the inactive cells. In coastal waters the range from minimum to maximum is about two-fold.

## References


Azam, F. and Holm-Hansen, O. (1973). *Mar. Biol.* **23,** 191–196.
Ferguson, R. L. and Sunda, W. G. (1984). *Limnol. Oceanogr.* **29,** 258–274.
Fuhrman, J. A. and Azam, F. (1982). *Mar. Biol.* **66,** 109–120.
Fuhrman, J. A. and Ferguson, R. L. (1986). *Mar. Ecol. Prog. Ser.* **33,** 237–242.
Griffith, P. C. (1988). *Limnol. Oceanogr.* **33,** 632–638.
Hobbie, J. E. and Crawford, C. C. (1969). *Limnol. Oceanogr.* **14,** 528–532.
Hobbie, J. E. and Rublee, P. (1977). *In* "Aquatic Microbial Communities" (J. Cairns Jr, Ed.), pp. 441–476, Garland Publishing Co., New York and London.
Hopkinson, C. S. Jr, Sherr, B. and Wiebe, W. J. (1989). *Mar. Ecol. Prog. Ser.* **51,** 155–166.
Jenkins, W. J. (1977). *Science* **196,** 291–292.
Kirchman, D. L., K'Ness, E. and Hodson, R. (1985). *Appl. Environ. Microbiol.* **49,** 599–607.
Moriarty, D. W. (1986). *Adv. Microbiol. Ecol.* **9,** 245–292.
Parsons, T. R. and Strickland, J. D. H. (1962). *Deep-Sea Res.* **8,** 211–222.
Pomeroy, L. and Deibel, O. (1986). *Science* **233,** 359–361.
Riley, G. A. (1951). *Bull. Bingham Oceanogr. Coll.* **13,** 1–126.
Simon, M. and Azam, F. (1989). *Mar. Ecol. Prog. Ser.* **51,** 201–213.
Sorokin, Yu. I. (1965). *Mem. Ist. Ital. Idrobiol.* **18,** 187–205.
Tranvik, L. J. and Höfle, M. G. (1987). *Appl. Environ. Microbiol.* **53,** 482–488.
Williams, P. J. LeB. (1981). *Oceanologica Acta* **4,** 359–364.
Wright, R. T. and Hobbie, J. E. (1966). *Ecology* **47,** 447–464.

# 8

# Methods for Studying Adhesion and Attachment to Surfaces

MADILYN FLETCHER

*Center of Marine Biotechnology, Maryland Biotechnology Institute, University of Maryland, Baltimore, MD, USA*



## I. Introduction

The attachment of bacteria to solid surfaces has significant and often serious implications in a number of industrial, clinical, and ecological areas. Examples are the initial fouling of manmade structures such as pipelines, heat exchangers, and ship hulls, of biomaterials used for medical implants or prosthetic devices, and of particulates or larger surfaces in natural aquatic and soil environments.

In the 1930s, early attempts to evaluate attachment to surfaces used test surfaces (usually glass slides) which were suspended in the environment of interest, retrieved some time later, and examined for attached microorganisms with a microscope (Cholodny, 1930; Henrici, 1933). However,

METHODS IN MICROBIOLOGY
VOLUME 22 ISBN 0-12-521522-3

the emergence and application of new techniques have advanced considerably since those early days. For example, the development of epifluorescence microscopy has allowed studies of attachment to surfaces to be extended to a wide range of surface materials and environments. Also, computer-enhanced image analysis has enabled the processing of many more samples than is possible for a single microscopist, and moreover has made possible long-term, real-time studies of surface colonization. The identification of biochemical 'markers' has allowed us to measure and characterize attached populations by biochemical analysis, whereas radioisotopes can now be used to label bacteria and 'track' them as they attach to surfaces. Moreover, further developments are visible on the horizon with the emergence of genetic techniques, particularly the application of nucleic acid hybridization probes. Each of these technologies has its value and limitations, and it is through the combination of approaches that most progress is made in measuring attachment, determining attachment mechanisms, and characterizing attached populations.

There are generally two basic types of attachment studies. First, there are those aimed at evaluating adhesion in natural environments. Test surfaces are placed in the environment of interest (water, soil, water delivery system, cooling tower, etc.), left for a period, retrieved, and the attached bacteria evaluated either microscopically or biochemically. The purpose of such studies is to determine the types of bacteria which are likely to become attached and to make up the natural attached population, as well as the probable time-scale of surface colonization.

Most attachment studies, however, fall into a second category, that is those which are laboratory based and designed to measure some specific aspect of the adhesion process. They are generally designed to measure (a) the adhesion ability of particular organisms, (b) the effect of particular environmental conditions on adhesion, or (c) the degree to which bacteria attach to a particular surface. Some laboratory techniques, e.g. microscopy (III.A), analysis of biochemical markers (III.B.3), and nucleic acid hybridization (III.D), may be applied to *in situ* studies, but must first be validated by laboratory experiments.

## II. Laboratory studies – experimental design

### A. Introduction

The extent of bacterial attachment to surfaces and rate of cell deposition is heavily dependent upon not only the species of bacteria, but also on the strain and nutrient phenotype. Also attachment can be considerably influenced by environmental conditions. These include nutrient sources, concentration, and flux (Ellwood *et al.*, 1974; Brown *et al.*, 1977; Molin *et al.*, 1982; Knox *et al.*, 1985; McEldowney and Fletcher, 1986), electrolyte concentration (Marshall *et al.*, 1971b; Orstavik, 1977; Gordon and Millero, 1984; Knox *et al.*, 1985), pH (Gordon *et al.*, 1981; Harber *et al.*, 1983), and temperature (Fletcher, 1977; Harber *et al.*, 1983). Particularly important are characteristics of the substrata, such as hydrophobicity or surface topography. Rinsing procedures can also affect the result, especially where the strength of binding is low.

Thus, the adhesion measurement obtained will depend very much on the design of the system. The organism, environmental conditions, solid surfaces, and rinsing procedures which are chosen will depend very much on the question to be answered, and if the ultimate objective is to understand a natural process, they must be appropriate to the natural environment being modelled.

## B. Selection of organisms

The choice of organism and its laboratory history can have an enormous effect on the degree of adhesion observed. If the experimental question relates to attachment in a natural environment, then it is clearly important to choose organisms representative of the natural populations. For example, studies aimed towards understanding adhesion in aquatic environments have frequently used *Pseudomonas* or *Vibrio* species. However, it is also extremely important to use adhesion phenotypes characteristic of the natural population. Experiments with *Escherichia coli* (cf. Isaacson, 1985) and *Streptococcus salivarius* (Weerkamp *et al.*, 1986) have clearly shown how mutations altering specific cell surface components can alter non-specific adhesion to non-biological surfaces. For example, the adhesion of *S. salivarius* to hydrophobic surfaces was related to the density of a fibrillar layer on the cell surface and to the properties and surface exposure of specific types of fibril (Weerkamp *et al.*, 1987). Significant surface changes can occur with natural isolates upon culture in the laboratory, possibly because of selection of non-adhesive strains by repeated transfer from liquid culture. However, recently it has also become apparent that there are a number of strains able to undergo phase transition and to demonstrate two or more adhesive phenotypes. Examples are *Pseudomonas atlantica* (Bartlett *et al.*, 1988) and *P. fluorescens* (Pringle *et al.*, 1983), which have both been shown to produce progeny of three separate colony (and adhesion) morphotypes, i.e. mucoid, smooth, and crenated. Interestingly, the mucoid phenotype of *P. atlantica* is apparently the most adhesive morphotype (Corpe *et al.*, 1976), while with *P. fluorescens* H2, the mucoid variant is much less adhesive than the smooth or crenated variants

(Pringle *et al.*, 1983; Pringle and Fletcher, 1983). Thus, great care must be taken to be certain that the strain or variant being used is the one of interest or that results obtained with one adhesive phenotype are not extrapolated erroneously to a different strain.

The production of cell surface polymers can be influenced considerably by growth conditions. Carbon source, carbon : nitrogen ratio, carbon flux, and nutrient concentration all have been shown to influence surface polymers and adhesive ability of some bacteria. However, not all bacteria are affected, and those affected are not influenced in the same or a predictable way (Marshall *et al.*, 1971b; Wardell *et al.*, 1980; Molin *et al.*, 1982; McEldowney and Fletcher, 1986). In batch culture, adhesion is affected by the growth phase (Powell and Slater, 1982; Harber *et al.*, 1983; Rosenberg and Rosenberg, 1985). However, the effect varies with the organism; with some bacteria adhesion is greatest with log phase cultures, whereas with others it is greatest at stationary phase. This emphasizes the importance of selecting the experimental conditions which suit the question being asked. Otherwise, essentially valid yet totally irrelevant adhesion results may be obtained.

Another factor that may considerably affect the adhesion of particular bacteria is interaction among bacterial species or with higher organisms. Very little is known about the frequency or mechanisms of such interaction, because the subject is practically untouched experimentally. However, there are clear indications that adhesion of given bacteria can be influenced by the presence of other species (Murchison *et al.*, 1985; McEldowney and Fletcher, 1987). This is possibly due to the production of dissolved substances which affect the adhesive interaction at the bacterium/solid interface (Fletcher, 1976; Velez *et al.*, 1976; Gordon, 1987). Alternatively, competitive or mutualistic interactions could influence cell surface properties (hence adhesives) indirectly, by mechanisms similar to those involved with the effect of nutrients on adhesion.

The preparation of the bacteria for the experiment may also have an influence on the results obtained. Centrifugation and washing is often used to standardize cell densities in buffer solution before adhesion assays. However, bacteria may have a capsule or extracellular polysaccharide which influences adhesion but is easily removed. Consequently, if the cells are centrifuged and washed before adhesion tests, results may be modified. Thus, it might be preferable to minimize such preparation when testing cells with easily removed surface polymers.

### C. Selection of surfaces

The number of cells which attach to available surfaces depends considerably on the physicochemical properties of the surface. These properties include electrostatic charge (Feldner *et al.*, 1983; MacRae and Evans, 1983), surface free energy (van Pelt *et al.*, 1985; Busscher *et al.*, 1986a, b) and a related parameter, hydrophobicity (Gerson and Scheer, 1980; Kawabta *et al.*, 1983; Hogt *et al.*, 1983; Ludwicka *et al.*, 1984), as well as texture (Baker, 1984). Thus, the outcome will depend significantly on the choice of surface. A given strain may attach in quite high numbers to one type of surface and very little to another. Thus, if an experiment is designed to evaluate the adhesiveness of the bacteria, then several types of substrata should be used to obtain a representative result.

In general, materials can be divided into two main classes, i.e. (1) high-surface energy materials, which are also hydrophilic, frequently negatively charged, and usually inorganic materials such as glass, metals, or minerals, and (2) low-energy surfaces, which are relatively hydrophobic and low in negative and positive electrostatic charge and are generally organic polymers, such as plastics. Because of their high surface activity, high-energy substrata readily adsorb dissolved solutes or atmospheric contaminants and thus are rarely clean. Therefore, if the material itself is of special interest, it must be scrupulously cleaned and used as soon as possible, or stored in non-contaminating conditions, e.g. under double-distilled, deionized water.

Glass is a commonly used surface which is quickly contaminated by adsorbents in the atmosphere or solution, and it should be cleaned thoroughly immediately before use. One cleaning protocol is to first degrease glass in acetone, soak it in 1:1 concentrated HCl: concentrated $HNO_3$, rinse three times in double-distilled water from an all-glass still, and steam before use (Abbott *et al.*, 1983).

Without special cleaning, metals will have an oxide coating. However, since such rigorously clean surfaces are virtually never encountered by bacteria in the real world, for most purposes, cleaning should probably be consistent with the conditions of interest. Thus, for example, if the key question relates to how bacteria in natural waters attach to stainless steel, the steel should not be specially cleaned to remove the oxide layer and detergent washing should probably be adequate. However, if the objective is to understand the physicochemical nature of bacterial adhesion to metal, then a chemically defined surface is required and the metal should be rigorously cleaned.

Low-energy materials do not adsorb contaminants as readily as high-energy surfaces. Therefore, detergent washing followed by thorough rinsing is generally suitable (Fletcher and Loeb, 1979; Pringle and Fletcher, 1983). Bacteriological or tissue culture plastic ware can be used as supplied, if precise knowledge of substratum composition is not required. In all cases, high-grade materials should be used, to minimize the amount of leachable plasticizers and non-polymerized monomers.

Sometimes surfaces with the optical properties of glass but with different chemical properties are desired. For this purpose, glass surfaces can be modified by chemical treatments. Satou *et al.* (1988) prepared four types of surface-derivitized glass:

(i) aminopropyl glass, prepared by alkylsilylation of glass slides with gamma-aminopropyltriethoxysilane, which carried positively charged amino groups and non-polar ethoxyl groups, and was considered to be amphipathic (able to enter into both polar and hydrophobic interactions);

(ii) aminopropyl glass subsequently treated to bind hexanal to the amino groups, thus introducing a hexyl group and rendering the surface hydrophobic;

(iii) aminopropyl glass treated to introduce carboxyl groups, so that both positively and negatively charged groups were present;

(iv) aminopropyl glass treated to introduce pentahydroxymethylene residues, which increased the hydrogen-bonding potential of the glass, making it more hydrophilic.

Glass may also be treated with a commercial siliconizing agent to provide a microscopically thin hydrophobic layer on the surface. The risk with such derivitization procedures is that the surface may not be completely modified, producing a heterogeneous surface on the microbial scale. This, however, could be checked by surface analysis techniques, such as X-ray photoelectron spectroscopy (XPS or ESCA), which gives information on elemental composition and chemical bonding.

The transparency and roughness of the surface will be an important consideration when selecting the method for measurement of attachment. Optically flat surfaces can be examined by microscopy and numbers of bacteria counted directly. However, rough or porous surfaces must be assessed by other means, such as scintillation counting of radiolabelled bacteria (III.C) or analysis of signature chemicals (III.B.3). Transparent surfaces, such as glass and some plastics, have the advantage that they can be evaluated by either light microscopy (III.A.2) or spectrophotometry (III.B.2), whereas opaque surfaces require assessment by epifluorescent or electron microscopy (III.A.3–4) or by biochemical (III.B.3) or radiotracer (III.C) techniques.

## D. Position of the surface

When determining the adhesiveness of a particular bacterium or comparing adhesiveness of strains, it is important that the position of all substrata in the bacterial suspension is such that the surfaces are exposed to the same cell densities, nutrient concentration, aeration, and water circulation. Also, whether the surface is vertical or horizontal can affect attachment numbers with non-motile cells, as sedimentation will bring the bacteria into contact with horizontal surfaces but not vertical ones. A number of devices have been designed by different workers to ensure even exposure of surfaces to bacteria and to minimize the influence of variable environmental factors. These range from the simple 'O' rings used to hold coverslip surfaces vertically in 25-ml bottles (Pringle and Fletcher, 1983) to more complicated devices designed to maintain controlled laminar flow.

### *1. Static or random-flow systems*

*(a) Horizontal surfaces.* The simplest method used to position the attachment surface is to use the suspension container itself as the substratum or to lay a flat test surface on the bottom of the container. Petri dishes have been used as the attachment substrata in this way (Fletcher, 1976) and have the advantage that they can be obtained in both non-wettable (bacteriological) and water-wettable (tissue culture) forms, allowing a comparison of surface characteristics. They are also inexpensive and can be used as they are supplied, without cleaning or sterilization. However, they have several possible disadvantages. They can only be used with small quantities of suspension (e.g. 8 ml in a 5-cm diameter dish). Also, the precise compositions of their surfaces are not known because of variations in commercial manufacturing procedures. In some cases it is desirable to eliminate the sedimentation of cells onto the surface, and thus not use a horizontally placed surface.

*(b) Surface-holding devices.* A number of devices can be improvised to hold test surfaces vertically in the bacterial suspension. An example is to use circular glass coverslips or discs cut from other materials as test surfaces and to insert these perpendicularly into 'O' rings cut from flexible tubing. One ring can be used to hold a single surface vertically in a small vessel, which has an internal diameter only slightly larger than the ring. Thus, the ring lies horizontally near the bottom of the vessel, and the test surface is in a vertical position. Alternatively, several surfaces can be placed in a flask. With shaking, the positions of the test surfaces are constantly changed, but the rings prevent them from abrading one another or adhering to the inside wall of the glass and facilitate their removal with forceps. A possible combination of dimensions for test surfaces and for tubing 'O' rings is 16 mm diameter for the surface and 14 mm (internal

diameter) × 20 mm (external diameter) × 10 mm (height) for the 'O' ring (Pringle and Fletcher, 1983). A high-quality, flexible tubing, e.g. silicone rubber or Tygon, is suitable and may be leached in distilled water before use.

A more complex device was used by van Pelt *et al.* (1985) to ensure that bacterial contact with vertical test surfaces was standardized (Fig. 1). They

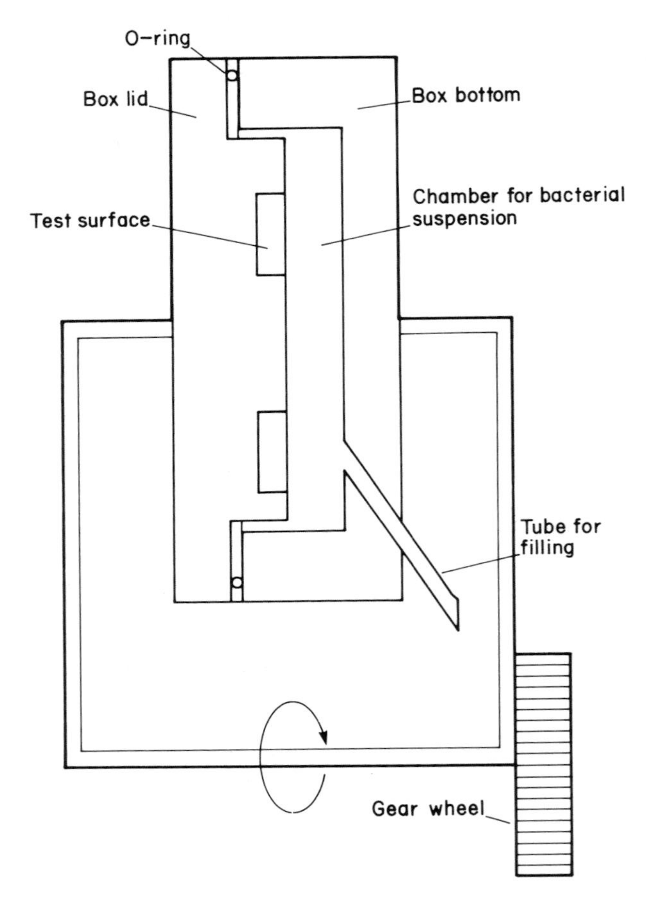


**Fig. 1.** Schematic diagram showing the basic features of a device developed by van Pelt *et al.* (1985). The box was one of six boxes which were clamped in a wheel rotating at 4 rpm. Each box contained six test surfaces, and the internal chamber of each contained 9 ml of bacterial suspension. The suspension was introduced through a narrow tube. In such a device, the substrata are always in a vertical position, and constant rotation of the sample chamber tends to prevent sedimentation of the bacteria.

placed test surfaces in recessed areas in the lids of flat boxes filled with the bacterial suspensions. The boxes were mounted on a wheel which was rotated vertically at 4 rpm. Thus, the surfaces maintained their vertical position and the bacterial suspension was agitated throughout the attachment period. Each wheel carried six boxes, which in turn held six substrata.

*(c) Adhesion screening of many samples.* Sometimes the attachment ability of a large number of strains must be quantified, so that microtechniques are appropriate. For this purpose, the wells of microtiter plates can be used as the attachment substrata. These were used to measure the adhesion of transposon mutants of *P. fluorescens* to identify strains with altered attachment ability (Cowan and Fletcher, 1987).

To prepare bacterial suspensions, single colonies of each strain are inoculated into 3-ml aliquots of medium and grown at the appropriate temperature with shaking. All strains should be grown up to the same stage of growth and the turbidity of each culture standardized to the same value, e.g. an $A_{600}$ of 0.60, with the appropriate buffer. A 1.5-ml sample of each culture is placed in an Eppendorf microcentrifuge tube and centrifuged at $7500 \times g$ for 3 min in a microcentrifuge. The supernatants are poured off and pellets resuspended in 1.0 ml of buffer. They are recentrifuged and resuspended in a final volume of 220 μl buffer. Duplicate 50-μl samples of each suspension are placed in wells of the microtiter plate. After evaluation of a number of commercially supplied plates, it has been found that Corning untreated 96-well microtiter plates and Costar tissue culture-treated plates resulted in reproducible and relatively high numbers of attached cells, and that different strains attached to these surfaces in noticeably different numbers (Cowan and Fletcher, 1987). The cells are allowed to attach for an appropriate time (e.g. 2 h), and at an appropriate temperature.

After attachment, unbound cells are removed by pouring distilled deionized water over the surface of the entire microtiter plate, directing the stream of water towards a few empty wells which are not used for the assay, and then allowing the water to flood into the wells containing the bacteria. The dish is then inverted to drain water, and the process repeated twice. The entire dish is immersed in a glass dish containing Bouin's fixative [saturated aqueous picric acid (75%), formalin (25%), and glacial acetic acid (5%) (v/v)] for 4 min. The dish is flushed with double-distilled water until no colour remains and is then stained with a general purpose stain. Congo red was previously used (Cowan and Fletcher, 1987), but crystal violet would also be suitable. For Congo red staining, after rinsing off the fixative, aspirate the water from each well and replace with 45 μl cetyl pyridinium chloride (0.34%). Add Congo red (10 μl) to the cetyl pyridinium chloride in each well and agitate the plate for 5 min. Rinse the plate exhaustively with double-distilled water and invert to dry. Just before reading, add 50 μl of 95% ethanol to each well to elute stain from the cells, cover the plate, and agitate for 10 min. Read absorption of the eluted stain at 490 nm on a microtiter plate reader. Crystal violet may be used in the same way as Congo red, except cetyl pyridinium chloride is not used during staining, and absorbance is read at 590 nm (Fletcher, 1976).

### 2. *Rotating disc*

A rotating disc apparatus was developed by Abbot *et al.* (1980), so that the hydrodynamics at the attachment surface could be controlled and the collision frequency between cells and surface could be calculated. The apparatus, described in detail elsewhere (Rutter and Abbot, 1978; Abbot *et al.*, 1980), consisted of a Perspex (polymethylmethacrylate) disc in the shape of a truncated cone and 5.5 cm in diameter. It was mounted on a stainless steel shaft and held inverted in a Perspex dish containing the bacterial suspension. The shaft extended through the base of the dish via two Teflon bearings and was connected to a variable speed motor. There was a recess in the base of the disc in which coverslips could be placed and secured by high-vacuum silicone grease; these served as the attachment substrata.

The theoretical number of bacterial collisions with the surface can be calculated from the Levich equation (Levich, 1962)

$$j = 0.621 D^{2/3} v^{-1/6} w^{1/2} C_0$$

where $D$ is the diffusion coefficient $(kT)/(6\pi na)$, in which $k$ is Boltzmann's constant, $T$ is the absolute temperature, $n$ is the viscosity of the medium, and $a$ is the radius of the particle. $v$ is the kinematic viscosity of the medium, which is the ratio of its viscosity and its density ($v = n/p$). $w$ is the angular velocity of the disc in radians per second, and $C_0$ is the initial single particle concentration.

This predicted number of collisions can be compared to the measured number of attached cells, in order to estimate the proportion of collisions which actually result in cell attachment to the surface. Abbott and co-workers (1980, 1983) used the technique to evaluate the influence of substratum properties and solution electrolyte concentration and pH on the number of 'successful' collisions. Such measurements are extremely valuable in attempts to understand the dynamics of the adhesion process, because they have the precision which allows some interpretation in terms of physicochemical theory.

### 3. *Flow-cell surface holder*

A second approach for standardizing flow over the surface is to place the surface in a tube through which the fluid with suspended bacteria flows, and to construct the device so that flow over the surface is laminar. An example is the polymethylmethacrylate flow cell apparatus devised by Pratt-Terpstra *et al.* (1987) (Fig. 2). The surfaces were 0.1 to 1.0 mm thick and approximately 1 by 1.5 cm in size and were held in a position parallel to the direction of flow. Liquid from a reservoir held at a position above the flow cell was allowed to pass through the cell by hydrostatic pressure at a continuous laminar flow rate of 60 ml $min^{-1}$ (equivalent to a shear rate of 21 $s^{-1}$ at the centre of the substratum).

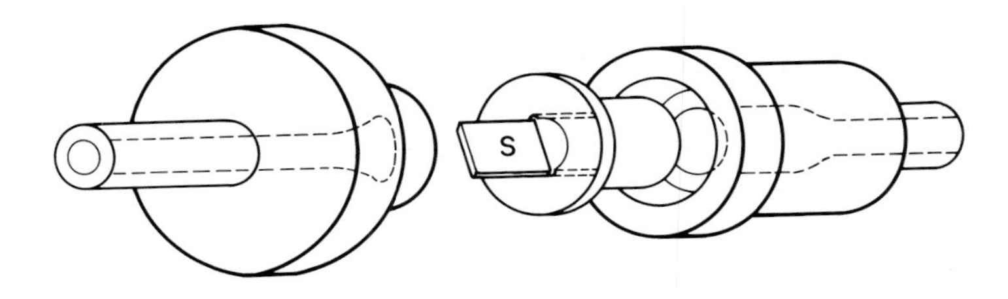


**Fig. 2.** Schematic diagram of a flow cell designed by Pratt-Terpstra *et al.* (1987). This expanded view illustrates the way in which the test surface (S) is positioned parallel to the direction of flow. Pratt-Terpstra *et al.* used a chamber diameter of 1 cm.

## E. Rinsing procedures

After the experimental attachment period, surfaces must be rinsed to remove unattached cells in surface-adsorbed liquid and cells which are only loosely associated with the surface. Removal of loosely attached cells will vary depending upon the force applied by the rinsing procedure. Thus, it is very important to standardize the rinsing procedure within a given laboratory, and in so far as possible make certain that it is simple enough and described well enough to be reproduced in other laboratories.

In many attachment experiments, the rinsing has been done by washing the surface with a stream of water or buffer. This is probably satisfactory for strongly attached cells, but on the whole should be avoided because strong jets of water can cause uneven removal of cells from a surface. Also, such rinsing is difficult for workers in other laboratories to reproduce.

Another factor which can introduce variability into attachment data is surface tension forces caused by passing the attachment surface through an air/water interface. This is particularly important with low energy, or hydrophobic, surfaces. Water will 'bead up' on these surfaces, often lifting up and carrying along attached cells, so they become deposited again at the site of the water bead. Thus, the fewer passages through an air/water interface, the better. It is generally easy to maintain the surface in liquid while rinsing, and while fixing and staining if required, by using an apparatus such as that shown in Fig. 3. Rinse water can be run through continuously for several volume changes, and it is not necessary to apply

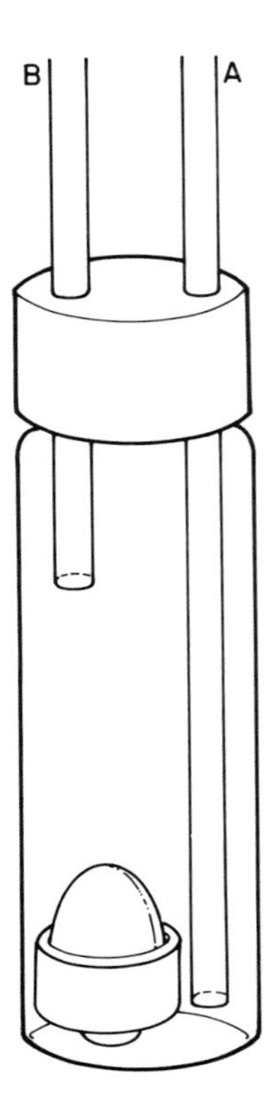


**Fig. 3.** Schematic diagram of an apparatus used to rinse and fix a coverslip with attached bacteria without exposing the surface to an air–water interface. It consists of a screw-cap bottle (*ca* 25 ml), with glass tubes inserted through drilled holes in the plastic cap and secured with silicone adhesive. A coverslip is held in an 'O' ring [II.D.1(b)] and is placed in the bottle containing the test bacterial suspension. After the attachment period, the suspension is flushed out with the rinse solution by introducing the solution under pressure through tube A. The suspension and rinse solution are expelled through tube B. After rinsing, the surface remains immersed in rinse solution, and fixation solution can then be introduced in the same way.

strong rinsing forces directly to the surface. After rinsing, most of the rinse water can be removed, still leaving the surface under water, and an appropriate volume and concentration of fixative can be added. Similarly, the surface can remain in liquid during subsequent rinsing and staining. Thus, the surface does not need to be taken through the air/water interface until after fixation, or even staining. In our experience, fixation with Bouin's fixative does appear to stabilize adhesion, so that cells become less likely to be removed by surface tension forces.

## F. Application of controlled shear stress

An important aspect of adhesive interactions is the strength of the adhesion bond. Various tests have been developed to test adhesive strength of attached eukaryotic cells by determining the applied force necessary to remove cells. A variety of forces have been used, ranging from gravitation (Weiss, 1967), a water jet (Christie *et al.*, 1970), centrifugation away from the surface (George *et al.*, 1971), and shear stress via liquid flow (Weiss, 1961a). The problem with all such measurements is quantification of the force at the cell–surface interface. This is complicated in that applied forces may be perpendicular tensile stress (e.g. centrifugation) or sideways shear (most washing procedures). Many fluids will form a good adhesion against tension (e.g. water adhering two opposing glass slides), but resistance to shear stress requires a viscous adhesive. Furthermore, the removal of attached cells in such tests is rarely if ever due to a break in the adhesive joint, but instead is due to a cohesive failure within the adhesive or the cell itself (Weiss, 1961b).

An alternative approach is to impose shear stress while cells are being exposed to the test surface and during attachment and to measure the shear stress which prevents attachment from occurring. This does not result in a measurement of adhesive strength, but rather reflects the residence time required by the cell at the surface to establish a stable adhesive interaction. The mean residence time at the surface is the net result of two opposing forces, i.e. the attractive forces between the cell and surface and fluid shear. If the time at the surface required for stabilization of adhesion is greater than mean residence time, then the cell will be washed away before adhesion can occur.

On the basis of this theoretical background, Fowler and McKay (1980) developed the radial flow growth chamber for determination of the maximum shear at which firm adhesion can take place. The apparatus consists of two parallel discs separated by a narrow space. A cell suspension is pumped at a constant volumetric flow rate through the centre of one disc, and it flows radially out between the two discs to a collection manifold. The cross-sectional area of the flow duct increases with increasing radius. Consequently, the linear fluid velocity and surface shear stress decrease radially across the disc. Thus, the shear stresses at which bacterial adhesion occurs can be identified. In addition, the influence of environmental conditions, e.g. nutrients, electrolyte concentration, pH, on attachment at different shear stress can be evaluated. For further details on construction of the radial flow chamber and its application see Fowler and McKay (1980).

## III. Methods for measurement of cell biomass or numbers

### A. Microscopy

#### *1. Introduction*

The first studies of bacterial attachment to surfaces used light microscopy, and microscopy is still the most valuable single technique for evaluating numbers and distribution of attached cells. Although surfaces must be optically transparent for transmitted light microscopy, the application of epifluorescent microscopy and scanning electron microscopy has enabled the examination of opaque surfaces. Consequently, adhesion to a wide range of surfaces can be visualized and evaluated. However, microscopy also has its disadvantages. Surfaces must be relatively flat. Also, because of large variations in distribution, it may be necessary to analyse large surface areas to obtain statistically sound data. Without computer-enhanced digital image analysis, this can be quite time consuming and tedious. The possibility of operator bias is another significant disadvantage of microscopy (Mackowiak and Marling-Cason, 1984) that can be minimized by image analysis. Nevertheless, microscopic examination is invaluable for the information it can give on cell distribution and relationships between location of cells and surface topography.

#### *2. Transmitted light*

In its simplest form, transmitted light microscopy is used to evaluate attachment by using glass slides or coverslips as attachment surfaces. Transparent plastics (e.g. polystyrene, polycarbonate) or mica can also be used, but as light is lost due to internal reflection, the image is poorer. With glass, phase contrast is suitable, but with other surfaces it is usually necessary to stain with a general purpose stain, e.g. crystal violet. In our experience, attachment should be stabilized before staining by fixation, such as by treatment with Bouin's fixative [III.D.1(c)] for approximately 3–5 min, although the time is not critical.

A number of workers have extended the basic light microscope system to allow more detailed analysis of various features of bacterial attachment. Examples are the adaptation of microscopes for *in situ* microscopy, the optimization of staining techniques, the construction of surface holders which allow the effects of fluid flow to be investigated, and the application of computer enhanced microscopy.

*(a)* In situ *light microscopy.* Staley (1969) and Hirsch (1972) used partially submerged microscopes and water immersion objectives to observe the colonization of glass surfaces by bacteria and algae. This allows some characterization of indigenous surface colonizers but is hampered by the morphological similarity among many aquatic bacteria and constrained by the resolving power of the microscope and practical considerations, e.g. risky use of expensive equipment.

*(b) Staining attached cells and polymers.* Attached cells are readily visualized with standard stains such as crystal violet or carbol fuchsin. However, the extracellular polymers associated with attached cells are frequently of interest, but are rarely or only faintly stained. Allison and Sutherland (1984) developed a staining method which demonstrated polymers associated with attached cells and allowed a visual estimation of polymer production.

With this method, the surface, e.g. glass slide, with attached bacteria is removed from the bacterial suspension, covered with 10 mM cetyl pyridinium chloride, and air-dried for 20–30 min. The slide is then fixed by gentle heating, allowed to cool, and stained for 15 min with a 2:1 solution of saturated aqueous Congo red and 10% (v/v) Tween 80. The slide is rinsed, stained with 10% (v/v) Ziehl carbol fuchsin, rinsed again, and dried at 37°C. The purpose of the cetyl pyridinium chloride is to precipitate the extracellular polysaccharide, which is then stained by Congo red and subsequently intensified by Tween 80. The cells are stained with carbol fuchsin.

*(c) Flow cells.* Frequently, real-time observations of living, attaching cells are required, and it is thus necessary to place the bacteria in a chamber through which fresh liquid can be passed to replenish oxygen, as well as nutrients in long-term studies. A coverglass, or some other optically transparent material, forms part of the wall of the chamber, so that cells attaching to the underside of the coverglass can be observed directly. Such

flow cells may also be designed to maintain laminar flow within the chamber, making it possible to calculate rates of bacterial collisions with the surface (Sjollema *et al.*, 1988).

The flow chamber devised by Szewzyk and Schink (1988) consisted of a central well in a Plexiglass (polymethylmethacrylate) slide covered by a glass coverslip, which also served as the attachment surface. Two holes were drilled through the side of the Plexiglass chamber to provide an inlet and outlet for flow of liquid. Flow was controlled by maintaining the liquid reservoir under $N_2/CO_2$ at a pressure of approximately 50 mbar and by regulating a setscrew on the tubing that connected the chamber and reservoir.

A temperature-controlled microperfusion chamber in which bacteria were separated from flowing medium by a dialysis membrane was described by Duxbury (1977). With this system, the influence of diffusible substances on bacterial behaviour, including attachment, can be observed. It has been used to test the influence of metabolic inhibitors on the surface gliding behaviour of *Flexibacter* BH3 (Duxbury *et al.*, 1980) and the influence of starvation on attached cells of *Vibrio* DW1 (Kjelleberg *et al.*, 1982).

More simple flow cells have been devised from capillary tubes (Rutter and Leech, 1980; Leech and Hefford, 1980) and glass chambers constructed from coverslips and microscope slides (Caldwell and Lawrence, 1986; Fletcher, 1988).

*(d) Computer-enhanced microscopy and image analysis.* Counting bacteria on surfaces and evaluating the area covered can be considerably facilitated by using computer-enhanced microscopy. The degree of image analysis depends upon the sophistication of the instrument, and there are image analysers on the market representing a wide range of capabilities and prices. Image analysis systems have been used to determine the numbers of cells attached, area coverage, and biovolume of attached cells, as well as the much more complicated evaluation of attachment, detachment, and cell growth on the surface in real time (Caldwell and Germida, 1985). The simplest and least expensive systems connect a microcomputer to a video camera via a video digitizer (Costello and Monk, 1985) and are suitable for pure culture studies in which high-contrast images can be obtained. When counting of lower contrast images or small cells is required, more sophisticated equipment may be necessary. Sieracki *et al.* (1985) describe the application of a commercial image analysis system to the enumeration and sizing of planktonic bacterial populations, and satisfactorily tested the system at sea. However, they make the important and generally applicable point that accuracy and precision of analysis depends on the quality of the

primary image. For example, in situations where bacteria occur in clusters or chains, it is necessary to evaluate attachment in terms of area coverage rather than numbers attached, unless time-consuming interactive cell-separation programs are applied. However, area coverage may be the more significant parameter, as it can be more readily related to biomass. For a discussion of the conversion of image area to biomass, see Bjornsen (1986).

Methods for analysis of time-dependent microbial colonization of surfaces using computer-enhanced microscopy have been developed by D. E. Caldwell and co-workers using the Kontron IBAS II microcomputer (Caldwell and Germida, 1985; Caldwell and Lawrence, 1986; Lawrence and Caldwell, 1987; Lawrence *et al.*, 1987). The attachment of cells in real time was monitored by capturing images at intervals (e.g. 1.2 min; Lawrence and Caldwell, 1987) and processing them for numerical data, e.g. surface area coverage, microcolonies per field. In experiments testing the effect of fluid flow, the system was used to determine the orientation of the long axis of attached bacteria relative to direction of flow by measuring the feret diameter in the $x$ and $y$ direction. The feret diameter in the $x$ or $y$ direction is a parameter describing the length of a projection of the object in the $x$ or y direction, respectively. The direction having the highest mean diameter was the orientation of the long axis (Lawrence *et al.*, 1987).

### 3. *Incident light*

*(a) Fluorescence.* Often the surfaces of interest are opaque, for example metals, minerals, anti-fouling coatings, macroorganisms. In such cases, epifluorescent microscopy can be used for direct observation of bacteria.

*(i) Fluorochrome dyes.* In general, attached bacteria are easily stained with fluorochromes, so they can be examined visually. The two most commonly used stains are acridine orange (AO) and DAPI (4′6-diamidino-2-phenylindole). Hoechst dyes have also been used (Paul, 1982). DAPI and Hoechst dyes have the advantage over AO when counting bacteria attached to surfaces which bind AO, e.g. polystyrene. Also, with AO staining, if the surfaces have been submerged in natural environments, non-specific AO adsorption to detritus can obscure bacteria.

In a comparison of AO, DAPI, and Hoechst dyes for counting of *Alteromonas citrea* and *Aeromonas* sp. attached to glass or polystyrene, Paul (1982) obtained similar counts with all three techniques. A variety of staining protocols has been used, and representative procedures are given below.

*(ii) Acridine orange (3,6-tetramethyldiaminoacridine).* AO binds to DNA and RNA and fluoresces when excited with light at a wavelength of 436 or

490 nm. Generally, the DNA–AO complex fluoresces greenish-yellow while the RNA–AO complex fluoresces red. However, factors such as dye concentration, pH, or fixative used can also influence the action of the dye (Chayen *et al.*, 1973). It has been used successfully at a wide range of concentrations, and some preliminary testing may be necessary to determine the optimum conditions for a particular situation. For example, Austin (1988) recommends a concentration of 5 mg $l^{-1}$, whereas Paul (1982) used 0.5 mg $l^{-1}$. Also, there is no dissolving medium of choice; various buffers (Meyer-Reil, 1978; Fletcher, 1979), sea water (Paul, 1982), or water are all suitable under appropriate conditions.

Examples of suitable staining solutions are:

0.02% (w/v) AO in 0.07 M phosphate buffer, pH 6.6 (Fletcher, 1979);
0.005% (w/v) AO in sea water (Paul, 1982);
0.5% (w/v) AO in distilled water (Fletcher and Loeb, 1979).

Staining for 2–3 min is generally sufficient, followed by thorough rinsing with the corresponding type of water or buffer, and finally distilled water.

Stock solutions, e.g. 0.1 g $l^{-1}$, may be stored at 4 °C for several months and should be filtered (0.22 μm) when made and just before use.

For examination, surfaces should be air-dried, and then a drop of low fluorescence immersion oil (e.g. Cargille type B) should be applied, followed by a coverslip. However, after application of oil, surfaces should be evaluated as soon as possible. There is frequently detachment of cells and their release into the oil, particularly with hydrophobic surfaces.

If the attachment substratum is plastic then fluorescence of the surface may be a complication. Ladd *et al.* (1985) resolved this problem when visualizing bacteria attached to catheter surfaces by staining the specimen with 0.01% AO for 2 min, followed by 1% (w/v) aqueous malachite green for 10 min. The cells stained dark green, while the background surface fluoresced orange.

*(iii) DAPI.* DAPI is specific for DNA under a variety of conditions, and fluoresces at a wavelength of >390 nm when excited with light at a wavelength of 365 nm. Unbound DAPI or DAPI bound to non-DNA material may fluoresce a weak yellow, and thus can be easily distinguished from bacteria which fluoresce blue.

Samples can be stained with a DAPI solution of 0.01 μg $ml^{-1}$ distilled water (0.029 μM) for 5 min or longer (Porter and Feig, 1980). They should be rinsed with distilled water and examined in the same way as described for AO. Paul (1982) used a higher DAPI concentration of 10 μM.

A concentrated stock solution of 1.0 mg $ml^{-1}$ may be stored indefinitely at 0 °C.

*(iv) Hoechst dyes 33258 and 33342.* Hoechst 33258 binds to DNA specifically, and there is an increase in its fluorescence when it binds to adenine-thymine-rich portions of DNA (Latt and Statten, 1976). Upon excitation with light of a wavelength of 360 nm, the DNA–dye complex fluoresces at a wavelength of 450 nm. Hoechst 33342 has an identical structure to that of Hoechst 33258, except that an hydroxyl group is replaced by an ethoxy group (Paul, 1982). Paul (1982) found that counts of cultures of estuarine bacteria attached to polystyrene were the same when stained with Hoechst 33258 or AO.

Surfaces should be rinsed with a buffer appropriate to the experiment and fixed in buffer containing 1.86% (w/v) formaldehyde for 1 h. They are then stained with 10 μM Hoechst 33258 or 33342 in buffer for 1 h, and rinsed with buffer.

*(v) Immunolabelling.* Often, there is a need to know not only the number of bacteria attached to a surface but also the types of species or strains attached from a mixed culture or natural community. When bacteria cannot be distinguished by their morphology or differential staining, then it may be possible to use immunofluorescent staining. Antibodies may be generated for the strains of interest, or in the case of a natural environment, antibodies may be generated from strains representing genera likely to be found in the environment being studied (Dahle and Laake, 1982; Zambon *et al.*, 1984). Generally, fluorescein isothiocyanate is the fluorochrome which has been used for labelling microorganism-specific antibodies. However, it should be possible to use dual labelling, whereby one antibody is conjugated with one fluorochrome, e.g. fluorescein isothiocyanate, whereas another is conjugated with a second fluorochrome with different fluorescence characteristics, e.g. rhodamine.

Generation of antibody, conjugation with fluorescein isothiocyanate, and establishment of a working titre should be done by standard immunological techniques (Clausen, 1981; Zambon *et al.*, 1984). For determination of numbers of cells attached to a surface, fix the cells in 2% glutaraldehyde, rinse with distilled water, and air-dry (Zambon *et al.*, 1984). If several antibodies are available, replicate surfaces can be stained sequentially, so that one surface is stained with one antibody, a second is stained with two antibodies, etc., so that the number of bacteria in different groups can be determined by subtraction. Total numbers of bacteria can be determined by epifluorescence with AO [III.A.3(a)(ii)].

Bacteria can also be specifically stained by treating bacteria on surfaces with polyvalent rabbit antibodies, followed by fluorescein isothiocyanate-conjugated swine anti-rabbit antibodies (Hoff, 1988). With this method, rabbit antiserum is diluted 1:100 in 0.01 M phosphate buffer, pH 7.2, supplemented with 20 g of NaCl, 4.3 g of $MgCl.6H_2O$, and 1 g of $NaN_3$

to give a final volume of 1000 ml. Filter antiserum (0.22 μm), and cover attached bacteria with it for 20 min. Gently wash surface with distilled water and allow it to drain. Dilute commercially prepared FITC-conjugated swine anti-rabbit immunoglobulins 1:100 with phosphate buffer, filter (0.22 μm), and add to antibody-treated bacteria for 20 min. Gently wash and drain. The sample can also be stained with DAPI (1 $\mu g\ ml^{-1}$) for 5 min and washed, to allow bacteria without bound antibody to be detected. DAPI-stained cells will be visible with excitation at 365 nm, while cells with bound fluoroscein isothiocyanate conjugate will be visible with excitation at 495 nm.

*(b) IRM.* Although microscopy is almost always used to determine numbers and distribution of attached cells, interference reflection microscopy (IRM) can be applied to investigate the interfacial area between an attached or attaching bacterium and the substratum, which is a glass coverslip (Fletcher, 1988). The bacterium is viewed by incident illumination which is reflected by both the bacterial surface and the adjacent surface of the coverglass substratum. The degree of interference between the reflected light from the two surfaces is a function of the distance separating them. This is visualized as differences in image darkness, so that at closest approach to the surface, an image of maximum darkness is obtained. As the cell–substratum distance is increased the image becomes progressively brighter. Maximum brightness is achieved at a distance that is a function of the wavelength of light used and the refractive indices of the phases through which the light travels. For green light of 546 nm, this is approximately 100 nm. Thus, the separation distance of living cells in real time can be evaluated, and the influence of changes in environmental factors, e.g. electrolyte concentration, nutrients, can be monitored. For more detail on the theory and complicating features of the technique, refer to Curtis (1964), Gingell and Todd (1979) and Izzard and Lochner (1976).

The basic technique is relatively simple to set up, but high precision is required for quantitative work (Gingell and Todd, 1979; Izzard and Lochner, 1976). A Zeiss Standard 14 microscope can be adapted for IRM, as described by Abercrombie and Dunn (1975), by fitting the epifluorescent condenser (IV FL) with an interference green filter (BP 546/9) in the exciter position and a dichroic reflector (LT 510). The objective must be a ×100 of high optical quality, and the iris field stop must be closed to form a small opening to collimate the light.

IRM observations can be alternated with phase contrast by blocking off the epi-illumination and turning on transmitted light. Images should be recorded on a high-contrast, high-resolution film, such as Technical Pan (Eastman Kodak). The film should be developed for high contrast, and it is important that developing conditions are highly standardized so that image density is not influenced by variations in development. Alternatively, images can be observed on a video monitor (Preston and King, 1984), but this does not allow quantitative comparison with other IRM images published in the literature (Gingell and Vince, 1982).

In practice, it can be difficult to obtain well-focused micrographs of IRM images, because the dark area is frequently so small it is hard to focus by eye. In such cases, it may help to focus on the edge of the iris diaphram, which is in approximately the same focal plane as cells attached to the underside of the coverslip.

### *4. Electron microscopy (EM)*

A number of attachment studies have used EM to assess numbers and distribution of attached cells. This can also give some information on the mechanism of attachment as structures such as fimbriae, flagella, or polymeric adhesives may be resolved.

EM may also give some information on attachment in natural environments. An example was the use of formvar- or collodion-coated electron microscope grids as attachment substrata, which were submerged in different aquatic habitats at various depths (Hirsch and Pankratz, 1970). These were then retrieved after submersion times ranging from 0.5 to 24 days, air-dried, shadowed at 11–18° with platinum–palladium, and examined by transmission electron microscopy. This provided more morphological detail than light microscope studies, but gave no information on attachment behaviour.

*(a) Scanning electron microscopy (SEM).* SEM is frequently used to visualize attachment to surfaces because it can be applied to a wide range of surfaces, such as metals, minerals, and plant and animal tissues. It can be used to count cells, determine topographical features of the surface, and observe distributional relationships among cells and substratum features. Advances in preparation techniques have led to improved preservation of surface polymers and appendages. However, it is important to realize that preparation can denature polymers, and artefacts are certain to occur without precautions which eliminate dehydration effects. When conventional chemical and fixation methods, followed by critical point drying or argon replacement induced drying (Lamb and Ingram, 1979), are used, bacteria typically have associated 'polymer' or 'fibril' material. Frequently, in the literature this material is attributed to be a presumptive adhesive. However, Fraser and Gilmour (1986) demonstrated that polymer fibrils and condensates were likely to be artefacts. They were eliminated by

cryofixation, that produced bacterial images which had apparently not contracted and on which the polymer presumably remained as a thin coat on the cell. Thus, great care must be taken when interpreting SEM images, and information on adhesive polymers is best obtained by other methods.

The method used by Fraser and Gilmour (1986) is as follows. Rinse surfaces with attached cells in distilled water and lay on sterile filter paper moistened with sterile distilled water in a Petri dish. When the substratum surface is free from excess moisture, it is attached to a specimen stub of a Sputter Cryo coating chamber system (e.g. Emscope Laboratories Ltd, Ashford, Kent) with contact adhesive and plunged in 'sub-cooled' liquified nitrogen (reliquified nitrogen slush) for *ca* 3 min. The stub is then transferred to the coating chamber and warmed to −80 °C for 10 min and recooled to −170 °C. It is sputter-coated with gold and transferred to the SEM cold stage (−165 °C) for viewing.

*(b) Transmission electron microscopy (TEM).* There are two types of TEM preparation for studying attached cells: (i) negative-stained or shadow preparations and (ii) thin sectioning.

The first method is analogous to light microscopy or SEM in that the attached bacterium is viewed from above, rather than edge-on. It has the advantage over SEM of increased resolution, so that attachment polymers or structures such as pili can be observed. Also, like SEM, it gives information on the attached cell numbers and distribution. The disadvantage of negative staining or shadowing and TEM is that the choice of substratum is limited to coated EM grids.

To prepare specimens for negative staining or shadowing, coated EM grids are used as the attachment surfaces and are immersed in the bacterial suspension. Nickel or gold grids are suitable for seawater suspensions. They can be coated with formvar or collodion, but formvar may give better results (Marshall *et al.*, 1971a). After the attachment period, the grid is removed, fixed in 2.5% formaldehyde for 30 min, rinsed, air-dried, and shadowed, e.g. with gold–palladium alloy (60% gold–40% palladium) (Marshall *et al.*, 1971a). Alternatively, grids may be negatively stained with 1.0% phosphotungstic acid, pH 6.8 (Watson and Remsen, 1969).

To view the interface between the attached cell and the underlying substratum, a sample can be prepared for TEM by thin-sectioning. Combined with histochemical staining, TEM can thus provide some information on the nature of the adhesive polymers. The most common procedure for examination of attached cells uses the polyanionic dye ruthenium red, which has demonstrated apparently adhesive material in a number of systems colonized by attached bacteria (Costerton, 1980).

Since the substratum must be sectioned as well as the bacterium, it is necessary to use either a prepolymerized piece of resin (Marshall and Cruickshank, 1973) or a material which can be embedded as the attachment surface (Fletcher and Floodgate, 1973). For example, cellulose ester membrane filters can be used as a substratum, and subsequently the filter with attached bacteria can be processed by standard preparation techniques. The actual details of the method can vary, but many are similar to the following example, which uses a cellulose ester membrane filter as the attachment substratum.

Place a sterile piece of cellulose ester membrane filter (0.22 μm porosity, *ca* 2 $cm^2$) in bacterial suspension, and allow the bacteria to attach for an appropriate time. Remove filter, and cut into pieces small enough to fit into the embedding capsule. Fix for 1 h at 4 °C in a solution consisting of equal parts of 3.6% (v/v) glutaraldehyde, 0.2 M sodium cacodylate buffer, pH 7.4, and ruthenium red stock solution (1.5 g $l^{-1}$ distilled water). Rinse three times in 0.15 M sodium cacodylate buffer over a period of 15 min. Post-fix for 3 h at room temperature in a solution consisting of 2 parts 2% (w/v) osmium tetroxide in 0.1 M sodium cacodylate buffer (pH 7.4) to 1 part ruthenium red stock solution. Rinse briefly in 0.15 M sodium cacodylate buffer, and dehydrate appropriately before embedding.

## B. Measurement of biomass by culture and biochemical techniques

### *1. Introduction*

Although microscopy can frequently provide valuable information on numbers, morphological types, and distribution of attached cells, as well as some information on attachment structures, it has its limitations. Its application to topographically complex surfaces is limited, and apart from histochemical observations, gives little information on the chemical composition of attached populations. Microscopy is also perhaps more subject to 'sampling error', as proportionally smaller areas are examined. Thus, where complex surfaces are involved and/or analysis of a large surface area is required, it is preferable to measure a chemical component, a biomarker, which is an indicator of biomass. A disadvantage of biomarker measurement is that it is not possible to determine cell numbers without a suitable calibration, and calibrations of numbers to specific cellular components can vary with nutritional status of the bacteria. A simpler alternative is to use culturing techiques to determine numbers of attached cells.

### 2. *Removal and culture of attached cells*

When attached bacteria are difficult or impossible to observe by microscopy, an alternative approach has been to remove the bacteria quantitatively from the surface and determine numbers by direct or viable counts. This is successful when cells are easily removed from the surface and has the advantage that attached bacterial strains may be identified to some extent through viable counts and the use of differential media. The greatest drawbacks are incomplete removal of the cells without loss of viability and their inefficient dispersal before direct counting or plating out. There is virtually always incomplete removal of cells and incomplete disruption of aggregates is probable, so numbers will be underestimated by this technique. However, if the underestimation is by a consistent amount, numbers can be reliably estimated by applying a separately determined calibration (Mackowiak and Marling-Cason, 1984).

For removal of bacteria from sediment particles, various combinations of homogenization, sonication, and treatment with detergents, cation sequestering agents, and adhesive denaturants have been applied with mixed success (Litchfield *et al.*, 1975; McDaniel and Capone, 1985). Thus, although this approach can be applied quantitatively (Mackowiak and Marling-Cason, 1984), no one procedure can be recommended, and an optimum protocol should be worked out empirically.

A related approach is to leave the bacteria attached to the surface and to evaluate numbers of attached cells by using the surface to imprint solid media which are subsequently incubated. The methylotrophic bacteria colonizing green plant leaves were evaluated in this way using methanol–ammonium salts agar (Corpe, 1985). A relatively low density of culturable bacteria on the surface is necessary. Otherwise crowding of colonies and confluent growth on the solid medium will prevent the possibility of distinguishing colonies.

### 3. *Spectrophotometric techniques*

Bacterial coverage can be measured semi-quantitatively by staining cells with a dye and then measuring the dye absorption in a spectrophotometer. This usually requires the improvisation of some sort of holder for the surface to place it in the light path in the spectrophotometer.

The bacteria should be fixed with a fixative such as Bouin's fixative, and after rinsing stained with a dye such as crystal violet or Congo red [II.D.1(c)]. The surface is then placed in the spectrophotometer, and absorption is read at the wavelength of optimum absorbance for the dye used: 590 nm for crystal violet or 470 for Congo red. If 50-mm-diameter polystyrene dishes are used as the attachment surface [III.D.1(a)], these can be inserted directly in the sample chamber of a number of spectrophotometers. Random readings can then be obtained by rotating the dish in the chamber.

### 4. *Biochemical markers*

A number of cell constituents, e.g. protein, carbohydrate, ATP, DNA, lipopolysaccharide (LPS), can be used as biochemical markers. Thus, the amount of a particular cell component is determined and then related to biomass using a separately determined calibration. The disadvantage of such analyses is that the numbers of attached bacteria per surface area may be small, so that large surface areas may be needed to make accurate measurements. In addition, the relative proportion of a given constituent may vary with different organisms or nutritional status, so that calibration to cell biomass is difficult.

*(a) ATP.* The biomass of attached cells has been determined indirectly by measuring extractable ATP by the luciferin/luciferase reaction. Harber *et al.* (1983) described a simple and rapid bioluminescence method for evaluating cell attachment to polystyrene tubes which served as the attachment substrata. Attachment was measured as an adherence ratio, in which the amount of ATP in bacteria which attached to the tube surface was related to ATP in the original bacterial suspension which was exposed to the tubes. The technique can be easily applied to various other substrata.

For the assay, the internal surface of tubes designed for use with a luminometer can be used as attachment surfaces. Harber *et al.* (1983) used polystyrene tubes for use with the LKB Luminometer 1250. With this method, samples (300 μl) of the bacterial suspension are dispensed into tubes and incubated at an appropriate temperature for attachment. The samples are then removed and the tubes washed twice with 1 ml of buffer. Bacterial ATP is then extracted by adding 40 μl of a mixture of equal volumes of Lumac NRB reagent (nucleotide-releasing reagent) and phosphate-buffered saline (PBS: $KH_2PO_4$, 3.45 g $l^{-1}$; $Na_2HPO_4$, 4.45 g $l^{-1}$; NaCl, 5.00 g $l^{-1}$; KCl, 0.20 g $l^{-1}$, pH 6.8). The extracts are then vortexed for 10 s and treated with 160 μl 25 mM HEPES buffer (pH 7.75, containing 2 mM EDTA). Extracted ATP is measured in a luminometer after adding 100 μl LKB ATP-monitoring reagent (firefly luciferin/luciferase) to a 200-μl sample. ATP from attached cells is then calibrated against bacterial suspensions of known density. The ATP is extracted from suspensions of bacteria in PBS by vortexing 100 μl of bacterial suspension with 100 μl of

Lumac NRB reagent for 10 s followed by the immediate addition of 800 μl HEPES buffer.

*(b) DNA.* Numbers of attached cells have also been evaluated by measuring their DNA with a fluorometric technique using Hoechst staining (Paul and Loeb, 1983). With this method, bacteria are allowed to attach to hydrophobic Petri dishes (60 mm × 15 mm), and non-attached cells are removed by rinsing. Bacteria attached to the side walls can also be removed by wiping with a cotton-tipped probe. Four μl of 5% (v/v) Triton X-100 are added to each dish, which is then put on a bed of crushed ice. After the addition of 4 ml of an ice-cold solution of 0.154 M NaCl and 0.015 M trisodium citrate, pH 7.0, to each dish, the attached cells are broken by sonication for 45 s at 100 W. The dish is rotated during sonication to ensure even exposure. The authors point out that longer sonication times may be necessary for more resistant bacteria. After sonication, 2 ml of the liquid is added to 1 ml of either $1.5 \times 10^{-6}$ M (for 0.5–10 μg DNA) or $1.5 \times 10^{-7}$ M (for 50–1500 ng DNA) Hoechst 33258 in the NaCl/trisodium citrate solution (above). The fluorescence of the solution is measured in a fluorimeter to determine DNA concentration, by using calf thymus DNA standards (Paul and Meyers, 1982).

*(c) Other biochemical markers.* If the attached organisms are Gram-negative bacteria, then they may be evaluated by determination of LPS concentration. This approach was used to measure attachment to surfaces *in situ* by utilizing *Limulus* amoebocyte lysate (Dexter *et al.*, 1975). This extract from *L. polyphemus* (the horseshoe crab) forms a gel or turbid solution in the presence of the protein–LPS endotoxin of Gram-negative bacteria, and the optical density of the solution is related to bacterial density. The technique has been infrequently used, probably because of the availability of methods with less exacting protocols and broader specificity.

A number of additional biochemical markers have been used to measure the biomass of multicellular layers of attached bacteria, or biofilms. Some of these may be effective for evaluating monolayers of bacteria, but their suitability for such low numbers of cells is largely untried. Cell constituents that could be measured include lipid phosphate (White *et al.*, 1979; Gehron and White, 1983), lipid A of lipopolysaccharides (Parker *et al.*, 1982), teichoic acids in Gram-positive bacteria (Gehron *et al.*, 1984), and muramic acid (Findlay *et al.*, 1983). Such techniques have the advantage that they may provide indications of constituent organisms and/or their physiological condition (White, 1984; McKinley *et al.*, 1988).

## C. Radiotracer labelling

### *1. Introduction*

A relatively rapid method for measuring attachment to surfaces is the prelabelling of cells with a radiotracer, allowing attachment to test surfaces, and then determining numbers of attached cells indirectly by scintillation counting. This method has the advantage that it is sensitive and very accurate and allows rapid processing of a large number of surfaces. The disadvantage is that preincubation with labelled substrate is required, and this may influence the attachment properties of the organisms. Supplementary experiments are required to calibrate scintillation counts to cell biomass, confirm that the radiolabel is stably retained within the cells for the required period of time, and ensure that radiation is not absorbed by test surfaces resulting in spuriously low counts. Two substrates which have been commonly used to label cells are leucine and thymidine.

### *2. Leucine*

Leucine is a substrate which is readily assimilated by many microorganisms and is often stably retained over periods of hours. With this technique, bacterial suspensions are radiolabelled (Pringle and Fletcher, 1983) by incubation with $^{14}$C-leucine at 1 μCi $ml^{-1}$ for 2 h at appropriate incubation temperature. The cells are then washed three times and resuspended in 0.01 M $KH_2PO_4$ buffer (adjusted to pH 7.4 with 1 M NaOH) to a concentration of $2.5 \times 10^9$ to $5 \times 10^9$ $ml^{-1}$. The attachment substrata are then exposed to the bacterial suspension at an appropriate temperature and for an appropriate time (2 h is generally suitable) to allow attachment. The cell suspension should be removed by flushing with buffer (100 ml) under constant flow conditions and the surfaces transferred to scintillation vials with only one passage through a liquid–air interface (the apparatus illustrated in Fig. 3 would be appropriate). Before addition of the liquid scintillation solution, the surfaces are air-dried for 1 h at 40°C and positioned horizontally in the counting vial.

When the radioactive sample is located primarily on the surface, there can be a loss in efficiency of counting $^{14}$C due to local 'self-absorption' by the substratum, often referred to as β-absorption. In this case, the surface absorbs β-emissions, so they do not produce a measurable excitation of the scintillant. This results in a decrease in sample count which is not the same as quenching and thus is not accounted for by a quench correction. An accurate calibration curve must therefore compensate for the relative level of β-absorption to measure the relative attachment to each type of surface.

Thus, a calibration curve relating liquid scintillation counts to cell numbers for each substratum used is produced. The washed radiolabelled cells are suspended in absolute ethanol and diluted to ten concentrations between $2.5 \times 10^7$ and $5 \times 10^9$ bacteria $ml^{-1}$. The suspensions (100 μl) are then either put in scintillation vials (controls) or allowed to spread on the substrata, dried, and counted as described for attached cells. The accuracy of the calibration curve can be verified by comparing direct microscopic counts of cells attached to selected test surfaces with the corresponding numbers derived from the calibration curve.

### 3. *Thymidine*

Ørstavik (1977) prelabelled streptococci by incubation with radiolabelled thymidine. With this method, 10 ml of an overnight culture is added to 90 ml of trypticase soy broth and incubated for 2 h. Then (6-$^3$H-thymidine (25.4 Ci $mmol^{-1}$, 105 mCi $mg^{-1}$) is added to give a final activity of 1 μCi $ml^{-1}$ and incubated for 45 min. The cells are harvested, washed three times, and resuspended in the buffer or salt solution used for the subsequent attachment experiment. After exposure of the labelled cells to glass coverslips, the surfaces are transferred to scintillation vials, and 500 μl of formamide is added to each vial and incubated for 2 h at 37°C with continuous stirring. Ten ml of a dioxane–xylene-based scintillation liquid is added to each vial and the radioactivity in counts per minute (cpm) determined by scintillation counting.

To convert cpm to bacterial density, a calibration curve is prepared by placing different amounts (10–50 μl) of the original bacterial suspension of known density onto coverslips, allowing these to dry, and then treating them with formamide and determining cpm as described above. Ørstavik (1977) determined that there was no loss of $^3$H activity from cells over 21 h, but there was a gradual decrease in cell numbers. The counting efficiencies of attached cells compared to those of suspended cells ranged from 101% to 107%, and the relationship between cpm and cell density (dried on coverslips) was linear.

## D. Nucleic acid hybridization probes

When primary interest is in identification of the species or physiological types attached to the surface, e.g. in characterization of environmental samples, suitable methods are few. The traditional technique of using differential growth media is severely limited because of the difficulty in removing cells from the attachment surface, at the same time maintaining culturability (III.B.2). Thus, in recent years, attempts are being made to develop methods to enable identification of cells while they are still on the surface. Specific fluorescence antibody techniques are one option, while more recently the possibility of nucleic acid probes has emerged.

Although DNA–DNA probes have been applied successfully to environmental isolates (Sayler *et al.*, 1985; Holben *et al.*, 1988), oligonucleotide probes for detection of 16S rRNA sequences probably hold the most immediate promise for identification of attached bacteria (Stahl *et al.*, 1988). The use of probes to identify naturally attached cells is not documented. However, Giovannoni *et al.* (1988) used oligodeoxynucleotide probes complementary to 16S rRNA to characterize bacteria artificially attached to a glass slide by first dipping the slide in gelatin and applying the bacteria to the gelatin-coated slide. Labelling of individual cells was then visualized by micro-autoradiography. It may be possible to apply a similar protocol to naturally attached cells, although it would be necessary to reinforce their attachment to the surface in some way (e.g. coating with a thin layer of gelatin) so cells would not be removed by the various chemical treatments involved in probe hybridization. Although the application of nucleic acid probes to adhesion studies is not yet established, it is clearly a potentially valuable technique for future studies.

## IV. Concluding remarks

Possibly the most important lesson to be learned from the past 15 years of research on bacterial adhesion is that there is a great deal of variability and diversity among bacterial adhesion mechanisms. The numbers or biomass of attached bacteria which are measured depend heavily upon the bacteria and environmental conditions, and it is rarely possible to extrapolate directly from one system to a different one.

Most of the techniques described in this chapter have been used to measure adhesion in laboratory systems, and little attention has been paid to adhesion measurements *in situ*. However, verification of techniques in the laboratory is necessary before transporting them to the field. It is clear from the care which must be taken in choosing appropriate laboratory approaches to address specific questions on adhesion that equal or even greater care is necessary to evaluate adhesion in natural environments. Similarly, realistic extrapolation from laboratory experiments to natural situations requires sufficient knowledge of both systems and adequate control over the laboratory model.

## References

Abbott, A., Berkeley, R. C. W. and Rutter, P. R. (1980). *In* "Microbial Adhesion to Surfaces" (R. C. W. Berkeley, J. M. Lynch, J. Melling, P. R. Rutter and B. Vincent, Eds), p. 117, Ellis Horwood, Chichester, UK.
Abbott, A., Rutter, P. R. and Berkeley, R. C. W. (1983). *J. Gen. Microbiol.* **129,** 439.
Abercrombie, M. and Dunn, G. A. (1975). *Exp. Cell Res.* **92,** 57.
Allison, D. G. and Sutherland, I. W. (1984). *J. Microbiol. Methods* **2,** 93.
Austin, B. (1988). "Methods in Aquatic Bacteriology", John Wiley, Chichester, UK.
Baker, J. H. (1984). *Can. J. Microbiol.* **30,** 511.
Bartlett, D. H., Wright, M. E. and Silverman, M. (1988). *Proc. Natl. Acad. Sci., USA* **85,** 3923.
Bjornsen, P. K. (1986). *Appl. Environ. Microbiol.* **51,** 1199.
Brown, C. M., Ellwood, D. C. and Hunter, J. R. (1977). *FEMS Microbiol. Lett.* **1,** 163.
Busscher, H. J., Uyen, M. H. W. J. C., van Pelt, A. W. J., Weerkamp, A. H. and Arends, J. (1986a). *Appl. Environ. Microbiol.* **51,** 910.
Busscher, H. J., Uyen, M. H. W. J. C., Weerkamp, A. H., Postma, W. J. and Arends, J. (1986b). *FEMS Microbiol. Lett.* **35,** 303.
Caldwell, D. E. and Germida, J. J. (1985). *Can. J. Microbiol.* **31,** 35.
Caldwell, D. E. and Lawrence, J. R. (1986). *Microb. Ecol.* **12,** 299.
Chayen, J. Bitensky, L. and Butcher, R. G. (1973). "Practical Histochemistry", John Wiley, London.
Cholodny, N. (1930). *Arch Mikrobiol.* **1,** 620.
Christie, A. O., Evans, L. V. and Shaw, M. (1970). *Ann Bot.* **34,** 467.
Clausen, J. (1981). "Immunochemical Techniques for the Identification and Estimation of Macromolecules", 2nd Edn, Elsevier/North Holland, Amsterdam.
Corpe, W. A. (1985). *J. Microbiol. Meth.* **3,** 215.
Corpe, W. A., Matsuuchi, L. and Armbruster, B. (1976). *In* "Proceedings of the 3rd International Biodegradation Symposium" (J. M. Sharpley and A. M. Kaplan, Eds), p. 433, Applied Science Publishers, London.
Costello, P. J. and Monk, P. R. (1985). *Appl. Environ. Microbiol.* **49,** 863.
Costerton, J. W. (1980). *In* "Adsorption of Microorganisms to Surfaces" (G. Bittton and K. C. Marshall, eds), p. 403, John Wiley, New York.
Cowan, M. M. and Fletcher, M. (1987). *J. Microbiol. Meth.* **7,** 241.
Curtis, A. S. G. (1964). *J. Cell Biol.* **21,** 199.
Dahle, A. B. and Laake, M. (1982). *Appl. Environ. Microbiol.* **43,** 169.
Dexter, S. C., Sullivan, J. D. Jr., Williams, J. III and Watson, S. W. (1975). *Appl. Microbiol.* **30,** 298.
Duxbury, T. (1977). *J. Appl. Bacteriol.* **43,** 247.
Duxbury, T., Humphrey, B. A. and Marshall, K. C. (1980). *Arch. Microbiol.* **124,** 169.
Ellwood, D. C., Hunter, J. R. and Longyear, V. M. C. (1974). *Arch. Oral. Biol.* **19,** 659.
Feldner, J., Bredt, W. and Kahane, I. (1983). *J. Bacteriol.* **153,** 1.
Findlay, R. H., Moriarty, D. J. W. and White, D. C. (1983). *Geomicrob. J.* **3,** 133.
Fletcher, M. (1976). *J. Gen. Microbiol.* **94,** 400.
Fletcher, M. (1977). *Can. J. Microbiol.* **23,** 1.

Fletcher, M. (1979). *Arch. Microbiol.* **122,** 271.
Fletcher, M. (1988). *J. Bacteriol.* **170,** 2027.
Fletcher, M. and Floodgate, G. D. (1973). *J. Gen. Microbiol.* **74,** 325.
Fletcher, M. and Loeb, G. I. (1979). *Appl. Environ. Microbiol.* **37,** 67.
Fowler, H. W. and McKay, A. J. (1980). *In* "Microbial Adhesion to Surfaces" (R. C. W. Berkeley, J. M. Lynch, J. Melling, P. R. Rutter and B. Vincent, Eds), p. 143, Ellis Horwood, Chichester, UK.
Fraser, T. W. and Gilmour, A. (1986). *J. Appl. Bacteriol.* **60,** 527.
Gehron, M. J. and White, D. C. (1983). *J. Microbiol. Methods* **1,** 23.
Gehron, M. J., Davis, J. D., Smith, G. A. and White, D. C. (1984). *J. Microbiol. Methods* **2,** 165.
George, J. N., Reed, R. L. and Reed, C. F. (1971). *J. Cell Physiol.* **77,** 57.
Gerson, D. F. and Scheer, D. (1980). *Biochim. Biophys. Acta* **602,** 506.
Gingell, D. and Todd, I. (1979). *Biophys. J.* **26,** 507.
Gingell, D. and Vince, S. (1982). *J. Cell Sci.* **54,** 299.
Giovannoni, S. J., DeLong, E. F., Olsen, G. J. and Pace, N. R. (1988). *J. Bacteriol.* **170,** 720.
Gordon, A. S. (1987). *Appl. Environ. Microbiol.* **53,** 1175.
Gordon, A. S. and Millero, F. J. (1984). *Appl. Environ. Microbiol.* **47,** 495.
Gordon A. S., Gerchakov, S. M. and Udey, L. R. (1981). *Can. J. Microbiol.* **27,** 698.
Harber, M. J., Mackenzie, R. and Asscher, A. W. (1983). *J. Gen. Microbiol.* **129,** 621.
Henrici, A. T. (1933). *J. Bacteriol.* **25,** 277.
Hirsch, P. (1972). *Z. Allg. Mikrobiol.* **12,** 203.
Hirsch, P. and Pankratz, St. H. (1970). *Z. Allg. Mikrobiol.* **10,** 589.
Hoff, K. A. (1988). *Appl. Environ. Microbiol.* **54,** 2949.
Hogt, A. H., Dankert, J., Vries, J. A. de and Feijen, J. (1983). *J. Gen. Microbiol.* **129,** 2959.
Holben, W. E., Jansson, J. K., Chelm, B. K. and Tiedje, J. M. (1988). *Appl. Environ. Microbiol.* **54,** 703.
Isaacson, R. E. (1985). *In* "Bacterial Adhesion" (D. C. Savage and M. Fletcher, Eds), p. 307, Plenum Press, New York.
Izzard, C. S. and Lochner, L. R. (1976). *J. Cell Sci.* **21,** 129.
Kawabata, N., Hayashi, T. and Matsumoto, T. (1983). *Appl. Environ. Microbiol.* **46,** 203.
Kjelleberg, S., Humphrey, B. A. and Marshall, K. C. (1982). *Appl. Environ. Microbiol.* **43,** 1166.
Knox, K. W., Hardy, L. N., Markevics, L. J., Evans, J. D. and Wicken, A. J. (1985). *Infect. Immun.* **50,** 545.
Ladd, T. I., Schmiel, D., Nickel, J. C. and Costerton, J. W. (1985). *J. Clin. Microbiol.* **21,** 1004.
Lamb, J. C. and Ingram, P. (1979). *Scanning. Elec. Microsc.* **3,** 459.
Latt, S. A. and Statten, G. (1976). *J. Histochem. Cytochem.* **24,** 24.
Lawrence, J. R. and Caldwell, D. E. (1987). *Microb. Ecol.* **14,** 15.
Lawrence, J. R., Delaquis, P. J., Korber, D. R. and Caldwell, D. E. (1987). *Microb. Ecol.* **14,** 1.
Leech, R., and Hefford, R. J. W. (1980). *In* "Microbial Adhesion to Surfaces" (R. C. W. Berkeley, J. M. Lynch, J. Melling, P. R. Rutter and B. Vincent, Eds), p. 544, Ellis Horwood, Chichester, UK.

Levich, V. G. (1962). "Physicochemical Hydrodynamics", Prentice Hall, New York.
Litchfield, C. D., Rake, J. B., Zindulis, J., Watanabe, R. T. and Stein, D. J. (1975). *Microb. Ecol.* **1,** 219.
Ludwicka, A., Jansen, B., Wadstrom, T. and Pulverer, G. (1984). *Zbl. Bakt. Hyg. A* **256,** 479.
Mackowiak, P. A. and Marling-Cason, M. (1984). *J. Microbiol. Methods* **2,** 147.
MacRae, I. C. and Evans, S. K. (1983). *Water Res.* **17,** 271.
Marshall, K. C. and Cruickshank, R. H. (1973). *Arch. Microbiol.* **91,** 29.
Marshall, K. C., Stout, R. and Mitchell, R. (1971a). *Can. J. Microbiol.* **17,** 1413.
Marshall, K. C., Stout, R. and Mitchell, R. (1971b). *J. Gen. Microbiol.* **68,** 337.
McDaniel, J. A. and Capone, D. G. (1985). *J. Microbiol. Meth.* **3,** 291.
McEldowney, S. and Fletcher, M. (1986). *J. Gen. Microbiol.* **132,** 513.
McEldowney, S. and Fletcher, M. (1987). *Arch. Microbiol.* **148,** 57.
McKinley, V. L., Costerton, J. W. and White, D. C. (1988). *Appl. Environ. Microbiol.* **54,** 1383.
Meyer-Reil, L.-A. (1978). *Appl. Environ. Microbiol.* **36,** 506.
Molin, G., Nilsson, I. and Stenson-Holst, L. (1982). *Eur. J. Appl. Microbiol. Biotechnol.* **15,** 218.
Murchison, H., Larrimore, S. and Curtiss, R. III (1985). *Infect. Immun.* **50,** 826.
Ørstavik, D. (1977). *Acta Path. Microbiol. Scand. Sect. B* **85,** 38.
Parker, J. H., Smith, G. A., Fredrickson, H. L., Vestal, J. R. and White, D. C. (1982). *Appl. Environ. Microbiol.* **44,** 1170.
Paul, J. H. (1982). *Appl. Environ. Microbiol.* **43,** 939.
Paul, J. H. and Loeb, G. I. (1983). *Appl. Environ. Microbiol.* **46,** 338.
Paul, J. H. and Meyers, B. (1982). *Appl. Environ. Microbiol.* **43,** 1393.
Porter, K. G. and Feig, Y. S. (1980). *Limnol. Oceanogr.* **25,** 943.
Powell, M. S. and Slater, N. K. H. (1982). *Biotech. Bioeng.* **24,** 2527.
Pratt-Terpstra, I. H., Weerkamp, A. H. and Busscher, H. J. (1987). *J. Gen. Microbiol.* **133,** 3199.
Preston, T. M. and King, C. A. (1984). *J. Gen. Microbiol.* **130,** 2317.
Pringle, J. H. and Fletcher, M. (1983). *Appl. Environ. Microbiol.* **45,** 811.
Pringle, J. H., Fletcher, M. and Ellwood, D. C. (1983). *J. Gen. Microbiol.* **129,** 2557.
Rosenberg, M. and Rosenberg, E. (1985). *Oil Petrochem. Poll.* **2,** 155.
Rutter, P. R. and Abbott, A. (1978). *J. Gen. Microbiol.* **105,** 219.
Rutter, P. and Leech, R. (1980). *J. Gen. Microbiol.* **120,** 301.
Satou, N., Satou, J., Shintani, H. and Okuda, K. (1988). *J. Gen. Microbiol.* **134,** 1299.
Sayler, G. S., Shields, M. S., Tedford, E. T., Breen, A., Hooper, S. W., Sirotkin, K. M. and Davis, J. W. (1985). *Appl. Environ. Microbiol.* **49,** 1295.
Sieracki, M. E., Johnson, P. W. and Sieburth, J. M. (1985). *Appl. Environ. Microbiol.* **49,** 799.
Sjollema, J., Busscher, H. J. and Weerkamp, A. H. (1988). *Biofouling* **1,** 101.
Stahl, D. A., Flesher, B., Mansfield, H. R. and Montgomery, L. (1988). *Appl. Environ. Microbiol.* **54,** 1079.
Staley, J. T. (1969). *Bacteriol. Proc.* G47.
Szewzyk, U. and Schink, B. (1988). *J. Gen. Microbiol.* **134,** 183.
van Pelt, A. W. J., Weerkamp, A. H., Uyen, M. H. W. J. C., Busscher, H. J., de Jong, H. P. and Arends, J. (1985). *Appl. Environ. Microbiol.* **49,** 1270.

Velez, B. J., Tosteson, T. R., Zaidi, B. R., Tsai, R. S. and Atwood, D. K. (1976). *FAO Fish. Rep. No. 200 Suppl.,* 363.
Wardell, J. N., Brown, C. M. and Ellwood, D. C. (1980). *In* "Microbial Adhesion to Surfaces" (R. C. W. Berkeley, J. M. Lynch, J. Melling, P. R. Rutter and B. Vincent, Eds), p. 221, Ellis Horwood, Chichester, UK.
Watson, S. W. and Remsen, C. C. (1969). *Science* **163,** 685.
Weerkamp, A. H., Handley, P. S., Baars, A. and Slot, J. W. (1986). *J. Bacteriol.* **165,** 746.
Weerkamp, A. H., van der Mei, H. C. and Slot, J. W. (1987). *Infect. Immun.* **55,** 438.
Weiss, L. (1961a). *Exp. Cell Res. Suppl.* **8,** 141.
Weiss, L. (1961b). *Exp. Cell Res.* **25,** 504.
Weiss, L. (1967). "The Cell Periphery, Metastasis and Other Contact Phenomena", North-Holland, Amsterdam.
White, D. C. (1984). *In* "Microbial Adhesion and Aggregation" (K. C. Marshall, Ed.), p. 159, Springer-Verlag, Berlin.
White, D. C., Davis, W. M., Nickels, J. S., King, J. D. and Bobbie, R. J. (1979). *Oecologia* **40,** 51.
Zambon, J. J., Huber, P. S., Meyer, A. E., Slots, J., Fornalik, M. S. and Baier, R. E. (1984). *Appl. Environ. Microbiol.* **48,** 1214.

# 9
# Methods for Studying Biofilm Bacteria

T. I. LADD*
*Biology Department, Saint Mary's University, Halifax, Nova Scotia B3H 3C3, Canada*

AND J. W. COSTERTON
*Biology Department, University of Calgary, Calgary, Alberta T2N 1N4, Canada*



## I. Introduction

Microorganisms will rapidly attach to almost any surface submerged in the aquatic environment. Similarly, plastic and metal prostheses introduced into the animal body provide a substratum for bacterial colonization. Firmly attached cells grow predominantly as microcolonies which often produce extracellular polymeric substances that enrobe the cells in a gelatinous-like matrix (Costerton *et al.*, 1981). The term biofilm refers to the combination of the cells with the surrounding matrix.

*Present address: Department of Biology, Millersville University, Millersville, PA 17551, USA.

METHODS IN MICROBIOLOGY
VOLUME 22 ISBN 0-12-521522-3

Despite the potential benefits associated with the removal of inorganic and organic contaminants by biofilms in natural streams and waste-water facilities, most of the recent interest in biofilms has been focused on their detrimental roles. Bacterial biofilms reduce the efficiency of heat exchangers, increase resistance to flow, and enhance corrosion. In medicine biofilms can become the foci of overt bacterial disease (Marrie and Costerton, 1982). Confluent biofilms have been shown to confer a degree of resistance to biocides (Marrie and Costerton, 1981) and antibiotics (Nickel *et al.*, 1985), thus making the control and management of nuisance biofilms difficult.

Numerous techniques are available for studying bacterial biofilms. Light and electron microscopy provide excellent means of examining the distribution of the bacteria on the surface. Scanning electron microscopy (SEM) can reveal some details about the mechanism(s) of attachment. Used in conjunction with intracellular formazan deposition techniques and a nalidixic acid sensitivity procedure, light microscopy can provide information about the percentage of actively metabolizing bacteria on a surface or concerning the sensitivity of the attached bacteria to biocidal agents. Procedures such as the Trinder (1969) glucose oxidase/peroxidase enzyme assay can be used to measure substrate utilization by intact biofilms. Radioisotopic techniques provide a versatile way of detecting bacteria on the surface and measuring their response to specific agents. These procedures lack the 'within-the-biofilm', resolution of techniques like nuclear magnetic resonance and Fourier transforming infrared spectroscopy, but provide a means of non-destructively measuring the activity of biofilm bacteria on virtually any type of surface material. The radioisotopic procedure can be modified to determine the activity of attached cells under continuous flow conditions as well. The evolution of radiolabelled carbon dioxide can be used to detect the presence of actively metabolizing bacteria (radiorespirometry) or calculate a rate of activity for the attached population (direct calculation and kinetic approaches). Changes in the activity of the biofilm bacteria in response to biocide challenges can be readily measured. The amount of substrate assimilated by the cells on the surface can also be measured to give a more complete understanding of the activity of the biofilm bacteria.

In this chapter methods are described that may be used to study adherent biofilms on inert surfaces. Many of the procedures allow analysis of attached microbial consortia with minimum perturbation to the biofilm. None require that the cells be removed from the surface. Antibiotic (biocide) sensitivity testing is an especially useful application of procedures which measure the activity of biofilm bacteria.

## II. Establishing surface-associated biofilms

### A. Colonization of surfaces

The growth and development of biofilms can be observed on surfaces deliberately exposed to a microorganism or on surfaces which have become unintentionally fouled by bacteria. The method selected will reflect the specific purpose of the study.

#### *1. Aquatic studies*

*(a) Procedure.* Cobble substrata can be cut with a high-speed diamond saw and ground with silica powders to the desired dimensions (1 cm × 1 cm × 1 mm). The samples are placed directly in the stream or lake or mounted in plastic or wooden platforms and exposed for the desired period of time (Ladd *et al.*, 1979). Samples are retrieved, rinsed at least three times with 30 ml filter-sterilized stream (lake) water (0.45 μm) and transferred to the appropriate vial for later analysis.

#### *2.* In vitro *studies*

A Robbins Device, a multiport sampling device with evenly spaced sampling ports, can be used to examine colonization of engineered surfaces (Ruseska *et al.*, 1982; McCoy *et al.*, 1981; Nickel *et al.*, 1985). The device can be constructed of metal or plastic and attached as a side spur to existing industrial systems or used as a model system for laboratory studies. Surfaces to be exposed are mounted into removable sampling plugs, the device is sterilized by ethylene oxide gas and then affixed to a heat exchanger, nutrient reservoir containing an appropriate medium, or other system as required. The surfaces are colonized over a period of time under continuous flow conditions (batch or continuous culture). Sampling coupons can be removed at intervals for analysis. Coupons are washed three times with 30 ml of sterile buffer to remove reversibly adsorbed cells and blotted with absorbing paper to remove excess rinse water. This procedure is repeated with a second 10-ml wash, and surfaces are transferred to vials for subsequent analysis.

Unmodified tubing can also be used for colonization studies (Ladd *et al.*, 1987). The tubing (glass, Silkolatex, polystyrene, polyethylene) is affixed to a nutrient reservoir via sterile Tygon tubing and flow is initiated through a peristaltic pump at 25–50 ml $h^{-1}$. Either batch culture or continuous culture flow conditions can be used. At the end of the exposure period, the entire length of tubing is rinsed at 200 ml $h^{-1}$ for 1 h with sterile buffer.

Samples which have been enclosed in sterile pipette wraps can be cut aseptically into specific lengths of 1–2 cm using alcohol-flamed forceps and scissors to provide surfaces for microscopy and activity measurements. Samples which cannot be sectioned (glass) can be examined using formazan extraction and radioisotopic procedures.

## III. Microscopic examination of surfaces

The growth and development of bacterial biofilms on a variety of surfaces can be studied using light and electron microscopy. The attachment of cells to transparent materials such as glass, Tygon tubing, and polyethylene tubing can be demonstrated by simple or negative staining procedures. Epifluorescence and scanning electron microscopy are used for both transparent and opaque surfaces.

### A. Bright-field light microscopy

#### *1. Simple staining*

*(a) Procedure.* Aseptically retrieved substrata are rinsed at least three times with 30 ml of sterile phosphate buffer (pH 6.8). Excess rinse water is removed by holding the colonized surface with alcohol-flamed forceps and gently touching the edge of the surface to a sterile 4 $cm^2$ piece of Whatman No. 1 filter paper. The filter paper sections are individually sterilized in glass Petri dishes and discarded after use. The surfaces are then rerinsed with 10 ml of buffer. Repeat with a second wash and transfer the samples to sterile Petri dishes. Fix in an atmosphere of 0.5% buffered glutaraldehyde for 1 h (Geesey *et al.*, 1978) and then stain for 2 min with a 1% solution of filter-sterilized crystal violet or similar basic stain. The stained samples are rinsed, blotted with absorbing paper, and allowed to air-dry. Tygon and other tubing must be flattened in order to view the surface. This can be accomplished by mounting the material on a glass slide with an appropriate glue.

*(b) Interpretation.* Cells on transparent surfaces will be deeply stained. Numbers are determined by counting the number of cells per oil immersion field (OIF) or per grid area if an ocular graticule is used.

*2. Negative staining*

*(a) Procedure.* A loopful of a 10% filtered solution of nigrosine containing 1 ml of 40% formaldehyde as a preservative is applied to the surface and removed with absorbing paper after 2 min. The preparation is allowed to air-dry and is examined under oil, with the oil placed directly on the stained surface.

*(b) Interpretation.* The cells appear as light shadows against a dark background. Determine the cell number and note the presence of capsular material around cells. When used in conjunction with the intracellular formazan procedure, actively metabolizing cells can be determined (Fig. 1).

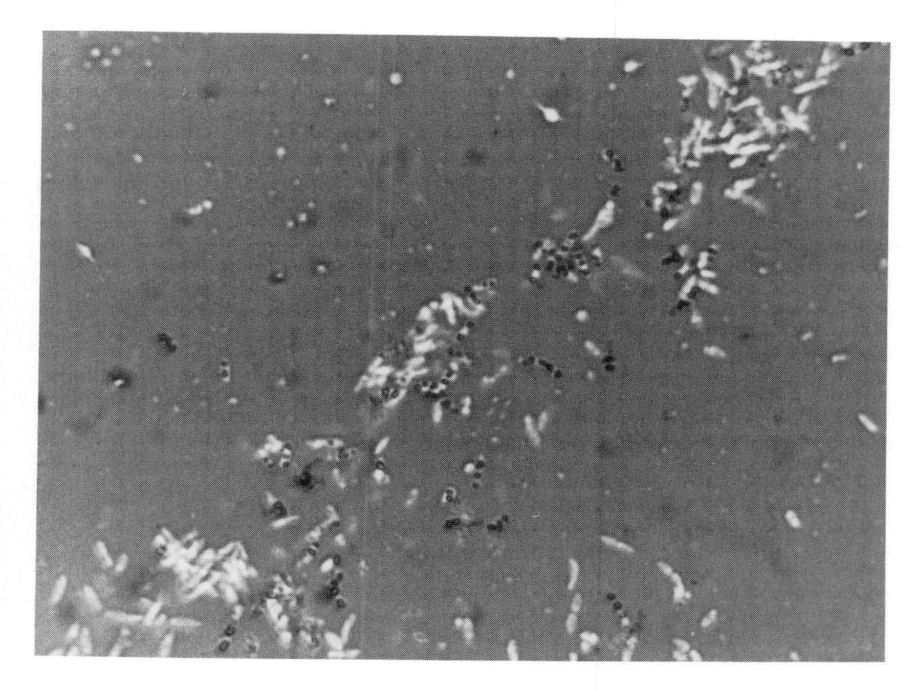

**Fig. 1.** Negatively stained bacteria attached to the surface of plastic discs. Formazan deposits are visible in many of the cells.

## B. Epifluorescence microscopy

### *1. Acridine orange staining*

*(a) Reagent.* Dissolve 10 mg of the fluorochrome acridine orange (AO) in 100 ml of sterile phosphate buffer (0.1 M; pH 7.2) and filter through a 0.2-μm Nucleopore filter (Geesey *et al.*, 1978). Prepare the stain fresh daily. Acridine orange is a potential carcinogen and must be handled with care.

*(b) Procedure.* Glutaraldehyde-fixed specimens are stained for 2 min and rinsed for 30 s with filter-sterilized 70% alcohol (Geesey *et al.*, 1978). The samples are air-dried and examined under oil using a microscope equipped for epifluorescence microscopy such as the Zeiss Standard 16 microscope illuminated with a HBO 50-W burner and fitted with a KP490 excitation and LP 520 barrier filter (Ladd *et al.*, 1982).

### *2. Counterstaining*

Surfaces which show bright background fluorescence such as Silokolatex catheters cannot be viewed unless the surface is counterstained with a basic dye such as crystal violet or malachite green (Ladd *et al.*, 1985).

*(a) Procedure.* Counterstain for 2 min, rinse with distilled water, blot, and air-dry. Examine under oil.

*(b) Interpretation.* Green, orange, and yellow–orange fluorescing cells are counted. For aquatic samples, bacteria can be distinguished from algae on the basis of size and morphology. A dye such as 4′6-diamidino-2-phenylindole provides a better distinction between bacteria and detritus (Porter and Feig, 1980). Counterstained catheter-associated bacteria appear as darkly stained cells against a bright background. Record cell density in cells $cm^{-2}$ (see Fig. 2).

## C. Scanning electron microscopy (SEM)

### *1. Procedure*

The sample is fixed in a solution containing 5% buffered glutaraldehyde (cacodylate buffer, 0.067 M; pH 6.2) for 24 h at 20°C. The sample is then dehydrated through a graded series of aqueous alcohol (30–100%) and

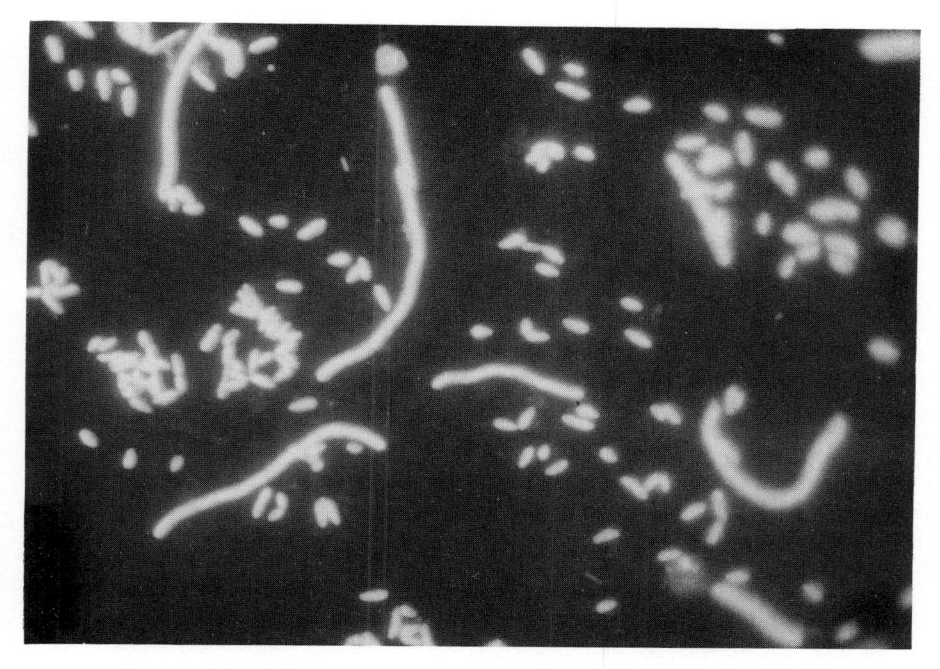

**Fig. 2.** Acridine orange-stained cells on surface of a stone. Note the presence of elongated cells in the absence of nalidixic acid.

alcohol-Freon 113 (30–100%) washes and critical point-dried (Nickel *et al.*, 1985). Dehydration times are 20–30 min duration. Specimens are coated in gold and viewed on a scanning electron microscope. Alternatively, the sample can be fixed in buffered glutaraldehyde containing 0.15% ruthenium red (Marrie and Costerton, 1984) and 'metallized' with osmium tetroxide and thiocarbohydrazide (Malick and Wilson, 1975). The samples are dehydrated in ethanol and Freon 113 before critical point-drying and examined with the scanning electron microscope.

## 2. *Interpretation*

Individual cells, microcolonies, fibrillar strands, and condensed polymeric material enrobing the cells may be seen (Figs 3A–C). Record the size, shape, and distribution of cells on the surface.

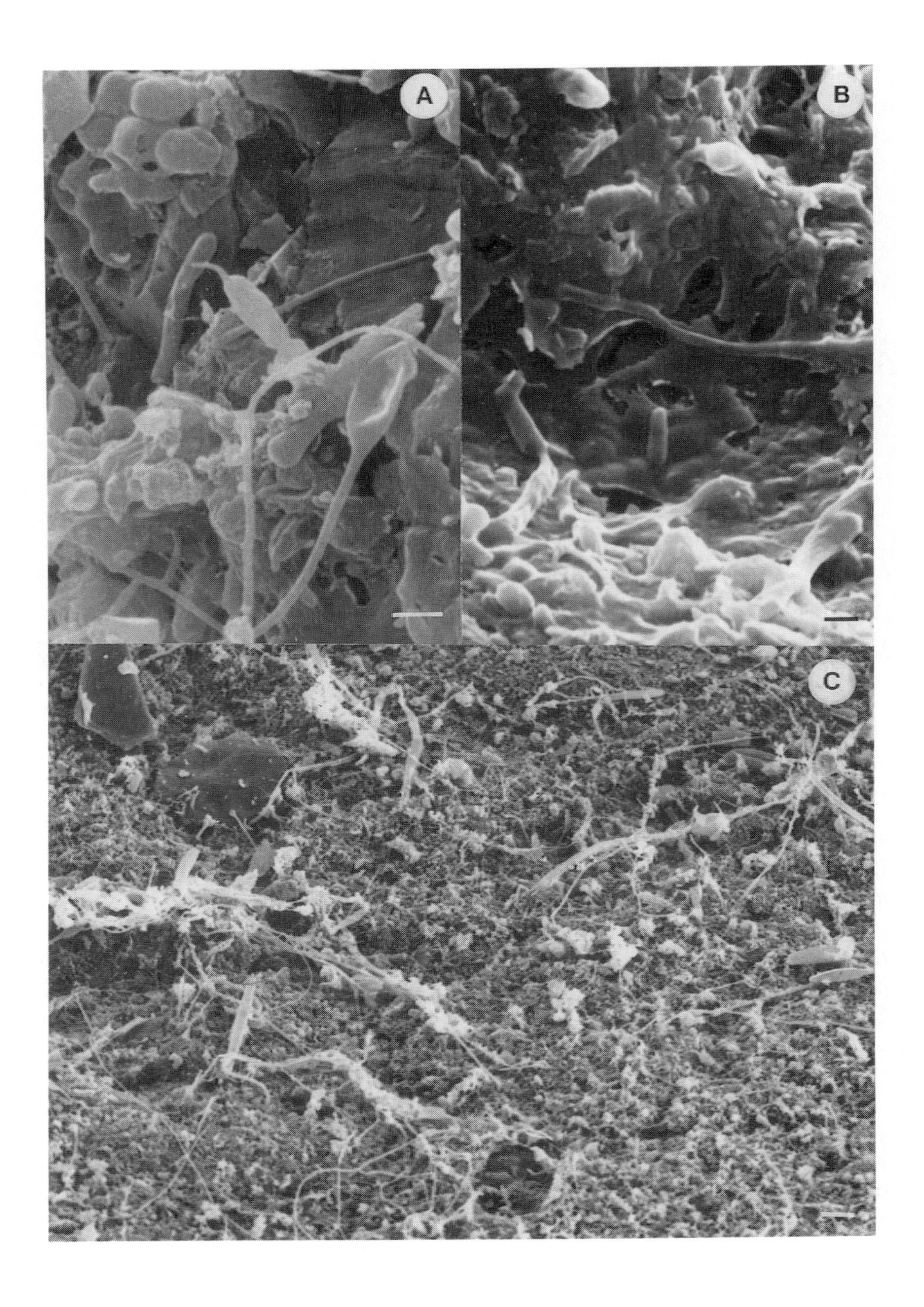
A
B
C

## IV. Methods for detecting actively metabolizing bacteria on surfaces

### A. Intracellular formazan deposition

#### *1. Microscopy*

The percentage of actively metabolizing microorganisms can be estimated on many surfaces using the reduction of INT as an indicator of electron transport activity (Zimmerman *et al.*, 1978).

*(a) Reagent.* The INT solution [2-(*p*-iodophenyl)-3-(*p*-nitrophenyl)-5-phenyl tetrazolium chloride (Eastman)] is prepared by dissolving 0.2 g of INT in 5 ml of 92–95% ethanol and bringing to volume with 95 ml water. Filter-sterilize the solution and store in an amber or dark bottle at 4°C for up to 1 week.

*(b) Procedure.* Rinse the surface and transfer to a sterile vial containing 5 ml of sterile phosphate buffer (pH 6.8, 0.2 M). Add 1 ml of the INT solution and incubate in the dark for 20–30 min at ambient temperatures. Remove the surface at the end of the incubation period and rinse with 20 ml of sterile buffer. Do not rinse the surface with alcohol. If required, the samples may be fixed with a 0.4% solution of formaldehyde. Fixation of the cells with glutaraldehyde may enhance the deposition of crystals and is not recommended. Cells attached to transparent materials can be readily visualized by negatively staining with nigrosine (Fig. 1). Spread a loopful of nigrosine over the surface and remove by gently blotting after 2 min. Add oil directly to the stained surface and examine under oil. Poisoned controls are used to check for abiotic reduction of the INT. Immersion oil may slowly extract formazan from the cells (Tabor and Neihof, 1984); therefore, long-term storage of samples is difficult.

*(c) Interpretation.* The dark crystals (red) are easily seen in the negatively stained cells (Fig. 1). Dark-staining crystals can also be seen on surfaces which have bright background fluorescence (e.g. Silkolatex catheters). The cells, however, cannot be seen. Light staining with a 0.1% solution of malachite green may provide a means of seeing the cells on these surfaces.

---

**Fig. 3.** (A) Stalked bacteria are visible in the biofilm on the surface of a stone. Bar equals 1.0 μm. (B) Both coccoidal and rod-shaped cells are enrobed in a confluent mucilage sheet. Bar equals 5.0 μm. (C) Comprehensive view of colonized stone after 42 days. Bar equals 1.0 μm.

Most basic stains, however, obscure the presence of formazan deposits in the catheter samples. When catheters are used, the INT–formazan procedure suggests the presence of active microorganisms on the surface and gives their relative distribution. It is possible to demonstrate the absence of actively metabolizing cells in the presence of germicidal agents. Opaque surfaces can be examined by epifluorescence microscopy; however, formazan crystals can usually only be seen in green fluorescing cells. The percentage of actively metabolizing bacteria is determined by counting the number of cells containing one or more crystals and dividing by the total number of cells. Record the results as percentage of active cells per area or OIF.

### 2. *Extraction of formazan*

Often a heavily colonized surface will show a distinct red colour following incubation with INT. This red colour can be partially extracted from a variety of surfaces with absolute alcohol, acetone, or methanol.

*(a) Procedure.* The formazan is extracted with 4 ml of absolute alcohol at 37°C for 1 h. More rapid extraction can be achieved at 60°C. The amount of alcohol-extracted formazan can be semi-quantitated by recording the optical density (OD) on a spectrophotometer at 495 nm (depending on the solvent used).

*(b) Interpretation.* OD readings between 0.05 and 1.25 can be obtained (Table I). Experience has shown that the extraction is incomplete and surface bacteria retain formazan. The presence of bacteria on a piece of C-flex or glass tubing, for example, can be ascertained by flowing INT over the surface and allowing the system to incubate at ambient temperature for 1 h. Subsequently, alcohol can be drawn through the system to extract the formazan on the surface. The collected extract can then be read spectrophotometrically. A number of ions such as copper and iron interfere with this assay. Surfaces containing these ions cannot be used. Abiotic and uncolonized surface controls must be used. Some red coloration may be observed on uncolonized sections incubated for periods greater than 24 h. Red blood cells do reduce INT, therefore surfaces which have erythrocytes may give erroneous results. Microscopic examination also shows that leucocytes contain formazan deposits. These cells could interfere with the assay.

**TABLE I**
Detection of INT–formazan on surfaces colonized by *E. coli*

| Surface colonized | Method of colonization | Visual appearance on surface[a] | | | | Extractable formazan[b] | | | | cfu $cm^{-2}$ ($n = 2$) |
|---|---|---|---|---|---|---|---|---|---|---|
| | | Trials | | | | Trials | | | | |
| | | A | B | C | D | A | B | C | D | |
| Rüsch catheters | | | | | | | | | | |
| 1.42 $cm^2$ | Batch (28 h) | +++ | +++ | +++ | | 0.18 | 0.13 | 0.18 | | $2.72 \times 10^6$ |
| 1.88 $cm^2$ | Batch (28 h) | + | + | + | | 0.04 | 0.10 | 0.16 | | $6.69 \times 10^5$ |
| 2.83 $cm^2$ | Batch (28 h) | ++ | ++ | ++ | | 0.23 | 0.16 | 0.20 | | $3.06 \times 10^6$ |
| Rüsch catheter 0.825 $cm^2$ | Continuous flow (18 h) | + | + | + | + | 0.02 | 0.01 | 0.02 | 0.03 | $5.2 \times 10^6$ |
| Tygon 22.23 $cm^2$ | Continuous flow (18 h) | +++ | | | | 0.23 | | | | – |

[a] +, faint red; ++, red or red blotches; +++, deep red; –, no detectable red.
[b] INT–formazan extracted with absolute alcohol for 1 h at 37°C.

## B. Nalidixic acid sensitivity

This antibiotic interferes with DNA replication without blocking other biosynthetic pathways (Goss *et al.*, 1964). Sensitive bacteria generally appear swollen and elongated following treatment with nalidixic acid (Fig. 4).

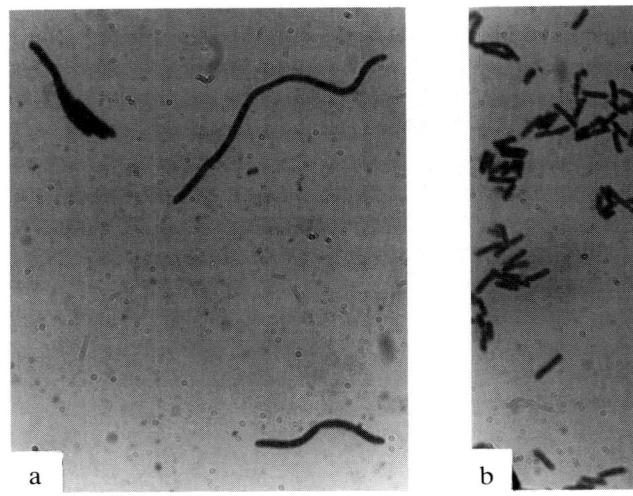
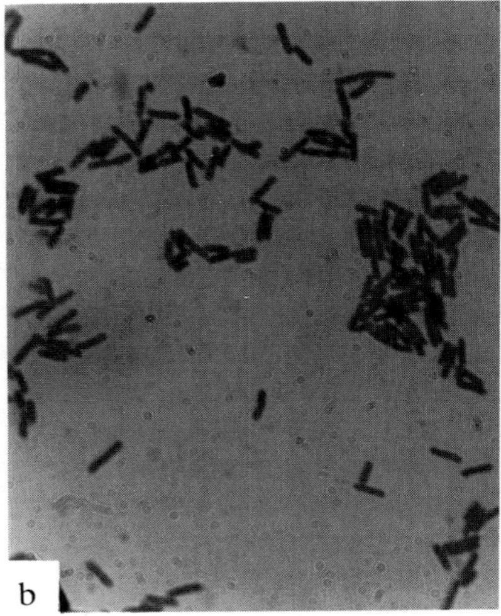


**Fig. 4.** (a) Surface-associated cells exposed to nalidixic acid. (b) Control group showing no elongated cells.

### *1. Procedure*

Colonized and rinsed substrata are transferred to a sterile vial containing 0.025% yeast extract (Kogure *et al.*, 1979) or other suitable substrate. Because bacteria may show substrate specificity, it is recommended that trials be done with other media preparations such as peptone, tryptone, and various carbohydrates. Freshly prepared, filter-sterilized 0.002% nalidixic acid is then added to each vial. The samples are then incubated for 8–12 h at the appropriate temperature. Environmental samples are incubated at 16°C; clinical samples at 37°C. At the end of the incubation period, the samples are removed from the incubation vials, rinsed with sterile buffer, and fixed with 0.5% glutaraldehyde. The fixed specimens are stained with acridine orange for 2 min, rinsed for 30 s with sterile alcohol, and examined under oil for the presence of elongated cells.

### 2. *Interpretation*

The percentage of actively metabolizing bacteria is estimated by dividing the number of elongated cells by the total number of bacteria present. It is important that a number of controls be conducted when using this method. Bacteria on the surface of some materials appear filamentous (McCoy *et al.*, 1981). An untreated, fixed sample can be used to estimate the number of elongated cells prior to antibiotic exposure. A substrate control is required to distinguish between cell growth on the surface and a response to the antibiotic. The biofilm may provide the bacteria with some degree of resistance to the antibiotic (Nickel *et al.*, 1985); therefore, antibiotic sensitivity over a range of concentrations (0.002–0.006%) should be examined. Although the procedure cannot enumerate Gram-positive bacteria or resistant strains of Gram-negative bacteria, the results are easier to interpret than those obtained from micro-autoradiography. The procedure can be adapted for use with simple staining or transparent or background fluorescing surfaces (Fig. 4).

## V. Activity measurements of biofilms

Both non-radioisotopic and radioisotopic methods can be used to measure the activity of biofilm bacteria non-destructively. The latter procedures require the safe and proper handling of radioactive materials by trained personnel.

### A. Non-radioisotopic procedures

The glucose oxidase/peroxidase assay (Trinder, 1969) is a relatively simple method which can be used to detect active bacteria and measure the rate of substrate utilization. This procedure is based on the oxidation of D-glucose to D-gluconic acid by glucose oxidase followed by the formation of a quinoneimine dye in the presence of peroxidase.

### 1. *Procedure*

Rinsed substrata are transferred to sterile 20-ml vials containing a 1 mg $ml^{-1}$ glucose solution. The samples are incubated with or without agitation at the appropriate temperature for up to 12 h. Glucose concentrations are monitored in solution over time and are compared to a standard curve. For routine assays 0.025 ml of glucose solution is removed from the vial and added to a cuvette capable of holding 3 ml. To this add 2.5 ml of the glucose reagent (Gilford Diagnostics). Mix vigorously, incubate for 10 min at 37°C (20–30 min at 26°C) and read at 505 nm on a spectrophotometer. Because glucose may serve as a growth substrate, it is necessary to monitor changes in the bacterial density on the surface and in the glucose incubation solution. This is done with a control set of vials which can either be used for direct microscopic examination or for vial plate counts. If vial plate counts are required, the surface is scraped with a sterile scalpel blade, rinsed, and vortexed for 30 s to dislodge as many surface-associated cells as possible (Ruseska *et al.*, 1982; Nickel *et al.*, 1985). Serially dilute the samples and plate on tryptic soy agar (Difco) or other general purpose medium.

### 2. *Interpretation*

Calculate the amount of substrate removed (Table II). In absence of significant changes in bacterial densities on the surface, the results can be used to estimate rates of substrate removal. The effect of biocides on biofilm bacteria can also be demonstrated.

**TABLE II**

Uptake of β-D-glucose by *S. epidermidis* attached to whole Foley (Rüsch) catheters as measured with the Trinder glucose oxidase/peroxidase enzyme assay[a,b]

| Sample | cfu $cm^{-2}$ ($n = 2$) | Amount of glucose consumed in mg $dl^{-1}$ ($\bar{X} \pm SD$) | ($n = 4$) |
|---|---|---|---|
| Control[c] | – | – | |
| Experimental: reading at 7 h | $2.4 \times 10^8$ | $6.35 \pm 1.98$ | |

[a]3.2 $cm^2$ sections cut from the whole catheter were exposed to 90 mg $dl^{-1}$ of glucose solution for 7 h.

[b]A continuous flow of the organism in nutrient medium was allowed to colonize three whole Foley (Rüsch) catheters for 20 h. The system was rinsed out with phosphate buffer for 1 h at 250 ml $h^{-1}$.

[c]Control sections were boiled for 1 h.

## B. Radioisotopic techniques

A number of procedures can be used to estimate the relative rates of heterotrophic activity of attached bacterial populations. Some techniques measure the amount of $^{14}C$-labelled substrate mineralized to $^{14}CO_2$ during a short (hours) incubation period. An estimation of the total activity of the biofilm population can be obtained by measuring both $CO_2$ evolution and assimilation. The latter determination requires a method for assessing the amount of labelled substrate associated with the cells on the surface. A modification of the above techniques involve measuring the disappearance of a known amount of labelled substrate from solution over time.

### *1. Radiorespirometric and direct calculation approach*

*(a) Procedure.* Add the colonized and rinsed substrata to sterile stoppered serum vials containing at least 5–10 ml sterile buffer to which between 0.10 and 30.0 $\mu g\ l^{-1}$ of uniformly labelled $^{14}C$-substrate (e.g. glutamic acid, glucose) has been added. Incubate the vials for several hours at ambient temperatures and terminate the reaction by acidification with 0.3 ml 5 N $H_2SO_4$. Acid-killed samples are used as controls.

*(b) $^{14}CO_2$ collection.* The released $^{14}CO_2$ can be trapped in a number of different ways. The trapping efficiency for each method is determined by adding a known amount of radioactively labelled sodium bicarbonate to a buffered solution (pH 10.5), acidifying the solution and collecting the evolved $^{14}CO_2$. The most rapid methods for collecting the carbon dioxide involve flushing the system (Wyndham and Costerton, 1981; Ladd *et al.*, 1987). One method of flushing involves pumping dehumidified, $CO_2$-free air into each vial and collecting the $^{14}CO_2$ in a cocktail of ethanolamine and methanol (1:3 v/v) for at least 20 min. Trapping efficiencies exceeding 85% can be expected. Alternatively, the $^{14}CO_2$ can be collected on fluted-filter wicks placed in a centre well and soaked with 0.2 ml of phenethylamine or 0.3 ml ethanolamine. The trapping efficiency of the ethanolamine is lower (50–55%) than that of the phenethylamine (83%); however, it is a less noxious and safer chemical to work with for this procedure.

*(c) Interpretation.* Results are scored as positive if the amount of $^{14}CO_2$ evolved (in cpm or dpm) is twice that of the acid-killed controls (Lehmicke *et al.*, 1979; Ladd *et al.*, 1987). The method is useful for detecting the presence of actively metabolizing bacteria which are capable of mineralizing the added substrate (Fig. 5) and assessing the activity of biofilm bacteria in different areas of a stream or lake ecosystem (Table III), or determining the effects of increasing flow rate (in rpm) on the heterotrophic activity of attached bacteria (Fig. 6). The results can also be

expressed as a rate of mineralization as suggested by Kadota *et al*. (1966) using the equation (direct calculation)

$$v = \frac{f(A)}{t}$$

where

$v$ = the velocity of mineralization by the microorganisms

$f$ = the fraction of available isotope respired during the incubation period

$A$ = the concentration of the compound added to the reaction vial in $\mu g\ l^{-1}$

$t$ = the incubation time in h.

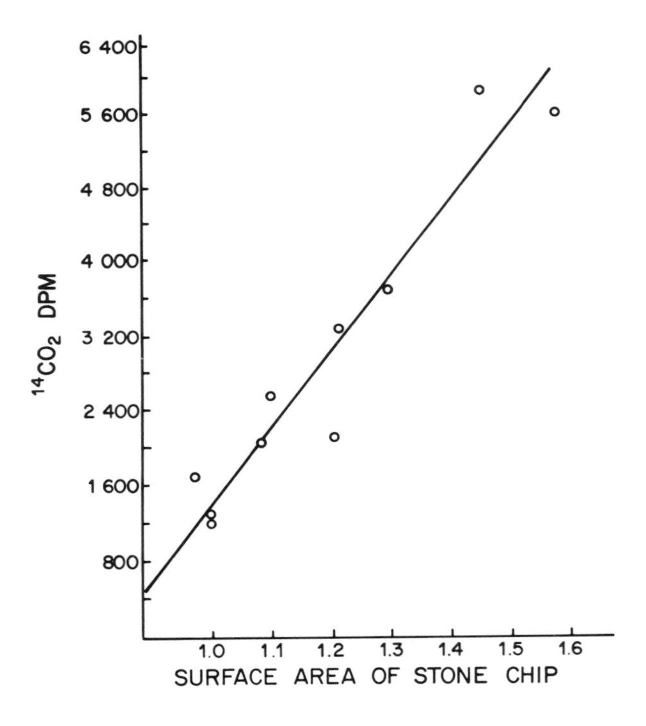


**Fig. 5.** The relationship between the amount of radioactively labelled substrate mineralized and surface area of a colonized stone section.

**TABLE III**

Heterotrophic activity of 12 individual stone squares removed from three different areas in Middle Creek, Alberta, Canada, $^{14}$C-glutamic acid (6.3 μl $l^{-1}$) used as the substrate[a,b]

| | Radioactivity dpm ± SD | | |
|---|---|---|---|
| | Respired | Assimilated | Total |
| Area 1 | | | |
| c.v.% (replicates) | 4055.0 ± 1363.9<br>33.6 (6) | 1973.5 ± 806.2<br>40.9 (6) | 6028.5 ± 1149.1<br>19.1 (6) |
| Area 2 | | | |
| c.v.% (replicates) | 5197.3 ± 1494.1<br>29.0 (6) | 1455.6 ± 579.9<br>40.0 (6) | 6652.9 ± 1596.1<br>24.0 (6) |
| Area 3 | | | |
| c.v.% (replicates) | 3334.2 ± 1595.9<br>47.7 (6) | 2531.9 ± 1236.2<br>48.8 (6) | 5876.1 ± 2108.5<br>35.9 (6) |

[a] Controls for total uptake between 64.5 and 210.0 dpm.
[b] Thirty-minute incubation period at 4°C. 2 cm$^2$ surface for each sample.

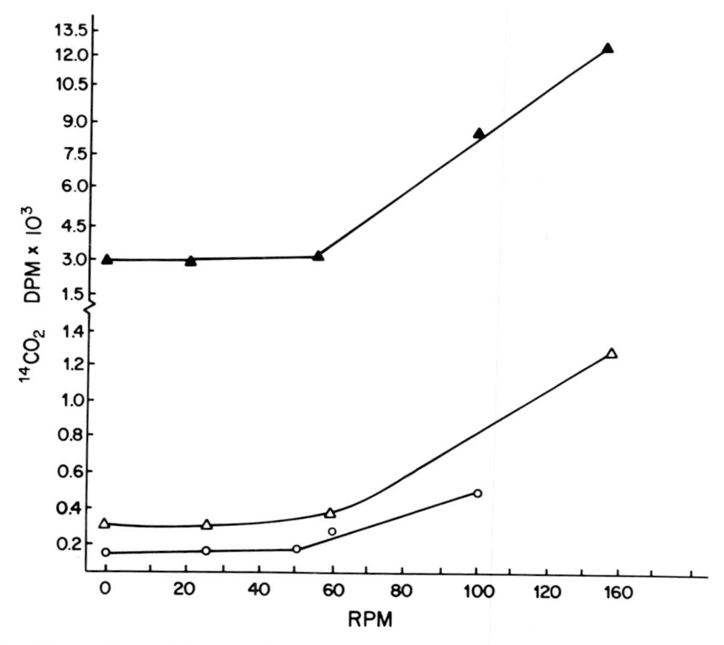


**Fig. 6.** The effect of increasing revolutions per minute (rpm) on the mineralization and assimilation (filled triangle) of added substrate by stone-associated biofilm bacteria.

The use of the single concentration–single incubation procedure for determining the rates of activity of natural microbial communities has been questioned (Wright and Burnison, 1979); nevertheless, the values are useful in demonstrating the response of attached cells to varying concentrations of antimicrobic agents over different exposure periods and evaluating the effect of increased flow rates (measured in rpm) on biofilm activity (Table IV). The values are not absolute and may change when different substrate concentrations are used (Lewis *et al.*, 1988). In addition, it should be noted that respiration may only account for a fraction of the total activity of the attached bacterial population.

**TABLE IV**
The effect of gentamycin sulphate (500 $\mu g\ ml^{-1}$) on the mineralization of $^{14}C$-glutamic acid (evolution of $^{14}CO_2$) by *P. aeruginosa* with respect to rpm[a]

| | rpm | Velocity of uptake ($\mu g\ l^{-1}\ h^{-1}$) ($n$ = 3) |
|---|---|---|
| Controls[b,c] | | |
| (no antibiotic) | 0 | 2.49 ± 0.27 |
| | 25 | 2.23 ± 0.11 |
| | 50 | 2.23 ± 0.06 |
| Antibiotic-treated samples[a] | | |
| (500 $\mu g\ ml^{-1}$) | 0 | 0.68 ± 0.34 |
| | 25 | 0.64 ± 0.58 |
| | 50[c] | 0.16 ± 0.17[c] |

[a]Samples were exposed to 500 $\mu g\ ml^{-1}$ of antibiotic for 14.5 h. The uptake and mineralization of the substrate was measured at a single concentration (5.3 $\mu g\ l^{-1}$) for 1 h.
[b]ANOVA showed no significant difference between shaken (25 and 50 rpm) and non-shaken (0 rpm) samples of the control.
[c]There is a significant difference if all groups are considered together (degrees of freedom = 17, $F$ = 7.17, $P \leq 0.05$). A Student–Newman–Keuls (SNK) test indicated the uptake at 50 rpm for the antibiotic exposed samples was significantly less than that of the other groups ($P$ = 0.05).

## 2. *Kinetic approach*

*(a) Procedure.* Colonized and rinsed substrata are transferred to several sterile serum vials containing an appropriate buffer (or filter-sterilized stream water) and varying concentrations of $^{14}C$-labelled substrate. The labelled substrate can be added to sterile, non-labelled substrate to conserve reagent. It is advisable to test the system over a range of substrate concentrations (11–15 concentrations ranging from 0.1 to 64.0 $\mu g\ l^{-1}$) to determine if the relationship between microbial uptake and substrate

concentration can be described by Michaelis–Menton kinetics (Wright and Hobbie, 1965). A time-course study at the highest and lowest concentrations of substrate is done to determine the appropriate incubation period for the kinetic parameters. No more than 5% of the substrate should be consumed during the incubation period. Periods exceeding 24 h should not be used due to the increased likelihood of cell replication. Kinetic measurements are generally made over a single time period at 5–6 concentrations using triplicate samples at each concentration for either mineralization, assimilation, or both. A set of acid-killed mineralization and assimilation controls at each concentration is required to test for abiotic release of carbon dioxide and binding of the substrate to the substratum. Choquet *et al.* (1987) have recommended using multiple-time, multiple-substrate concentration experiments for measuring kinetic parameters. This procedure is useful for systems which show a periodic failure to give statistically acceptable parameters. It will require 3–4 times as many available substrata sections to complete an experiment, however. Evolved $^{14}CO_2$ is collected using a method which has a high trapping efficiency (flushing). Substrata which can be heated at elevated temperatures without undergoing chemical modification can be oxidized in an auto-oxidizer (Packard model 306) to determine assimilated uptake. Plastics and other synthetic compounds should not be placed in this instrument. For these substrata, kinetic parameters are measured for mineralization alone. Prior to oxidizing the samples, the substrata are removed from the vials, rinsed three times with 30 ml of sterile buffer, and allowed to air-dry. Do not add acid or chemical fixatives such as glutaraldehyde or formaldehyde. Heavy metal poisons such as mercuric chloride can be used; however, rinsing the samples adequately is generally sufficient to terminate the reaction.

*(b) Interpretation.* The data can be analysed using a Hanes plot transformation of the Michaelis–Menton equation:

$$t/f = A/V_{max} + K + S_n/V_{max}$$

where

$f$ = the fraction of available isotope assimilated or mineralized
$t$ = the incubation time in h
$S_n$ = the natural concentration of a given substrate in $\mu g\ l^{-1}$
$A$ = the concentration of the added substrate in $\mu g\ l^{-1}$
$V_{max}$ = the maximum velocity of uptake or mineralization
$T_n$ = the turnover time in hours of the added substrate by the population.

The $t/f$ values are calculated by dividing the incubation time by the determined fraction, $f$, and plotting the values against $A$, the substrate concentration. If the plot is linear, $V_{max}$, the maximum rate of uptake, is determined by taking the reciprocal of the slope of the line. The $y$ intercept, $(K + S_n/V_{max}) = T_n$, represents the turnover time, the time for complete mineralization or assimilation of an amount of substrate equal to the *in situ* concentration.

The procedure can be used to determine the response of the microbial assemblages to germicides (Fig. 7) or enrichment or to compare the activity of biofilm bacteria attached to different surfaces (Table V). If data plotted according to the Hanes plot give a line which is parallel to the $x$ axis, this suggests that the data fit a first-order model, i.e. the system does not become saturated and uptake or respiration shows a direct proportionality to substrate concentration. This may indicate that the uptake is due primarily to diffusion (Wright and Hobbie, 1965). Inspection of the graph may reveal that there is more than one best line for the plotted data. This indicates that the kinetics of the system cannot be represented by a single set of kinetic parameters. Thus, the saturation model may not provide a satisfactory model for predicting $V_{max}$ for kinetically heterogeneous microbial assemblages which display a diversity of kinetic parameters over a wide range of concentrations (Lewis *et al.*, 1988). The approach used by Choquet *et al.* (1987), which uses non-linear regression analysis, provides a means of analysing data which do not fit the saturation model.

**TABLE V**

Uptake of $^{14}C$-glutamic acid by stream bacteria attached to stone, glass, and metal surfaces. $V_{max}$ values corrected to 1 $cm^2$ area

| Surface | AODC $\times 10^7$ $cm^{-2}$ | $^{14}CO_2$ $V_{max}$ (ng $cm^{-2}$ $h^{-1}$ $\times 10^{-3}$) | Total $V_{max}$[a,b,c] (ng $cm^{-2}$ $h^{-1}$ $\times 10^{-1}$) | Per cent respired |
|---|---|---|---|---|
| Quartzite | 6.77 | 8.50 | 3.31 | 4.90% |
| Glass | 5.85 | 4.70 | 0.62 | 2.71% |
| Metal (tempered steel) | 4.32 | 2.70 | – | – |

[a]Turnover times for the stone and glass surfaces were 163.2 and 71.2 h, respectively.
[b]Specific activity indexes for stone and glass were 4.89 and 1.06 μg $cell^{-1}$ $h^{-1}$ $\times 10^{-12}$.
[c]Correlation coefficients for $^{14}CO_2$ $V_{max}$ and total uptake were greater than $R = 0.86$.

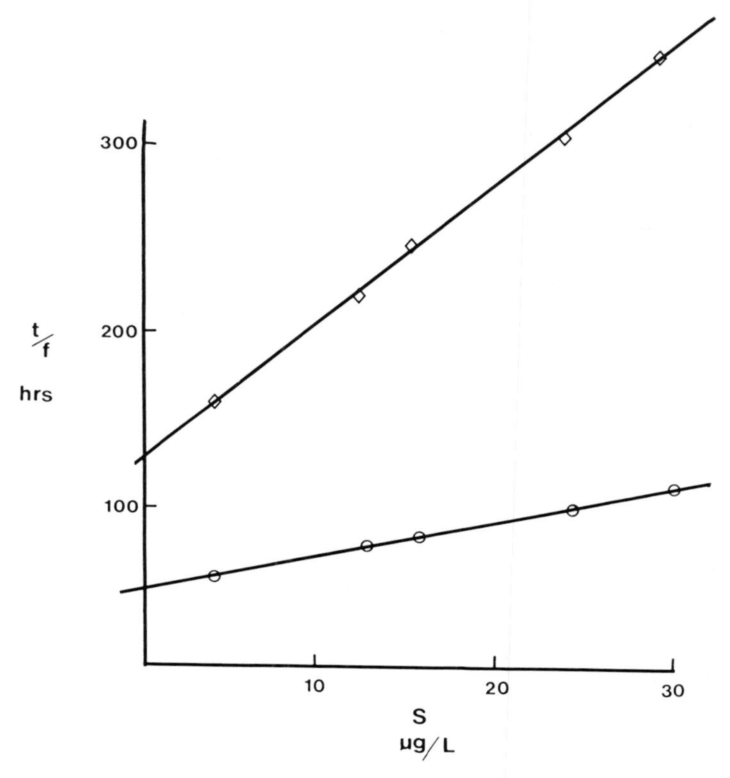


**Fig. 7.** The effect of tobramycin sulphate (120 µg $ml^{-1}$) on the heterotrophic activity (mineralization) of surface-associated *Pseudomonas aeruginosa*. the turnover time for the control sample (○) is 42 h and for the treated sample (◇) is 125 h.

### *3. Continuous uptake model for closed tubing systems*

*(a) Procedure.* The colonized tubing is washed for 1 h at 200 ml $h^{-1}$ with sterile buffer. The tubing connecting the colonized surface to the rinse reservoir is replaced with new sterile tubing which is then aseptically affixed to the substrate reservoir containing 5.0–20.0 µg $l^{-1}$ radioactively $^{14}$C-labelled substrate. Flow is adjusted to between 30 and 200 ml $h^{-1}$ with a 1-h acclimatization period between each change in flow rate. Four to six 5-ml fractions are collected on ice at each flow rate and 1-ml samples are added to a scintillation cocktail (Ready Safe, Beckman Corp.) for reading in a scintillation counter. The radioactivity measured at each speed is compared to the amount of radioactivity in the substrate reservoir at the

beginning and end of the experiment. Controls include uncolonized tubing, samples treated with 0.5% glutaraldehyde, and samples packed in ice.

*(b) Interpretation.* The velocity of uptake at the different flow rates is calculated using the following formula (LaMotta, 1976)

$$v = Q/A(S_i - S_e)$$

where

$v$ = the velocity of uptake in μg cm$^{-2}$ h$^{-1}$
$Q$ = surface area colonized in cm$^2$
$S_i$ = substrate concentration in the substrate reservoir (μg l$^{-1}$)
$S_e$ = substrate concentration in the effluent (μg l$^{-1}$)

Plots of velocity versus flow rate may show an increase in uptake at the higher flow rates. This suggests that the activity of the attached population is diffusion limited (Fig. 8).

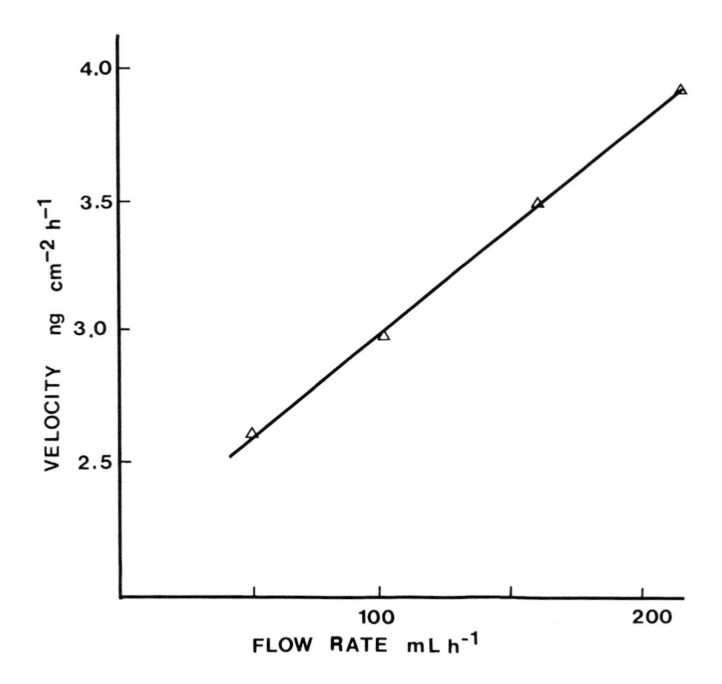


**Fig. 8.** The effect of increasing flow rate on the velocity of uptake of radioactively labelled substrate by catheter-associated pseudomonads measured under continuous flow conditions. Measurements made using intact tubing which was rinsed for 1 h at 250 ml h$^{-1}$.

## Acknowledgements

The assistance of D. Nahri and A. Berih is gratefully acknowledged. This research was funded in part by the Kidney Foundation of Canada.

## References

Choquet, C. G., Ferroni, G. D. and Leduc, L. G. (1987). *FEMS Microbiol. Ecol.* **45,** 59–64.

Costerton, J. W., Irwin, R. T. and Cheng, K. J. (1981). *Ann. Rev. Microbiol.* **35,** 299–324.

Geesey, G. G., Mutch, R., Costerton, J. W. and Green, R. B. (1978). *Limnol. Oceanogr.* **23,** 1214–1223.

Goss, W. A., Deitz, W. H. and Cook, T. M. (1964). *J. Bacteriol.* **88,** 1112–1118.

Kadota, H., Hata, Y. and Miyoshi, Y. (1966). *Mem. Res. Inst. Food Sci. Kyoto Univ.* **27,** 28–30.

Kogure, K., Simidu, U. and Taga, N. (1979). *Can. J. Microbiol.* **26,** 415–420.

Ladd, T. I., Costerton, J. W. and Geesey, G. G. (1979). *In* "Native Aquatic Bacteria: Enumeration, Activity, and Ecology" (J. W. Costerton and R. R. Colwell, Eds), pp. 180–195, ASTM Press, Philadelphia.

Ladd, T. I., Ventullo, R. M., Wallis, P. M. and Costerton, J. W. (1982). *Appl. Environ. Microbiol.* **44,** 321–329.

Ladd, T. I., Schmeil, D., Nickel, J. C. and Costerton, J. W. (1985). *J. Clin. Microbiol.* **21,** 1004–1007.

Ladd, T. I., Schmiel, D., Nickel, J. C. and Costerton, J. W. (1987). *J. Urol.* **138,** 1451–1456.

LaMotta, E. J. (1976). *Biotech. Bioeng.* **18,** 1359–1370.

Lehmicke, L. G., Williams, R. T. and Crawford, R. L. (1979). *Appl. Environ. Microbiol.* **38,** 644–649.

Lewis, D. L., Hodson, R. E. and Hwang, H. (1988). *Appl. Environ. Microbiol.* **54,** 2054–2057.

Malick, L. E. and Wilson, B. W. (1975). *Stain Technol.* **50,** 265–270.

Marrie, T. J. and Costerton, J. W. (1981). *Appl. Environ. Microbiol.* **42,** 1093–1102.

Marrie, T. J. and Costerton, J. W. (1982). *Circulation* **66,** 1339–1341.

Marrie, T. J. and Costerton, J. W. (1984). *J. Clin. Microbiol.* **19,** 687–693.

McCoy, W. F., Bryers, J. D., Robbins, J. and Costerton, J. W. (1981). *Can. J. Microbiol.* **27,** 910–917.

Nickel, J. C., Ruseska, R., Wright, B. and Costerton, J. W. (1985). *Antimicrob. Agents Chemother.* **27,** 619–624.

Porter, K. G. and Feig, Y. S. (1980). *Limnol. Oceanogr.* **25,** 943–948.

Ruseska, I., Robbins, J., Lashen, E. S. and Costerton, J. W. (1982). *Oil Gas J.* March 8, 253–264.

Tabor, P. S. and Neihof, R. A. (1984). *Appl. Environ. Microbiol.* **48,** 1012–1019.

Trinder, P. (1969). *Ann. Clin. Biochem.* **6,** 24–25.

Wright, R. T. and Burnison, B. K. (1979). *In* "Native Aquatic Bacteria: Enumeration, Activity and Ecology" (J. W. Costerton and R. R. Colwell, Eds), pp. 140–155, ASTM Press, Philadelphia.

Wright, R. T. and Hobbie, J. E. (1965). *Limnol. Oceanogr.* **10,** 22–28.

Wyndham, R. C. and Costerton, J. W. (1981). *Appl. Environ. Microbiol.* **41,** 783–790.

Zimmerman, R., Iturriaga, R. and Becker-Birck, J. (1978). *Appl. Environ. Microbiol.* **36,** 926–935.

# 10

# Methods for Studying the Microbial Ecology of Soil

T. R. G. GRAY

*Department of Biology, University of Essex, Colchester, Essex CO4 3SQ, UK*



METHODS IN MICROBIOLOGY
VOLUME 22 ISBN 0-12-521522-3

## I. Introduction

Soil may be characterized as a habitat for microorganisms which has a low average rate of input of energy-yielding substrates, is subject to diurnal, seasonal and random fluctuations in external environmental factors, contains solid, liquid and gaseous phases in varying proportions and, as a result of interactions between all of these factors, is highly heterogeneous. Heterogeneity of the environment in both space and time in turn leads to heterogeneity of the microbial communities at different levels or scales, e.g. the gross level such as in soil profiles, the microhabitat level such as the rhizosphere or the molecular level such as at clay–liquid interfaces (Nedwell and Gray, 1987).

The extreme heterogeneity of the environment has led to the development of two fundamentally different approaches to studying the ecology of these microbial communities. On the one hand, soil may be treated as a functional black box with emphasis being placed on the rates of input of materials, the size and turnover of the total microbial biomass involved and the rates of output of materials. This synecological approach can operate at different levels, depending upon how widespread or restricted the process being studied is. Thus carbon dioxide output might be a useful indicator of the activity of most aerobic microorganisms whilst rates of cellulolysis will be relevant to only a small subset of microorganisms. Many of these studies on rates of processes have been hampered by the slow rate at which metabolic transformations occur in soil. On the other hand, some microbiologists are interested in particular key species of microorganisms and may wish to study the ecology of these in detail. This autecological approach will involve answering many questions about the organisms including where are they and what do they look like in field samples, how can they be isolated, how many are there, what metabolic processes are they carrying out, what survival mechanisms do they have, do they interact or compete with other organisms and how active are they?

In the past 20 years, most emphasis has been placed on the development of synecological approaches to elucidate the functioning of terrestrial ecosystems. At the present time, some attention is being paid to autecological methods as these will be essential for the study of natural or genetically manipulated organisms released into the environment. Methods for the recognition of individual species and strains of organisms in soil and for following changes in their population sizes are being developed.

Some of these autecological approaches are dealt with in other chapters in this book. Here coverage will emphasize methods which attempt to measure the entire microflora of a soil, together with some methods appropriate for examining bacteria in soil which are not dealt with in other chapters. The methods described will inevitably be a very small selection of those available and readers are referred to the many reviews of methods that have already been published for more complete coverage (Parkinson *et al.*, 1971; Burns and Slater, 1982). The methods will be considered under four headings, activity, biomass, location and isolation of organisms.

## II. Methods for measuring microbial activity

Microbial activity is best measured by determining the growth rate of microbial cells, the rate of disappearance of a particular substrate or the rate of appearance of selected metabolites. Care must be taken to distinguish between the *actual* activity of the organisms which exists under field conditions and *potential* activity which shows how organisms may respond to a particular set of changed conditions, usually substrate saturation and optimal incubation conditions. However, application of methods designed to measure actual activity often induces changes in the samples being investigated and this possibility must be taken into account when interpreting results. Thus, disturbance of soils and consequent changes in the arrangement of particles, substrates and organisms may be sufficient to cause a temporary increase or flush of activity (Griffiths and Birch, 1961).

There are no accurate methods for determining the growth rate of soil organisms but a number of techniques have been designed to determine the quantity of soil organisms which are metabolically active or not. Two of these are described below.

### A. Detection of fluorescein diacetate (FDA)-metabolizing organisms (Söderström, 1977)

FDA is a fluorogen which is transformed to the fluorescent compound fluorescein when it is attacked by esterases. Organisms that are active in this respect are stained by the fluorescein and can be detected with short wavelength blue light (Söderström, 1977, 1979a, b).

Between 5.0 and 25.0 g soil, depending upon the size of the microbial population, are dispersed in 250 ml phosphate buffer (60 mM, pH 7.5) after removing roots thicker than 1.0 mm, using an MSE Atomix running at 6000 rpm (half-speed) for 2 min. If appropriate, the homogenate can be diluted with further phosphate buffer and FDA (Koch Light, Colnbrook, UK) is added to the final dilution. FDA may be dissolved in acetone to give a stock solution of 2 mg ml$^{-1}$ (store at −20°C) and this is added to the homogenate to give a final concentration of 10 μg ml$^{-1}$. The suspension is stained for 3 min and then filtered through a black 0.8 μm Millipore filter held in a Pyrex microanalysis filter holder. The filter is removed, mounted in non-fluorescent immersion oil and examined under an epifluorescent microscope fitted with an HB0200 mercury lamp and a 3 mm BG12 primary and 1 mm GG9 + 2 mm OG515 secondary filters. Organisms containing esterase activity are stained green.

A number of precautions are sensible. The objectives used should have as wide a numerical aperture as possible, the suspensions should be kept stirred on a magnetic stirrer run at a constant speed and sampled by pipette from the same spot in the beaker each time. The dilution factors should also be standardized (Söderström, 1979b). Microphotographs can be taken on Agfapan 100, exposing for 4 s and developing in Agfa Rodinal (1 + 25) for 10 min at 25°C.

The best results have been obtained for fungi, although some authors have used the technique for bacteria as well. Quantities of stained hyphae can be expressed as lengths or converted to biomass figures using suitable figures for mean hyphal diameter, hyphal water content and hyphal specific gravity.

The most serious source of error is the drastic effect of homogenization on FDA stainability which destroys the integrity of some hyphae, causing cytoplasmic leakage and failure to stain. It is desirable, therefore, to determine the optimum homogenization time for each new soil type used and to recognize that the results obtained may be underestimates (Söderström, 1979a). Samples should also be analysed as soon as possible although storage at 5–12°C minimizes change during the first 48 h (Söderström, 1979a).

### B. Autoradiographic methods for the detection of activity (Bääth, 1988)

Autoradiography is used to detect organisms using $^{14}C$-glucose as a substrate. Uniformly labelled $^{14}C$-D-glucose (Amersham; >230 mCi $mmol^{-1}$) is diluted aseptically in a mineral solution ($NaNO_3$, 3 g $l^{-1}$; $KH_2PO_4$, 1 g $l^{-1}$; $MgSO_4.7H_2O$, 0.5 g $l^{-1}$; KCl, 0.5 g $l^{-1}$; $FeSO_4$, 10 mg $l^{-1}$).

Ten grams of soil are homogenized in 200 ml mineral solution in an MSE Atomixer at 6000 rpm (half speed) for 2 min. 0.5 ml of this suspension is incubated with the isotope solution (final concentration 20 μCi $ml^{-1}$) at 22°C for 2 h and then an equal volume of formalin (final concentration 4%) is added. Aliquots are pipetted onto gelatin-subbed glass slides (Rogers, 1967), dried at room temperature and washed repeatedly in water. Ilford liquid nuclear emulsion K5 is melted in a darkroom (45°C) and diluted 1:3 with distilled water. The emulsion is pipetted onto the glass slides, which are then stood vertically to allow the excess emulsion to drain, and allowed to set on a cold steel strip. After drying for 1 h at 30°C, the slides are exposed in a light-proof box at 4°C for 4–8 days. The slides are processed at 20°C in Ilford Phen-x developer (5 min), distilled water rinse (2 min), Kodak F-24 fixative (10 min) and a distilled water rinse (30 min).

The method has been used for metabolically active fungi which can be detected at a magnification of ×400. Use of $^{3}$H-glucose might give greater resolution and allow discrimination of bacteria.

Controls using non-tagged glucose and/or autoclaved or formalin-treated preparations should be included and processed separately to avoid transfer of $^{14}$C-labelled volatiles (Preston and Waid, 1972). The duration of the incubation period is important as too short a time gives inadequate labelling and too long a period allows fungal growth. Use of more highly labelled glucose or prolonged exposure times can cause increased background labelling. Homogenization has the same affect as that noted for FDA staining (see above).

The amount of hyphae detected by this method is only about a quarter to a half of that detected by FDA, perhaps due to the rather slow growth and uptake of glucose by hyphae that occur in soil. At present, the technique should be used only for comparative purposes.

### C. Decomposition of $^{14}$C, $^{15}$N-labelled plant material (Ladd *et al.*, 1981)

There have been a variety of studies on the decomposition of isotope-labelled plant material in soil which generally show that the concentration of $^{14}$C-labelled material in soil decreases rapidly in the first few months but that then rates slow down considerably so that there are substantial residues still present after several years. The methods reported here utilize double-labelled legume tissue for following decomposition over a 4-year period.

#### *1. Preparation of labelled plant material*

Seeds of *Medicago littoralis* are germinated and grown in low organic N sand in pots. The pots are placed in a sealed growth cabinet 1 week after germination in an atmosphere enriched in $^{14}CO_2$. Radioactivity is maintained by automatic addition of labelled sodium bicarbonate to a phosphoric acid reservoir. Plant nutrients, including $(^{15}NH_4)_2SO_4$ are added through tubes to the individual pots at weekly intervals for 8 weeks. Water is added to keep the sand at *ca* 10% gravimetric water content. Plants are harvested at the flowering stage when they may be sorted into leaves, stems and roots and dried at 40°C for 4 days. The specific activity of plant C and the atom % enrichment of plant N can be determined.

*2. Incorporation of labelled material into soil*

Open galvanized steel cylinders (6.7 cm dia. × 85 cm) are pushed into the ground to a depth of 80 cm with a hydraulic ram. By offsetting the leading edge of the cylinder, soil compaction can be avoided. The top 10 cm of the soil is removed from the cylinder and sieved (<6 mm) to remove coarse fresh plant tops and roots and pooled with replicate samples. Sufficient labelled plant material is added as dried, unground leaves, stems and roots to give the equivalent of 44–67 kg N $ha^{-1}$ for a 0–10 cm depth. In the study reported this involved mixing 757–1135 mg plant material with *ca* 500 g moist soil. The particular amounts would depend upon the study in hand, the C and N contents of the plant material and the degree of labelling. All these would need to be determined before embarking on such a study. The amended soils are then replaced in the cylinders and the soil surfaces tamped gently.

*3. Subsequent sampling and analysis of amended soils*

Soils inside the cylinder are maintained free of vegetation. At suitable intervals of time, e.g. 4, 8, 16, 32, 53, 104 and 208 weeks, duplicate cylinders are removed and sawn into lengths. Soils from depths 0–10 cm and 10–20 cm are weighed moist, mixed thoroughly and subsampled. Samples may be analysed fresh within 24 h or stored as freeze-dried or deep-frozen material.

Residual $^{14}C$ and $^{15}N$ are determined on ground samples. 50 g of each soil are ground for 30 s in a Sieb Technik mill and 5 g subsamples given a second grinding (<53 μm). The radioactivities of triplicate samples (20–40 mg) are determined by scintillation counting using the methods described by Adu and Oades (1973). Total N can be determined by Kjeldahl techniques (Bremner, 1965a) and atom % excess $^{15}N$ calculated after analysis (Ross and Martin, 1970), using a mass spectrometer. Inorganic N is determined in duplicate KCl extracts of field moist soils (125 ml 2 M KCl: 50 g soil) which are steam distilled as described by Bremner (1965b). Atom % enrichments of the distilled $NH_4^+$–N are determined as above. Microbial biomass can also be determined using Jenkinson's fumigation technique (see below), modified so that $^{14}CO_2$ can be determined as well. Labelled and unlabelled ammonium- and nitrate-N in various extracts can be determined so that the amount of mineral-N derived from the biomass and released following fumigation can be calculated. Approximate soil biomass-N can be derived from this by multiplying the values by 1.6 (Amato and Ladd, 1980). A balance sheet for carbon and nitrogen added to soils can be constructed and the rates of decomposition (half lives and decomposition constants) determined in the initial rapid and subsequent slow phases. The labels can be tracked through the microbial biomass during the decomposition period.

### D. **Litter bag methods** (Gilbert and Bocock, 1962; McClaugherty *et al.*, 1985)

It is often necessary to determine the rate at which different leaves or twigs or components of them decay in different soils. In addition, information may be needed on whether decomposition is assisted or hindered when certain classes of decomposer are excluded from the samples. These questions are conveniently answered by use of the litter bag technique. Data can be analysed by comparing treatment means using parametric statistical procedures or mathematical descriptions of the data characterizing the observed changes with time. A valuable critique of these approaches to be read by anyone using these techniques is given by Wieder and Lang (1982).

Leaves are collected at leaf fall in nets or in screens suspended above the forest floor and used within 3 days of collection. The leaves are air-dried at 20–25°C for about 3 days to a constant mass when the water content is about 10–12% (fresh weight). Weighed samples ranging from 1 to 10 g are placed in nylon or polyester litter bags (*ca* 15 × 15 cm) with known mesh sizes. There is some variation in the mesh sizes used but Edwards and Heath (1963) suggested that a 7-mm mesh will allow colonization by all the microflora and the invertebrate fauna, a 1-mm mesh will do the same but exclude earthworms, a 0.5-mm mesh allows only microorganisms and mites, springtails, encytraeids and small insects to colonize the leaves while a very small mesh of 0.003 mm will exclude all animals and allow only microbial colonization. More recently, McClaugherty *et al.* (1985) used only two mesh sizes, 0.1 mm and 2.0 mm. Fine mesh nets may affect some environmental conditions around the leaves, e.g. moisture levels.

The litter bags are placed on the forest floor or buried in particular soil horizons, taking care to put them at random within pre-established grids in each study area. Replicate bags (about four) are collected at suitable intervals until decomposition is complete. This may take several months or even years depending upon leaf type.

Initially and upon sampling, leaves should be dried to a constant mass at 60°C for 48 h and weighed. If it is desired to follow the fate of leaf components, then the dried leaves can be ground to pass a 1-mm mesh screen and subsamples analysed for ash content (450°C for 8 h), moisture content (105°C for 48 h), total nitrogen, e.g. by a suitable Kjeldahl system, polar and non-polar extractable materials, acid-soluble (holocellulose) and acid-insoluble (lignin) residues.

An alternative though less accurate indicator of litter decomposition is the reduction in the surface area of leaves remaining. For this, 50 replicate leaf discs initially of 2.5 cm diameter can be placed in each bag (Edwards and Heath, 1963) and their areas reassessed upon sampling by gridding methods or with a leaf area meter.

### E. Rates of carbon dioxide output measured by infrared gas analysis (Ineson and Gray, 1980; Reiners, 1968)

Infrared gas analysers can measure the amount of carbon dioxide in gas mixtures specifically, continuously, accurately and non-destructively with the output being sent to a chart recorder or computer. The gas stream to be analysed is passed through an optical cell placed between a source of infrared radiation and a detector. Radiation of specific wavelengths is absorbed by the carbon dioxide and the change in energy level reaching the detector is amplified and forms the output signal. A useful description of the instrument and its application to photosynthetic measurements is given by Long (1986) but much of the information is of relevance to measurements of soil respiration.

Implicit in the use of this system is that a stream of gas flows over or through the soil sample being analysed. This gas stream will cause the oxygen supply rate to the soil, the water content of the sample and the rate of removal of carbon dioxide to change from that found naturally. This may be exacerbated if the soil has been homogenized and sieved which is often the case when the response of soils to various treatments is being examined. Since the air flow rate can be varied between 0 and 2.0 l $min^{-1}$, these changes can be quite considerable and so the assumption that field rates are being measured is not justified. Nevertheless, valuable comparative measurements can be made. It is certainly advisable to humidify the air being passed through the soil samples to prevent desiccation. Use of an air stream containing carbon dioxide is possible and so disturbance of the bicarbonate–carbon dioxide equilibrium and consequent release of abiotic carbon dioxide can be minimized in alkaline, carbonate-containing soils.

A suitable analyser for making such measurements is the Type 225 Mk 3 (Analytical Development Co. Ltd, Hoddesdon, Herts, UK). This is a non-dispersive, double-beam instrument which will measure differential increases in carbon dioxide concentration up to 50 ppm. A survey of all the analysers available currently with their specifications is given by Bingham and Long (1988). The analyser should be accompanied by a gas-handling device, fitted with a sequential timer circuit which routes gas through a series of separate channels either to the analyser or to waste. Gas handlers have been constructed which can switch between 30 or more channels, but commercially available gas handlers with six channels are adequate for most purposes.

Air can be supplied to respiration chambers, one of which is empty and acts as a reference, from a common compressed air supply after it has been passed through a humidifier. Excess pressure is dissipated and the gas-handler pump then draws the humidified air through the respiration chambers. Respiration chambers may be designed to suit the experimenter but glass cylinders (26 × 3 cm), fitted with a water jacket through which water circulates at a predetermined temperature, are convenient. To avoid condensation of the water in the air supply, the humidifier should be kept at the same temperature.

The signal will be read out in ppm carbon dioxide and to convert this to the volume of carbon dioxide evolved per hour per unit weight of soil, the temperature, barometric pressure, flow rate of the gas stream and the soil dry weight should all be recorded.

Soil may be placed as intact cores in the chambers or may be homogenized and sieved. Alternatively, Reiners (1968) has described a method for monitoring carbon dioxide output from soil in the field. A steel sleeve (20 × 50 × 20 cm) is inserted into the soil until the top edge is just above the litter surface. The sleeve is fitted with a plastic lid provided with inflow and outflow tubes. Air is pumped through the space above the soil surface and sampled for carbon dioxide with an infrared gas analyser. One of the problems with methods of this type is relating the gas exchange to a particular volume of soil. One has to assume that the carbon dioxide has come only from the volume of soil inside the cylinder but this remains an untested assumption. The development of field methods will be assisted by the relatively new portable IRGA systems (Bingham and Long, 1988).

There are many other methods that can be used to measure carbon dioxide evolution including absorption in alkali and titration, Wosthoff Ultragas analysis, gas chromatography and mass spectrometry; it is also possible to measure oxygen uptake and hence determine the respiratory quotient. Some of these methods have been described by Parkinson *et al.* (1971).

## III. Methods for measuring microbial biomass

When studying primary and secondary productivity in ecosystems, it is necessary to compare the amounts of microorganisms present in different samples. This must be done in a way that is directly comparable with determinations of the amount of plant and animal material in the system. In practice, this involves determining the weight of living material of each component (biomass) at chosen times and sometimes the weight of dead material as well (necromass). Until about 20 years ago, most methods for determining microbial biomass involved counting microbial propagules and converting these counts to weights by means of assumed or measured conversion factors. These conversion factors were often very imprecise as they were determined of necessity for cells grown in culture, rather than field-grown cells. They were also inappropriate where one was not counting uniformly sized propagules, e.g. fungal propagules range in size from small spherical spores to large and variable lengths of hyphae. Furthermore, many of the counting methods used were more or less selective and underestimated the numbers of cells present.

For these reasons, new methods were developed which attempted to measure the total microbial biomass (subsuming bacteria, fungi, algae, protozoa, etc.) based on supposedly universal properties of living organisms. A few of the most widely used methods are described below which are based on responses to fumigation, responses to added glucose, presence of ATP and measurements of biovolume. Methods typical of those for determining the biomass of bacteria based on the presence of characteristic wall polymers are also described and some comments made on the value of dilution plate counting. It should be noted that in many of the methods described, reliance is still placed on the use of conversion factors and it cannot be stressed too much that many of these conversion factors are not universal and may only be provisional estimates anyway.

### A. **The soil fumigation method** (Jenkinson and Powlson, 1976)

If soil is sieved and made relatively homogeneous, placed in a container and its carbon dioxide output measured over 20 days, there will be a period of rapid respiration (a flush), followed by a period in which a low stable respiration rate occurs. This flush is brought about by disturbance effects which include killing of some organisms and bringing oxygen, substrates and organisms into new relationships. Eventually the flush subsides and a rate commensurate with the normal rate of substrate oxidation obtains.

If, however, soil is prefumigated with chloroform, most of the biomass will be killed and the normal flush prevented. When the chloroform is removed, the few residual organisms (if necessary with some added at this stage) will have not only the original organic matter for a substrate but the microbial corpses derived from the biomass. The flush of respiration in the first 10 days will then be due to this additional organic matter. Jenkinson

(1966) suggested that the microbial biomass carbon ($B$) could be calculated as follows

$$B = \frac{X - y}{k}$$

where $X$ is the weight of carbon given off as carbon dioxide in the first 10 days, $y$ is the weight of carbon given off as carbon dioxide in the second 10 days and $k$ is the percentage of the biomass carbon mineralized to carbon dioxide in the first 10 days. It was shown that for a range of dead organisms decaying at 25°C, $k$ was 0.5 (Jenkinson, 1976), although Anderson and Domsch (1978) suggested a figure of 0.411 for soil incubated at 22°C and Jenkinson and Ladd (1981) a value of 0.45.

An alternative method of calculating biomass uses the formula

$$B = \frac{X - x}{k}$$

where $x$ is the weight of carbon given off as carbon dioxide from unfumigated soil during the first 10 days. As discussed above, disruption of the soil will induce a flush and so $y$ will represent the rate after both components of the flush have subsided in the fumigated soil and so overestimate biomass. On the other hand, since mechanical disruption kills part of the biomass in the unfumigated soil, use of $x$ will underestimate the biomass. The true value will lie between these limits.

There are some clear restrictions on the use of the technique. It cannot be used on previously air-dried soils as air-drying is partially biocidal, on soils containing a lot of chalk (without correcting for abiotic carbon dioxide) or for soils with large recent additions of organic matter. In practice it does not seem to work well with very acid soils. Its principal value is in comparing biomass levels in agricultural and some woodland soils.

Four 250 g portions of moist soil, previously passed through a 6.5-mm sieve, are used for each sample. They are placed in 400-ml beakers and two of them are fumigated with chloroform in desiccators (i.d. 30.5 cm) lined with moist paper. Each desiccator contains a separate beaker containing 50 ml alcohol-free chloroform, prepared by shaking Analar grade chloroform three times with 5% conc. sulphuric acid, washing it five times with water and drying it over anhydrous potassium carbonate before redistillation. Such chloroform can be kept for a few weeks over anhydrous potassium carbonate in the dark.

Each desiccator is evacuated until the chloroform boils, then closed and left in the dark at 25°C for 18–24 h. The beaker of chloroform and the paper are removed and the desiccator evacuated eight times for 3 min (three times with a water pump, five times with a high-vacuum oil pump) to remove the residual chloroform.

The respiration rates of the fumigated and unfumigated samples are measured over 20 days to determine the values of $X$ and $y$. In the original method, Jenkinson and Powlson (1976) placed the samples in 3.75-l glass jars with 100 ml 1 N NaOH in a beaker and 20 ml distilled water and estimated the respiration rate titrimetrically (Underwood, 1961) but Grisi and Gray (1986) used infrared gas analysis as described in an earlier section.

Biomass can be expressed as mg C per 100 g oven-dry soil so it is also necessary to determine the water content of each soil. Values between 20 and 120 mg C per 100 g oven-dry soil are common which represent between 1.7% and 3.7% of the soil organic carbon.

Amato and Ladd (1988) have shown that the release of ninhydrin-reactive compounds from soil directly following fumigation is also an indicator of biomass carbon. They found that for 25 different soils amended with plant material and incubated for 66 weeks, total biomass carbon was 21 times the total amount of ninhydrin-reactive material that could be measured in 2 M KCl extracts of fumigated soil (2.5:1 v/w). Biomass nitrogen was 3.1 times the total amount of ninhydrin-reactive material. Ninhydrin-reactive nitrogen was estimated by reacting 0.5 or 2.0 ml of the filtered extracts with a ninhydrin reagent (Moore and Stein, 1954). Absorbances were read at 570 nm. This may prove to be a more reliable indicator of biomass than carbon dioxide release.

### B. Relation between response to glucose and microbial biomass
(Anderson and Domsch, 1978)

When glucose is added to soil, it is metabolized and converted to biomass and carbon dioxide. If sufficient glucose is added to saturate the enzyme systems of the biomass, then the initial rate of carbon dioxide production (that which occurs before new biomass is formed) should always be the same and will be related to the biomass present. Anderson and Domsch (1978) calibrated such responses against biomass estimated by the fumigation technique (see above) and obtained a straight-line relationship between biomass and carbon dioxide production at 22°C according to the expression

$$x = 40.4y + 0.37$$

where $y$ is the initial rate of respiration in ml carbon dioxide per 100 g soil per hour and $x$ is the biomass in mg microbial carbon per 100 g soil. There was a correlation coefficient of 0.96.

The advantage of this method is that only one series of respiration measurements is needed but the calibration must be redone for each new soil type or incubation temperature used. All the errors associated with the fumigation technique are implicit in this technique. Additionally, one should recognize that the biomass may be underestimated because some organisms are unable to use glucose and because anaerobic organisms may behave differently.

Soils are sieved (2 mm) to remove roots and rootlets and stored moist at 22°C for 10 days. Samples of between 20 and 100 g are amended with sufficient glucose (see below). The glucose is put in a glass mortar with 0.5 g talcum, ground to a fine powder and blended with the soil in a 1-l beaker with a hand-operated electric mixer operating at 1600 rpm for 25–30 s. The amount of glucose required must be determined by experiment. Samples with excess glucose will show increases in carbon dioxide production after a short lag and samples with insufficient glucose will show a decrease from the initial maximal response. In soils with the correct amount of glucose, the initial maximal response will show an unchanged rate for several hours. Depending upon the soil type, such glucose concentrations will vary but values in the range 100–4000 $\mu g\ g^{-1}$ soil are typical.

Samples are placed in suitable chambers and their respiration measured hourly at 22°C with an Ultragas 3 carbon dioxide analyser (Wösthoff Co., Bochum, FRG) for up to 6 h. It should be possible to use an IRGA instead. Biomass is calculated using the above equation or one similar to it derived for other soils.

### C. **Relationship of biomass and biovolume of cells** (Jenkinson *et al.*, 1976)

It is possible to estimate the biovolume of cells in a soil suspension microscopically by making assumptions about the spherical and/or cylindrical shapes of the majority of microorganisms. These estimates can be converted to biomasses if the specific gravity and the water content of the cells is known or assumed; and to biomass carbon contents if the percentage carbon content of the cells is known. It is necessary to prepare a representative suspension of soil in a diluent and to make the organisms visible by means of phase-contrast microscopy or by staining. Much controversy surrounds the choice of stain and particularly whether some stains are capable of distinguishing living from dead cells. Amongst the stains used most frequently are phenol aniline blue, acridine orange, fluorescein isothiocyanate, europium chelate + fluorescent brighteners, ethidium or propidium bromide or various fluorescent antibodies. Intending users will find some useful comments on the techniques in Postgate

(1967) and Gray (1976). Another problem is ensuring that the preparations observed under the microscope are homogeneous. To date, the best method for preparing material is the agar film method pioneered by Jones and Mollison (1948) and modified by Jenkinson *et al.* (1976).

1–2.5 g moist soil are added to 60 ml warm (60°C) agar detergent solution (10 ml 12% Decon 75, 50 ml 0.9% Oxoid agar No. 1) that has been filtered through a No. 4 sinter. The agar solution may be filtered prior to use through a 0.22 μm Millipore filter to remove contaminating bacteria (Harris, 1969). The suspension is ultrasonified for 10 min in a 60°C cleaning bath (Dawe Instruments Ltd, model 1143A operating at 25 Kc $s^{-1}$), taking care to mechanically stir the suspension all the time to prevent the sand from settling. After ultrasonification, the stirring is maintained for a few minutes and only stopped 30 s before removing a 0.02 ml sample with a pipette (tip i.d. 1 mm) exactly 1 cm below the surface. The sample is run into a haemocytometer cell (0.1 mm depth) under a thick (0.47 mm) coverslip and pressed firmly but gently down with a 5 g weight (Thomas *et al.*, 1965). The film is allowed to set, floated off the haemocytometer, trimmed, positioned on a microscope slide, dried at room temperature and stained with phenol aniline blue (phenol, 5% aqueous, 15 ml; aniline blue, 6% aqueous, 1 ml; glacial acetic acid, 4 ml; filter 1 h after mixing) for 1 h. The stained films are rinsed in water and 3–4 changes of 98% alcohol and mounted in Euparal.

Stained spherical and nearly spherical organisms (dia. 0.3–19 μm) can then be grouped in 13 diameter classes and the number in each counted. The New Porton G12 eyepiece graticule (Graticules Ltd) is used for organisms up to 1.5 μm diameter; for larger organisms a graticule marked with a 10-mm side square divided into a 0.5-mm grid is appropriate. Hyphae can be grouped into seven diameter classes, ranging from 1 to 13 μm, and counted and their lengths determined on the same grid. Hyphae not lying parallel to the grid must have their lengths corrected by a factor of 1–1.4, depending upon the angle. Details of the diameter classes of use are given by Jenkinson *et al.* (1976). Fields can be spaced equally so that 20 can be counted in a single traverse, a second traverse being made if less than 40 organisms are counted. Only clear blue or purplish-blue objects are counted. Unstained, brown hyphae and cells, objects with asymmetrical outlines and indefinite boundaries and faint, eroded and decomposing hyphae should be ignored. Organisms viewed between crossed polarizers are isotropic, which helps to differentiate them from birefringent soil particles that can obscure large organisms.

The biovolume of spheres in a given size class is obtained by multiplying number by class volume calculated from the nominal diameter of the class. Hyphal volumes are calculated by multiplying length by cross sectional area.

Jenkinson *et al.* (1976) concluded from using this technique on a number of soils that a class of rare large organisms contained as much biovolume as a class of numerous small organisms. Thus, careful observation of the slides is essential. It is also noteworthy that the biovolumes of hyphae and spherical organisms in all the soils they studied were roughly equal.

Phenol aniline blue and some of the other stains mentioned above can be used in techniques used for locating organisms in soil (see below).

### D. Determination of ATP content of soils (Jenkinson and Oades, 1979; Oades and Jenkinson, 1979)

ATP is a universal energy carrier in biological systems which degenerates rapidly on death and could thus be a good indicator of biomass. Although it is easy to analyse, there are considerable problems associated with extraction of ATP from soil colloids and with its hydrolysis during extraction. The method described below aims to minimize these problems, not eliminate them. It is therefore necessary to estimate the efficiency of extraction by using internal standards, usually either known organisms or pure ATP. Neither are completely satisfactory (Jenkinson and Oades, 1979). ATP levels in cells are also related to the physiological state of the cells and so comparisons between different populations may be of doubtful value; it may be useful to determine the adenylate energy charge (Brookes *et al.*, 1987). Samples containing large amounts of roots or animals which also contain ATP may be unsuitable for application of these methods.

Four portions of sieved (<2 mm), moist soil consisting of 2.5 g oven-dry soil (80°C, 24 h) in 50 ml polypropylene centrifuge tubes are stood in ice. Twenty-five ml extractant A (25.8 g paraquat hydrochloride in 100 ml water added to previously dissolved 81.6 g trichloroacetic acid and 89.6 g $Na_2HPO_4.12H_2O$ in 600 ml water, then all made up to 1 l with water; can be stored at −15°C) is added to tubes 1 and 3 and 25 ml extractant B (5 ml 0.1 mM ATP made up to 1 l with extractant A; can be stored at −15°C) added to tubes 2 and 4. The soil is then dispersed for 1 min with an ultrasonic probe (12.5 mm dia.) attached to a 150 W sonifier at full power. The probe tip must be 1 cm below the liquid surface. The tubes are cooled in ice for 5 min, filtered (Whatman No. 44) and frozen quickly at −15°C (storage time up to 6 weeks). The pH of the extracts must be below 2; if it is not, the extraction must be repeated with less soil. Only 0.5–1.0 g soil is required for soils with >5% organic matter.

Fifty μl extract is pipetted into 5 ml arsenate buffer in a dry, acid-washed, 25 ml scintillation vial, 100 μl enzyme–luciferin solution added, the vial sealed and then placed in a liquid scintillation spectrometer (gain set at 100%, window setting of 50–90 scale divisions, photomultiplier tubes set out of coincidence), allowed to stand for 1 h and the vials counted for 0.1 min. The enzyme–luciferin solution can be made from Sigma pre-prepared enzyme FLE-50 by suspending the extract from 50 mg dried firefly tails in 10 ml ice-cold water and adding 10 ml ice-cold buffer solution (31.2 g $Na_2HAsO_4.7H_2O$ in 800 ml water mixed with 10 ml 0.2 M tetra-sodium salt of EDTA to which is then added 2.46 g $MgSO_4.7H_2O$ in 100 ml water; pH adjusted to 7.4 with 1 M sulphuric acid and solution made up to 1 l). Alternatively, a slightly better preparation can be obtained directly from firefly tails as described by Jenkinson and Oades (1979).

Freshly prepared standards are run in triplicate, each consisting of 5 ml buffer, 50 µl extractant A, 100 µl enzyme–luciferin solution and 50 µl arsenate buffer containing 0, 1.5, 2.5, 5.0, 10, 25, 40 and 50 pmol ATP (Sigma, di-sodium salt). Counts per 0.1 min less background count per 0.1 min can be plotted against ATP concentration on log 2 cycle: log 2 cycle paper. A straight line is obtained from 1 to 25 pmol ATP. If counts from samples exceed those of the 50 pmol standard, the extraction should be repeated with less soil.

Let $a$ be the ATP content (pmol on the calibration curve) of 50 µl extract from $n$ g moist soil ($m$ g oven-dry soil) with 25 ml extractant A. Let $b$ be the corresponding value with extractant B. Then

$$\% \text{ recovery of ATP} = 4(b - a)[1 + (n - m)/25]$$

The ATP content of the soil (as the anhydrous di-Na salt, MW 551.2) will be

$$6.89\, a/m(b - a) \ \mu\text{g ATP g}^{-1} \text{ oven-dry soil}$$

Jenkinson and Oades (1979) tested the precision of this method and found that the mean percentage recovery of added ATP from one of their soils was 67.9% (SD ± 3.4) and the corresponding ATP content was 1.57 µg ATP $g^{-1}$ oven-dry soil (SD ± 0.13 µg). However, the percentage recovery varies from soil to soil in the range 45–84% and the ATP content from 0.64 to 9.03 µg $g^{-1}$ oven-dry soil. Precision is affected greatly by soil sample size and by the accuracy of pipetting enzyme and sample into the scintillation vials.

Oades and Jenkinson (1979) proposed that ATP content could be related to biomass by the formula

$$\text{biomass C content} = 120(\text{ATP content})$$

but they pointed out that this relationship might not hold for all soils and was probably not relevant to soils in which microbes were actively proliferating and which had received large additions of substrate within a few days of measurement. Depending upon assumptions made, there are about 10.0 μmoles ATP $g^{-1}$ resting biomass carbon which is surprisingly close to the ATP concentration found in exponentially growing prokaryotic and eukaryotic organisms (Brookes *et al.*, 1987).

## E. Determination of the amounts of cell wall components of bacteria in soil (Moriarty, 1975, 1977, 1980)

Muramic acid is released from bacterial walls by acid hydrolysis and analysed by HPLC (Moriarty, 1983) or is hydrolysed with alkali to release D-lactic acid (Tipper, 1968) which is then assayed (Noll, 1974; Stanley, 1971). Amounts of muramic acid estimated in this way in cultured bacterial cells vary from 17 to 115 μg $mg^{-1}$ carbon (Moriarty, 1975), with Gram-negative bacteria having about 12 μg $mg^{-1}$ carbon and Gram-positive bacteria 44 μg $mg^{-1}$ (Moriarty, 1978). Thus, it is improbable that the estimates of biomass from use of these methods will be very accurate unless the proportions of these two groups are known. If necessary, this can be determined by Gram staining of bacteria released from soil (Moriarty, 1980). A formula for calculating biomass in marine sediments from muramic acid is

$$C = 1000M/12n + 40p$$

where $C$ is the biomass carbon (mg), $M$ is the muramic acid in μg, $n$ is the proportion of Gram-negative and weakly Gram-positive bacteria and $p$ is the proportion of Gram-positive bacteria (Moriarty, 1977). The formula needs modification for soil. Allowance must also be made if bacterial endospores are present as they contain 3.5 times as much muramic acid as vegetative cells of Gram-positive bacteria and for Gram-positive bacteria with thick walls. Blue-green bacteria may contain 500 times more muramic acid than Gram-negative bacteria. Variations of this type may have to be checked by examining samples microscopically (Moriarty, 1978). In principle, this method developed for marine sediments could be applied to soil and should represent an improvement on the method originally described by Millar and Casida (1970). Some of the problems associated with this technique will also apply to methods used to quantify fungi by determining their chitin content (see Frankland *et al.*, this Volume).

A weighed sample is hydrolysed with 3 M HCl (1 ml per 200–500 mg soil or 10–20 mg bacteria) in a sealed tube at 100°C for 8 h. However, if the soil contains calcium carbonate this should be removed first by treatment with glacial acetic acid (2 ml for every 1 g calcium carbonate), overnight at room temperature and the sediment recovered by centrifugation (3000 *g* for 5 min) and washed with distilled water (Moriarty, 1980). After hydrolysis, the hydrochloric acid is removed *in vacuo* and the sample dissolved in distilled water. 0.2 ml of 0.5 M di-sodium hydrogen phosphate is added and the pH adjusted to 8.0 with 5 M and then 1 M NaOH. The sample is then centrifuged to remove the precipitate which contains heavy metals and humic compounds. After centrifugation, the pellet is washed and the washings added to the supernatant from the centrifugation. This material can be analysed directly for muramic acid using HPLC or it can be hydrolysed and analysed biochemically for D-lactate. HPLC methods depend upon the instrumentation available and will not be described but the methods are simpler and faster. They are based on an assay of *o*-phthaldialdehyde derivatives of primary amines which can discriminate muramic acid from amines (Moriarty, 1983).

*Biochemical analysis.* The material can be freeze-dried first after adjusting the pH of the material to 1–2 with hydrochloric acid. The dried samples can then be extracted at any time with diethyl ether to remove D-lactate liberated by acid hydrolysis and other compounds such as gly-collic acid (Moriarty, 1978). Samples are vortexed for 1 min with 10 ml ether acidified with concentrated HCl (0.2% v/v). Excess ether is removed by drying and the samples redissolved in water and the pH adjusted to 7.5–8.0 with sodium hydroxide to give a known volume, e.g. 2 ml. Half of each sample is frozen until needed for assay of residual D-lactate.

The other half of the material is subjected to alkaline hydrolysis to liberate D-lactate from muramic acid. The pH of the sample is raised to 12.5 with more NaOH. After incubation at 35°C for 2 h, the pH is reduced to 9.0 with HCl to make the volume up to 2 ml. It may be necessary to confirm the presence of muramic acid by thin layer chromatography if soil samples are known to contain a large amount of carbohydrate as some carbohydrates can liberate D-lactate after alkali hydrolysis.

Known volumes of the hydrolysate (50 μl acid hydrolysate after ether extraction; 100 μl alkali hydrolysate; 100 μl alkali hydrolysate + 50 μl of a 1 ng $\mu l^{-1}$ solution of D-lactate as an internal standard) are put in three separate tubes and made up 150 μl with distilled water. A set of standards, usually in the range 0–200 ng D-lactate in 150 μl, are also prepared. These mixtures may be slightly inhibitory to the subsequent assay and if this is so, then the samples can be diluted further. Samples can be frozen until ready for analysis.

A lactate reaction mixture is prepared (0.1 M, pH 9.4 glutamate buffer, 0.5 ml; 33 mg $ml^{-1}$ NAD, 10 μl; 5 mg $ml^{-1}$ D-lactic acid dehydrogenase,

4 μl; 10 mg $ml^{-1}$ (80 units $mg^{-1}$) glutamate pyruvate transaminase, 2 μl). If the pH is below 9.0, it should be raised to 9.0 with 1 M NaOH. 0.5 ml of this mixture is added to small test tubes containing the sample to be analysed and mixed rapidly. After incubation at 30°C for 20–30 min, the tubes are transferred to crushed ice.

Two ml phosphate mixture (0.1 M, pH 7.5 phosphate buffer, 2 ml; 2-mercaptoethanol, 14 μl; 0.5 mg $ml^{-1}$ water flavin mononucleotide, 10 μl; saturated solution of tetradecylaldehyde in ethanol, 10 μl) are placed in a small polypropylene vial and mixed with 20 μl luciferase solution. Luciferase is prepared by dissolving 2.0 mg $ml^{-1}$ phosphate buffer and adding an equal weight of bovine serum albumen. The solution is kept on ice and centrifuged for 15 min at 3000 *g* at 0°C.

After 15 s or similar known time interval, 200 μl of the lactate reaction mixture is mixed with the phosphate mixture plus luciferase and the tube is placed in a scintillation vial. After exactly 30 s, the counting sequence is initiated. It is usual to measure the standards first keeping the interval between standards to about 15 min. If it is longer, computed corrections must be applied to the standard curve (Moriarty, 1980).

Variations in the procedure may be required depending upon the precise amount of lactate in the samples, variations in enzyme activity and the presence of divalent or heavy metals during ether extraction (Moriarty, 1980).

## F. Some comments on dilution plate counts

The dilution plate count has probably been the most widely used method for determining numbers of microorganisms in soil and biomasses are often derived from the counts obtained. Brock (1987) has recently restressed the drawbacks which have discredited the method. These drawbacks relate principally to the selectivity of the procedure and its inability to distinguish dormant and active organisms; the organisms which are counted are not necessarily those which are functioning in the ecosystem. To these draw-backs may be added the philosophical difficulty of deciding what an individual is. This is hard enough for clumps of cells of bacteria but is impossible for variously sized, branching, anastomosing and fragmenting mycelia. Despite these difficulties, it is doubtful that use of the technique will be abandoned completely because versions of it are often incorporated into procedures designed to isolate organisms, where it may be useful to have a crude idea of the numbers of propagules present which will respond to particular plating and environmental factors. It will continue to be used where selective methods for particular biochemical or physiological groups are needed. However, those using it should resist the temptation to believe that total bacterial or fungal biomasses can be estimated from it. For these reasons, the technique is not described here but descriptions of it can be found in Parkinson *et al.* (1971).

## IV. Methods for locating organisms in soil

Brock (1987) has said that all ecological studies of natural environments should start with the microscope. Certainly, it is true that in soil, above all other habitats, the spatial relationships of organisms and their habitats are of very great importance. Furthermore, endless speculation on the ecological relationships of different organisms which have never been seen growing together is unprofitable. However, microscopic observations of soil are hampered by the opacity of many soil particles, by the ease with which the interrelationships of these particles can be disrupted and the lack of morphological characteristics enabling one to identify the strains and species of organisms present. We do not always know if organisms growing in soil have the same morphology as those growing in culture and the mere presence of microbially sized objects in soil is no guarantee that they are organisms at all. There are well-known, embarrassing examples of pieces of larger organisms being mistaken for microorganisms. Resolution and depth of focus are also limiting factors for many techniques, as well as the non-specificity of some of the stains used to enhance contrast between organisms and their backgrounds.

The microscopic techniques available are legion but here I shall describe only representative examples in which some of the above problems are minimized. Only techniques which look at soil itself are included and not those which involve burial of glass slides, capillaries, organic materials, etc. from which information on the colonization of model substances can be obtained. Such techniques have been described already by Parkinson *et al.* (1971).

### A. Maintenance of the integrity of soil structure (particle and section analysis)

If one is interested only in the distribution of organisms on the surface of individual, large soil particles, then careful staining and mounting of them is all that is required. Because particles are variable in diameter and irregular in shape, it can help if the particles, following staining, are pressed lightly into a bed of agar on a microscope slide with a coverslip. In this way, the upper surfaces are aligned and made easier to focus without grinding the lens into larger particles. If the particles are opaque, it will be necessary to observe them with transmitted white or short wavelength light.

However, many soil microorganisms form colonies which cover or penetrate particles in a relatively large volume of soil and under appropriate conditions may change their growth patterns (Rayner *et al.*, 1987). To observe such organisms, it is necessary to retain the relationships between the different soil particles while processing them. This can be achieved by impression methods in which an adhesive-coated support, e.g. a slide, or SEM stub is pressed against an exposed soil surface, removed, dried and stained (Brown, 1958; Parkinson *et al.*, 1971). It is also possible to impregnate soil and prepare sections and two methods are described to achieve this for light and transmission electron microscopy.

In Tippkotter *et al.*'s method (1986), samples of soil containing roots in Kubiena boxes (8 × 6 × 4.5 cm) are placed in a polyethylene dish inside a desiccator and the pressure reduced to 20 kPa for 15 min. Glutaraldehyde [2.5% (v/v) in 0.1 M Sorensen's phosphate buffer, pH 7.2] is then added to the dish until the soil is immersed and left for 2–3 h. Fixed samples are dehydrated for 1–2 h in each of 70, 90, 100 and 100% acetone, again at a reduced pressure of 20 kPa. The lids of the Kubiena boxes should be replaced with muslin for efficient dehydration.

The Kubiena boxes are then placed in a 2-l tank, the pressure reduced to 20 kPa for at least 10 min and Crystic 17449 resin (B & K Resins, Bromley, UK) added slowly to the bottom of the tank so that it rises at $<1$ mm min$^{-1}$ and totally immerses the soils after 1 h. The most suitable formulation of resin is resin, 800 ml; acetone, 200 ml; cobalt accelerator 1% (w/v), 2.5 ml; catalyst MEKP, 5.0 ml. The resin-impregnated blocks are maintained at low pressure for 1 day. The tanks containing the samples are transferred to a fume cupboard for 2 days to allow the acetone to evaporate and then to a spark-free embedding oven at 25°C for 10 days to allow polymerization. Polymerization is completed by raising the temperature to 50°C for one day.

Slices (8–10 mm thick) are cut from the resin-impregnated blocks with a diamond saw and ground flat with a surface grinding machine. They are lapped on diamond-impregnated bronze plates (240, 320 and 400 grit size) (Robertson and Normington, 1976). The resulting polished blocks are stained (see below) and mounted on clean glass plates using Crystic 17449 resin as an adhesive [resin, 100 ml; cobalt accelerator (1% w/v), 2 ml; catalyst MEKP, 2 ml]. After 1 day, excess material is removed by cutting with a diamond saw 800 μm above the glass. The thickness of the section is reduced to 25–30 μm with a surface grinder and then to 15–20 μm with diamond lapping plates. The section may be mounted in further adhesive resin mixture under a coverslip and cured at 40°C for 1 h.

Staining of thin sections is difficult because the sections may crack and become detached from the slides but the surfaces of polished blocks of resin can be stained with basic fuchsin (10 g $l^{-1}$), methylene blue (1 g $l^{-1}$ in 0.1 N NaOH) or the fluorochrome Mg-ANS (the magnesium salt of 8-anilino-1-napthalene sulphonic acid; 3.5 g $l^{-1}$) for 5–10, 1–3 or 3–5 min, respectively. Prestaining of soil before fixation may be possible but this is likely to disturb the arrangement of cells. Earlier work (Burges and Nicholas, 1961) using similar techniques has shown that fungal hyphae, bacteria, faecal pellets, etc. can be seen in soil sections of about 50 μm thickness, often without staining. These methods have not been used widely enough to assess what proportion of different cell types can be seen and until combinations of stains that can improve visibility of both roots and microbes can be tested, interpretation of results must be cautious.

For examination of soils under the electron microscope, it is necessary to cut sections with a diamond knife and for this reason samples containing large sand grains must be avoided. The methods are thus only appropriate for examining root/soil interfaces and leaf litter. Foster and Rovira (1973) reported a method where roots are dissected out of the soil mass and cut into 1 mm segments. They are fixed in 3% (v/v) glutaraldehyde + 3% acrolein in 0.1 M sodium phosphate buffer for 3–4 h at 4°C. Excess fixative is removed with several changes of buffer and the roots fixed overnight in 2% osmium tetroxide in 0.1 M sodium phosphate buffer at 4°C, taking care not to wash off adhering soil. The samples are dehydrated through tertiary butyl alcohol, rinsed in epoxy propane and embedded in araldite. The blocks are trimmed for ultra-thin sectioning of 30 nm thickness. Sections may be stained in Reynolds' lead citrate solution (Reynolds, 1963).

Transmission electron microscopy can also be used to examine micro-organisms extracted from soil in aqueous suspensions by mounting them on formvar-coated grids and staining or shadowing them by conventional means. Scanning electron microscopy (Gray, 1967) can be used to examine the relationship between organisms and their habitat but the preparative techniques are quite drastic and special steps must be taken to prevent collapse of cells or extrusion of protoplasm, e.g. the use of critical point drying (Cohen *et al.*, 1968). Even so, where cells are situated on delicate mucilaginous surfaces, the surfaces may collapse and interpretation is difficult (Costerton and Cheng, 1982). Scanning electron microscopy has the supreme advantage of coupling great depth of focus with relatively good resolution, so much so that three-dimensional images can be obtained. Its disadvantage, compared with optical microscopy, is that the contrast between cells and their background is less good and at low magnifications, cells can easily be missed. However, it can be combined with X-ray microanalysis to plot the distribution of particular elements in cells and their surroundings (Heldal *et al.*, 1985).

## B. Biological stains for light microscopy

Most of the stains used for soil organisms are cytoplasmic stains and it has been claimed that the presence of cytoplasm is an indicator of viability since it disappears rapidly following death. Thus, many microbiologists will record only deeply stained cells although this does introduce a subjective decision. Typical stains include phenol aniline blue (see above), protein stains like fluorescein isothiocyanate and nucleic acid stains like acridine orange, europium chelate and ethidium bromide.

### *1. Fluorescein isothiocyanate* (Babiuk and Paul, 1970)

The following are mixed and allowed to stand for 10 min: 1.3 ml of 0.5 M carbonate–bicarbonate buffer (pH 7.2); 5.7 ml 0.85% physiological saline; 5.3 mg crystalline fluorescein isothiocyanate. Preparations are stained for 1 min at 37°C, rinsed for 10 min in 0.5 M carbonate–bicarbonate buffer (pH 9.6) and mounted in buffered glycerol (pH 9.6). Observations are best made under short wavelength blue light. Dead cells lose their ability to be stained after 2 days in soil, compared with 4–5 days for loss of stainability by europium chelate or fluorescent antibody (see below) (Gray and Deaney, 1983).

### *2. Acridine orange* (Trolldenier, 1973)

A solution of acridine orange in water (1 : 15 000) is put on the specimen for 2 min and removed by rinsing twice in tap water. Specimens are best irradiated with ultraviolet light. Bacteria may stain either green or orange but although this was thought originally to distinguish living from dead bacteria as only bacteria lacking a cytoplasmic membrane took up the stain in large quantities, it is now known that this is not so for colour also depends upon the Gram reaction of the bacteria and the precise concentration of the stain. The idea that the cytoplasmic membrane disappears at the moment of death is also suspect (Postgate, 1967).

### *3. Europium chelate/fluorescent brightener differential stain (DFS)* (Anderson and Slinger, 1975a)

Europium (iii) thenoyltrifluoroacetonate (Eastman organic Chemicals, Rochester, NY, USA) known as europium chelate or $Eu(TTA)_3$, complexes with nucleic acids. The fluorescent brightener (FB) di-sodium 4,4′-bis[4-anilino-6-bis(2-hydroxyethyl) amino-5-triazin-2-ylamino] 2,2′-stilbene disulphonate (Cyanamid Co., Bound Brook, NJ, USA) stains microbial cell walls. When the two are mixed and used as a differential stain (DFS), living or recently dead cells stain red while cells lacking nucleic acids and which are dead stain green (Anderson and Slinger, 1975a). Problems may be encountered as some recently dead organisms fluoresce a deep shade of pink. Anderson and Slinger (1975b) suggested that DFS stained greater numbers of cells (although some of them were green and presumed dead) than fluorescein iso-thiocyanate (FITC) and differentiated cells from organic matter more efficiently than phenol aniline blue. However, Jenkinson *et al.* (1976) still preferred phenol aniline blue, even though $Eu(TTA)_3$- and FITC-stained organisms were easier to see and less tiring to count because it was the only stain that detected organisms in all the size classes they examined (see above).

$Eu(TTA)_3$ and FB are stirred and dissolved in absolute ethanol in a volumetric flask and diluted with sterile distilled water to give 50% (v/v) ethanol, 2 mM $Eu(TTA)_3$ and 25 μM FB. Agar films (see above) are stained by immersing the films for 18 h at room temperature. These are rinsed with a fixed volume of ethanol (12.5–20.0 ml) by running it down the slide sloped at 30° to the horizontal and dried in a sterile air cabinet. Soil smears are stained for 90–120 min and rinsed in only 5 ml ethanol. Specimens are mounted in Gurr Fluoromount (BDH, Poole, Dorset, UK) and examined with a microscope fitted with an HBO 200 lamp, 4 mm red minus BG38 and 1 mm UG2 filters (300–400 nm wavelength light transmitted) and a 1.5 mm clear glass filter overlaid with a yellow Wratten gelatin filter (Kodak No. 12) which enhances the red and green fluorescence.

#### 4. *Ethidium bromide* (Roser, 1980)

Soil smears can be stained with a solution of ethidium bromide (100–500 mg $l^{-1}$ water) and then washed for 1 min and mounted in distilled water. Jones and Mollison agar films (see above) can be stained with a 10 mg $l^{-1}$ solution for 10 min after air-drying the films for 4 h. The fluorochrome can be excited with 390–490 nm light when it fluoresces red, although some soil particles may autofluoresce. Less autofluorescence is obtained if the specimen is irradiated with 520–560 nm green light and this enables double staining and detection of proteins with fluorescein isothiocyanate to be carried out.

### C. Use of fluorescent antibodies for identification

Some biochemical transformations in soil are carried out by a narrow range of organisms, e.g. autotrophic nitrification and nitrogen fixation. Some other organisms have important properties, e.g. production of endospores and thus long survival times or an ability to decompose a wide range of litter types, e.g. some basidiomycetes. Symbiotic organisms may also have a free-living phase of existence in the environment. In all these cases, it is desirable to study the autecology of the organisms involved and for this one requires specific identification techniques. It is generally possible to devise a selective isolation procedure and subsequently identify organisms thus cultured (see below) but the only methods for identifying and locating specific organisms in soil samples are immunological and genetic ones.

The fluorescent antibody technique has been used to identify some soil organisms though the ecological investigations in which the technique has been applied (e.g. Siala and Gray, 1974; Siala *et al.*, 1974), are fewer in number. The reasons for this include the inability of the stain to distinguish living from dead organisms, insufficient or too great specificity of the antisera for the organisms being studied, autofluorescence of some soil particles and organisms and the ability of some soil particles to adsorb antisera and mask the stained organisms. This latter problem was overcome by Bohlool and Schmidt (1968) by pretreating specimens with a gelatin–rhodamine conjugate which blocked non-specific adsorption and acted as a counterstain for FITC conjugates.

A 2% aqueous solution of gelatin (adjusted to pH 10–11 with 1 N NaOH) is autoclaved for 10 min at 121°C and then readjusted to the same pH. Rhodamine isothiocyanate is dissolved in a minimum volume of acetone and added to the gelatin to give 8 μg dye to each mg gelatin. After stirring the mixture overnight, it is passed through a Sephadex column (G25, coarse) equilibrated with phosphate-buffered saline, pH 7.2. The conjugate separates from the uncombined dye and can be collected and preserved at −20°C with the addition of 1:10000 merthiolate.

Specimens (on slides) are immersed in the conjugate and placed at 60°C until the preparation is dry. After cooling, the specimen is stained with an FITC conjugate specific for the organism being detected. Cells will fluoresce green against an orange background. It is preferable to use a microscope fitted with epifluorescence and wide aperture, achromatic lenses. Immersion oil must be non-fluorescent.

Protocols for the production of fluorescent antisera for particular organisms and the staining techniques used are varied and will not be described here. Some suitable ones for soil organisms are referred to by Bohlool and Schmidt (1980) or given by Malajczuk *et al.* (1975) and Chard *et al.* (1983, 1985b). Specificity of the conjugates will be improved if antisera are raised against particular cell fractions (Chard *et al.*, 1985a) or if monoclonal antibodies are prepared (Wong *et al.*, 1988).

## V. Methods for isolating organisms from soil

Media for the isolation of microorganisms from soil have usually been divided into selective and non-selective types. However, all media are selective and so are the incubation conditions used. In this brief summary of isolation methods, media except for soil extract will not be described as it is impossible to make a rational selection. Formulae for media are available in Parkinson *et al.* (1971) and Burns and Slater (1982). Instead, reference will be made to direct methods of isolation which provide cells in the condition they were found in soil (micromanipulation and centrifugation) and suggestions made concerning the relevance of open (continuous culture) and closed (batch) culture procedures.

### A. Micromanipulation (Warcup, 1955; Casida, 1962; Skerman, 1968)

Many fungal hyphae are associated with the heaviest soil particles which may come out of suspension during dilution procedures. Soil crumbs can be saturated and broken with a jet of sterile water and the heavy particles allowed to sediment. The fine particles are decanted off and washing continued until only the large particles remain. These may then be spread in a film of water in a Petri dish and using a stereo-dissecting microscope, fungal hyphae are picked off with sterile fine forceps or needles before plating on a suitable isolation medium. Adherent contaminating organisms are often removed by drawing each hyphae through semi-solid agar. Plates are incubated and examined daily to prove that growth originates from the plated hyphae and not from any spores that remain adhering to the hyphae. Warcup (1955) showed that this method was especially successful for isolating fungi that grew slowly on plates but which were important in decomposition of soil organic matter, e.g. basidiomycetes and ascomycetes. The method is selective for these fungi and for fungi with large diameter, non-fragmenting, dark-coloured hyphae which extend into the soil spaces from their substrates. It is time consuming and requires much practice but is one of the few methods available for isolation of these particular fungal types.

Casida (1962) was able to isolate bacteria in a similar way but the bacteria had to be stained first. He applied 10 ml acridine orange (1:5000–1:10 000) to 1 g soil for 10 min and spread the mixture thinly on a slide coated with agar. Any other vital stain could be used. By incubating the slides for 4 h, cells grew to their maximum size before division and they could then be observed with a fluorescence microscope and removed with micromanipulators. The method is useful if one wishes to relate the morphology of bacteria seen in soil to their morphology in culture.

Skerman (1968) designed a much improved micromanipulation system for isolating single cells of bacteria from colonies of bacteria that had just initiated growth when placed on water agar films. This could be used in conjunction with the above methods.

## B. Centrifugation

If it is necessary to obtain a large sample of organisms in the form in which they were growing in soil, e.g. for biochemical, genetic or morphological analysis, it is necessary to extract them rapidly under conditions which will not alter the cells. It may be necessary to free the cells from as much of the clay-humus fraction as possible as this could interfere with subsequent analyses.

The following method (Faegri *et al.*, 1977) is relatively simple. About 10 g (wet weight) soil are homogenized with 90 ml 1:20 Winogradsky's salt solution [1.43 mM $K_2HPO_4$, 1.0 mM $MgSO_4.7H_2O$, 2.14 mM NaCl, 4.75 μM $Fe_2(SO_4)_3.7H_2O$, 14.8 μM $MnSO_4.4H_2O$] in a Waring blender in a cold room, placing the samples in an ice bath between runs. After diluting the homogenate to 500 ml, it is centrifuged at 1000 *g* in a refrigerated centrifuge. The sediment is rehomogenized for 60 s, diluted and centrifuged twice more and the combined supernatants from the three centrifugations centrifuged at 10 000 *g* for 30 min in an angled head at low temperature. The supernatant from this centrifugation contains a few bacteria which may be counted and is then usually discarded. The sediment contains 50–80% of the bacteria and can be resuspended and stored or used appropriately. The sediment from the third low-speed centrifugation contains most of the fungi, some soil debris and a few bacteria. The whole procedure takes about 3 h. A greater proportion of bacteria might be obtained with more elaborate procedures but speed is of the essence if unaltered organisms are required.

Holben *et al.* (1988) have modified the above procedure by removing contaminating humic acid with polyvinylpolypyrrolidone (PVPP). Fifty g soil samples and 10 g acid-washed PVPP are homogenized in a Waring blender for three 1-min intervals in Winogradsky's salt solution diluted 1:20 to which is added 0.2 M sodium ascorbate just before use to achieve a final concentration of 0.2 M. The homogenates are cooled in an ice bath for 1 min between intervals. PVPP is prepared by suspending 300 g in 4 l 3 M HCl for 12–16 h at room temperature. The suspension is filtered through MIRACLOTH (Chicopee Mills, Milltown, NJ, USA) and the PVPP is suspended in 4 l 20 mM potassium phosphate (pH 7.4) and stirred for 1–2 h. This process is repeated until the pH reaches 7.0 after which it is refiltered and air-dried overnight.

The homogenate is transferred to a 250 ml centrifuge bottle and centrifuged at 1000 *g* for 15 min at 4°C which sediments the fungi and soil debris. The supernatant is recentrifuged at 23 000 *g* for 20 min at 4°C to sediment the bacteria. The fungal/soil debris pellet can be subjected to two further rounds of homogenization and centrifugation to provide more bacteria. The bacterial fractions are combined, suspended in 200 ml 2% (w/v) sodium hexametaphosphate, pH 8.5, collected by centrifugation at 23 000 *g* and washed twice in 200 ml buffer [33 mM Tris (pH 8.0) and 1 mM EDTA]. After final centrifugation, the bacteria can be resuspended in buffer if desired.

The bacteria may be used for lysis and isolation of DNA by equilibrium density gradient methods. This DNA can be digested and probed for the presence of specific DNA base sequences with $^{32}P$-labelled DNA. As little as 0.1 pg of the sequence of interest can be detected and if this sequence is unique to a particular organism (either naturally or through insertion by genetic engineering), then that organism can also be detected. Holben *et al.* (1988) were able to detect as few as $4.3 \times 10^4$ cells $g^{-1}$ oven-dry soil of *Bradyrhizobium japonicum* using these methods and research is progressing currently to improve their sensitivity through use of the polymerase chain reaction (Sarki *et al.*., 1986).

Macdonald (1986a) considered that too many bacteria were lost in the Faegri *et al.* (1977) procedure. He attempted to isolate non-filamentous organisms from soil, dispersing soil and microorganisms by replacement of polyvalent cations on ion-exchanging soil surfaces with $Na^+$ from an ion-exchange resin in the presence of the detergent sodium cholate. Soil was removed from the suspensions by sieving and elutriation and the microorganisms concentrated with a Millipore Pellicon casette system. They were subsequently fractionated and separated from clay particles by density gradient centrifugation in linear gradients of Percoll (limiting densities 1.101–1.139 g $cm^{-3}$) (Macdonald, 1986b). To allow subsequent chemical analysis of the cells, Percoll was removed by gel chromatography on columns of Sephacryl S-1000 or by centrifugation. The whole procedure is lengthy and uses a density gradient medium which is unstable with ions and autoclaving and relatively difficult to remove from cell fractions. The presence of some capsulated cells and associations of cells and non-microbial material in flocs of gummy material can complicate interpretation of the results.

## C. Preparation of soil extract media

Soil extract media are useful for supporting the growth of a wide range of soil organisms. However, they are not defined and contain many uncharacterized compounds and growth factors. An easy extraction method (James, 1958) is as follows.

One kg soil is autoclaved with 1 l water for 20 min at 121°C, strained and restored to a volume of 1 l. If the extract is cloudy, a little calcium sulphate should be added to it in order to flocculate the clays. It should be allowed to stand for a while and then filtered. The extract may be used as a medium, with or without agar, or be supplemented with 0.025% $K_2HPO_4$ or 1% glucose, 0.5% yeast extract and 0.02% $K_2HPO_4$, or incorporated into other media.

Alternatively, a cold extract (Hill and Gray, 1967) can be made by ball-milling 500 g soil with 1 l water for 7 days in a cold room. The extract is repeatedly filtered through cellulose filter pulp or centrifuged at 2500 *g* for 3 h to clarify it. Subsequent sterilization must be by membrane filtration. This type of extract preserves thermolabile growth-promoting materials. Inhibitory humic materials may be extracted if very acid forest soils are used.

The media can be made selective for bacteria or fungi by incorporation of antibiotics. Fungi are suppressed by addition of 50 $\mu g\ ml^{-1}$ actidione (cycloheximide) and/or 50 $\mu g\ ml^{-1}$ nystatin and bacteria are inhibited by 30 $\mu g\ ml^{-1}$ aureomycin. Gram-positive bacteria can be suppressed with 1.0 $\mu g\ ml^{-1}$ penicillin and Gram-negative bacteria inhibited by 5.0 $\mu g\ ml^{-1}$ polymixin B. Further selectivity can be obtained by adjusting the pH of the medium to 5.0–5.5 which will depress the growth of most bacteria.

### D. Open and closed systems of enrichment

Traditionally, microorganisms present in samples in low numbers have been enriched prior to their isolation on selective media. Enrichments are made by applying a selective environmental pressure, e.g. high temperature or by adding a substance to soil which is required by the target organisms and no others. The additives are supplied in large amounts to maximize growth but as enrichment proceeds, the concentrations of the additives will decrease, the concentration of some metabolites and organisms increase and thus the conditions and selectivity of the enrichment culture will alter.

It has been argued that additions of nutrients at high concentrations will select organisms which are not necessarily those which are active in the soil where growth-limiting nutrients are usually present in very low concentrations and where the growth parameters of the organisms ($\mu_{max}$ and $K_s$) may result in the occurrence of a rather different community. To overcome these problems, it is possible to use a chemostat as an enrichment device where conditions are defined and stable and it is possible to select growth conditions to favour bacteria with particular growth parameters.

*1. Closed systems*

Any container can be used to hold soil amended with particular nutrients and incubated under set conditions. Wide-necked, conical flasks are popular. Incubation conditions suitable for the target organisms must be chosen, e.g. aeration, temperature, pH, illumination, and suitable nutrients (together with appropriate trace elements and/or inhibitors) added.

Soil perfusion (Lees and Quastel, 1946) is a more sophisticated system in which dissolved nutrients are continually circulated from a reservoir (any suitable flask) through a vertical glass column containing soil and back to the reservoir. Circulation may be achieved by pumping, by suction or by driving the solution round by means of an air supply. The latter method has the advantage that the soils remain reasonably aerated but even here, the soil is in a more waterlogged condition than normal. Indeed, many soils must be opened out by addition of coarse sand in order to permit free circulation of the nutrients. Samples may be taken from a side arm on the reservoir without disturbing the soil column, though this assumes that the organisms being sought will be washed out of the soil. It is also possible to connect the air stream which powers the system to a gas analyser and monitor the oxygen uptake (Greenwood and Lees, 1959) or carbon dioxide output from the soil and thus estimate the rate at which the enrichment is taking place.

A much-simplified perfusion apparatus was suggested by Cripps and Norris (1969). Soil (0.5 g) is placed in a glass tube (6 × 1 cm), plugged at both ends with glass wool. The two ends of the tube are connected by polyvinyl chloride tubing in which 8–10 ml medium is placed. The completed circular systems are fitted around a cylinder which is rotated by a motor. Many of them can be placed on a single cylinder and operated for days or weeks. There is no aeration in this system; indeed it could easily be made anaerobic and hence useful for organisms which thrive under reduced oxygen tensions. Sampling is by means of a syringe pushed through the tubing wall.

*2. Open systems*

Any type of chemostat may be used for enrichment but a simple system with a temperature-controlled, aerated fermenter (volume, 500 ml or less) which is agitated by a mechanical stirrer, fitted with sampling and medium input and output ports and supplied with medium by a cheap peristaltic pump will do. Antifoam sensors are rarely required and pH and oxygen control are only needed in special cases.

More important is the setting up of the fermenter. If the fermenter is filled with a complete medium and inoculated and the organisms allowed to grow for some time before the medium pump is switched on, then the initial enrichment will be a closed enrichment of the type reviewed above. Some organisms may even die out before the pump is switched on and a steady state is achieved. Even if the pump is switched on at the moment of inoculation, the growth-limiting substrate will still be at a high concentration and will decrease and only reach a low steady-state concentration after about five times the fermenter volume of medium has flowed through the system. A more sensible strategy would be to fill the fermenter with medium minus the growth-limiting substrate, inoculate it and switch the pump on together and thus start to supply the complete medium. In this way, the growth-limiting nutrient will never be present at high concentration and the enrichment will favour organisms competing best at low growth-limiting substrate concentrations throughout.

Single-stage chemostats have been used most commonly but multistage systems may be of use where one is concerned with the sequential breakdown of complex substances. Breakdown of the substrate by organisms in one vessel will supply new substrates for different organisms which may be enriched in a subsequent vessel. It is common to find that some substrates cannot be degraded completely by pure cultures of organisms; rather they are attacked by consortia of organisms, each component of which is responsible for a separate stage. Attempts to isolate a single organism responsible for degradation will fail (Slater and Hardman, 1982).

## References

Adu, J. K. and Oades, J. M. (1973). *Soil Biol. Biochem.* **10,** 109–115.
Amato, M. and Ladd, J. N. (1980). *Soil Biol. Biochem.* **12,** 405–411.
Amato, M. and Ladd, J. N. (1988). *Soil Biol. Biochem.* **20,** 107–114.
Anderson, J. P. E. and Domsch, K. H. (1978). *Soil Biol. Biochem.* **10,** 215–221.
Anderson, J. R. and Slinger, J. M. (1975a). *Soil Biol. Biochem.* **7,** 205–209.
Anderson, J. R. and Slinger, J. M. (1975b). *Soil Biol. Biochem.* **7,** 211–215.
Bääth, E. (1988). *Soil Biol. Biochem.* **20,** 123–125.
Babiuk, L. A. and Paul, E. A. (1970). *Can. J. Microbiol.* **16,** 57–62.
Bingham, M. J. and Long, S. P. (1988). "Equipment for crop and environmental physiology: specifications, sources and costs". Dept of Biology, University of Essex, Colchester.
Bohlool, B. B. and Schmidt, E. L. (1968). *Science, NY* **162,** 1012–1014.
Bohlool, B. B. and Schmidt, E. L. (1980). *Adv. Microb. Ecol.* **4,** 203–241.
Bremner, J. M. (1965a). *In* "Methods of Soil Analysis" (C. A. Black, Ed.), Vol. 2, pp. 1149–1178.
Bremner, J. M. (1965b). *Ibid.*, pp. 1179–1237.

Brock, T. D. (1987). *In* "Ecology of Microbial Communities" (M. Fletcher, T. R. G. Gray and J. G. Jones, Eds), Symp. Soc. Gen. Microbiol. Vol. 41, pp. 1–17.
Brookes, P. C., Newcombe, A. D. and Jenkinson, D. S. (1987). *Soil Biol. Biochem.* **19,** 211–217.
Brown, J. C. (1958). *J. Ecol.* **46,** 641–664.
Burges, A. and Nicholas, D. P. (1961). *Soil Sci.* **92,** 25–29.
Burns, R. G. and Slater, J. H. (1982). "Experimental Microbial Ecology", Blackwells, Oxford.
Casida, L. E. Jr (1962). *Can. J. Microbiol.* **8,** 115–119.
Chard, J. M., Gray, T. R. G. and Frankland, J. C. (1983). *Trans. Br. Mycol. Soc.* **81,** 503–511.
Chard, J. M., Gray, T. R. G. and Frankland, J. C. (1985a). *Trans. Br. Mycol. Soc.* **84,** 235–241.
Chard, J. M., Gray, T. R. G. and Frankland, J. C. (1985b). *Trans. Br. Mycol. Soc.* **84,** 143–249.
Cohen, A. L., Marlow, D. P. and Garner, G. E. (1968). *J. Microsc.* **7,** 331–342.
Costerton, J. W. and Cheng, K-J. (1982). *In* "Experimental Microbial Ecology" (R. G. Burns and J. H. Slater, Eds), pp. 275–290, Blackwells, Oxford.
Cripps, R. E. and Norris, J. R. (1969). *J. Appl. Bacteriol.* **32,** 259–260.
Edwards, C. A. and Heath, G. W. (1963). *In* "Soil Organisms" (J. Doeksen and J. van der Drift, Eds), pp. 76–84, North Holland, Amsterdam.
Faegri, A., Torsvik, V. L. and Goksöyr, J. (1977). *Soil Biol. Biochem.* **9,** 105–112.
Foster, R. C. and Rovira, A. D. (1973). *In* "Modern Methods in the Study of Microbial Ecology" (T. Rosswall, Ed.), pp. 93–102, NFR, Stockholm.
Gilbert, O. J. and Bocock, K. L. (1962). *In* "Progress in Soil Zoology" (P. W. Murphy, Ed.), pp. 348–352, Butterworth, London.
Gray, T. R. G. (1967). *Science, NY* **155,** 1668–1670.
Gray, T. R. G. (1976). *In* "The Survival of Vegetative Microbes" (J. R. Postgate and T. R. G. Gray, Eds), Symp. Soc. Gen. Microbiol., Vol. 26, pp. 327–364.
Gray, T. R. G. and Deaney, N. B. (1983). *Abstr. Int. Symp. Mic. Ecol.* **3,** 63.
Greenwood, D. J. and Lees, H. (1959). *Plant Soil* **11,** 87–92.
Griffiths, E. and Birch, H. F. (1961). *Nature, Lond.* **189,** 424.
Grisi, B. M. and Gray, T. R. G. (1986). *R. Bras. Ci. Solo* **10,** 109–115.
Harris, P. J. (1969). *Soil Biol. Biochem.* **1,** 103–104.
Heldal, M., Norland, S. and Tumyr, O. (1985). *Appl. Environ. Microbiol.* **50,** 1251–1257.
Hill, I. R. and Gray, T. R. G. (1967). *J. Bacteriol.* **93,** 1888–1896.
Holben, W. E., Jansson, J. K., Chelm, B. K. and Tiedje, J. M. (1988). *Appl. Environ. Microbiol.* **54,** 703–711.
Ineson, P. and Gray, T. R. G. (1980). *In* "Microbial Growth and Survival in Extremes of Environment" (G. W. Gould and J. E. L. Corry, Eds), pp. 21–26, Academic Press, London.
James, N. (1958). *Can. J. Microbiol.* **4,** 363–370.
Jenkinson, D. S. (1966). *J. Soil Sci.* **17,** 280–302.
Jenkinson, D. S. (1976). *Soil Biol. Biochem.* **8,** 203–208.
Jenkinson, D. S. and Ladd, J. N. (1981). *In* "Soil Biochemistry", Vol. 5 (E. A. Paul and J. N. Ladd, Eds), pp. 415–471, Marcel Dekker, New York.
Jenkinson, D. S. and Oades, J. M. (1979). *Soil Biol. Biochem.* **11,** 193–199.
Jenkinson, D. S. and Powlson, D. S. (1976). *Soil Biol. Biochem.* **8,** 209–213.
Jenkinson, D. S., Powlson, D. S. and Wedderburn, R. W. M. (1976). *Soil Biol. Biochem.* **8,** 189–202.

Jones, P. C. T. and Mollison, J. E. (1948). *J. Gen. Microbiol.* **2,** 54–69.
Ladd, J. N., Oades, J. M. and Amato, M. (1981). *Soil Biol. Biochem.* **13,** 119–126.
Lees, H. and Quastel, J. H. (1946). *Biochem. J.* **40,** 803–815.
Long, S. P. (1986). *In* "Advanced Agricultural Instrumentation: Design and Use" (W. G. Gensler, Ed.), pp. 39–91. Martinus Nijhoff, Dordrecht.
Macdonald, R. M. (1986a). *Soil Biol. Biochem.* **18,** 399–406.
Macdonald, R. M. (1986b). *Soil Biol. Biochem.* **18,** 407–410.
Malajczuk, N., McComb, A. J. and Parker, C. A. (1975). *Aust. J. Bot.* **23,** 289–309.
McClaugherty, C. A., Pastor, J. and Aber, J. D. (1985). *Ecology* **66,** 266–275.
Millar, W. N. and Casida, L. E. Jr (1970). *Can. J. Microbiol.* **16,** 299–304.
Moore, S. and Stein, W. H. (1954). *J. Biol. Chem.* **211,** 907–913.
Moriarty, D. J. W. (1975). *Oecologia* **20,** 219–229.
Moriarty, D. J. W. (1977). *Oecologia* **22,** 317–323.
Moriarty, D. J. W. (1978). *In* "Microbial Ecology" (M. W. Loutit and J. A. R. Miles, Eds), pp. 31–33, Springer Verlag, Berlin.
Moriarty, D. J. W. (1980). *In* "Biogeochemistry of Ancient and Modern Environments" (P. A. Trudinger, M. R. Walter and B. J. Ralph, Eds), pp. 131–138, Australian Academy of Science, Canberra.
Moriarty, D. J. W. (1983). *J. Microbiol. Methods* **1,** 111–117.
Nedwell, D. B. and Gray, T. R. G. (1987). *In* "Ecology of Microbial Communities" (M. Fletcher, T. R. G. Gray and J. G. Jones, Eds), Symp. Soc. Gen. Microbiol., Vol. 41, pp. 21–54.
Noll, F. (1974). *In* "Methods of Enzymatic Analysis" (H. U. Bergmeyer, Ed.), pp. 1475–1479, Academic Press, New York.
Oades, J. M. and Jenkinson, D. S. (1979). *Soil Biol. Biochem.* **11,** 201–204.
Parkinson, D., Gray, T. R. G. and Williams, S. T. (1971). "Methods for Studying the Ecology of Soil Micro-organisms", Blackwells, Oxford.
Postgate, J. R. (1967). *Adv. Microbial Phys.* **1,** 1–23.
Preston, K. J. and Waid, J. S. (1972). *Trans. Br. Mycol. Soc.* **59,** 151–153.
Rayner, A. D. M., Boddy, L. and Dowson, C. G. (1987). *In* "Ecology of Microbial Communities" (M. Fletcher, T. R. G. Gray and J. G. Jones, Eds), Symp. Soc. Gen. Microbiol., Vol. 41, pp. 83–123.
Reiners, W. A. (1968). *Ecology* **49,** 471–483.
Reynolds, E. S. (1963). *J. Cell. Biol.* **17,** 208.
Robertson, L. and Normington, J. H. (1976). *Lab. Practice* **25,** 470–471.
Rogers, A. W. (1967). "Techniques of Autoradiography", Elsevier, Amsterdam.
Roser, D. J. (1980). *Soil Biol. Biochem.* **12,** 329–336.
Ross, P. J. and Martin. A. E. (1970). *Analyst, Lond.* **95**, 817–822.
Sarki, R. K., Scharz, S., Faloona, F., Mullis, K. B., Horn, G. J., Ehrlich, H. A. and Arnheim, N. (1986) *Science*, **230,** 1076–1079.
Siala, A. and Gray, T. R. G. (1974). *J. Gen. Microbiol.* **81,** 191–199.
Siala, A., Hill, I. R. and Gray, T. R. G. (1974). *J. Gen. Microbiol.* **81,** 183–190.
Skerman, V. B. D. (1968). *J. Gen. Microbiol.* **54,** 287–297.
Slater, J. H. and Hardman, D. J. (1982). *In* "Experimental Microbial Ecology" (R. G. Burns and J. H. Slaten, Eds), pp. 255–274.
Söderström, B. (1977). *Soil Biol. Biochem.* **9,** 59–63.
Söderström, B. (1979a). *Soil Biol. Biochem.* **11,** 147–148
Söderström, B. (1979b). *Soil Biol. Biochem.* **11,** 149–154.
Stanley, P. E. (1971). *Anal. Biochem.* **39,** 441–453.
Strugger, S. (1948). *Can. J. Res. Ser. B* **26,** 188–193.

Thomas, A., Nicholas, D. P. and Parkinson, D. (1965). *Nature, Lond.* **205,** 105.
Tipper, D. J. (1968). *Biochemistry* **7,** 1441–1449.
Tippkotter, R., Ritz, K. and Darbyshire, J. F. (1986). *J. Soil Sci.* **37,** 681–690.
Trolldenier, G. (1973). *In* "Modern Methods in the Study of Microbial Ecology" (T. Rosswall, Ed.), pp. 67–75, NFR, Stockholm.
Underwood, A. L. (1961). *Anal. Chem.* **33,** 955–956.
Warcup, J. H. (1955). *Nature, Lond.* **175,** 953–954.
Wieder, R. K. and Lang, G. E. (1982). *Ecology* **63,** 1636–1642.
Wong, W. C., White, M. and Wright, I. G. (1988). *Lett. Appl. Microbiol.* **6,** 39–42.

# 11

# Methods for Studying Fungi in Soil and Forest Litter

JULIET C. FRANKLAND,* J. DIGHTON* AND LYNNE BODDY†

**Institute of Terrestrial Ecology, Merlewood Research Station, Grange-over-Sands, Cumbria LA11 6JU, UK*

*†School of Pure and Applied Biology, University of Wales, College of Cardiff, Cardiff CF1 3TL, UK*



METHODS IN MICROBIOLOGY
VOLUME 22 ISBN 0-12-521522-3

## I. Introduction

The history of methods for studying fungi in soil stretches back into the last century, but Garrett's statement in 1952 that we see what we cannot identify, and identify what we cannot see still rings true. Mycologists continue to grapple with inadequate techniques in a notoriously complex milieu, where fungal mycelium entwines and interacts with plant roots, animals, and many different types of microorganism. The hyphae, which may be active, dormant or dead are often featureless, and many species have not been grown successfully on laboratory media. The review chapters in Vol. 4 of this series (Booth, 1971a) and the excellent manual on methods for use in the International Biological Programme (Parkinson *et al.*, 1971), although published nearly two decades ago, are rightly still much in use. Rather than repeating, therefore, all the classic recipes they describe, this chapter contains our personal selection of old and new techniques with emphasis on recent progress, particularly in forest studies. Pathogens and fungi of the root region, other than mycorrhizal species, are excluded.

Advances in the last 10 years have been due partly to highly sophisticated technology (e.g. in microscopy and analytical procedures), and partly to different approaches to fungal ecology using the simplest of techniques. In particular, recognition of genetic individuals in natural populations of fungal mycelia in soil and litter has greatly increased the scope of field investigations (see Cooke and Rayner, 1984; Jennings and Rayner, 1984). The concept of 'soil' has been gradually widened too. After 1960, more attention was given to leaf litter, once conveniently scraped away before sampling the soil. Now, with the focus on nutrient cycling by fungi in whole ecosystems, bulky dead wood is also treated as an important constituent of a forest floor. The current upsurge in mycorrhizal research has also generated new approaches to integrated studies of fungi in soil, litter and roots. In the international symposium volume on *The Ecology of Soil Fungi* (Parkinson and Waid, 1960), a landmark at the time, only two pages were devoted to endotrophic fungi, yet now *Glomus* might well be claimed to be the most abundant mycelium in many soils.

Whatever the shortcomings of old and new methods of studying soil fungi, investigators' claims have tended to become more reliable, with wider recognition that the use of sound statistical techniques when designing experiments and handling numerical data is essential. Statistical analyses are not discussed in this chapter, but some useful references are mentioned.

The methods considered here have been divided into four broad categories: Observation, Isolation, Quantification and Measurement of Activity. Identification is also discussed.

## II. Observation

The identity of species, location and spatial arrangement of mycelia, resource relationships, and involvement with other organisms are all of ecological significance. These patterns of occurrence can be observed and analysed either by direct examination with minimal disturbance, or indirectly by removal (isolation) of the mycelium from its microhabitat. The two approaches are complementary and both are necessary in the attempt to bridge the immense divide between field and laboratory. This will be obvious to anyone who has seen the straggling, anaemic growth in soil of, say, a *Penicillium*, and compared it with its well-defined, dense and often brightly coloured colony on an artificial, nutrient-rich medium.

As Brock (1987) cogently reminded bacteriologists, however essential it may be to experiment in the laboratory, most studies in microbial ecology should begin with direct observation. The importance, and also the difficulties, of examining a mycoflora *in situ* cannot be overemphasized. A great number of observational, non-quantitative techniques and their modifications are described for this purpose in the literature, but it is questionable whether more than a handful have advanced the ecology of soil fungi significantly. Most of the techniques referred to here, as in subsequent sections, are those we consider to be particularly useful or promising, but the researcher's choice must depend on the objectives. Useful earlier accounts of relevant methods are given by Parkinson *et al.* (1971), Johnson and Curl (1972), Waid (1979), and Macauley and Waid (1981).

### A. Macroscopic features

Whenever fungi produce macroscopic structures, such as the fruit bodies and vegetative cords of basidiomycetes, or distinctive patterns of discoloration and decay, valuable information can be gained from observations just with the naked eye. Otherwise, recourse must be made to light or even electron microscopy, but minimum disturbance of the system should be the aim. Owing to masking by soil particles, opacity of organic matter, and sheer bulk of roots and wood, some disruption of the microhabitats is inevitable with microscopy, even with modern developments in equipment such as fibre optics.

The location of macroscopic fruit bodies of many Basidiomycotina and Ascomycotina can immediately indicate likely nutrient resources or mycorrhizal hosts. Further investigation of the more ephemeral of these species in soil and litter, by repeated recording and mapping, can also reveal unexpected and often fascinating non-random distribution patterns associated with variations in resources or host roots (Mason *et al.*, 1982; Swift, 1982; Frankland, 1984). Likewise, fruit bodies on wood can be indicative of the organisms effecting decay. However, although distribution of fruit bodies provides a useful starting point, it can be misleading, since it may not always reflect the distribution and activity of the mycelium. For example, wood-decay fungi are often found fruiting on the surface of logs when the vegetative mycelium is no longer in occupation. Similarly, *Boletinellus merulioides*, among other species, fruits on or in rotten wood, but it is almost certainly not a wood decomposer (Cotter and Bills, 1985).

The detection, tracking or mapping of the larger vegetative structures, such as sclerotia, cords and rhizomorphs, obviates the above problem and does not require elaborate equipment, only a little ingenuity and manipulative skill. Some particularly successful ecological investigations in recent years have been of this type (see Cooke and Rayner, 1984; Thompson, 1984). For example, Thompson and Rayner (1982), by repeatedly excavating the cord system of *Phanerochaete laevis* in the surface soil and litter of a *Fagus* woodland, were able to follow changes in its distribution and development over a period of 13 months. Patterns and rates of colonization of cord-forming fungi have also been studied by placing small (8 $cm^3$), precolonized wood blocks in the litter layer of woodlands (Dowson *et al.*, 1988a, b).

It is more difficult to track undifferentiated mycelium of soil and litter saprotrophs in the field without some form of labelling (see pp. 348-350), but sometimes it can be located by tracing fruit bodies to their origin. In this way, Newell (1984a) demonstrated a vertical zonation of the mycelia of two basidiomycetes in the organic horizons of a Sitka spruce forest.

In solid, bulky substrata, such as wood, considerable information can be gained by direct examination of the arrangement of discoloured decay columns and boundary zone lines in longitudinal or transverse sections, a few centimetres in thickness, (e.g. Rayner and Boddy, 1988a, b). These patterns can be recorded by drawing, photography and computerized image analysis (Mercer, 1979; Blanchette, 1982). Further information can be obtained by incubating the sections in a moist atmosphere (polythene bags are frequently successful) to allow mycelia to grow out, often luxuriantly, from the different regions. Samples can then be removed for identification, avoiding the selective properties of laboratory media. If surface growth is sparse, momentary immersion in a simple nutrient medium, such as 2% malt extract broth, is often a remedy. Checks for contaminants can be made by successively correlating the position of the indigenous fungi on separately incubated, serial sections.

Similar direct incubation can be employed for leaves and other organic substrata. The damp chamber technique (Keyworth, 1951; Frankland

and Latter, 1973) is one of the simplest and most effective means of inducing sporulation on such material. The material is exposed to a moist atmosphere in a Petri dish with a base lining of filter paper and an inner rim of compressed, highly absorbent paper pulp. To avoid undue bacterial growth, the paper is kept just moist but not sodden by damping the rim with sterile water; free water should not reach the specimen. Microfungi usually start to spore in 2 days at room temperature, whereas some agarics require several months incubation. The arrangement of reproductive structures can be seen without disturbance, and spores held above the substratum are easily removed to culture media with an agar-tipped dissecting needle.

Direct observation of intact mycorrhizas in the field has been limited. Only recently have rhizotrons been set up to observe root growth, and their full potential for examining mycorrhizas has yet to be realized. In most instances, the distribution of mycorrhizas in the field is revealed only by destructive sampling, the roots being washed out of soil cores. By this method, the aggregated distribution of Douglas fir (*Pseudotsuga menziesii*) ectomycorrhizas in organic fractions of soil was demonstrated by Harvey *et al.* (1976).

## B. Light microscopy

### *1. Direct examination*

Initial investigation of field material removed to the laboratory need not involve elaborate procedures. Soil particles, or soil smears in water, agar or gelatine can be examined for mycelium under bright-field epi-illumination, with or without a metallurgical objective and fixatives. Similarly, the presence and type, if not the identity, of mycelium can be detected in organic matter and surface litter picked from soil and mounted under a microscope. Phase-contrast will reveal more detail of the physiological state of the hyphae, such as the extent of wall lysis, cytoplasmic content and vacuolation, than bright-field optics. In one ecological investigation, up to 42% of hyphae that had grown onto glass slides buried in leaf litter were ghost-like or invisible under bright-field illumination, even when stained, but were clearly seen by phase contrast (Frankland, 1974, 1975a).

Many attempts have been made to examine soil and litter microscopically with least possible disturbance to the organisms and their microhabitats, so as to elucidate spatial relationships. The elegant techniques of Kubiena (1932, 1938) using a 'soil microscope', which was actually attached to soil profiles in the field, were followed by the development of a great variety of methods for making impressions, peels and thick sections of soil and litter. A three-dimensional picture of a soil profile can be obtained by sectioning a soil block or core after impregnating it with resin and grinding with Carborundum (Nicholas *et al.*, 1965), or after freezing the sample in the field, embedding in gelatine and slicing with a knife (Anderson, 1978). The latter method is not recommended for mineral soils lacking sufficient air spaces for penetration by the gelatine. In turn, these approaches have provided fundamental information, but their potential is limited, particularly in proportion to the labour involved.

The distribution of hyphae in wood that is only slightly decayed can be ascertained from sections cut by hand or a sledge microtome (Wilcox, 1964, 1968). More decayed specimens may have to be embedded before sectioning. Paraffin embedding is useful for production of serial sections on a rotary microtome, but it is time consuming, involving as it does alcohol dehydration, and it may lead to distortion. Celloidin embedding followed by cutting with a sledge microtome has given excellent results with very decayed samples (even with white-rotted wood that has lost 70% of its original weight), but the procedure takes several months to complete (Wilcox, 1964, 1968). Polyethylene glycol provides a rapid alternative which can be used on moist material without dehydration (Pearce and Rutherford, 1981). Finally, the sections can be mounted in water or lactic acid without staining, and examined directly by transmitted light.

Ectomycorrhizas, unlike mycorrhizas formed by most endotrophic fungi, often have natural features, such as distinctive and diagnostic colouration, branching, or mantle surface characters, readily observed with the naked eye or by low-power microscopy. Vesicular–arbuscular mycorrhizal (VAM) infection cannot usually be detected without clearing and staining the roots, but Arias *et al.* (1987) found that some VAM fungi were associated with yellow pigmentation. This can be seen microscopically in fresh intact roots of certain hosts, but the colour disappears after exposure to light.

### 2. *Staining and 'labelling'*

Failing direct observation of the fungus in its nearly natural state, 'labelling' of the mycelium *in situ* or staining of microscope preparations is often necessary, although various artefacts will be introduced. Mycologists have used numerous stains and mountants over the years, and made many unsubstantiated claims for them. For a good selection of stains and procedures, see Gurr (1965) and Clark (1981). Rapid techniques for staining nuclei and conidial appendages in Coelomycetes have been described more recently by Punithalingam (1989). Watling's publication

(1981) is also especially recommended for methods of preparing fruit bodies and spores of agarics for microscopy. As a general purpose stain, one of the best is phenolic aniline blue, permanent microscope preparations of which can be sealed with Glyceel (British Drug Houses) or nail varnish (Dring, 1971; Baker, 1981, 1988).

It is often difficult to distinguish hyphae within plant tissues, and heavily pigmented material must be decolorized before staining. Hering and Nicholson (1964) recommended a sodium chlorite/methyl salicylate procedure for various leaf litters including oak (*Quercus*) with its high tannin content, but Visser and Parkinson (1975) found that hydrogen peroxide caused less damage to the more delicate leaves of aspen poplar (*Populus tremuloides*). Simple overnight methods of decolourizing plant tissues, using methanol or chlorine from domestic bleaching agents, have been described by Baker (1988).

Safranin–picroaniline blue stain has been found to be particularly useful for wood, giving good differentiation of hyphae (dark blue) and cell walls (pink). Pianese III B, thionin-orange G, toluidine blue and rhodamine B/methyl green also give good differentiation (Pearce and Rutherford, 1981; Pearce, 1984), and, recently, fluorescent-labelled lectins, proteins which selectively bind with various carbohydrates and glycoproteins, have been employed to enhance hyphal visibility in wood (Morrell *et al.*, 1985).

For some years, the most popular procedure for microscopic examination of mycorrhizal roots has been clearance of cytoplasm in potassium hydroxide followed by trypan blue staining in lactophenol (Phillips and Hayman, 1970). The less toxic lactic acid is recommended in place of phenol for large-scale operations (Kormanik *et al.*, 1980), and has been successfully combined with acid fuchsin (Kormanik and McGraw, 1982), but comparisons have shown that chlorazol black E can be superior to trypan blue for revealing details of VAM hyphae and arbuscules (Brundrett *et al.*, 1984).

The principal use of fluorescent stains (see also p. 351) is for specimens viewed against an opaque background with epi-U.V. or B.V. illumination (e.g. Roser *et al.*, 1982). Otherwise their advantages are limited owing to uneven adsorption of the stain and autofluorescence of soil particles and plant material. Claims that they distinguish living from dead hyphae are mostly unreliable. Fluorescein diacetate, which fluoresces when enzymatically hydrolysed within living cells, has been the most successful and thoroughly tested vital stain for mycelium in soil, but, nevertheless, appears to underestimate the living component (Söderström, 1977, 1979).

Brighteners, a group of fluorescent stilbene compounds, have been applied as hyphal labels when stable binding and lack of toxicity were important (Darken, 1961, 1962). If translocated by the fungus, they can be used to detect new growth and spore germination (Tsao, 1970), but in the case of one saprotrophic agaric, *Mycena galopus*, the brightener Calcofluor White ST was found to be neither translocated nor leached out by rain. This enabled an estimate of the decay rate of hyphae on leaf litter placed in the field to be made by recording residual fluorescence (Frankland, 1975b; Frankland *et al.*, 1979).

Autoradiography is an alternative approach to the problem of distinguishing between metabolically active and non-active hyphae in the field situation. Waid *et al.* (1971, 1973), for example, described a procedure in which hyphae on leaf litter were labelled with $^{14}$C-glucose and removed in polystyrene peels. Autoradiographs were prepared, using liquid photographic emulsion to give close contact. The radioactive labelling of bacteria associated with hyphae, particularly when they are dead or moribund, is a serious drawback to this method; $^{14}$C-labelled volatiles may also be a source of error (Preston and Waid, 1972). Autoradiographic detection of metabolically active soil hyphae can correlate well with FDA-stained hyphal length, but considerably higher values have been obtained by the latter method (Bååth, 1988a). Isotopic labelling has, however, been used successfully to follow mycelium into a microarthropod food chain (e.g. Coleman and McGinnis, 1970, with $^{65}$Zn), and it is valuable for measuring metabolic processes (see Measurement of Activity, pp. 377-391).

### C. Electron microscopy

Transmission and scanning electron microscopy (SEM) are used to achieve higher resolution and depth of focus than light microscopy. SEM has the advantage that the organisms do not have to be removed from their substrata and irregular surfaces can be surveyed in focus (Gray, 1967). Many striking SEM photographs of the surface features of fungi with a three-dimensional effect have been published, including some showing fungal structures such as arbuscules inside plant cells (e.g. Kinden and Brown, 1975). Their interpretation is often difficult due to artefacts, but these can be reduced by metal coatings (Draggan, 1976), and by low-temperature SEM in which the material is examined in a frozen hydrated state instead of freeze-dried (Campbell and Porter, 1982).

SEM was hailed as a revolutionary tool, but again the labour and, especially in this case, the expense must he weighed against the gain in information. The samples are necessarily small, and the fine detail of, say, spore walls aid the taxonomist rather than the ecologist. A three-dimensional image does, however, offer opportunities for assessing interactions between different hyphal systems and their substrata. For example, electron microscope studies of decaying wood have done much to elucidate modes of colonization, patterns of enzyme production and spatial distribution of hyphae on and within woody cell walls (e.g. Messner and Stachelberger, 1984; Hale and Eaton, 1983, 1985a,b,c). Similarly, both types of electron microscopy have proved valuable in investigations of root fungi (Greaves and Campbell, this volume, Chapter 14). SEM has also, in a more specific problem, helped to explain the lectin-mediated, adhesive capture mechanism of nematode-trapping fungi (Nordbring-Hertz and Mattiasson, 1979; Nordbring-Hertz, 1984). Another important ecological application of SEM is its use in combination with electron microprobe analysis for determining the elemental composition of a fungus *in situ* on its substratum (p. 382).

### D. Immunological techniques

The value of examining fungi in field material is greatly extended if the organisms can be identified. Immunological techniques hold out the greatest promise of recognizing mycelia in their microhabitats (Schmidt, 1973a; Bohlool and Schmidt, 1980). They should not, however, be embarked upon in the hope of a quick answer. Each new system can be fraught with technical problems in establishing specificity, and success has been achieved in relatively few autecological studies of filamentous fungi, several of these being animal or plant pathogens in well-defined microhabitats. Perhaps ecologists need that extra drive shown by the more applied scientist in using new techniques. The methods are essentially simple and the preliminaries and standardizations, rather than their application, are time consuming. The need for rigorous controls cannot be overemphasized, but with continuing, fast developments on this front, it is not unrealistic to assume that eventually 'banks' of specific antisera or monoclonal antibodies will be available to the fungal ecologist.

The fluorescent antibody technique (FAT), radioimmunoassay (RIA), and the enzyme-linked immunosorbent assay (ELISA) all depend on the specificity of an antigen–antibody union which is exhibited by some visual or measurable feature of a 'label' coupled with the antibody, i.e. the fluorescence of a fluorescent compound such as fluorescein isothiocyanate, or the activity of a radioisotope or enzyme. The FAT applied to microscopical preparations is the least disruptive method and most applicable to field studies of intact hyphae, but many users have had difficulties in removing cross reactions with unrelated species, and non-specific adsorption by soil organic matter. Success appears to have depended partly on the genera concerned.

Early investigators using fluorescent antibodies succeeded in following the growth of *Aspergillus flavus* (Schmidt and Bankole, 1962, 1963,

1965), *Arthrobotrys conoides* (Eren and Pramer, 1966) and *Arthroderma uncinatum* (Ibbotson and Pugh, 1975) through unsterile soil. The potential use of FAT in leaf litter in an autecological study of a saprotrophic basidiomycete, *Mycena galopus*, has also been investigated (Frankland *et al.*, 1981; Chard *et al.*, 1983, 1985a, b; see also p. 375). A partially purified antigenic protein fraction of the *Mycena* mycelium was used in place of a crude 'whole' cell preparation for producing the final antiserum. Most non-specific fluorescence was reduced to acceptable levels by absorbing the antiserum with mycelium from cross-reacting fungi. Non-specific binding sites can also be masked by incubating the specimens with pre-immune serum and unconjugated commercial protein as described by Zollfrank *et al.* (1987) in their qualitative assay of hyphae and rhizomorphs of *Armillaria* and *Heterobasidion* in wood. The mycelium of another wood-rotter, *Phaeolus schweinitzii*, has been detected in both natural and artificially infested soil samples by the FAT (Dewey *et al.*, 1984).

The immunofluorescence technique has been particularly effective for mycorrhizal fungi. It has enabled *Thelephora terrestris*, for example, to be distinguished from *Pisolithus tinctorius* on mycorrhizal roots, and from 31 diverse non-mycorrhizal species (Schmidt *et al.*, 1974). Vesicular–arbuscular fungi have also been detected in roots by FAT. Kough *et al.* (1983), for example, developed an antiserum specific to *Glomus*. Antigenic extracts of chlamydospores were used to overcome the failure to culture these fungi in the absence of living roots. The method was later adapted for monitoring extraradical hyphae of a single *Glomus* species in mineral soil and sand (Kough and Linderman, 1986), but it was unsuccessful in peat owing to non-specific staining and autofluorescence of organic matter. Hyphae collected from germinating spores of *Gigaspora* have also been a source of VAM antigens, the resulting antiserum distinguishing between hyphae of VAM and non-endogonaceous fungi (Wilson *et al.*, 1983).

RIA and ELISA are quantitative assays in which the material containing the target antigens can be homogenized and bound to a solid phase. In RIA the 'label' is usually an isotope of iodine ($^{125}$I or $^{131}$I), whereas in ELISA it is usually alkaline phosphatase or peroxidase, the enzyme substrate giving a coloured reaction product. The cheaper ELISA has been the more frequent choice of the two methods, but they are similarly powerful. Antigen concentrations of the Dutch elm disease pathogen as low as 500 pg ml$^{-1}$, and beyond the limit of visible staining, have been detected by ELISA (Dewey and Brasier, 1988; Breuil *et al.*, 1988).

Lung-Escarment *et al.* (1985) described ELISA as a simple and quick method of directly characterizing two *Armillaria* species from crude extracts of their fruit bodies, rhizomorphs or mycelia. Preliminary trials of the assay on the wood-decay fungus *Poria placenta* have also been promising (Goodell *et al.*, 1988; Jellison and Goodell, 1988). ELISAs have been performed, in preference to FAT, on VAM hyphae washed from roots in inoculated soils, but the primary objective was, again, to locate a particular fungus rather than study its mycelium *in situ* (Aldwell *et al.*, 1983; Aldwell and Hall, 1986).

Conventional polyclonal antisera were used in the immunological investigations mentioned above. Monoclonal antibodies (MAbs), the current alternative, produced by a technique devised in the mid-1970s by Köhler and Milstein, have had a revolutionary impact on immunobiology. The crux of the procedure is the endowment of 'immortality' on a normal antibody-producing lymphocyte by fusion with a myeloma cell (Köhler and Milstein, 1975; Köhler *et al.*, 1978). A useful, albeit much simplified, introduction to the principles involved from the viewpoint of forestry research has been published by Mitchell (1985); see also Tiffin (1987) for an outline of the techniques.

Specificity, overcoming cross-reactivity, is the great advantage of MAbs, although in some circumstances it can be so extreme as to be a disadvantage compared with that of a broad-spectrum antiserum. MAbs bind to a single antigenic site on the antigen and can differentiate between protein molecules differing only in a single amino acid. In addition, unlimited quantities of the pure 'reagent' can be obtained from relatively small amounts of impure antigen. MAbs could be valuable ecological tools both as species probes and for detecting metabolites and extracellular enzymes, but much fundamental work remains to be done on the serology of fungi.

Again, initiative in using the MAb technique is seen in pathology rather than ecology. Hardham *et al.* (1986), for example, have used MAbs in combination with immunofluorescence to identify components of zoospores and cysts of the pathogenic *Phytophthora cinnamomi* specific to isolates, species and genera. Wright *et al.* (1987) and Wright and Morton (1989), who claimed they were the first users of MAbs for identifying VAM fungi, developed an ELISA with MAbs to identify hyphae and spores of *Glomus occultum*. Specificity was such that a single spore of this species was detectable among high numbers of spores of other mycorrhizal fungi.

## III. Isolation

For all their value, direct methods of examining fungi still give us only a partial view of the composition of a soil mycoflora. Even at the descriptive stage, further information must usually be obtained indirectly by some method of fungal isolation, be it a 'trap', 'bait' or culture medium in a Petri dish. Owing to the selectivity of these methods, however, bias and distortion of the true situation are particular risks, and the results must be interpreted cautiously. As already mentioned, a high proportion of microorganisms will not grow on laboratory culture media at all (Macdonald, 1986), and glass capillaries (pedoscopes) containing a film of nutrient or water have revealed, when buried in soil, many strange fungi never seen in culture (Aristovskaya and Parinkina, 1961; Perfiliev and Gabe, 1969; Parkinson *et al.*, 1971).

### A. Traps and baits

Fungal traps of many types have been inserted in soil. They range from simple non-nutrient structures, such as glass fibres or nylon mesh, to containers with restricted apertures leading to nutrient media separated from the soil by an air gap, such as Sewell's slide traps and Chester's tubes with small orifices (see Warcup, 1967; Parkinson *et al.*, 1971). At the time, each one seemed to be a novel method of isolating fungi from soil, but there have always been lingering doubts about the extent to which they stimulate germination of dormant spores nearby and isolate only fast-growing species. Their main contribution has been information on seasonal differences in the growth of some common species.

Many types of intentionally selective bait have been used to isolate particular fungi from soil, usually a physiological group, more occasionally single species (see Booth, 1971a; Gray and Williams, 1971a). Examples include cellulose film for cellulolytic fungi (Tribe, 1957); human hair for keratinophilic species (Griffin, 1960); chitin strips from *Sepia* shells (Gray and Bell, 1963); and tubes of a modified polysulphide medium, with an inhibitor of fungal spore germination and antibacterial antibiotics, for fungi capable of utilizing reduced sulphur compounds (Wainwright and Killham, 1980). Small seeds are a favourite bait for certain water moulds found in soil. Fungi can be stained and examined *in situ* on these baits or transferred to culture media.

### B. Macroscopic fungal samples

Whenever mycelium occurs in the form of macroscopic aggregations, such as sheets, fans, cords, rhizomorphs or fruit bodies, attempts can be made to transfer it to an artificial culture medium after cleaning or surface sterilizing. Usually the thicker the tissue the better is the chance of obtaining an uncontaminated isolate. Caps of agaric fruit bodies should be surface sterilized in alcohol and/or by flaming the non-hymenial surface before excising small pieces of the inner tissue. If, as in population studies, it is important to obtain only secondary mycelia (heterokaryons), then care should be taken to avoid the hymenium with its homokaryotic spores. Cords and rhizomorphs (in 3–4-cm lengths) can be cleaned of extraneous material by scraping the surface under sterile distilled water. The cords should then be agitated mechanically in several changes of water. Many other useful tips on the isolation of basidiomycetes, including use of reproductive propagules, have been given in excellent and detailed accounts by Watling (1971, 1981).

### C. Hyphal extraction

Removal of single hyphae from soil, litter or roots is far more difficult than obtaining isolates from aggregated mycelium, and to be worthwhile appears to need the mycological equivalent of 'green fingers'. Warcup (1955) did describe a method of picking individual hyphae from soil particles that had been separated from a soil suspension. His remarkable studies of wheat-field and pasture soils followed, and hyphal isolations of many species, including basidiomycetes, absent from the more usual species lists were recorded, altering our whole concept of these mycofloras (Warcup, 1957, 1959). Unfortunately, the success rate per number of hyphal transfers, especially by later workers (e.g. Williams and Parkinson, 1964), was very low, and many hyphae intimately associated with organic material were not extracted. Some improvement in the success rate was achieved with the organic layer of a coniferous forest soil by selecting only living hyphae after staining with fluorescein diacetate (Söderström and Erland, 1986; p. 349). However, the number of species isolated did not differ significantly from that obtained by less laborious techniques.

### D. Dilution plating

If hyphae cannot be dissected from field material for identification, they must be induced to grow out as visible colonies onto an artificial culture medium. The dilution plate method, in which a dilution series is prepared from a soil suspension and the selected dilution incorporated in an agar medium, continues to be used for this purpose despite criticism. It was designed for isolation of bacteria, is time consuming, and above all favours sporulating fungi, so its use for ecological purposes rather than as a neat exercise for student classes is rarely justified (see Parkinson *et al.*, 1971 and Schmidt, 1973b, for discussion of the technique, also p. 363).

### E. Soil plates

The soil plate method (Warcup, 1950, 1951a) is a simple, quick alternative to dilution plating for isolating fungi from soil samples. It obviates both loss of inoculum in residues discarded in the dilution procedure, and excessive fragmentation of hyphae. The advantages are obvious from the longer species lists it produces, with less emphasis on heavy sporers. A small amount of soil (less than 1 mg per 9-cm Petri dish) is dispersed over the base of the culture plate and covered with an agar medium, which should be as non-selective as possible in a general study of a mycoflora (see media p. 358). Sufficient growth for obtaining a pure subculture, without contamination by, or suppression of, neighbouring colonies should be the aim. Bacteria are retarded by the depth of agar, and early excision of rapidly growing fungi with a sterile scalpel allows more slowly emerging species, including basidiomycetes, to develop. Various antibiotics and inhibitors of fungal growth may be added to the culture medium. The method can also be adapted for selective isolation of particular fungi by pretreating the soil or by using selective culture media (p. 359). For example, the soil can be heat treated to kill most species and to stimulate germination of dormant ascomycete spores (Warcup, 1951b; Warcup and Baker, 1963).

In this and similar methods, conditions of culture should be chosen to mimic as far as possible the natural physical environment. There is usually no advantage, for instance, in choosing the traditional incubation temperature of 25°C when the mean temperature of the woodland, if temperate, may be 11°C, and when the optimum temperature for growth of several basidiomycetes is around 22°C.

### F. Soil washing

A more intensive study of the composition of a soil mycoflora can be made by washing whole soil samples to remove spores, and to separate out component organic and inorganic particles (Parkinson and Williams, 1961). Various modifications of this soil-washing technique have been introduced to automate the system, but the principle has remained the same (Williams *et al.*, 1965; Hering, 1966a; Bissett and Widden, 1972). The soil is serially washed by bubbling compressed air through a water suspension, and the particles are collected on graded sieves for aseptic removal to culture media. A wide range of species can be obtained by this method, but, more importantly, information on the origin of the propagules is retained. Some decrease in efficiency must be expected, however, with increase in humus and colloidal clay, and the effect of particle size must also be taken into account. Bååth (1988b) advocated use of relatively small particles, after finding that the species composition was affected by the number and identity of species isolated from each particle.

### G. Macroscopic organic components

Fungi in some of the larger organic components of soil and litter, such as fine roots and leaves, can be isolated by the technique of Harley and Waid (1955), originally designed solely for roots. The material is cut into small fragments (1–3 $mm^3$), which are washed repeatedly to remove spores by a predetermined procedure, blotted dry to restrict bacterial growth, and plated equidistantly (about 5 pieces per 9-cm-diameter Petri dish). The particles are pressed into a solidified agar medium, and produce discrete centres of growth, more certainly of hyphal origin than those on Warcup plates. Serial water washing is usually preferable to surface sterilization, since the concentration and contact time of the chemical can be critical (Kirby, 1987).

Bulkier roots and litter can be separated into different tissues before plating (e.g. roots, Waid, 1956; *Pteridium* petioles, Frankland, 1966). Some degree of selective sampling of wood is also often more rewarding than a completely objective approach. For example, by sectioning wood containing a mature community of decomposers, sampling from different decay columns (distinguishable by colour, texture and interaction zone lines), and then plating, it is possible not only to determine the species composition but also to map the actual three-dimensional structure of the community (Rayner and Boddy, 1988a, b). To ensure isolates come from *within* bulky tissues, it is often helpful to surface sterilize the material by immersing it briefly in 5% domestic bleach or a similar solution. This is not recommended for well-decayed tissues as bleach is readily absorbed and could kill the internal fungus.

### H. Mycorrhizal fungi

Ectomycorrhizal species can often be isolated by plating fruit-body tissue on agar media as described for other basidiomycetes (p. 354). Isolates may also be obtained from roots that have been surface sterilized (Zak and Marx, 1964); sodium hypochlorite or mercuric chloride is commonly used as the sterilant. Success of this method depends again on the time the sterilant is in contact. Zak and Marx found that only about 13% of root tips yielded viable cultures. An additional drawback to the technique is that it is often difficult to be sure which species has been isolated unless identification of the mycorrhizal root has been attempted.

An effective medium for the culture of VAM fungi has yet to be developed. Consequently, the isolation and characterization of these fungi relies on the extraction of spores from soil (see p. 363).

## I. Analysis of mycelial populations

Collection of floristic data (species lists) by one of the methods described above can provide valuable comparisons between fungal communities of different habitats and resources but little information on the structure of populations of individual species. As mentioned (p. 344), recent ecological research, on the higher fungi in particular, has gained by a more genetical approach with emphasis on *individual* mycelia. It is now clear that in genetically variable populations of Ascomycotina and Basidiomycotina different genotypes are delimited as a result of somatic/vegetative incompatibility mechanisms (Rayner and Todd, 1979; Rayner *et al.*, 1984; Rayner and Boddy, 1988a, b). The traditional concept of a mycelium in the field being an anastomosing, intraspecific mosaic is no longer tenable.

Interactions between genetically different mycelia within organic substrata in the field are often evident from pigmented zone lines. These are particularly common in decomposing wood (see also p. 357, 380) but also quite frequent in leaf litter. Similar interaction zones form between colonies of somatically incompatible mycelia on agar plates, including formation of mycelial barrages and troughs, whereas strains which are genetically identical (as regards somatic compatibility) intermingle without signs of antagonism. Pairing of isolates of known origin on culture media provides, therefore, a means of distinguishing genotypes and mapping the population structure. Typically, inocula (approx. 4 mm$^2$) are cut from the margin of the colonies and plated 1–2 cm apart on an agar medium.

This genetical approach has been used successfully to analyse populations of several wood decomposers, including species confined within individual logs, branches and twigs (e.g. Boddy and Rayner, 1983, 1984; Chapela *et al.*, 1988) and extensive cord formers (Thompson and Rayner, 1982; Thompson and Boddy, 1983). Some of the latter were found to grow through soil between roots of suppressed trees up to 50 m apart. Similar studies have been made of a more diffusely spreading leaf-litter fungus (Frankland, 1984), 'fairy rings' of *Clitocybe nebularis* (Dowson *et al.*, 1989a), and ectomycorrhizal species (Fries, 1987).

## J. Choice of media

Recipes for a wide range of culture media, together with methods for maintenance of fungi once they have been isolated, are well documented elsewhere (see Parkinson *et al.*, 1971; Johnson and Curl, 1972; Gams *et al.*, 1987; Smith and Onions, 1983; Kirsop and Snell, 1984).

Choice of media depends upon the taxa being sought. For general purposes, a number of broad-spectrum media, such as Czapek-Dox, malt extract, potato dextrose and corn meal agars are commonly used. To these, antibiotics/inhibitors (e.g. chloramphenicol, novobiocin, streptomycin and Rose Bengal) or organic acids (e.g. lactic acid or malic acid) are often added to suppress bacterial activity (see Martin, 1950; Bakerspigel and Miller, 1953; Vaartaja, 1960; Tsao, 1970; Booth, 1971c). Czapek-Dox agar with addition of 0.05% yeast extract and, if necessary, Rose Bengal (1:15 000) is recommended for isolation of a soil or litter community by Warcup soil plates or a similar technique. It contains sufficient trace elements and vitamins, such as thiamine, for the more nutritionally demanding species, and yet is more standardized than media prepared from plant extracts. Weaker media, such as potato extract agar without dextrose, must often be used to maintain the fungus in a sporulating state after isolation.

Selective media are available for the isolation of particular fungal taxa. For example, benomyl will inhibit many microfungi, particularly species of *Penicillium* and *Trichoderma*, giving advantage to others (Maloy, 1974). Cellulolytic fungi may be isolated using the cellulose agar formulated by Eggins and Pugh (1962) and modified by Park (1973) . Basidiomycetes can be selected out using Hunt and Cobb' s (1971) medium, while for white-rot species an orthophenyl phenol medium is often successful (Russell, 1956); see also Watling (1981). Czapek-Dox medium in which sulphur has been precipitated by the reaction between polysulphide and hydrochloric acid can be used to isolate sulphur-oxidizing fungi, important in pollution studies (Wainwright, 1978).

Ectomycorrhizal fungi can be grown in a variety of agar media, liquid cultures or combinations of liquid media with a solid component such as vermiculite/peat. Modified Melin–Norkrans and Hagem's recipes are used regularly and other basic media are outlined by Molina and Palmer (1982). Most mineral contain essential elements, glucose and trace elements and are often supplemented with the vitamins biotin and thiamine. However, further research on the suitability of different culture media for the growth of these fungi is needed. Some species grow readily on the standard media, some very slowly, and certain well-known species of, for example, *Russula, Inocybe* and *Cortinarius* not at all. Again, it has been found that media required for the rapid production of mycelium and for long-term maintenance of ectomycorrhizal fungi may differ.

For the isolation of individual species of fungi of all categories, even more selective media can be devised by examining the tolerance of the species to a range of inhibitory substances, and then incorporating various combinations of them in the medium at sublethal concentrations (e.g. Papavizas, 1967; Barrett, 1978; Thomas, 1985).

Whilst much research has been done using media of the types described, they have many drawbacks and pose problems of interpretation. As Jennings (1987) emphasized in his Presidential Address to the British Mycological Society, the time is ripe for mycologists to develop new media, more relevant to fungal growth in the field (see p. 381). Some workers have appreciated this point and have used media lacking carbon to isolate fungi which are oligocarbotrophic. Agar itself can be a nutrient source, but more oligocarbotrophic media can be prepared by using silica gel as the solidifying agent (Tribe and Mabadeje, 1972; Payton *et al.*, 1976).

A different approach, which mimics situations often occurring in the field, is to use living mycelium as the selective agent for a particular species, the rationale being that living homokaryons used as baits can be dikaryotized only by spores of their own species. This method has been successful in studies of airborne basidiospore populations of the heterothallic wood-decay fungi *Coriolus versicolor* and *Flammulina velutipes* (Adams *et al.*, 1984; Williams *et al.*, 1984), and the leaf-litter agaric *Mycena galopus* (J. C. Frankland and J. Poskitt, unpublished).

## IV. Identification

Many keys, monographs and general mycofloras have been published for the identification of temperate fungi sporulating in the field or in culture after isolation from soil and litter. Guides to the literature, applicable mainly to the British Isles and north-western Europe, are given by Holden (1982) and Sims *et al.* (1988); see also Hawksworth (1974) and Hawksworth *et al.* (1983). Ectomycorrhizal fungi are included in the standard works on agarics and boleti such as Moser's (1983) key. Even in well-worked temperate habitats, however, it is not unusual to find undescribed species in all taxonomic groups, and comparison with material in a Type Culture Collection, or confirmation by a taxonomist, is often essential for the correct identification of many species and some genera.

Several methods can be employed successfully to induce microfungi to sporulate if they remain vegetative in the laboratory (see Booth, 1971b,c; Frankland and Latter, 1973). Basidiomycetes, however, cause more problems than most fungi, since many species do not fruit readily in culture. Much research has been done on the culture conditions required by this group for fruiting (e.g. Badcock, 1943; Cartwright and Findlay, 1958; Tamblyn and da Costa, 1958; Gramms, 1979), but whether the lengthy procedures are often worthwhile is questionable, particularly since the fruit bodies frequently possess abnormalities. Hence, identification of these fungi must often be based solely on vegetative features.

Praiseworthy attempts have been made to produce vegetative keys for wood-decaying basidiomycetes (e.g. Nobles, 1965; Stalpers, 1978), but their coverage is limited. For example, Stalper's key, although the most comprehensive one so far produced, considers only Aphyllophorales. Its use, involving various tests and recording of characters, is extremely laborious, although construction of computer data bases, as by the UK Forestry Commission (D.A. Rose, personal communication), should lessen this problem eventually.

There is no need to despair completely over the naming of mycelia! Many higher fungi, in particular, are highly selective for certain ecological situations, so the range of species likely to be encountered can be narrowed down considerably. As a first step, mycelium can be isolated from fruit bodies collected from the habitat concerned, and some background information on intrinsic cultural variability obtained. Secondly, vegetative characters of known and unknown mycelia can be compared. Despite popular opinion, many mycelia exhibit a wide range of morphological and physiological features which are relatively easy to recognize. Some of these characters, useful for identifying wood-decay fungi in particular, are listed below; more detailed accounts are given by Stalpers (1978) and by Rayner and Boddy (1988a). Field and/or macroscopic characters include: colour; odour; texture; and plectenchymatous (tissue-like) structures such as rhizomorphs, cords, sclerotia and pseudosclerotial plates. Vegetative hyphae can often be characterized by: diameter; clamp connections and septum formation; nuclear behaviour; wall thickness; angle and frequency of branching; straightness; compartment length; inclusions; surface deposits and terminal structures. Some of these may be so distinctive as to allow immediate identification of a genus, or even a species, from a particular site. For example, species of *Coniophora, Phanerochaete* and *Stereum* produce particularly large vegetative hyphae with verticillate clamps. Rapid extension rates; the so-called 'cardinal' points, and enzyme tests (e.g. Taylor, 1974; Poppe and Welvaert, 1983) may also be diagnostic. With all these parameters, however, it is essential to be aware of potential variability within a species, 'turned on' by physiological switches and changes in environmental conditions. Finally, the ability to mate with a known strain provides a means of confirming the identification of a mycelium, or discriminating between several possibilities (see, for example, Rayner and Boddy, 1988a). These tests often involve simply the juxtapositioning of strains on a culture plate, and with Basidiomycotina are possible even when one isolate is a secondary mycelium.

Methods for the identification of both ectomycorrhizas and VAM are outlined in Schenck (1982). Classification of ectomycorrhizal roots relies on the colour and morphological characteristics of the fungal sheath, emergent hyphae, and other surface features identifiable by light (Agerer, 1986, 1987; Haug, 1988) or scanning electron (Seviour *et al.*, 1978) microscopy. A comprehensive key to VAM spores based on spore size, colour and wall construction (murographs) is given by Schenck and Pérez (1988).

Biochemical techniques have also been used to identify both ectomycorrhizal and VAM fungi. Pyrolysis of the stipes of three *Suillus* species, followed by analysis of thermal degradation products, was an elaborate approach tested by Söderström *et al.* (1982). Species differences were detected using principal component analysis and a pattern recognition program, but the complex chemical analysis and computing procedures obviously count against this method. In an investigation of a single host genus, Seviour and Chilvers (1972) used polyacrylamide gel electrophoresis to produce protein separation patterns for identifying ectomycorrhizal fungi of *Eucalyptus*. Similar techniques using the separation patterns of the enzymes esterase, glutamate oxaloacetate transaminase, hexokinase, malate dehydrogenase, peptidase and phosphoglucomutase have been employed to separate *Glomus* species by their spores (Sen and Hepper, 1986). Resolution by this method can be effective on extracts of only 10 spores, and it is independent of soil type or host. Restriction fragment length polymorphism electrophoresis has also been reported as successful for identifying VAM fungi (Millner, 1987), and use of immunological techniques has already been mentioned (p. 352). These biochemical methods are relative newcomers to the mycorrhizal field and their full potential has yet to be realized. They are all more likely to be of value in a circumscribed project involving relatively few species than in a broad ecological survey.

## V. Quantification

All methods of quantifying soil fungi are imperfect. The heterogeneity of soil and the difficulties of separating the organisms from this complex medium, or of differentiating living from dead, cause major problems. Success depends to a great extent on soundly based sampling and cross checks between methods rather than reliance on a single procedure. Useful check lists of questions to be answered when designing experiments and sampling have been drawn up by Jeffers (1978, 1979). He concluded: 'There is usually little that a statistician can do to help you once you have committed yourself to a particular sampling scheme'.

The number of samples required to satisfy statistical validity and the number which can be handled adequately are often at variance with one another. Detailed discussion of such sampling problems can be found elsewhere (e.g. Butcher, 1971). Briefly, sampling patterns may be subjective or objective, the former being appropriate when distinct patterns in fungal distribution are evident, whilst an objective approach, with random or stratified random sampling, is applicable when there is considerable microspatial heterogeneity. In contrast to random or semi-random procedures, ordered sampling patterns have been used by many researchers. Chapela and Boddy (1988a), for example, found them to be particularly useful for quantifying fungi colonizing wood at early stages of decay.

Because they are so intimately associated with roots, enumeration of mycorrhizal fungi is particularly fraught with statistical and practical problems. Most quantitative techniques require the whole or a fraction of the root system to be recovered from soil, and cleaned of soil particles before an estimate of infection can be made.

### A. Absolute numbers

The counting of absolute numbers of organisms is, in general, the prerogative of bacteriologists rather than mycologists, although theoretically this is less so since the discovery that higher fungi often occur in the field as populations of genetically distinct mycelia (p. 358). Since the latter are recognizable by simple pairing experiments in the laboratory, it is sometimes possible to estimate the number of individual mycelia belonging to a single species within, say, a certain area of the forest floor or in a particular branch of a tree. However, naturally occurring mycelia are dynamic entities that cannot be treated as countable units (A.D.M. Rayner, personal communication). Counts of mycelia, if related to size, can be clues to their life span and colonization strategies but no more. Dilution plate counts of fungal colonies, arising as they do from hyphal fragments and spores, are also meaningless in terms of either abundance or biomass of a species. Such colony counts are useful only in limited circumstances, as in comparative studies of certain species on selective media, or as a guide to the distribution of spores. Burges (1950), for example, used dilution counts to follow movement of spores down a soil profile after adding a spore suspension to the soil.

The VAM 'inoculum potential' of soils can be evaluated by counting infective propagules. It is usually based on the number of spores contained in a unit volume of soil, although infected root fragments may also act as sources of inoculum. The spores, because of their relatively large size, shape, and wall characteristics, can be quantitatively recovered from soil and identified.

Two methods of determining VAM propagule numbers can be used: extraction of spores by wet sieving and indirect baiting of soil with a host plant. Techniques for wet extraction are given by Daniels and Skipper (1982). The extraction is usually carried out over a final sieve grid size of 38–350 μm, and a secondary separation of spores of different sizes achieved by sucrose density centrifugation. Flotation bubbling through sieves, and direct sucrose centrifugation from soil have also been employed and are discussed by Daniels and Skipper.

In the baiting technique, a test host is planted in the soil and the number of infection 'units' on the root system assessed after a period of growth. The most probable number (MPN) technique, where two- and tenfold dilutions of the test soil are made with sterile diluent soil, is commonly used (Porter, 1979). Presence or absence of VAM infection is scored for each dilution and the number of infective propagules determined from MPN tables. Porter found that results from this technique agreed well with spore counts by wet sieving for one soil but gave significantly higher estimates of propagule density for another. An and Hendrix (1988), however, suggested that vital staining of VAM spores, using MTT [3-(4,5-dimethylthiazol-yl)-2,5-diphenyl-2H-tetrazolium bromide], would give an even higher estimate of viable spores than the plant bioassay.

Counts of fruit bodies of higher fungi visible above ground can be useful in a variety of investigations. For example, Hering (1966b) related total numbers of agaric fruit bodies to their weights in permanent quadrats, when comparing mull and mor sites of some UK woodlands. Evidence of the existence of perennial or renewable mycelia of *Mycena galopus* in a coniferous plantation has been obtained by determining the numbers of fruit bodies that occupied identical locations in successive years (Frankland, 1984). Again, enumeration of ectomycorrhizal fruit bodies can be a method of assessing the relative contribution of different species to a mycorrhizal succession (e.g. Chu-Chou, 1979; Chu-Chou and Grace, 1981; Mason *et al.*, 1982; Dighton et al., 1986; Ammirati *et al.*, 1987), but then it is essential to demonstrate that the fruit bodies observed are related to the mycorrhizas formed on the roots. Numbers again were the basis of Arnolds' (1985, 1988) surveys of the effects of long-term pollution on this group of fungi. Occasional attempts have also been made to relate fruit body numbers or weight to the live mycelial biomass of the same species below ground, including a litter saprotroph (Frankland, 1975b) and some ectomycorrhizal fungi (Miller, 1982).

In general, numerical surveys of fruit bodies give unreliable results unless there is repeated sampling both within a season and also over a period of several years to offset the bias due to the ephemeral nature of many species and vagaries of fruiting. An extensive 6-year study of the influence of soil and litter properties on the occurrence of the fruit bodies of macrofungi in 95 plots (500 $m^2$) along a mull/mor gradient in a Swedish beech forest (Hansen, 1988a,b, 1989) has demonstrated the value of a large data base. Both univariate and multivariate statistical methods, including principal component, regression and gradient analyses, were applied to both accumulated numbers and presence/absence data, and occurrence predicted from models of soil and litter variables.

## B. Frequency of occurrence

For simple comparisons, assessments of percentage frequency of occurrence can be obtained by several of the methods of observation and isolation already described, but they cannot be an absolute measure unless related to unit amounts of the substratum. The basis for measurements of frequency of occurrence and abundance varies from one investigation to another, so the terms should always be clearly defined. For example, in a study of the fungal colonization of wood (Chapela and Boddy, 1988b), branches were cut into standard lengths termed 'units', and chips of wood were taken from random sections. The percentage of units colonized was termed fungal frequency and the percentage of chips yielding isolates as abundance.

Presence and absence data obtained from Warcup soil plates and by similar methods are usually expressed as the percentage number of plates on which the fungus has grown. Frequency of occurrence has also been expressed on an area basis when comparing amounts of mycelium in soil sections and on impression slides (Nicholas and Parkinson, 1967); reasonably similar results were obtained by the two methods.

More discerning comparisons of fungal communities can be made by mathematical analysis of the distribution of species-abundance data. This approach is particularly useful in studies of the development and stability of communities after disturbance. Zak's (1988) investigation of strip-mine spoils, although it was concerned with fungi of root surfaces rather than soil, is a good example of the use of logarithmic functions to characterize the mycoflora of disturbed habitats. Cluster analysis to evaluate non-parametric data in a comparative study of fungal endophytes in *Pinus* and *Fagus* tissues has been described by Petrini and Fisher (1988).

Ecologists have developed a variety of mathematical indices to describe species diversity that take account of the composition of communities and abundance of individual species. As pointed out by Holder-Franklin (1986) in a critical evaluation of these indices, to be of ecological value an index must be capable of distinguishing between closely similar environments. A limiting factor, particularly in soil, is the labour required for adequate sampling. Christensen (1969) was able to discern distribution patterns of soil microfungi in forest stands of northern Wisconsin by three-dimensional ordination, based on coefficients of similarity, but few have emulated her intensive studies. Swift (1976), in an investigation of communities much less complex than those of most soils, employed Simpson's index, based on relative frequencies of isolation, to describe changes in species diversity during progressive fungal decomposition of branch wood.

Most methods of determining the frequency of ectomycorrhizas involve extraction of a subsample of the root system by washing the roots over sieves which collect root fragments of different sizes (see Grand and Harvey, 1982). An arbitrary scale may be used or the number of short mycorrhizal roots counted to obtain the percentage of root tips that are mycorrhizal. Identification of species, or at least of morphotypes, from surface features (see p. 362) will then allow the calculation of the percentage frequency of occurrence (or % 'contribution') of different mycorrhizal types in that mycorrhizal community. Where the number of mycorrhizal 'units' of each morphotype are determined, care must be exercised in defining the 'unit'. Multibranching and coralloid types have numerous root tips within one mycorrhizal unit, whereas monopodal and bifurcate morphs may have only one or two root tips associated with each mycorrhizal unit.

Determination of the frequency or 'intensity' of infection (i.e. colonization) of VAM fungi in a root usually involves clearing and staining of the root (p. 349). The procedure can be streamlined by processing the roots in capsules that allow large numbers to be stained at once (Kormanik and McGraw, 1982). Having stained the fungal structures (vesicles, arbuscules, spores and hyphae) within or on the roots, the intensity of infection can be assessed by a number of measures. A grid-line intersect method (Newman, 1966; Tennant, 1975; Giovanetti and Mosse, 1980) can be used to determine total root length and the percentage length of root infected. Alternatively, l-cm root segments can be mounted on microscope slides, gently squashed if necessary, and the length of cortical colonization measured and expressed again as a percentage of total root length.

Biermann and Linderman (1981) attempted to find a standard technique for VAM roots, by comparing estimates based on percentage length of root containing VAM structures with percentage of root segments showing such evidence of VAM infection. They found that the former measure was more accurate and concluded that, with their material, a minimum of seven samples per plant, each consisting of 25 randomly selected, 0.5–1.0-cm root segments, were needed for the confidence level to be within 10% of the mean. A modification of this approach was adopted by Toth and Toth (1982) who superimposed a grid of points (dots) over a root squash and determined the number of points overlying colonized and uncolonized cortical cells. Using morphometric techniques, they calculated an index of infectivity:

$$P \text{ fungus}/P \text{ root} = Pp \text{ for the fungus in the root}$$

or,

$$P \text{ cells with arbuscules}/P \text{ cortical cells} = Pp \text{ for invaded cells}$$

the number of invaded cells which must be counted to satisfy a certain standard error being calculated from:

$$P = Pp\,(1 - Pp) \times 1/\sigma^2 Pp$$

where $P$ = number of points over all cortical cells; $Pp$ = the point fraction ($P$ fungus/$P$ root); and $\sigma^2 Pp$ = the variance.

Determination of the degree of mycorrhizal infection by assessing the proportion of roots containing VAM structures, or the percentage number of short ectomycorrhizal roots on a root system, may not be an adequate measure of the *physiological* capacity of the mycorrhizas, as has been stressed by several workers in this field. Frankland and Harrison (1985) found significant relationships between ectomycorrhizal infection of *Betula* (as % of total root tips) and certain host plant variables or soil factors, by examining 'immature' mycorrhizas (i.e. without external features typical of infection) rather than mature mycorrhizal roots. This suggested that the mycorrhizas were physiologically active at a relatively early stage in their development, which could have important practical implications when selecting quantitative methods in other mycorrhizal investigations. This consideration apart, since the basic assumption regarding the role of mycorrhizas (Harley, 1969; Harley and Smith, 1983) is that they allow greater exploitation of the soil than uninfected roots and extend the nutrient depletion zone around the root (Nye and Tinker, 1977), it is important to look beyond the root and obtain estimates of the extent of the extraradical hyphae in the rooting zone. Attempts to do this by measuring hyphal lengths or by biochemical assays in laboratory or pot experiments are discussed on pp. 371 and 373.

## C. Biomass

Absolute measures of biomass (equivalent to standing crop or mycelial concentration) and production are basic requirements of many ecological studies (see Frankland, 1982; Kjøller and Struwe, 1982). Fungal biomass, although small relative to the soil mass, usually exceeds that of bacteria and is a major influence on plant nutrition through uptake, immobilization and mineralization of nutrients. Again, definitions are important. On etymological grounds the term 'biomass' should be restricted to living matter, but owing to its frequent misuse is specified here as living or dead unless total mass is the intended meaning. 'Production' which depends on rate of turnover is the live biomass produced over a stated period of time; it implies change in biomass.

Complete separation of hyphae from soil for the determination of fungal biomass has not been achieved, and use of more drastic extraction procedures tends to result in excessive fragmentation of the hyphae. In one attempt, 20% of hyphae were estimated to have been recovered satisfactorily from mineral soil by wet sieving and density gradient centrifugation (Bingle and Paul, 1986). Such bulk samples of naturally grown mycelium can be useful for determining nutrient concentrations. Otherwise, chemical analysis of field biomass depends on the availability of sufficient quantities of mycelium in sheets or fans which can be stripped from litter and wood, or as macroscopic cords, rhizomorphs and fruit bodies (see Frankland, 1982; Lodge, 1987).

### *1. Measurement of hyphal length*

*(a) Agar film technique.* The most direct method of obtaining an absolute measure of fungal biomass in soil and the softer litters, on the basis of either weight or volume of soil or litter, is by measuring hyphal length using, for example, the Jones and Mollison (1948) agar film technique as modified by Thomas *et al.* (1965). Stained films of known thickness are prepared on a haemocytometer slide from a standardized molten agar suspension of ground soil or homogenized litter. The procedure for soil is described by Parkinson *et al.* (1971), and for litter by Frankland *et al.* (1978). Hyphal lengths are determined by using a map measurer (tachometer) on a projected image of a microscope field, or from the number of intersections with a grid system on an eyepiece graticule (Olson, 1950); the former method appears from comparisons between workers to be more subjective but less tedious. Biovolume and biomass are then calculated from length by assuming that the hyphae are cylindrical and by using conversion factors for the diameter, moisture content and relative density of the hyphae. Problems of 'masking' and focusing, which markedly reduce the accuracy of measuring hyphal lengths in soil sections and on impression slides (Nicholas and Parkinson, 1967), are overcome to a great extent by the thinness of the films and dispersion of the material. Errors due to masking are usually more of a problem with litter than soil.

The agar film technique has stood the test of time in comparison with alternative methods remarkably well, and improved standardization has increased its value. The time factor is the usual reason for rejecting it. Measurement of variable hyphal lengths unlike that of bacterial shapes does not lend itself easily to automation. Careful choice of the experimental design can, however, reduce considerably the labour involved. Frankland *et al.* (1978) examined the sources of variation when the technique was applied to measurement of the biomass of a basidiomycete in leaf litter. By comparing the contributions of various stages in the procedure to the final variance, an optimum design for minimum variance with an acceptable amount of labour was obtained (see Gore, 1970). Optimum replication is usually achieved by increasing the number of primary (field) samples of soil and litter relative to the number of agar films and microscope fields, with consequent reduction in much of the tedium.

The duration and intensity of any maceration needed to disperse the hyphae are among critical factors in preparing agar films. Again, optima should be predetermined to obtain a satisfactory balance between reduction of fragment size and retention of recognizable lengths of hyphae (Swift, 1973a; Leonard and Anderson, 1981). Hutchinson and King (1989) and Hutchinson *et al.* (1990) reduced this problem by combining data from successive extractions of relatively short duration. Removal of detached mycelium after each extraction reduces the risk of its destruction during further maceration. A factor can then be calculated for deriving total yield of mycelium from the yield of the first extraction, if the yield per extraction declines in a consistent way.

Choice of optical system for the agar film technique is also important. Bååth and Söderström (1980), for example, doubled their values for hyphal length by increasing the magnification from 800× to 1250×, although Ingham and Klein (1984a) obtained maximum values at relatively low magnifications when the films were freshly prepared; the latter also stressed that inappropriate dilutions could cause large errors. Considerable increases in estimates of hyphal length have also been obtained by examining the films for unstained 'ghost' hyphae under phase contrast instead of bright-field illumination (Frankland, 1974), and this is now the system usually adopted.

The frequent use of factors borrowed from the literature for converting hyphal length to biomass has been the most unsatisfactory aspect of the agar film technique. As much attention should be paid to the conversion factors as to the primary measurements. Moisture content and density are known to vary widely with age of the mycelium, growth conditions and species (van Veen and Paul, 1979; Newell and Statzell-Tallman, 1982). Lodge (1987) recommended expression of density on a dry weight basis to avoid the use of the highly variable moisture factor. Ideally measurements should be made on mycelium collected from the field, as by Lodge, or the cultures should at least mimic the low nutrient conditions of soil and litter. Density can be measured by volume displacement in water (e.g. Lodge, 1987) or by isopycnic centrifugation in a gradient of silica gel (Bakken and Olsen, 1983).

As Bååth and Söderström (1979) pointed out, hyphal diameter is the most crucial conversion factor used in the agar film technique because it is squared in the calculation of biomass. They compared the effects of different methods of estimating mean hyphal cross sections, and showed how greater accuracy could be achieved by taking account of differences in the distribution and breakage of thick and thin hyphae. Errors due to shrinkage during film preparation have been discussed by Jenkinson *et al.* (1976) and Jenkinson and Ladd (1981).

Some estimate of the proportion of living hyphal biomass on agar films can be obtained by assuming that cell contents are indicative of life and that their loss during preparation has been minimal (Frankland, 1975a). The hyphae are divided into categories according to the volume of cell contents visible under phase contrast, and the *proportion* of the total hyphal lumen containing cytoplasm calculated on a volume basis. More realistic figures have been obtained by this method than by recording presence and absence of cell contents per unit length of hyphae, or by use of vital stains (p. 349). It is not claimed that the amount of cytoplasm can be equated with metabolic activity, and the method does not distinguish between dead, functionless hyphae and 'empty' hyphae which may be forming a transport link between the resource and active hyphal tips.

*(b) Membrane filtration.* The membrane filter technique (Hanssen *et al.*, 1974) is more rapid than the agar film method of obtaining hyphal length measurements. Again, the objective is to obtain well-dispersed hyphae in thin preparations of soil or litter on a quantitative basis. Dilute homogenized suspensions are passed under vacuum through membrane filters, which are stained and mounted in a 'clearing' agent appropriate to the type of filter chosen (e.g. immersion oil for Millipore filters) so that the hyphae retained on the filter surface can be measured. Fungal stains (e.g. 0.05% trypan blue) can be added before or after filtration. Addition of a dispersion agent, such as Winogradsky's solution (Pochon and Tardieux, 1962) or Calgon (Albright and Wilson Ltd, UK), may also be necessary to remove excessive colloidal matter. The procedures for measuring hyphal lengths and converting length to biomass are the same as those used in the agar film technique – likewise the scope for introducing errors! Sundman

and Sivelä (1978) somewhat simplified the original method and gave some advice on appropriate dilutions and size of membrane pore, both of which should be selected after preliminary trials. A useful description of the general procedure, as applied to field soils, has also been given by Elmholt and Kjøller (1987).

Membrane filtration has also been used to obtain measurements of the length of extraradical hyphae of certain mycorrhizal fungi. Abbot *et al.* (1984), for example, separated a VAM fungus from soil by flotation before filtration, stained the hyphae with trypan blue, and determined their length by a grid-line intersect method. Schubert *et al.* (1987), after flotation, used a 50% sucrose solution followed by a 2-min sedimentation time before centrifuging and filtering off the supernatant to extract the hyphae. They stained with trypan blue or with fluorescein diacetate to obtain an estimate of total and active hyphal lengths.

Good correlation between agar film and membrane filter data from several soils has confirmed the latter method's comparative value, but its lower estimates of biomass when conventional stains are used suggest that it is less accurate (Bååth and Söderström, 1980). Membrane filters have, however, the advantage that fluorescent stains can be applied, as mentioned above. Thus, good recovery of mycelium added to soil was reported by Paul and Johnson (1977) using fluorescent aniline blue, and West (1988) obtained significantly greater estimates of mycelium in pasture soils with a fluorescent brightener on membrane filters than with phenolic aniline blue and phase contrast microscopy on agar films. Further comparative trials of the two methods, both with litter and a wider range of soil types, are needed.

### 2. *Substrate-induced respiration*

A different approach to the assessment of fungal biomass in soil can be made by measuring respiration. Methods of estimating the soil microbial biomass in its entirety via respiration are discussed in detail by Gray (Chapter 10); also by Jenkinson and Ladd (1981) and by Sparling (1985). Only the fungal contribution to 'soil respiration' in relation to biomass is considered here.

The carbon dioxide flush from a soil amended with an excess of a non-selective carbon source, such as glucose, reflects the metabolism of the total *active* microflora. If the duration of the incubation period is relatively brief (1–3 h), an initial maximum respiration value can be recorded on increasing the added substrate level. This can be 'equated' with the metabolic activity of the total living microbial biomass, before new microbial communities develop, and it can be calibrated against total live microbial biomass measurements obtained by the so-called soil fumigation technique, since, for many soils, a close linear relationship has been found between them (Jenkinson, 1966; Jenkinson and Powlson, 1976; Martens, 1987; see Gray, Chapter 10). This substrate-induced respiration (SIR) method has been adapted to obtain a direct measure of the respiratory contributions of the fungal and bacterial components of a soil microflora by adding selective inhibitors, usually actidione (= cycloheximide) and streptomycin, that block the major anabolic processes leading to the formation of fungal or bacterial biomass (Anderson and Domsch, 1973a, b, 1974, 1975, 1978). Various assumptions must be made. In mycological studies, the bactericide should not affect the fungi or be deactivated, and the incubation period should be brief enough to avoid effects from microbial utilization of either the inhibitors or the dead organisms. It also has to be assumed that the initial respiration rate is the same for different sections of the mycoflora. West (1986) found from an investigation of pasture soils that the timing of additions of antibiotics and substrate in respect to the assay periods was an important factor.

The advantages of the SIR method are that it is simple, rapid and objective, but it does require more specialized equipment than the techniques involving microscopy. Domsch *et al.* (1979) compared it with the agar film and membrane filter (fluorescein diacetate-active mycelium; Söderström, 1977) methods on six soil types. Good reproducibility was reported, but live biomass estimated from respiration was relatively very high despite the fact that dead hyphae were included in the agar film measurements. Ineson and Anderson (1982) also compared SIR with the agar film technique, using selective fluorescent stains (europium chelate and a brightener; Anderson and Slinger, 1975) to distinguish the living components of the microbial biomass in thin agar films of leaf litter from microcosms. Again, there was poor agreement between the biomass estimates. They suggested that the chelate might have stained DNA retained in old hyphae, or that basidiomycetes, a major fungal group of leaf litter, might not have been stimulated by the glucose substrate. The latter factor could, they suggested, have a bearing on the low biomass values often obtained from acid organic soils by the respiratory method (Jenkinson *et al.*, 1976; Domsch *et al.*, 1979). However, reasonable agreement has been obtained between SIR and direct microscopy for estimates of live biomass in larch humus and in an acid heathland soil (Sparling and Williams, 1986). In this case, estimation of live fungal biomass by microscopy was based on the volume of cytoplasm in the hyphae (Frankland, 1975a).

### *3. Hyphal components and products*

Hyphal constituents may appear promising as determinants of fungal biomass, but even in monoxenic conditions they have severe limitations. They must be quantitatively extractable and proportional to the amount of fungal material, and therefore present in all parts of the biomass. Methods of analysis have become more sensitive with the use of such techniques as high-performance liquid chromatography (HPLC), but variations in concentration with species, physiological state and age, like those recorded by Sharma *et al.* (1977) and others, are a continuing problem.

The relative sensitivity of assays for Kjeldahl nitrogen, nucleic acids, ATP and $^{15}NH_4$ (incorporation) when determining fungal biomass in solid substrata under axenic conditions has been discussed by Matcham *et al.* (1984); see also Wissler *et al.* (1983) on use of $^{15}N$ mass spectrometry for measuring biomass of *Chaetomium* in solid-state fermentations. In a mixed microflora only more specific constituents of mycelium have potential as indices of biomass. Of these, chitin, ergosterol and the growth-related laccase enzymes have been tested most thoroughly.

*(a) Chitin.* Chitin (a polymer of N-acetyl glucosamine) is a component of the cell walls of most fungi other than Oomycetes, and almost absent from prokaryotes and higher plants but present in invertebrate exoskeletons. Its use in soil and litter as an index of fungal biomass is therefore limited. Assays of fungal chitin in plant tissues by measuring hydrolytic release of glucosamine colorimetrically have been described by Ride and Drysdale (1972) and Swift (1973a, b). Swift used this method to estimate the biomass of fungi growing in wood as monocultures. The determination of a factor for converting glucosamine content to mycelial dry weight which took account of shifting phases of growth was the principal difficulty.

From a detailed comparison of the chitin assay and the agar film technique (with regard to precision, efficiency as man-hours, sensitivity and accuracy), when used to measure the biomass of a basidiomycete in leaf litter culture, the assay was found to be less sensitive but at least as accurate as the method based on hyphal length (Frankland *et al.*, 1978). Since then, the speed and sensitivity of chitin assays have been increased by using HPLC in place of the hydrolytic and colorimetric procedure (Zelles, 1988).

The potential value of quantifying mycorrhizal fungi by chitin assay of either roots or extraradical hyphae in laboratory or pot experiments has been explored by several workers. Hepper (1977) estimated glucosamine colorimetrically after heating VAM roots in potassium hydroxide. Her results correlated well with estimates of infected lengths of root using a staining technique. Similarly, Bethlenfalvay *et al.* (1981) found a significant linear relationship between chitin estimates and intensity of infection determined microscopically. The correlation was two-phased, and indicated a greater rate of chitin accumulation above 60% infection than below. Chitin assay has also shown good correlation with measurements of a yellow pigmentation characteristic of VAM in unsuberized roots of onion (Becker and Gerdemann, 1977; see also p. 348).

The biomass of VAM extraradical hyphae has been estimated by chitin determinations on soil washed from roots by Bethlenfalvay *et al.* (1982) and Pacovsky and Bethlenfalvay (1982), but they showed that chemicals present in soil and roots can interfere with the colorimetric estimates of glucosamine. They also advised particular care in standardizing procedures, since colloidal clay particles and humic acids can protect hexosamines from biodegradation. Their soil was washed over 43-μm mesh to reduce the effect of colloidal suspensions. In pot culture experiments, Bethlenfalvay and Ames (1987) were able to show a linear relationship between the length of VAM-fungal hyphae (i.e. hyphae >5 μm dia.) and VAM-fungal biomass by assaying soils with and without VAM plants to obtain a chitin standardization curve.

Choice of conversion factors for determining the biomass of mycorrhizal fungi from chitin estimates is again a problem. Whipps (1987) measured the chitin content of ectomycorrhizal fungi grown on cellophane membranes to facilitate estimation of chitin content per unit dry weight of mycelium of different ages. This assumes, however, that the chitin content of a fungus growing on a laboratory substrate will be the same as that under natural conditions.

Whipps *et al.* (1982) listed several potentially useful biochemical 'markers' besides chitin for determining growth and/or biomass of symbiotic fungi. Most require further investigation, but chitin stood out as the cell wall component most widely applicable to fungi in the presence of vascular plant tissue. In the same investigation, the chitin assay was compared with that of another cell wall polymer, mannan, for estimating infection by ectomycorrhizal and VAM fungi, mannose from mannan being measured by gas-liquid chromatography. It was concluded that chitin analysis was the better of the two assays because of the high concentration of mannan in cell walls of the host.

*(b) Ergosterol.* Ergosterol is the most abundant sterol in fungi and relatively specific, being only a minor constituent of higher plants. The general consensus is that its assay is more specific to fungi than that of glucosamine (Seitz *et al.*, 1979; Newell *et al.*, 1988). A rapid HPLC assay for ergosterol has been described by Grant and West (1986), Zill *et al.*

(1988), and Salmanowicz and Nylund (1988; ectomycorrhizal infection). It can also be measured by gas chromatography or U.V. spectrophotometry; the latter is convenient because plant sterols absorb poorly at the absorption maxima for ergosterol (Matcham *et al.*, 1984). West *et al.* (1987) considered it was a sensitive indicator of changes in live fungal biomass in agricultural soils, but, again, before absolute measurements can be made confidently, further evaluation of the effect of growth conditions and hyphal physiology on ergosterol content is required (Nout *et al.*, 1987).

*(c) Extracellular enzymes.* In certain limited situations extracellular enzymes can be used as a measure of live fungal biomass. For example, live biomass of *Agaricus bisporus* in liquid culture has been found to be directly proportional, at least in the early stages of growth, to the quantity of extracellular laccases. These enzymes have also given realistic estimates, without background interference, of the living biomass of some closely related species of *Agaricus* growing as mixed populations in composted straw (Wood, 1979). Specificity of the enzyme assay was confirmed by antigen–antibody reactions.

### 4. *Quantitative immunology*

The use of fluorescent antibodies simply as detectors of particular fungi in ecological investigations has already been mentioned (p. 351–353). Preliminary trials with an antiserum raised against *Mycena galopus* showed that fluorescent-antibody staining combined with membrane filtration was also a feasible means of estimating mycelial biomass of a single species in leaf litter (Frankland *et al.*, 1981).The basidiomycete was the principal colonizer of the litter, selected from field material for the tests; for more complex fungal communities, more specific polyclonal antisera or perhaps monoclonal antibodies would be needed. Other methods with potential for quantitative ecological studies, although more laboratory orientated, are ELISA and RIA (see p. 351), but the antigen involved must be related to the amount of fungal biomass.

Electrophoretic separation of specific proteins can also be the basis for biomass determinations, but the lengthy assay procedures are likely to preclude its use for large numbers of samples.

## D. Production

The accuracy of methods for determining total fungal biomass and turnover in a field situation can only be assessed by examining the inputs and outputs of the ecosystem, and then calculating whether the annual supply of nutrients is sufficient to sustain the calculated production of fungi and other organisms (see definition p. 368; Gray, Chapter 10; Gray and Williams, 1971b; Gray *et al.*, 1974). The complexities of ecosystems and the possibilities for errors in such an assessment are obvious. Much quantitative fungal ecology of this type was stimulated by the International Biological Programme (IBP; 1967–1972), one of its aims being to compare globally the biological productivity of terrestrial ecosystems. Kjøller and Struwe (1982) have discussed in detail the attempts made during IBP to calculate fungal production based on measurements of biomass, turnover rates and substrate input or use.

The simplest estimate of minimum production can be made by measuring biomass at the beginning and end of a growing season, ignoring fluctuations (Visser and Parkinson, 1975). Improvements in the data can be obtained by consecutive sampling and summing of positive changes to obtain net production (Witkamp, 1974), but gross production includes the component of biomass lost by grazing and lysis, and it is difficult to estimate in a mixed community. A field value has been obtained, however, for the gross annual production of a single species (*Mycena galopus*) in naturally colonized *Quercus* litter, when it was the sole basidiomycete occupant of the leaves (Frankland, 1975b; Frankland *et al.*, 1979). Net change in biomass was estimated by measuring the length of the hyphae in agar films, and biomass loss in the field (from all causes) by recording the decrease in mycelium labelled with a fluorescent brightener (see p. 350). The rate of loss was determined by regression analysis.

Nagel-de Boois (1971) and Nagel-de Boois and Jansen (1971), approached the problem of mycelial production differently, using the often-quoted nylon mesh technique (Waid and Woodman, 1957). They measured the increase of live mycelium growing over nylon mesh buried in soil and litter during periods of slow decomposition, but the results were unconvincing owing to the variability in the growth rates and life span of mycelia. The disadvantages inherent in all methods involving buried substrata have already been mentioned (p. 354).

Hanssen and Goksøyr (1975) calculated the production of decomposer fungi in alpine-tundra sites indirectly from estimates of litter input, by assuming that 90% of the litter carbon was utilized and that the assimilatory efficiency was 40%. More sophisticated modelling of decomposition and energy transfer in ecosystems, from which fungal productivity could be predicted, has also been attempted (e.g. Heal and MacLean, 1975; Flanagan and Bunnell, 1976). A combination of field studies and laboratory analyses was used to obtain values for litter input, fungal biomass, respiration, and yield efficiency, and of substrate needed for cell maintenance (without growth). A similar approach was used by Hutchinson and King (1989) to calculate potential microbial production in litter of pasture plants.

## VI. Measurement of activity

The ultimate goal of many ecological studies is measurement of activity: what does or can the fungus do, and at what rate? How often projects, full of promise at the start, tail off at this stage as time runs out. The worker engrossed in studying the composition and concentration of populations and communities then leaves the subject to a mythical successor. This simply underlines the difficulties to be overcome and the need to improve techniques.

Available methods can be broadly divided into those that attempt to measure activity in the field, and those that assess only potential capabilities usually in laboratory systems (see Parkinson *et al.*, 1971; Taylor and Parkinson, 1988). A major problem is separation of the fungal contribution from that of many other organisms involved in the same metabolic processes, without operating in entirely unnatural conditions.

The four most fundamental aspects of fungal activity commonly assessed are: germination and growth (i.e. mycelial extension and/or biomass production); hyphal and mycelial interactions; uptake, translocation and release of carbon compounds, mineral nutrients and water; and, since fungi are heterotrophic for carbon, the utilization of organic substrata and production of degradative enzymes.

### A. Germination and growth

Some indication of the viability of fungal spores and hyphae can be obtained by using various stains and markers (p. 349), but this takes no account of dormancy. Actual germination of spores in field soil can be measured by burying and retrieving spores on non-nutrient substrata such as fibre-glass tape, water agar discs or membrane filters (see Johnson and Curl, 1972). In a direct assay which avoids introduction of these foreign materials, a spore suspension is applied to the smoothed surface of a moist, compact soil in a Petri dish, stained *in situ*, and recovered on collodion film (Lingappa and Lockwood, 1963; Chacko and Lockwood, 1966). Otherwise, germination in the laboratory is usually assessed on agar media, cellulose strips, or slivers of natural substrata. However, some fungal spores must be stimulated before they will germinate. Laboratory use of heat, low temperatures and chemicals for this purpose, and also methods of removing inhibitors have been described by Booth (1971b).

Laboratory studies of spore germination enable the influence of varying environmental parameters to be examined. A technique which enables the effect of multiple combinations of physical factors to be measured on the same agar plate has been devised by Boddy *et al.* (1989). The plates are constructed so that gradients of solutes (e.g. salt, sugar) and pH can be set up in one or two directions; the latter arrangement can produce about 64 solute combinations on a 10 cm × 10 cm square plate.

It is sometimes possible to measure the extension rate of mycelium in the field. The cord- and rhizomorph-forming basidiomycetes, with their macroscopic hyphal aggregations, are among fungi most easily measured by direct observation (p. 346). The extension of 'fairy rings', such as those produced by *Clitocybe nebularis, Collybia peronata* and *Marasmius wyneii* in the litter layer of deciduous woodlands, can be deduced by mapping the position of fruit bodies (p. 345), and direct measurements can be made if the production of mycelium is sufficiently prolific and the mycelial front clearly visible (Dowson *et al.*, 1989a). Where mycelia are less obvious, or where there is doubt as to their origin, systematic isolations can be made from litter along transects, and pairings made on agar to determine, on the basis of somatic incompatibility interactions, whether or not the isolates are from the same fungal individual (pp. 358, 380).

In unsterile soil, measurements of growth are usually indirect and based on the previously discussed methods of observing and quantifying fungi. They include: assessment of the rate of colonization of buried substrata, either inert or selective (p. 354); measurement of change in biomass by sampling after various time intervals (p. 364), and labelling techniques (p. 348). The procedures are destructive, or involve amendment of the soil/litter system so that potential rather than actual growth rates are measured.

Failing measurement in the field, growth rates can be studied on agar media inoculated with the fungus, but they should be viewed with some scepticism even if obtained under varying conditions of temperature, water potential, gaseous regime, nutrient status, etc. (see Griffin, 1977, 1981; Magan and Lacey, 1984a,b; Boddy *et al.*, 1985). Growth form and extension rates on artificial media are very often different from those in natural substrata. Some fungi grow more rapidly on the usually richer laboratory media, often branching profusely to form dense compact colonies, whereas many fairy-ring formers grow more slowly on agar (Dowson *et al.*, 1989a). Again, many fungi that form cords do not produce them on standard culture media, or only in ageing cultures.

Some form of laboratory microcosm that still allows easy manipulation of the microenvironment is a better alternative to agar media. Soil fungi can be grown in tubes of glass or Perspex containing soil and/or litter. Direct measurements of fungal growth in sterile soil have been made in 'soil-recolonization tubes', either by viewing the mycelium through the glass or by sequential sampling (Evans, 1955). Side arms on these tubes allow addition of water and soil sampling. The problem of viewing fungi that grow deeply in soil can be avoided or lessened by using trays of compacted soil (although alteration of the soil structure is undesirable), or by arranging only a shallow layer of soil in the colonization tubes. The latter, achieved by placing an inner tube within an outer tube and filling the space with soil, has been particularly successful with cord-forming species. The method has revealed not only differences between species, with varying temperature and soil matric potential, but also a remarkable switch in colony morphology and extension rate of *Steccherinum fimbriatum,* triggered apparently internally rather than by environmental factors (Dowson *et al.*, 1989b).

Extension rates within bulky materials such as lengths of wood (say, 1.5 × 1.5 × 20 cm) can be followed by inoculating them at one end with the fungus, and sampling destructively after various time intervals. Extent of the colony can then be determined by making isolations and/or microscopic observations of sections. Extension growth can also be assessed non-destructively by growing fungi in thin sections of wood. Hale and Eaton (1985a,b,c) employed this method very successfully to follow hyphal extension and cavity formation by soft-rot fungi.

Several techniques have been developed for studying the growth of mycorrhizas. Fortin *et al.* (1980) used polyester root pouches for observing two-dimensional growth of ectomycorrhizas in liquid culture. Yang and Wilcox (1984) devised a similar system in which the root system grew between chromatographic paper and the wall of a large test tube containing nutrient solution. Petri dishes have also been used to maintain a two-dimensional rooting zone for following growth (Finlay and Read, 1986). Indirect measurements of the rate of spread of VAM hyphae in soil have been obtained by Schüepp *et al.* (1987 ) using a ' trap' plant to pick up VAM infection from an inoculated plant, separated in cuvettes by a 2-cm root-free zone. This was delineated by 80-μm mesh that allowed free passage of the mycorrhizal hyphae. Different rates of spread, depending on soil type, were measurable by harvesting the 'trap' plants over a period of time.

## B. 'Inoculum potential' and 'competitive saprophytic* ability'

The success of a saprotrophic fungus in colonizing a resource in the initial stages, either ahead or in competition with other microorganisms, depends

*The wider term 'saprotrophic' has been adopted in this chapter, but the implications are the same.

not only on its environment but also on certain intrinsic factors, such as the energy available for growth, growth rate and hyphal age. Garrett (1956, 1970) and his co-workers have studied this aspect of fungal activity extensively, and the criteria for success were summed up in two terms, 'inoculum potential' and 'competitive saprophytic ability'. Various simple laboratory methods were devised for comparing the inoculum potential and colonizing abilities of root-infecting fungi in particular, but they are also applicable to obligate saprotrophs. For example, the degree of colonization by a test fungus can be assessed by isolating it on an agar medium, after exposing discrete units of the organic material to a range of soil/inoculum mixtures (see Parkinson *et al.*, 1971).

### C. Hyphal and mycelial interactions

Fungal interactions that have occurred in the field are usually difficult to unravel, the main exceptions being those that occur between basidiomycetes or xylariacious ascomycetes in wood and sometimes in litter. As already mentioned (p. 358), interaction zone lines demarcate individuals of different species and different genotypes of the same species. These zone lines are valuable evidence of present or past activity. If their position appears to be permanent, the outcome of the interaction can be said to be 'deadlock', but the occurrence of relic zone lines within an area of decay suggests replacement of one mycelium by another (see Rayner and Todd, 1979; Rayner and Boddy, 1988a, b). Evidence of replacement or intermingling can also be obtained by isolating mycelium from different locations in and beyond the area of confrontation.

Interspecific interactions, similar to those in the field, can be demonstrated in agar cultures, samples of wood or litter, and laboratory microcosms. The conditions are far from natural, but the responses of wood-decay fungi correspond broadly with those observed in the field, although such factors as water potential, temperature, and gaseous composition have been shown to affect the outcome of interactions of certain species of wood and grain fungi (Magan and Lacey, 1984a,b; Boddy *et al.*, 1985; Chapela *et al.*, 1988). Use of agar cultures can also be criticized because of differences in the concentration, availability and distribution of nutrients compared with natural resources (Dowding, 1978). In addition, differences in the outcomes between different species of cord formers, and between cord formers and non-cord formers, have been demonstrated in the laboratory, depending on the state of decay of wood blocks used in the experiments, and also on whether the interactions occurred between mycelia already established in wood or between cords growing through soil towards colonized wood blocks (Dowson *et al.*, 1988c; L. Boddy and

M. Warren, unpublished). Extrapolations from the laboratory to the field must therefore be made, as always, extremely cautiously.

The mechanisms involved in these interactions, including killing from a distance, mycoparasitism and hyphal interference, can also be observed in artificial culture, for example on agar, cellophane films or wood slivers (e.g. Ikediugwu and Webster, 1970a, b; Deacon and Henry, 1978). Timelapse photography has proved useful for watching the progress of these activities.

## D. Uptake and translocation

Until recently, uptake and translocation of nutrients and water by fungi have been studied in the laboratory by the physiologist rather than the ecologist. Much research remains to be done in testing fundamental theories under more natural conditions, and in relating them to nutrient cycling within an ecosystem.

Rhizomorphs and cords are particularly amenable to experimentation. In the laboratory, apices or segments can be used, or the fungus can be grown from wood blocks and other inoculated resources over an inert surface such as Perspex sheeting (Brownlee and Jennings, 1981; Jennings, 1982). As Jennings (1987) has suggested, more information might be gained in such experiments if more appropriate materials were used, such as films of silica gel to mimic soil. Particularly successful results have been obtained by growing cord systems from wood blocks in trays or tubes of unsterile soil (systems similar to those used by Dowson *et al.*, 1986). Using this approach, rates of translocation greater than 1 m $h^{-1}$ have been detected (J.M. Wells, L. Boddy and J. Dighton, unpublished)! By dividing the trays into sections, radiotracer can be added to different regions of the system, and reallocation of nutrients can be assessed. Addition of 'baits', such as blocks of colonized or uncolonized wood of different sizes or leaf litter, has revealed remarkable patterns of nutrient reallocation which might never have been observed in less realistic systems (Wells *et al.*, 1990).

Isotopic labelling provides many possibilities in the search for methods of investigating physiological processes. Specialized laboratory facilities are, however, required, and field experiments on nutrient cycling, using either stable or radioisotopes, are particularly prone to problems caused by background interference. For useful references to some of the earlier ecological studies, in which biologically compatible radioisotopes with suitable half-lives were employed, see Swift *et al.* (1979). Absorption of phosphate by intact mycelial cords of saprotrophic wood-decaying basidiomycetes has been determined in the field by containment of the fungi. Cords growing through the litter layer were exposed and lowered into a labelling vessel, which was sunk below the surface of the forest floor and contained a solution of $[^{32}P]KH_2PO_4$ (Clipson *et al.*, 1987; Jennings, 1990). Alternatively, labelled litter components can be placed in the vicinity of cord apices, and translocation following colonization assessed. Similarly, translocation from colonized woody resources can be followed by drilling a hole in the wood, adding isotope, plugging, and then monitoring movement of the tracer (J.M. Wells, L. Boddy and J. Dighton, unpublished).

$^{32}P$-uptake and translocation have been demonstrated in ectomycorrhizal strands of a *Pinus radiata*/*Rhizopogon luteolus* combination in root observation chambers in the laboratory, and by use of the tracer in the field (Skinner and Bowen, 1974). Translocation over distances of 12 mm in the chambers and 120 mm in the field was recorded. Similar experiments by Finlay and Read (1986) were also successful in detecting translocation in *Rhizopogon* species. In a genuine field experiment, differences in the uptake of phosphorus by various ectomycorrhizal fungi, associated with birch (*Betula*), have been recorded by applying radiotracer phosphorus to defined mycorrhizal zones under birch trees (Dighton *et al.*, 1990).

Procedures using $^{14}CO_2$ to trace movement of carbon through mycorrhizal mycelium, in root chambers and in the field, have been described by Read and his co-workers (e.g. Read *et al.*, 1985). In one field experiment, the crowns of tall forest 'donor' trees were enclosed in polythene sacks. Activity was recorded in 'receiver' roots, but variability was understandably high!

The uptake of nutrients and other mineral elements by mycorrhizas can be determined indirectly, by a variety of organic and inorganic methods of analysing fungal or host plant tissue. For example, polyphosphate accumulation has been measured in VAM fungi by phenol detergent extraction and polyacrylamide electrophoresis (Callow *et al.*, 1978). Electron probe/X-ray microanalysis or laser probe/mass spectrometry has been used to determine various mineral elements in ectomycorrhizas and VAM vesicles (Strullu *et al.*, 1983a, b; Denny and Wilkins, 1988a, b). Cytochemical and ultracytochemical techniques have also been employed to quantify polysaccharides and proteins in mycorrhizas (Gianninazi-Pearson *et al.*, 1981; Piché *et al.*, 1981). Similarly, several chemical tests for glycogen and other carbohydrates have been used to examine their release in gels secreted by strands of ectomycorrhizal fungi of *Pinus radiata* (Foster, 1981).

Investigation of nutrient release caused by fungal activity is generally confined to microcosm studies, in which provision is made for collection of leachates as in a field lysimeter (see p. 384).

## E. Utilization of organic resources

In utilizing a resource, a decomposer fungus alters the physical and chemical composition, takes up oxygen, evolves carbon dioxide and sometimes other gases, produces water and releases energy. Measurement of any of these changes could in theory provide an estimate of fungal activity, but some are more useful than others, and it is nearly always necessary to resort to Petri dishes and microcosms, i.e. the 'bottom-up' approach, building up from a simple monoxenic system to the more complex. The 'top-down' approach by selective removal of certain groups of organisms has met with less success (see Coleman, 1985).

### *1. Evidence from the field*

Decay rates (usually estimated from changes in weight, tensile strength and chemical composition) of individual litter components or pure substrates – bagged, tethered or buried in the field – have been studied extensively (see Dickinson and Pugh, 1974; Swift *et al.*, 1979; Harrison *et al.*, 1988). However, these methods measure activities of a decomposer community as a whole, and attempts to partition functions between specific groups of organisms are beset with problems.

Decomposition of leaf litter in the field, for example, can be estimated by the nylon hair-net technique (Gilbert and Bocock, 1962), using 30–40-μm mesh to exclude animals, but the involvement of bacteria and fungi has not been separated satisfactorily. Biocides are available but are of dubious value in the field (see p. 384). In addition, the fine mesh causes excessive retention of moisture. However, nylon mesh in sheet form was effective in an animal exclusion experiment, designed to show that selective grazing by a collembolan could affect the activities of two competing basidiomycetes in the field (Newell, 1984b).

In wood, which is much bulkier than most types of litter and where the decomposer community changes relatively slowly (i.e. after months rather than days), measurements of density (p. 385) and other changes can sometimes be made on individual decay columns so that decomposer activity is partitioned between species.

The principal value of field samples lies in the clues they can give as to the functions of predominant decomposers, especially when observations are backed up by laboratory tests on isolated species (e.g. Frankland, 1966, 1969). For instance, presence of hyphae in 'bore holes' of fibre cell walls and in bleached litter can indicate fungal attack on cellulose and lignin. Laboratory tests alone are no substitute for such field evidence, and the influence of competition and possibility of synergistic effects should always be borne in mind.

*2. Laboratory measurements*

The aim of laboratory experiments on decomposition by fungi, whether they be in a Petri dish or a more elaborate microcosm, is to obtain realistic measurements of some activity under controlled conditions. Microcosms, as defined by Taylor and Parkinson (1988), are 'simplified analogues of natural ecosystems'. They vary greatly in complexity and therein lies the dilemma – how to balance realism with simplification.

In partitioning activities between fungi and other organisms, the first stage could be measurement of changes in a flask of sterile litter inoculated with a single fungal species (e.g. Hering, 1967; Frankland, 1969). This can lead stepwise to more competitive situations, e.g. two competing fungi on precolonized litter with a grazing collembolan (Newell, 1984b), plant/mycorrhizal/saprotrophic fungal interactions (Dighton *et al.*, 1987), or Perspex chambers from which leachates can be extracted for analysis (Anderson and Ineson, 1982), containing unsterile litter with added microfauna (Ineson *et al.*, 1982). In the last example, chemical changes were related to fungal biomass by measuring hyphal length.

*(a) Sterilization of litter and soil.* In many microcosm studies, it is necessary to sterilize the litter or soil. A variety of procedures have been described, including, most recently, treatment with microwaves (Ferriss, 1984; Speir *et al.*, 1986; Jeng *et al.*, 1987; Gibson *et al.*, 1988). It may be tempting to use one of the ubiquitous, domestic microwave ovens, but the effects on litter and soil have not been investigated as thoroughly as those caused by gamma-irradiation. The latter is still the most effective, well-tested method of sterilizing samples with minimum chemical and physical change (Cawse, 1969; Howard and Frankland, 1974). Air-dry samples can be irradiated at a commercial plant and stored until required, allowing a 2-week period for cessation of free enzyme activity and carbon dioxide dispersion. For most leaf litters, a dosage of $2.5 \times 10^4$ Gy on 2–3 g samples has been adequate, but sterility tests should be carried out for exceptionally resistant organisms such as the *Bacillus* found in *Quercus* litter by Frankland and Collins (1974). A similar dosage is usually recommended for soil, but the samples should be kept relatively small to allow the radiation to penetrate adequately. Cawse (1975) discussed the effects on different soil types and the biochemical artefacts which may be produced. On a practical point, release of iron(?) causing dark discoloration of glass containers can be reversed by heating for several hours in an oven.

Partial sterilization of organic materials, using X-rays or naphthalene to reduce or eliminate microarthropods is not recommended, because these treatments are known to have adverse effects on some fungi (Newell *et al.*,

1987; Blair *et al.*, 1989). Modern pesticides that are claimed to be entirely specific in their action also need more testing by ecologists.

Sterilization must always be the last resort. Even by gamma irradiation, sterilization of soil can result in different fungal growth patterns. For example, *Hypholoma fasciculare, Phanerochaete velutina* and *Steccherinum fimbriatum* can grow significantly more slowly in sterile soil than in unsterile, and fail to form mycelial cords (Thompson and Rayner, 1983).

*(b) Weight loss.* A common index of the rate of decay is weight loss, which is usually expressed as a percentage of the original oven (105°C) dry weight to allow for moisture loss. Since oven-drying of litter can result in considerable structural and chemical changes, use of 'fresh' litter in decay-rate experiments is preferable. Consequently, since the original dry weight is usually unknown, a factor for converting fresh or air-dry weight to oven-dry weight must be calculated by drying representative samples of such materials as leaf litter or fine twigs. Weight loss of bulky, slowly decaying substrata such as wood is usually calculated by comparing density (dry mass per unit volume of *undried* wood) after decay with density of undecayed wood (Healey and Swift, 1971; Christensen, 1984; Rayner and Boddy, 1988a):

$$\%\text{ weight loss} = \frac{(\text{density undecayed wood} - \text{density decayed wood})}{\text{density undecayed wood}} \times 100$$

One source of error arises when estimating the volume of wood, owing to sample shape. If the samples are rectilinear or cylindrical, measurement of linear dimensions is appropriate. In other cases, displacement of sand or water can be measured, but if the latter is used the wood must be covered in wax or polythene film to prevent absorption of the water.

Whichever criterion is adopted (i.e. density or weight loss), the value obtained will underestimate decomposition, since no account is usually taken of the weight of mycelium present. Occasionally all external mycelium can be removed, as was done in the case of smooth bracken (*Pteridium*) petioles which were wiped clean (Frankland, 1969). The error due to internal mycelium is usually considered to be small, but this cannot be so in some well-decayed samples of wood riddled with cords or rhizomorphs. Correction for internal mycelium can of course be made by estimating biomass, using one of the methods described earlier in this chapter.

*(c) Tensile strength.* Measurement of loss of tensile strength is an alternative to weighing samples in laboratory decomposition experiments. Unlike weight loss, it is not directly proportional to utilization of substrate

but depends on the breaking of bonds in the cellulose fibres. Thus, severing of glycosidic linkages at early stages would result in a decline in tensile strength, which would not be accompanied by much loss of glucose units. Measurement of tensile strength might therefore be more useful than weight loss measurements at early stages of decomposition, since the latter would not detect fungal activity. However, at later stages the possibility of a reduction in tensile strength owing to localized decay, despite only small weight losses, is a disadvantage.

Instrumentation and tensile strength measurements are critically discussed in a symposium volume on the cotton strip assay (Harrison *et al.*, 1988). As described in that volume, standardized cotton fabric strips have been used extensively in soils all over the world to obtain a comparative index of decomposer activity. Cellulolytic fungi and pigmentation caused by fungi occur regularly on these strips, but the method is not recommended as an indicator of the presence or activity of particular species in the field (Gillespie *et al.*, 1988).

*(d) Chemical changes.* Gross chemical changes effected by fungi can be quantified by using standard methods of analysing ecological materials (Allen *et al.*, 1974; Allen, 1989), usually after the samples have been air-dried and ground. Chemical components can be estimated as percentages of the initial dry weight to show absolute rates of loss or accumulation.

Specific sites of chemical change can be revealed by a combination of various types of microscopy and staining. Bulky and recalcitrant wood is particularly suited to this form of analysis. Loss of cellulose and presence of soft-rot cavities can be seen using polarized light microscopy, since cellulose, unlike hemicellulose and lignin, is doubly refractive, so regions from which cellulose has been removed appear dark when viewed through crossed, Nicol polarizing prisms (Lohwag, 1937; Schulze and Theden, 1938). An indication of the lignin content can be obtained from the degree of opacity when viewed with UV light microscopy at 278 nm, since lignin strongly absorbs UV light at this wavelength (Wilcox, 1968). Useful stains include safranin and fast green, which stain cellulose green and lignin red (Wilcox, 1968; Pearce and Rutherford, 1981). Scanning electron microscopy (pp. 350–351) can also provide direct evidence of patterns of decay (e.g. Blanchette, 1980, 1983; Hale and Eaton, 1983, 1985a,b,c).

One method of measuring the rate of chemical change, which avoids inaccuracies in estimating weight loss, depends on attachment of dyes to particular substrates and measurement of their release (or residual retention) when the substrate is decomposed. Some success, for example, has been achieved with cellulose strips dyed with Remazol Brilliant Blue R, which is released upon hydrolysis of the glycosidic bonds (Poincelot and Day, 1972; Moore *et al.*, 1979), but some basidiomycetes decolourize the dye. The residual dye is measured spectrophotometrically, after extracting it by boiling the strips in sodium or potassium hydroxide. These cellulose strips have been used in litter bags placed in the field and as overlays on agar media. In both cases, the dye aids the detection of localized cellulase activity along fungal hyphae. Other dyes have been used to label organic substrates, but radioactive labelling (p. 381) has more potential as a technique. Moreover, simple substrates are no replacement for natural organic materials. It is well known, for example, that some cellulolytic fungi will not degrade pure cellulose, and it rarely occurs as such in the field (see discussion by Howard, 1988; Latter and Walton, 1988).

*(e) Gaseous exchange.* Most of the laboratory methods for quantifying fungal activity already described result in destruction of the sample. Also, in many of them, especially with substrata such as wood, long time intervals are necessary before accurate estimates of activity can be obtained. Measurement of oxygen uptake or carbon dioxide evolution can, however, provide an almost immediate assessment of activity. However, partitioning of activity between heterotrophic groups in soil has not usually been attempted except in Anderson and Domsch's method of determining biomass (p. 372).

The most inexpensive method of measuring $CO_2$ evolution is by absorbing the gas in concentrated alkali (e.g. KOH) and then back-titrating with acid (see Johnson and Curl, 1972), or by measuring changes in electrical conductivity when $CO_2$ is absorbed by an electrolyte (see Chapman, 1971). Various types of laboratory respirometer are available. These can be multi-channel, and have become increasingly more automated, more sensitive and easier to calibrate. They include two, now rather dated, manometric respirometers, the Warburg and the Gilson (Umbreit *et al.*, 1964; Parkinson *et al.*, 1971), and, more recently, an automatic, computer-controlled, electrolytic respirometer that measures $O_2$ uptake has been described (Tribe and Maynard, 1989).

Gas chromatography is a particularly useful, rapid and sensitive technique, which compares very favourably with other respirometric methods (Putman, 1976; Boddy, 1983). It simply involves injection of a small (about 0.5 ml) sample of gas from the respiration chamber into the analyser. The apparatus includes a katharometer detector and a glass column of beads (e.g. 150–180 μm Poropak Q) with helium as carrier gas for $CO_2$ measurements. Sensitivity is increased 12× by using helium instead of nitrogen. The former allows detection of $CO_2$ at levels of 3.5 μl in any injected sample, and other carbon-containing gases (e.g. $CH_4$) can also be measured. Simultaneous measurement of oxygen is made possible by connecting a second column of Poropak Q, held at −78°C, to the exhaust port of the primary column (Blackmer *et al.*, 1974), or by using a second column with molecular sieve at room temperature (Tadmor *et al.*, 1971).

Infrared gas analysis (IRGA; see Ineson and Gray, 1980, for details) is another sensitive technique for measuring $CO_2$ in gas samples, and can be used for continuous monitoring of the atmosphere in respiration chambers, such as microcosms of fungi on litter. The gas to be analysed is passed through an optical cell where infrared radiation in a particular band of the spectrum is absorbed by the $CO_2$, reducing the level of energy reaching a detector. The sensitivity of these differential instruments is enhanced by the ability to compare reference and $CO_2$-enriched air streams.

Mass spectrometry, a further means of simultaneous and continuous monitoring of many atmospheric and dissolved gases over a wide range of concentrations, is being developed for general field use (Boddy and Lloyd, 1990), but its potential for mycological studies has not, as yet, been realized.

The efficiency of fungi as decomposers can be assessed from respiration values and biomass data, and the validity of the methods can be checked by constructing a carbon-balance sheet for a laboratory system (see Frankland *et al.*, 1978). Care is, however, required when interpreting data on $CO_2$ evolution in decomposition studies, since 'instantaneous' measurements may not reflect only decay of the substratum at that time but could result partly from metabolism of storage materials. Similarly, measurement of fungal dry weight may not always reflect actual utilization of a carbon source (Taber and Taber, 1988).

Extrapolation from laboratory respiration experiments to a field situation is sometimes possible. For example, Ingham and Klein (1984b) and Bååth and Söderström (1988) found correlations between the length of fluorescein diacetate (FDA)-stained hyphae and rate of $CO_2$ evolution by fungi inoculated into sterile soil. The latter authors used this relationship to obtain a tentative estimate of the contribution of fungi to carbon cycling in a field soil, in which soil temperatures and FDA-fungal biomass had been measured.

A method for estimating the ratio of the respiratory activities of fungi and bacteria in homogenized soil, from which extrapolation to soil *in situ* is a possibility, was devised by Fægri *et al.* (1977). Respiration was measured on the fungal and bacterial fractions separated by centrifugation of a soil homogenate. The respiration rates of these fractions were constant for 1–2 h if the separation was carried out rapidly (less than 3–4 h) in the cold, and the sum of the rates agreed well with that of the unfractionated soil homogenate, making it possible to calculate the respiration ratio.

However, inhibitory conditions in intact soil compared with homogenates complicate extrapolation to the field.

Respirometry, as described, has also been used to determine the activity of mycorrhizal fungi. For example, Harley and McCready (1981) demonstrated increases in $O_2$ uptake of excised beech (*Fagus*) mycorrhizas on addition of phosphorus by using a Warburg apparatus. Direct IRGA respirometry of the extraradical and main root components of pine (*Pinus*) ectomycorrhizas was achieved in root observation chambers by Söderström and Read (1987). Saif (1981), studying a different aspect of gaseous exchange, showed that the growth and nutrient content of VAM-inoculated plants could be influenced by modifying the supply of soil oxygen through the base of plant pots.

*(f) Micro-calorimetry.* Energy (calorific) contents and energy flow resulting from metabolic activity can be measured by calorimetry in conjunction with estimates of biomass. The measurements are most conveniently made in a bomb calorimeter (containing oxygen under pressure) which completes combustion without loss of energy (see Chapman, 1986). A micro-bomb calorimeter designed by Phillipson (1964) is available commercially, and, although not as accurate as larger models, is convenient for small samples such as 5–100 mg of fungal mycelium (Allen *et al.*, 1974; Allen, 1989) Calorific values may be required to calculate, for example, an energy budget in a laboratory microcosm study. Micro-calorimeters have been used to examine microbial activity in soil, but, again, the measurements cannot be partitioned between heterotrophic groups (Ljungholm *et al.*, 1979).

## F. Enzyme activity

The enzyme activity of a natural microbial community but rarely that of saprotrophic fungi alone can be measured by various methods *in situ* (Tabatabai, 1982). There have been numerous investigations of the total enzyme activity of agricultural soils (e.g. Burns, 1978; Nannipieri, 1984) but relatively few detailed studies of forest soils, particularly in relation to other soil factors or over a period of time, with which to compare laboratory measurements. Notable exceptions include the research of Harrison and Pearce (1979) and of Rastin *et al.* (1988). Many tests for extracellular enzymes exist (see, for example, Table 3.4 in Rayner and Boddy, 1988a), but most involve addition of a substrate, so, although useful in laboratory studies, measure potential rather than actual activity.

Production of extracellular laccases which can be measured by an oxygen-electrode method (Wood and Goodenough, 1977; Wood, 1979) has already been mentioned in relation to estimation of biomass (p. 375). In that example, the enzyme complex was sufficiently specific to species of *Agaricus* for an autecological assay to be feasible in unsterile composted straw. One of the few natural situations where enzyme activity of a saprotrophic fungus alone can be assessed is in wood, where there are regions which are effectively occupied by 'pure cultures' of fungi. The location of esterase, dehydrogenase, laccase and other enzyme activity, for example, has been determined in Norway spruce (*Picea abies*) roots colonized by *Heterobasidion annosum* (Johansson and Stenlid, 1985). Detection of localized cellulase activity using Remazol Blue has already been mentioned (p. 386).

Mycorrhizal fungi provide several opportunities for measurement of enzyme activity in the laboratory. It can be examined in roots collected from the field, in pure culture, or in synthesized organs under axenic conditions. Surface phosphatase activity of ectomycorrhizas has been determined in beech (*Fagus*; Bartlett and Lewis, 1973), Sitka spruce (*Picea sitchensis*; Alexander and Hardy, 1981) and willow (*Salix*; Linkins and Antibus, 1981); also in pure cultures of the fungi (Ho and Zak, 1979; Dighton, 1983), using, for example, Hoffman's method as cited by Harrison and Pearce (1979). Proteinase activity of ectomycorrhizal plants has also been demonstrated in the laboratory by measuring the decomposition of bovine serum albumin (Abuzinadah and Read, 1986a, b).

### G. Interactions between mycorrhizal fungi and host plants

Apart from such functions as growth, nutrient uptake and respiration, the ability of a fungus to form mycorrhizas and the host response can be used as further criteria of physiological activity and possibly of 'success' in ecological terms.

The potential for forming mycorrhizas can be tested under axenic conditions. Synthesis has been performed in several artificial media and in a variety of containers. For ectomycorrhizas, the vermiculite/peat technique of Zak (1976) using Erlenmeyer flasks is recommended. This basic method has been modified for multiple plant synthesis by McKay (1982) and Mason *et al.* (1983).

Synthesis of VAM fungi can be performed just in sterile sand amended with the appropriate fungal spores, although more elaborate techniques have been developed. St John *et al.* (1981) used short-stemmed, glass filter funnels containing sand, kept moist by a cheesecloth wick dipping into nutrient solution, the whole being maintained in a sterile air flow. Macdonald (1981), in another method, used a small-scale hydroponic growth chamber, and grew plants on agar-coated glass strips inoculated with VAM spores. See also Hepper's (1981) methods for studying the effects of media on interactions between germinating spores of *Glomus* and clover roots, using agar slopes, strips of chromatography paper and Fahraeus slides.

Host response to mycorrhization can be determined from the usual growth parameters, such as height and weight of shoot and root/shoot ratios, in comparison with uninoculated plants. Comparisons of nutrient content present more problems, since control plants of the same size are not usually available. See Harley and Smith (1983) for further discussion of relevant methods.

As discussed by Fogel (1980) in his account of the role of mycorrhizas in nutrient cycling, there are considerable problems in attempting to convert measures of their activity and numbers or biomass into estimates of nutrient flux in a soil ecosystem. Roots have often been arbitrarily divided on the basis of diameter (coarse and fine roots) rather than on activity. Conversion of root tip numbers to biomass often involves the assumption that mycorrhizas produced by different fungi have the same weight(!), or the surface area is calculated (Muchovej *et al.*, 1987) without considering the biomass and functioning of the extraradical hyphae. These difficulties are, however, not easily overcome and present a challenge for the development of novel techniques.

It will have been obvious to the reader that this is not a chapter of mycological recipes but more a guide to the literature. As in other branches of ecology, a particularly flexible attitude is needed, and in most investigations each method has to be tailored to the circumstances. In addition, a range of comparative techniques must often be used to offset their many limitations when applied to field material. In both autecological studies and at the community level, evidence frequently has to be pieced together from several sources.

Although this chapter is confined to mycological methods, other organisms in soil and litter cannot be disregarded. A multidisciplinary approach to the interactions between fungi, fauna and other microorganisms may be crucial to the success of an ecological investigation (see Anderson *et al.*, 1984; Visser, 1985).

Owing to the fast pace of technology, more sophisticated instrumentation is continuously coming onto the market, speeding up and refining some of the analytical methods mentioned. Although bacteriologists are in advance, DNA probes and genetic engineering are also likely before long to have a major impact on methods in fungal ecology. However, alongside these advances to come, straightforward observation of individual hyphae and mycelia will surely continue to be of special value.

## References


Abbot, L. K., Robson, A. D. and de Boer, G. (1984). *New Phytol.* **97,** 437–446.
Abuzinadah, R. A. and Read, D. J. (1986a). *New Phytol.* **103,** 481–493.
Abuzinadah, R. A. and Read, D. J. (1986b) *New Phytol.* **103,** 495–506.
Adams, T. J. H., Williams, E. N. D., Todd, N. K. and Rayner, A. D. M. (1984). *Trans. Br. Mycol. Soc.* **82,** 359–361.
Agerer, R. (1986). *Mycotaxon* **26,** 473–492.
Agerer, R. (1987). "Colour Atlas of Ectomycorrhizae", Eindhorn-Verlag, Munich.
Aldwell, F. E. B. and Hall, I. R. (1986). *Trans. Br. Mycol. Soc.* **87,** 131–134.
Aldwell, F. E. B., Hall, I. R. and Smith, J. M. B. (1983). *Soil Biol. Biochem.* **15,** 377–378.
Alexander, I. J. and Hardy, K. (1981). *Soil Biol. Biochem.* **13,** 301–305.
Allen, S. E. Ed. (1989). "Chemical Analysis of Ecological Materials" 2nd edn., Blackwell Scientific Publications, Oxford.
Allen, S. E., Grimshaw, H. M., Parkinson, J. A. and Quarmby, C. (1974). "Chemical Analysis of Ecological Materials", Blackwell Scientific Publications, Oxford.
Ammirati, S., Ammirati, J. and Bledsoe, C. (1987). *In* "Mycorrhizae in the Next Decade: Practical Applications and Research Priorities", Proceedings 7th North American Conference on Mycorrhizae (D. M. Sylvia, L. L. Hung and J. H. Graham, Eds), p. 81, Institute of Food and Agricultural Science, University of Florida, Gainesville, Florida.
An, Z. Q. and Hendrix, J. W. (1988). *Mycologia* **80,** 259–261.
Anderson, J. M. (1978). *J. Biol. Educ.* **12,** 82–88.
Anderson, J. M. and Ineson, P. (1982). *Soil Biol. Biochem.* **14,** 415–416.
Anderson, J. M., Rayner, A. D. M. and Walton, D. W. H. (Eds) (1984). "Invertebrate–Microbial Interactions", Cambridge University Press, Cambridge.
Anderson, J. P. E. and Domsch, K. H. (1973a). *Ecol. Res. Comm. Bull. (Stockholm)* **17,** 281–282.
Anderson, J. P. E. and Domsch, K. H. (1973b). *Arch. Mikrobiol.* **93,** 113–127.
Anderson, J. P. E. and Domsch, K. H. (1974). *Ann. Microb.* **24,** 189–194.
Anderson, J. P. E. and Domsch, K. H. (1975). *Can. J. Microbiol.* **21,** 314–322.
Anderson, J. P. E. and Domsch, K. H. (1978). *Soil Biol. Biochem.* **10,** 215–221.
Anderson, J. R. and Slinger, J. M. (1975). *Soil Biol. Biochem.* **7,** 205–209.
Arias, I., Sainz, M. J., Grace, C. A. and Hayman, D. S. (1987). *Trans. Br. Mycol. Soc.* **89,** 128–131.
Aristovskaya, T. V. and Parinkina, O. M. (1961). *Soviet Soil Sci.* **1,** 12–20.
Arnolds, E. (Ed.) (1985). *Wetenschappelijke Mededeling van de Koninklijke Nederlandse Natuurhistorische Vereniging* **167,** Hoogwoud.
Arnolds, E. (1988). *Trans. Br. Mycol. Soc.* **90,** 391–406.
Bååth, E. (1988a). *Soil Biol. Biochem.* **20,** 123–125.
Bååth, E. (1988b) *Can. J. Bot.* **66,** 1566–1569.
Bååth, E. and Söderström, B. (1979). *Oikos* **33,** 11–14.
Bååth, E. and Söderström, B. (1980). *Soil Biol. Biochem.* **12,** 385–387.
Bååth, E. and Söderström, B. (1988). *Soil Biol. Biochem.* **20,** 403–404.
Badcock, E. C. (1943). *Trans. Br. Mycol. Soc.* **26,** 127–132.
Baker, J. H. (1981). *In* "Microbial Ecology of the Phylloplane" (J. P. Blakeman, Ed.), pp. 1–14, Academic Press, London.
Baker, J. H. (1988). *In* "Methods in Aquatic Bacteriology" (B. Austin, Ed.), pp. 171–191, John Wiley, Chichester.

Bakken, L. R. and Olsen, R. A. (1983). *Appl. Environ. Microbiol.* **45,** 1188–1195.
Barrett, D. K. (1978). *Trans. Br. Mycol. Soc.* **71,** 507–508.
Bartlett, E. M. and Lewis, D. H. (1973). *Soil Biol. Biochem.* **5,** 249–257.
Becker, W. N. and Gerdemann, J. W. (1977). *New Phytol.* **78,** 289–295.
Bethlenfalvay, G. J. and Ames, R. N. (1987). *Soil Sci. Soc. Am. J.* **51,** 834–837.
Bethlenfalvay, G. J., Pacovsky, R. S. and Brown, M. S. (1981). *Soil Sci. Soc. Am. J.* **45,** 871–875.
Bethlenfalvay, G. J., Pacovsky, R. S., Brown, M. S. and Fuller, G. (1982). *Plant Soil* **68,** 43–54.
Biermann, B. and Linderman, R. G. (1981). *New Phytol.* **87,** 63–67.
Bingle, W. H. and Paul, E. A. (1986). *Can. J. Microbiol.* **32,** 62–66.
Bissett, J. and Widden, P. (1972). *Can. J. Microbiol.* **18,** 1399–1404.
Blackmer, A. M., Baker, J. H. and Weeks, M. E. (1974). *Soil Sci. Soc. Am. J.* **38,** 689–691.
Blair, J. M., Crossley, D. A. Jr. and Rider, S. (1989). *Soil Biol. Biochem.* **21,** 507–510.
Blanchette, R. A. (1980). *Can. J. Bot.* **58,** 1496–1503.
Blanchette, R. A. (1982). *Plant Dis.* **66,** 394–397.
Blanchette, R. A. (1983). *Mycologia* **75,** 552–556.
Boddy, L. (1983). *Soil Biol. Biochem.* **15,** 501–510.
Boddy, L. and Lloyd, D. (1990). *In* "Nutrient Cycling in Terrestrial Ecosystems: Field Methods, Application and Interpretation" (A. F. Harrison, P. Ineson and O. W. Heal, Eds), pp. 139–152, Elsevier, London.
Boddy, L. and Rayner, A. D. M. (1983). *New Phytol.* **93,** 177–188.
Boddy, L. and Rayner, A. D. M. (1984). *Trans. Br. Mycol. Soc.* **82,** 501–505.
Boddy, L., Gibbon, O. M. and Grundy, M. A. (1985). *Trans. Br. Mycol. Soc.* **85,** 201–211.
Boddy, L., Wimpenny, J. W. T. and Harvey, R. D. (1989). *Mycol. Res.* **93,** 106–109.
Bohlool, B. B. and Schmidt, E. L. (1980). *In* "Advances in Microbial Ecology" (M. Alexander, Ed.), Vol. 4, pp. 203–236.
Booth, C. (Ed.) (1971a). "Methods in Microbiology", Vol. 4, Academic Press, London.
Booth, C. (1971b). *In* "Methods in Microbiology" (C. Booth, Ed.), Vol. 4, pp. 1–47, Academic Press, London.
Booth, C. (1971c). *In* "Methods in Microbiology" (C. Booth, Ed.), Vol. 4, pp. 49–94, Academic Press, London.
Breuil, C., Seifert, K. A., Yamada, J., Rossignol, L. and Saddler, J. N. (1988). *Can. J. For. Res.* **18,** 374–377.
Brock, T. D. (1987). *In* "Ecology of Microbial Communities" (M. Fletcher, T. R. G. Gray and J. G. Jones, Eds), pp. 1–17, Cambridge University Press, Cambridge.
Brownlee, C. and Jennings, D. H. (1981). *Trans. Br. Mycol. Soc.* **77,** 615–619.
Brundrett, M. C., Piché, Y. and Peterson, R. L. (1984). *Can. J. Bot.* **62,** 2128–2134.
Burges, A. (1950). *Trans. Br. Mycol. Soc.* **33,** 142–147.
Burns, R. G. (1978). "Soil Enzymes", Academic Press, London.
Butcher, J. A. (1971). *Mat. Org.* **6,** 209–232.
Callow, J. A., Capaccio, L. C. M., Parish, G. and Tinker, P. B. (1978). *New Phytol.* **80,** 125–134.
Campbell, R. and Porter, R. (1982). *Soil Biol. Biochem.* **14,** 241–245.
Bakerspigel, A. and Miller, J. J. (1953). *Soil Sci.* **76,** 123–126.

Cartwright, K. St. G. and Findlay, W. P. K. (1958). "Decay of Timber and its Prevention", 2nd Edn, HMSO, London.
Cawse, P. A. (1969). U.K.A.E.A. Research Report AERE-R 6061, HMSO, London.
Cawse, P. A. (1975). *In* "Soil Biochemistry" (E. A. Paul and A. D. McLaren, Eds), Vol. 3, pp. 213–267, Marcel Dekker, New York.
Chacko, C. I. and Lockwood, J. L. (1966). *Phytopathology* **56,** 576–577.
Chapela, I. H. and Boddy, L. (1988a). *New Phytol.* **110,** 39–45.
Chapela, I. H. and Boddy, L. (1988b). *New Phytol.* **110,** 47–57.
Chapela, I. H., Boddy, L. and Rayner, A. D. M. (1988). *FEMS Microbiol. Ecol.* **53,** 59–70.
Chapman, S. B. (1971). *Oikos* **22,** 348–353.
Chapman, S. B. (1986). *In* "Methods in Plant Ecology" (P. D. Moore and S. B. Chapman, Eds), 2nd Edn, pp. 1–59, Blackwell Scientific Publications, Oxford.
Chard, J. M., Gray, T. R. G. and Frankland, J. C. (1983). *Trans. Br. Mycol. Soc.* **81,** 503–511.
Chard, J. M., Gray, T. R. G. and Frankland, J. C. (1985a). *Trans. Br. Mycol. Soc.* **84,** 235–241.
Chard, J. M., Gray, T. R. G. and Frankland, J. C. (1985b). *Trans. Br. Mycol. Soc.* **84,** 243–249.
Christensen, M. (1969). *Ecology* **50,** 9–27.
Christensen, O. (1984). *Oikos* **42,** 211–219.
Chu-Chou, M. (1979). *Soil Biol. Biochem.* **11,** 557–562.
Chu-Chou, M. and Grace, L. (1981). *Soil Biol. Biochem.* **13,** 247–249.
Clark, G. (Ed.) (1981). "Staining Procedures", 4th Edn, Williams and Wilkins, Baltimore.
Clipson, N. J. W., Cairney, J. W. G. and Jennings, D. H. (1987). *New Phytol.* **105,** 449–457.
Coleman, D. C. (1985). *In* "Ecological Interactions in Soil. Plants, Microbes and Animals" (A. H. Fitter, D. Atkinson, D. J. Read and M. B. Usher, Eds), pp. 1–21, Blackwell Scientific Publications, Oxford.
Coleman, D. C. and McGinnis, J. T. (1970). *Oikos* **21,** 134–137.
Cooke, R. C. and Rayner, A. D. M. (1984). "Ecology of Saprotrophic Fungi", Longman, London.
Cotter, H. Van T. and Bills, G. F. (1985). *Trans. Br. Mycol. Soc.* **85,** 520–524.
Daniels, B. A. and Skipper, H. D. (1982). *In* "Methods and Principles of Mycorrhizal Research" (N. C. Schenck, Ed.), pp. 29–35, Am. Phytopath. Soc., St Paul, Minnesota.
Darken, M. A. (1961). *Science, N.Y.* **133,** 1704–1705.
Darken, M. A. (1962). *Appl. Microbiol.* **10,** 387–393.
Deacon, J. W. and Henry, C. M. (1978). *Soil Biol. Biochem.* **10,** 409–415.
Denny, H. J. and Wilkins, D. A. (1988a). *New Phytol.* **106,** 525–534.
Denny, H. J. and Wilkins, D. A. (1988b). *New Phytol.* **106,** 545–553.
Dewey, F. M. and Brasier, C. M. (1988). *Plant Pathol.* **37,** 28–35.
Dewey, F. M., Barrett, D. K., Vose, I. R. and Lamb, C. J. (1984). *Phytopathology* **74,** 291–296.
Dickinson, C. H. and Pugh, G. J. F. (Eds) (1974). "Biology of Plant Litter Decomposition" Vols 1 and 2, Academic Press, London.
Dighton, J. (1983). *Plant Soil* **71,** 455–462.
Dighton, J., Mason, P. A. and Harrison, A. F. (1990). *In* "Nutrient Cycling in

Terrestrial Ecosystems: Field Methods, Application and Interpretation" (A. F. Harrison, P. Ineson and O. W. Heal, Eds), pp. 389–399, Elsevier, London.
Dighton, J., Poskitt, J. M. and Howard, D. (1986). *Trans Br. Mycol. Soc.* **87,** 163–171.
Dighton, J., Thomas, E. D. and Latter, P. M. (1987). *Biol. Fert. Soils* **4,** 145–150.
Domsch, K. H., Beck, Th., Anderson, J. P. E., Söderström, B., Parkinson, D. and Trolldenier, G. (1979). *Z. Pflanzenernaehr. Bodenkd.* **142,** 520–533.
Dowding, P. (1978). *Ann. Appl. Biol.* **89,** 166–171.
Dowson, C. G., Boddy, L. and Rayner, A. D. M. (1989b). *Mycol. Res.* **92,** 383–391.
Dowson, C. G., Rayner, A. D. M. and Boddy, L. (1986). *J. Gen. Microbiol.* **132,** 203–211.
Dowson, C. G., Rayner, A. D. M. and Boddy, L. (1988a). *New Phytol.* **109,** 335–341.
Dowson, C. G., Rayner, A. D. M. and Boddy, L. (1988b). *New Phytol.* **109,** 343–349.
Dowson, C. G., Rayner, A. D. M. and Boddy, L. (1988c). *New Phytol.* **109,** 423–432.
Dowson, C. G., Rayner, A. D. M. and Boddy, L. (1989a). *New Phytol.* **111,** 699–705.
Draggan, S. (1976). *Appl. Environ. Microbiol.* **31,** 313–315.
Dring, D. M. (1971). *In* "Methods in Microbiology" (C. Booth, Ed.), Vol. 4, pp. 95–111. Academic Press, London.
Eggins, H. O. W. and Pugh, G. J. F. (1962). *Nature, London* **193,** 94–95.
Elmholt, S. and Kjøller, A. (1987). *Soil Biol. Biochem.* **19,** 679–682.
Eren, J. and Pramer, D. (1966). *Soil Sci.* **101,** 39–45.
Evans, E. (1955). *Trans. Br. Mycol. Soc.* **38,** 335–346.
Fægri, A., Torsvik, V. L. and Goksøyr, J. (1977). *Soil Biol. Biochem.* **9,** 105–112.
Ferriss, R. S. (1984). *Phytopathology* **74,** 121–126.
Finlay, R. D. and Read, D. J. (1986). *In* "Physiological and Genetical Aspects of Mycorrhizae" (V. Gianinazzi-Pearson and S. Gianinazzi, Eds), pp. 351–355, Inst. Nat. de Recherche Agron., Paris.
Flanagan, P. W. and Bunnell, F. L. (1976). *In* "The Role of Terrestrial and Aquatic Organisms in Decomposition Processes" (J. M. Anderson and A. Macfadyen, Eds), pp. 437–457, Blackwell Scientific Publications, Oxford.
Fogel, R. (1980). *New Phytol.* **86,** 199–212.
Fortin, J. A., Piché, Y. and Lalonde, M. (1980). *Can. J. Bot.* **58,** 361–365.
Foster, R. C. (1981). *New Phytol.* **88,** 705–712.
Frankland, J. C. (1966). *J. Ecol.* **54,** 41–63.
Frankland, J. C. (1969). *J. Ecol.* **57,** 25–36.
Frankland, J. C. (1974). *Soil Biol. Biochem.* **6,** 409–410.
Frankland, J. C. (1975a). *Soil Biol. Biochem.* **7,** 339–340.
Frankland, J. C. (1975b). *In* "Biodégradation et Humification" (G. Kilbertus *et al.*, Eds), pp. 33–40, Pierron: Sarreguemines.
Frankland, J. C. (1982). *In* "Decomposer Basidiomycetes: their Biology and Ecology" (J. C. Frankland, J. N. Hedger and M. J. Swift, Eds), pp. 241–261, Cambridge University Press, Cambridge.
Frankland, J. C. (1984). *In* "The Ecology and Physiology of the Fungal Mycelium" (D. H. Jennings and A. D. M. Rayner, Eds), pp. 241–260, Cambridge University Press, Cambridge.
Frankland, J. C. and Collins, V. G. (1974). *Soil Biol. Biochem.* **6,** 125–126.

Frankland, J. C. and Harrison (1985). *New Phytol.* **101,** 133–151.
Frankland, J. C. and Latter, P. M. (1973). Merlewood Research and Development Paper No. 43, The Nature Conservancy, Merlewood Research Sta., Grange-over-Sands, UK.
Frankland, J. C., Bailey, A. D. and Costeloe, P. L. (1979). Ann. Rep. Inst. Terres. Ecol. 1978, pp. 26–28., Inst. Terres. Ecol., Natural Environment Research Council, Cambridge.
Frankland, J. C., Lindley, D. K. and Swift, M. J. (1978). *Soil Biol. Biochem.* **10,** 323–333.
Frankland, J. C., Bailey, A. D., Gray, T. R. G. and Holland, A. A. (1981). *Soil Biol. Biochem.* **13,** 87–92.
Fries, N. (1987). *New Phytol.* **107,** 735–739.
Gams, W., van der Aa, H. A., van der Plaats-Niterink, A. J., Samson, R. A. and Stalpers, J. A. (1987). "CBS Course of Mycology", 3rd Edn, Centraalbureau voor Schimmelcultures, Baarn, Netherlands.
Garrett, S. D. (1952). *Sci. Progr. London* **40,** 436–450.
Garrett, S. D. (1956). "Biology of Root-infecting Fungi", Cambridge University Press, Cambridge.
Garrett, S. D. (1970). "Pathogenic Root-infecting Fungi", Cambridge University Press, Cambridge.
Gianninazi-Pearson, V., Morandi, D., Dexheimer, J. and Gianninazi, S. (1981). *New Phytol.* **88,** 633–639.
Gibson, F., Fox, F. M. and Deacon, J. W. (1988). *New Phytol.* **108,** 189–204.
Gilbert, O. J. W. and Bocock, K. L. (1962). *In* "Progress in Soil Zoology" (P. W. Murphy, Ed.), pp. 348–352, Butterworths, London.
Gillespie, J., Latter, P. M. and Widden, P. (1988). *In* "Cotton Strip Assay: an Index of Decomposition in Soils" (A. F. Harrison, P. M. Latter and D. W. Walton, Eds), pp. 60–67, Institute of Terrestrial Ecology, Grange-over-Sands, UK.
Giovanetti, M. and Mosse, B. (1980). *New Phytol.* **84,** 489–500.
Goodell, B. S., Jellison, J. and Hösli, J. P. (1988). *Forest Prod. J.* **38,** 59–62.
Gore, A. J. P. (1970). Merlewood Research and Development Paper No. 18, The Nature Conservancy, Grange-over-Sands, UK.
Gramms, G. (1979). *Z. Mykol.* **45,** 195–208.
Grand, L. F. and Harvey, A. E. (1982). *In* "Methods and Principles of Mycorrhizal Research" (N. C. Schenck, Ed.), pp. 157–164, Am. Phytopath. Soc., St Paul, Minnesota.
Grant, W. D. and West, A. W. (1986). *Microbiol. Methods* **6,** 47–53.
Gray, T. R. G. (1967). *Science* **155,** 1668–1670.
Gray, T. R. G. and Bell, T. F. (1963). *In* "Soil Organisms" (J. Doeksen and J. van der Drift, Eds), pp. 222–230, North Holland Publishing Co., Amsterdam.
Gray, T. R. G. and Williams, S. T. (1971a). "Soil Micro-organisms", Oliver and Boyd, Edinburgh.
Gray, T. R. G. and Williams, S. T. (1971b). *In* "Microbial and Biological Productivity" (D. E. Hughes and A. H. Rose, Eds), Symp. Soc. Gen. Microbiol., Vol. 21, pp. 255–286.
Gray, T. R. G., Hisset, R. and Duxbury, T. (1974). *Rev. Ecol. Biol. Sol.* **10,** 15–26.
Griffin, D. M. (1960). *Trans. Br. Mycol. Soc.* **43,** 583–596.
Griffin, D. M. (1977). *Ann. Rev. Phytopathol.* **15,** 319–329.
Griffin, D. M. (1981). *Adv. Microbiol. Ecol.* **5,** 91–136.

Gurr, E. (1965). "The Rational Use of Dyes in Biology", Leonard Hill, London.
Hale, M. D. and Eaton, R. A. (1983). *In* "Biodeterioration" (T. A. Oxley and S. Barry, Eds), Vol. 5, pp. 54–63, Wiley-Interscience, Chichester.
Hale, M. D. and Eaton, R. A. (1985a). *Trans. Br. Mycol. Soc.* **84,** 277–288.
Hale, M. D. and Eaton, R. A. (1985b). *Mycologia* **77,** 447–463.
Hale, M. D. and Eaton, R. A. (1985c). *Mycologia* **77,** 594–605.
Hansen, P. A. (1988a). Ph.D. Dissertation, University of Lund.
Hansen, P. A. (1988b). *Vegetatio* **78,** 31–44.
Hansen, P. A. (1989). *Vegetatio* **82,** 69–78.
Hanssen, J. F. and Goksøyr, J. (1975). *In* "Fennoscandian Tundra Ecosystems", Part 1 (F. E. Wielgolaski, Ed.), pp. 239–243, Ecological Studies 16, Springer, Berlin.
Hanssen, J. F., Thingstad, T. F. and Goksøyr, J. (1974). *Oikos* **25,** 102–107.
Hardham, A. R., Suzaki, E. and Perkin, J. L. (1986). *Can. J. Bot.* **64,** 311–321.
Harley, J. L. (1969). "The Biology of Mycorrhiza", 2nd Edn, Leonard Hill, London.
Harley, J. L. and McCready, C. C. (1981). *New Phytol.* **88,** 675–681.
Harley, J. L. and Smith, S. E. (1983). "Mycorrhizal Symbiosis", Academic Press, London.
Harley, J. L. and Waid, J. S. (1955). *Trans. Br. Mycol. Soc.* **38,** 104–118.
Harrison, A. F. and Pearce, H. T. (1979). *Soil Biol. Biochem.* **11,** 405–410.
Harrison, A. F., Latter, P. M. and Walton, D. W. (Eds) (1988), "Cotton Strip Assay: an Index of Decomposition in Soils", Institute of Terrestrial Ecology, Grange-over-Sands, UK.
Harvey, A. E., Larsen, M. J. and Jurgensen, M. F. (1976). *For. Sci.* **22,** 393–398.
Haug, I. (1988). *In* "Ectomycorrhiza and Acid Rain" (A. E. Jansen, J. Dighton and A. H. M. Bresser, Eds), pp. 32–46, Commission European Community, Bilthoven, Air Pollution Research Report 12.
Hawksworth, D. L. (1974). "Mycologists' Handbook", Commonwealth Mycological Institute, Kew.
Hawksworth, D. L., Sutton, B. C. and Ainsworth, G. C. (1983). "Ainsworth and Bisby's Dictionary of the Fungi", 7th Edn, Commonwealth Mycological Institute, Kew.
Heal, O. W. and MacLean, S. F., Jr (1975). *In* "Unifying Concepts in Ecology" (W. H. van Dobben and R. H. Lowe-McConnell, Eds), pp. 89–108, Proc. 1st Int. Congr. Ecology, Centre Agric. Publishing and Documentation, Wageningen.
Healey, I. N. and Swift, M. J. (1971). *In* "Organismes du Sol et Production primaire", Proc. 4th Coll. Soil Zool., pp. 417–430, Inst. Nat. de Recherche Agron., Paris.
Hepper, C. M. (1977). *Soil Biol. Biochem.* **9,** 15–18.
Hepper, C. M. (1981). *New Phytol.* **88,** 641–647.
Hering, T. F. (1966a). *Plant Soil* **25,** 195–200.
Hering, T. F. (1966b). *Trans. Br. Mycol. Soc.* **50,** 267–273.
Hering, T. F. (1967). *Trans. Br. Mycol. Soc.* **50,** 267–273.
Hering, T. F. and Nicholson, P. B. (1964). *Nature, Lond.* **201,** 942–943.
Ho, I. and Zak, B. (1979). *Can. J. Bot.* **57,** 1203–1205.
Holden, M. (1982). *Bull. Br. Mycol. Soc.* **16,** 36–55, 92–112.
Holder-Franklin, M. A. (1986). *In* "Microbial Autecology. A Method for Environmental Studies" (R. L. Tate III, Ed.), pp. 93–132, John Wiley, New York.
Howard, P. J. A. (1988). *In* "Cotton Strip Assay: an Index of Decomposition in

Soils" (A. F. Harrison, P. M. Latter and D. W. H. Walton, Eds), pp. 34–42, Institute of Terrestrial Ecology, Grange-over-Sands, UK.
Howard, P. J. A. and Frankland, J. C. (1974). *Soil Biol. Biochem.* **6,** 117–123.
Hunt, R. S. and Cobb, F. W., Jr (1971). *Can. J. Bot.* **49,** 2064–2065.
Hutchinson, K. J. and King, K. L. (1989). *Aust. Ecol.* **14,** 157–167.
Hutchinson, K. J., King, K. L., Nicol, G. R. and Wilkinson, D. R. (1990). *In* "Nutrient Cycling in Terrestrial Ecosystems: Field Methods, Application and Interpretation" (A. F. Harrison, P. Ineson and O. W. Heal, Eds), pp. 291–314, Elsevier, London.
Ibbotson, R. and Pugh, G. J. F. (1975). *Mycopathologia* **56,** 119–123.
Ikediugwu, F. E. O. and Webster, J. (1970a). *Trans. Br. Mycol. Soc.* **54,** 180–204.
Ikediugwu, F. E. O. and Webster, J. (1970b). *Trans. Br. Mycol. Soc.* **54,** 205–210.
Ineson, P. and Anderson, J. M. (1982). *Soil Biol. Biochem.* **14,** 607–608.
Ineson, P. and Gray, T. R. G. (1980). *In* "Microbial Growth and Survival in Extremes of Environment" (G. W. Gould and J. E. L. Corry, Eds), pp. 21–26, Academic Press, London.
Ineson, P., Leonard, M. A. and Anderson, J. M. (1982). *Soil Biol. Biochem.* **14,** 601–605.
Ingham, E. R. and Klein, D. A. (1984a). *Soil Biol. Biochem.* **16,** 279–280.
Ingham, E. R. and Klein, D. A. (1984b). *Soil Biol. Biochem.* **16,** 273–278.
Jeffers, J. N. R. (1978). "Design of Experiments. Statistical Checklist 1", Institute of Terrestrial Ecology, Grange-over-Sands, UK.
Jeffers, J. N. R. (1979). "Sampling. Statistical Checklist 2", Institute of Terrestrial Ecology, Grange-over-Sands, UK.
Jellison, J. and Goodell, B. S. (1988). *Biomass* **15,** 109–116.
Jeng, D. K. H., Kazmarek, K. A., Woodworth, A. G. and Balasky, G. (1987). *Appl. Environ. Microbiol.* **53,** 2133–2137.
Jenkinson, D. S. (1966). *J. Soil Sci.* **17,** 280–302.
Jenkinson, D. S. and Ladd, J. N. (1981). *In* "Soil Biochemistry" (E. A. Paul and J. N. Ladd, Eds), Vol. 5, pp. 415–471, Marcel Dekker, New York.
Jenkinson, D. S. and Powlson, D. S. (1976). *Soil Biol. Biochem.* **8,** 209–213.
Jenkinson, D. S., Powlson, D. S. and Wedderburn, R. W. M. (1976). *Soil Biol. Biochem.* **8,** 189–202.
Jennings, D. H. (1982). *In* "Decomposer Basidiomycetes: their Biology and Ecology" (J. C. Frankland, J. N. Hedger and M. J. Swift, Eds), pp. 91–108, Cambridge University Press, Cambridge.
Jennings, D. H. (1987). *Trans. Br. Mycol. Soc.* **89,** 1–11.
Jennings, D. H. (1990). *In* "Nutrient Cycling in Terrestrial Ecosystems: Field Methods, Application and Interpretation" (A. F. Harrison, P. Ineson and O. W. Heal, Eds), pp. 233–245, Elsevier, London.
Jennings, D. H. and Rayner, A. D. M. (Eds) (1984). "The Ecology and Physiology of the Fungal Mycelium", Cambridge University Press, Cambridge.
Johansson, M. and Stenlid, J. (1985). *Eur. J. For. Pathol.* **15,** 32–45.
Johnson, L. F. and Curl, E. A. (1972). "Methods for Research on the Ecology of Soil-borne Plant Pathogens", Burgess Publishing Co., Minneapolis.
Jones, P. C. T. and Mollison, J. E. (1948). *J. Gen. Microbiol.* **2,** 54–69.
Keyworth, W. G. (1951). *Trans. Br. Mycol. Soc.* **34,** 291–292.
Kinden, D. A. and Brown, M. F. (1975). *Phytopathology* **65,** 74–76.
Kirby, J. J. H. (1987). *Trans. Br. Mycol. Soc.* **88,** 559–562.

Kirsop, B. E. and Snell, J. J. S. (Eds) (1984). "Maintenance of Microorganisms. A Manual of Laboratory Methods", Academic Press, London.
Kjøller, A. and Struwe, S. (1982). *Oikos* **39,** 389–422.
Köhler, G. and Milstein, C. (1975). *Nature* **256,** 495–497.
Köhler, G., Hengartner, H. and Shulman, M. J. (1978). *Eur. J. Immunol.* **8,** 82–88.
Kormanik, P. P. and McGraw, A. C. (1982). *In* "Methods and Principles of Mycorrhizal Research" (N. C. Schenck, Ed.), pp. 37–45, Am. Phytopath. Soc., St Paul, Minnesota.
Kormanik, P. P., Bryan, W. C. and Schultz, R. C. (1980). *Can. J. Microbiol.* **26,** 536–538.
Kough, J. L. and Linderman, R. G. (1986). *Soil Biol. Biochem.* **18,** 309–313.
Kough, J., Malajczuk, N. and Linderman, R. G. (1983). *New Phytol.* **94,** 57–62.
Kubiena, W. L. (1932). *Arch. Mikrobiol.* **3,** 507–542.
Kubiena, W. L. (1938). "Micropedology", Collegiate Press, Inc., Ames, Iowa.
Latter, P. M. and Walton, D. W. H. (1988). *In* "Cotton Strip Assay: an Index of Decomposition in Soils" (A. F. Harrison, P. M. Latter, and D. W. H. Walton, Eds), pp. 7–10, Institute of Terrestrial Ecology, Grange-over-Sands, UK.
Leonard, M. A. and Anderson, J. M. (1981). *Soil Biol. Biochem.* **13,** 547–549.
Lingappa, B. T. and Lockwood, J. L. (1963). *Phytopathology* **53,** 529–531.
Linkins, A. E. and Antibus, R. K. (1981). *In* "Arctic and Alpine Mycology" (G. A. Laursen and J. F. Ammirati, Eds), pp. 509–531, University of Washington Press, Washington.
Ljungholm, K., Noren, B., Sköld, R. and Wadsö, I. (1979). *Oikos* **33,** 15–23.
Lodge, D. J. (1987). *Soil Biol. Biochem.* **19,** 727–733.
Lohwag, K. (1937). *Microchemie* **23,** 198–202.
Lung-Escarment, B., Mohammed, C. and Dunez, J. (1985). *Eur. J. For. Pathol.* **15,** 278–288.
Macauley, B. J. and Waid, J. S. (1981). *In* "The Fungal Community. Its Organisation and Role in the Ecosystem" (D. T. Wicklow and G. C. Carroll, Eds), pp. 501–531, Marcel Dekker, New York.
Macdonald, R. M. (1981). *New Phytol.* **89,** 87–93.
Macdonald, R. M. (1986). *Biol. Agric. Hortic.* **3,** 361–365.
Magan, N. and Lacey, J. (1984a). *Trans. Br. Mycol. Soc.* **82,** 83–93.
Magan, N. and Lacey, J. (1984b). *Trans. Br. Mycol. Soc.* **82,** 305–314.
Maloy, O. C. (1974). *Plant Dis. Reporter* **58,** 902–904.
Martens, R. (1987). *Soil Biol. Biochem.* **19,** 77–81.
Martin, J. P. (1950). *Soil Sci.* **69,** 215–232.
Mason, P. A., Dighton, J., Last, F. T. and Wilson, J. (1983). *For. Ecol. Manage.* **5,** 47–53.
Mason, P. A., Last, F. T., Pelham, J. and Ingleby, K. (1982). *For. Ecol. Manage.* **4,** 19–37.
Matcham, S. E., Jordan, B. R. and Wood, D. A. (1984). *In* "Microbiological Methods for Environmental Biotechnology" (J. M. Grainger and J. M. Lynch, Eds), pp. 5–18, Soc. Appl. Bacteriol., Technical Series. Academic Press, London.
McKay, H. M. (1982). *Plant Soil* **66,** 257–262.
Mercer, P. C. (1979). *Ann. Appl. Biol.* **91,** 107–112.
Messner, K. and Stachelberger, H. (1984). *Trans. Br. Mycol. Soc.* **83,** 113–130.
Miller, O. K. (1982). *Holarct. Ecol.* **5,** 125–134.

Millner, P. D. (1987). *In* "Mycorrhizae in the Next Decade: Practical Applications and Research Priorities", Proceedings 7th North American Conference on Mycorrhizae (D. M. Sylvia, L. L. Hung and J. H. Graham, Eds), p. 211, Institute of Food and Agricultural Science, University of Florida, Gainesville, Florida.

Mitchell, L. A. (1985). "Monoclonal Antibodies and Immunochemical Techniques: Applications in Forestry Research", Information Report, Pacific Forest Research Centre, No. BC-X-258, Canadian Forestry Service, Victoria, B.C.

Molina, R. and Palmer, J. G. (1982). *In* "Methods and Principles of Mycorrhizal Research" (N. C. Schenck, Ed.), pp. 115–129, American Phytopath. Soc., St Paul, Minnesota.

Moore, R. L., Bassett, B. B. and Swift, M. J. (1979). *Soil Biol. Biochem.* **11,** 311–312.

Morrell, J. J., Gibson, D. G. and Krahmer, R. L. (1985). *Phytopathology* **75,** 329–332.

Moser, M. (1983). *In* "Kleine Kryptogamenflora", 5th Edn (H. Gams, Ed.), Vol. 26, pp. 1–533, G. Fischer, Stuttgart. English translation to 4th Edn, "Keys to Agarics and Boleti" by S. Plant (G. Kibby, Ed.), R. Phillips, London.

Muchovej, J. J., Muchovej, R. M. C. and Marx, D. H. (1987). *In* "Mycorrhizae in the Next Decade: Practical Applications and Research Priorities", Proceedings 7th North American Conference on Mycorrhizae (D. M. Sylvia, L. L. Hung and J. H. Graham, Eds), p. 213, Institute of Food and Agricultural Science, University of Florida, Gainesville, Florida.

Nagel-de Boois, H. M. (1971). "Proceedings of the 4th Colloquium of the Soil Zool. Comm. of the ISSS", pp. 447–454, Inst. Nat. de la Recherche Agronomique, Dijon.

Nagel-de Boois, H. M. and Jansen, E. (1971). *Rev. Ecol. Biol. Sol.* **8,** 509–520.

Nannipieri, P. (1984). *In* "Current Perspectives in Microbial Ecology" (M. J. Klug and C. A. Reddy, Eds), pp. 515–521, American Society of Microbiology, Washington.

Newell, K. (1984a). *Soil Biol. Biochem.* **16,** 227–233.

Newell, K. (1984b). *Soil Biol. Biochem.* **16,** 235–239.

Newell, K., Frankland, J. C. and Whittaker, J. B. (1987). *Biol. Fert. Soils* **3,** 11–13.

Newell, S. Y. and Statzell-Tallman, A. (1982). *Oikos* **39,** 261–268.

Newell, S. Y., Arsuffi, T. L. and Fallon, R. D. (1988). *Appl. Environ. Microbiol.* **54,** 1876–1879.

Newman, E. I. (1966). *J. Appl. Ecol.* **3,** 139–145.

Nicholas, D. P. and Parkinson, D. (1967). *Pedobiologia* **7,** 23–41.

Nicholas, D. P., Parkinson, D. and Burges, N. A. (1965). *J. Soil Sci.* **16,** 258–269.

Nobles, M. K. (1965). *Can. J. Bot.* **43,** 1097–1139.

Nordbring-Hertz, B. (1984). *In* "The Ecology and Physiology of the Fungal Mycelium" (D. H. Jennings and A. D. M. Rayner, Eds), pp. 419–432, Cambridge University Press, Cambridge.

Nordbring-Hertz, B. and Mattiasson, B. (1979). *Nature, London* **281,** 477–479.

Nout, M. J. R., Bonants-van Laarhoven, T. M. G., de Jongh, P. and de Koster, P. G. (1987). *Appl. Microb. Biotech.* **26,** 456–461.

Nye, P. H. and Tinker, P. B. (1977). "Solute Movement in the Soil–Root System", Blackwell, Oxford.

Olson, F. C. W. (1950). *Trans. Am. Microscop. Soc.* **59,** 272–279.

Pacovsky, R. S. and Bethlenfalvay, G. J. (1982). *Plant Soil* **68,** 143–147.

Papavizas, G. C. (1967). *Phytopathology* **57,** 848–852.
Park, D. (1973). *Trans. Br. Mycol. Soc.* **60,** 148–151.
Parkinson, D. and Waid, J. S. (Eds) (1960). "The Ecology of Soil Fungi", Liverpool University Press, Liverpool.
Parkinson, D. and Williams, S. T. (1961). *Plant Soil* **13,** 347–355.
Parkinson, D., Gray, T. R. G. and Williams, S. T. (1971). "Methods for Studying the Ecology of Soil Micro-organisms", International Biological Programme Handbook, No. 19, Blackwell Scientific Publications, Oxford.
Paul, E. A. and Johnson, R. L. (1977). *Appl. Environ. Microbiol.* **34,** 263–269.
Payton, M., McCullough, W. and Roberts, C. F. (1976). *J. Gen. Microbiol.* **94,** 228–233.
Pearce, R. B. (1984). *Trans. Br. Mycol. Soc.* **82,** 564–567.
Pearce, R. B. and Rutherford, J. (1981). *Physiol. Plant Pathol.* **19,** 359–369.
Perfiliev, B. V. and Gabe, D. R. (1969). "Capillary Methods of Studying Micro-organisms", Oliver and Boyd, Edinburgh.
Petrini, O. and Fisher, P. J. (1988). *Trans. Br. Mycol. Soc.* **91,** 233–238.
Phillips, J. M. and Hayman, D. S. (1970). *Trans. Br. Mycol. Soc.* **55,** 158–161.
Phillipson, J. (1964). *Oikos* **15,** 130–139.
Piché, Y., Fortin, J. A. and Lafontaine, J. G. (1981). *New Phytol.* **88,** 695–703.
Pochon, J. and Tardieux, P. (1962). "Techniques d'Analyse en Microbiologie du Sol", Editions de la Tourelle, St Mande, Seine.
Poincelot, R. P. and Day, P. R. (1972). *Appl. Microbiol.* **23,** 875–879.
Poppe, J. and Welvaert, W. (1983). *Med. Fac. Landbouww. Rijksuniv. Gent.* **48/3,** 901–912.
Porter, W. M. (1979). *Aust. J. Soil. Res.* **17,** 515–519.
Preston, K. J. and Waid, J. S. (1972). *Trans. Br. Mycol. Soc.* **59,** 151–153.
Punithalingam, E. (1989). *In* "Novel Approaches to the Systematics of Fungi" (S. L. Jury and P. F. Cannon, Eds), *Bot. J. Linn. Soc.* **99,** 19–32.
Putman, R. J. (1976). *J. Appl. Ecol.* **13,** 445–452.
Rastin, N., Rosenplänter, K. and Hüttermann, A. (1988). *Soil Biol. Biochem.* **20,** 637–642.
Rayner, A. D. M. and Boddy, L. (1988a). "Fungal Decomposition of Wood: its Biology and Ecology", John Wiley, Chichester.
Rayner, A. D. M. and Boddy, L. (1988b). *Adv. Microb. Ecol.* **10,** 115–166.
Rayner, A. D. M. and Todd, N. K. (1979). *Adv. Bot. Res.* **7,** 333–420.
Rayner, A. D. M., Coates, D., Ainsworth, A. M., Adams, T. J. H., Williams, E. N. D. and Todd, N. K. (1984). *In* "The Ecology and Physiology of the Fungal Mycelium" (D. H. Jennings and A. D. M. Rayner, Eds), pp. 509–540, Cambridge University Press, Cambridge.
Read, D. J., Francis, R. and Finlay, R. D. (1985). *In* "Ecological Interactions in Soil. Plants, Microbes and Animals" (A. H. Fitter, D. Atkinson, D. J. Read and M. B. Usher, Eds), pp. 193–217, Blackwell Scientific Publications, Oxford.
Ride, J. P. and Drysdale, R. B. (1972). *Physiol. Plant Path.* **2,** 7–15.
Roser, D. J., Keane, P. J. and Pittaway, P. A. (1982). *Trans. Br. Mycol. Soc.* **79,** 321–329.
Russell, P. (1956). *Nature, Lond.* **177,** 1038–1039.
Saif, S. R. (1981). *New Phytol.* **88,** 649–659.
Salmanowicz, B. and Nylund, J. E. (1988). *Eur. J. For. Pathol.* **18,** 291–298.
Schenck, N. C. (Ed.) (1982). "Methods and Principles of Mycorrhizal Research", Am. Phytopath. Soc., St Paul, Minnesota.

Schenck, N. C. and Pérez, Y. (1988). "Manual for the Identification of VA Mycorrhizal Fungi", 2nd Edn, International Culture Collection of VA Mycorrhizal Fungi (INVAM), Gainesville, Florida.
Schmidt, E. L. (1973a). *In* "Modern Methods in the Study of Microbial Ecology" (T. Rosswall, Ed.), *Ecol. Res. Comm. Bull. (Stockholm)* **17,** 67–76.
Schmidt, E. L. (1973b). *Ecol. Res. Comm. Bull. (Stockholm)* **17,** 453–454.
Schmidt, E. L. and Bankole, R. O. (1962). *Science* **136,** 776–777.
Schmidt, E. L. and Bankole, R. O. (1963). *In* "Soil Organisms" (J. Doeksen and J. van der Drift, Eds), pp. 197–203, North Holland Publishing Co., Amsterdam.
Schmidt, E. L. and Bankole, R. O. (1965). *Appl. Microbiol.* **13,** 673–679.
Schmidt, E. L., Biesbrock, J. A. and Bohlool, B. B. (1974). *Can. J. Microbiol.* **20,** 137–139.
Schubert, A., Marzachi, C., Mazzitelli, M., Cravero, M. C. and Bonfante-Fasolo, P. (1987). *New Phytol.* **107,** 183–190.
Schüepp, H., Miller, D. D. and Bodmer, M. (1987). *Trans Br. Mycol. Soc.* **89,** 429–435.
Schulze, B. and Theden, G. (1938). *Holz als Roh-Werkstoff* **1,** 548–554.
Seitz, L. M., Sauer, D. B., Burroughs, R., Mohr, H. E. and Hubbard, J. D. (1979). *Phytopathology* **69,** 1202–1203.
Sen, R. and Hepper, C. M. (1986). *Soil. Biol. Biochem.* **18,** 29–34.
Seviour, R. J. and Chilvers, G. A. (1972). *New Phytol.* **71,** 1107–1110.
Seviour, R. J., Hamilton, D. and Chilvers, G. A. (1978). *New Phytol.* **80,** 153–156.
Sharma, P. D., Fisher, P. J. and Webster, J. (1977). *Trans. Br. Mycol. Soc.* **69,** 479–483.
Sims, R. W., Freeman, P. and Hawksworth, D. L. (Eds) (1988). "Key Works to the Fauna and Flora of the British Isles and North-western Europe", 5th Edn, Clarendon Press, Oxford.
Skinner, M. F. and Bowen, G. D. (1974). *Soil Biol. Biochem.* **6,** 53–56.
Smith, D. and Onions, A. H. (1983). "The Preservation and Maintenance of Living Fungi", CAB International Mycological Institute, Kew.
Söderström, B. E. (1977). *Soil Biol. Biochem.* **9,** 59–63.
Söderström, B. E. (1979). *Soil Biol. Biochem.* **11,** 147–148.
Söderström, B. and Erland, S. (1986). *Trans. Br. Mycol. Soc.* **86,** 465–468.
Söderström, B. and Read, D. J. (1987). *Soil Biol. Biochem.* **19,** 231–236.
Söderström, B., Wold, S. and Blomquist, G. (1982). *J. Gen. Microbiol.* **128,** 1773–1784.
Sparling, G. P. (1985). *In* "Soil Organic Matter and Biological Activity" (D. Vaughan and R. E. Malcolm, Eds), pp. 223–262, Martinus Nijhoff/Dr W. Junk, Dordrecht.
Sparling, G. P. and Williams, B. L. (1986). *Soil Biol. Biochem.* **18,** 507–513.
Speir, T. W., Cowling, J. C., Sparling, G. P., West, A. W. and Speir, D. M. (1986). *Soil Biol. Biochem.* **18**, 377–382.
St John, T. V., Hayes, R. I. and Reid, C. P. P. (1981). *New Phytol.* **89,** 81–86.
Stalpers, J. A. (1978). *Studies in Mycology, Baarn* **16,** 1–248.
Strullu, D. G., Chamel, A., Eloy, J. F. and Gourret, J. P. (1983a). *New Phytol.* **94,** 81–88.
Strullu, D. G., Harley, J. L., Gourret, J. P. and Garrec, J. P. (1983b). *New Phytol.* **94,** 89–94.
Sundman, V. and Sivelä, S. (1978). *Soil Biol. Biochem.* **10,** 399–401.

Swift, M. J. (1973a). *In* "Modern Methods in the Study of Microbial Ecology", *Ecol. Res. Comm. Bull.* **17,** 323–328.
Swift, M. J. (1973b). *Soil Biol. Biochem.* **5,** 321–332.
Swift, M. J. (1976). *In* "The Role of Terrestrial and Aquatic Organisms in Decomposition Processes" (J. M. Anderson and A. Macfadyen, Eds), pp. 185–222, Blackwell Scientific Publications, Oxford.
Swift, M. J. (1982). *In* "Decomposer Basidiomycetes: their Biology and Ecology" (J. C. Frankland, J. N. Hedger and M. J. Swift, Eds), pp. 307–337, Cambridge University Press, Cambridge.
Swift, M. J., Heal, O. W. and Anderson, J. M. (1979). "Decomposition in Terrestrial Ecosystems", Blackwell Scientific Publications, Oxford.
Tabatabai, M. A. (1982). *In* "Methods of Soil Analysis, Part 2, Chemical and Microbiological Properties" (A. C. Pace, R. H. Miller and D. R. Keeney, Eds), 2nd Edn, pp. 903–947, American Society of Agronomy, Madison.
Taber, W. A. and Taber, R. A. (1988). *Mycologia* **80,** 855–858.
Tadmor, U., Applebaum, S. W. and Kafir, R. (1971). *J. Exp. Biol.* **54,** 437–440.
Tamblyn, N. and da Costa, E. W. B. (1958). *Nature, London* **181,** 578–579.
Taylor, J. B. (1974). *Ann. Appl. Biol.* **78,** 113–123.
Taylor, B. and Parkinson, D. (1988). *Can. J. Bot.* **66,** 1933–1939.
Tennant, D. (1975). *J. Ecol.* **63,** 995–1001.
Thomas, R. J. (1985). *Trans. Br. Mycol. Soc.* **84,** 519–526.
Thomas, A., Nicholas, D. P. and Parkinson, D. (1965). *Nature, London* **205,** 105.
Thompson, W. (1984). *In* "The Ecology and Physiology of the Fungal Mycelium" (D. H. Jennings and A. D. M. Rayner, Eds), pp. 185–214, Cambridge University Press, Cambridge.
Thompson, W. and Boddy, L. (1983). *New Phytol.* **93,** 277–291.
Thompson, W. and Rayner, A. D. M. (1982). *Trans. Br. Mycol. Soc.* **78,** 193–200.
Thompson, W. and Rayner, A. D. M. (1983). *Trans. Br. Mycol. Soc.* **81,** 333–345.
Tiffin, A. I. (1987). *J. App. Bact., Symp. Supplement,* 127S–139S.
Toth, R. and Toth, D. (1982). *Mycologia* **74,** 182–187.
Tribe, H. T. (1957). *In* "Microbial Ecology" (C. C. Spicer and R. E. O. Williams, Eds), Symp. Soc. Gen. Microbiol. Vol. 7, pp. 287–298, Cambridge University Press, Cambridge.
Tribe, H. T. and Mabadeje, S. A. (1972). *Trans. Br. Mycol. Soc.* **58,** 127–137.
Tribe, H. T. and Maynard, P. (1989). *Mycologist* **3,** 24–27.
Tsao, P. (1970). *Soil Biol. Biochem.* **2,** 247–256.
Umbreit, W. W., Burris, R. H. and Stauffer, J. F. (1964). "Manometric Techniques", 4th Edn, Burgess Pub. Co., Minneapolis.
Vaartaja, O. (1960). *Phytopathology* **50,** 870–873.
van Veen, J. A. and Paul, E. A. (1979). *Appl. Environ. Microbiol.* **37,** 686–692.
Visser, S. (1985). *In* "Ecological Interactions in Soil. Plant, Microbes and Animals" (A. H. Fitter, D. Atkinson, D. J. Read and M. B. Usher, Eds), pp. 297–317, Blackwell Scientific Publications, Oxford.
Visser, S. and Parkinson, D. (1975). *Can. J. Bot.* **53,** 1640–1651.
Waid, J. S. (1956). *Nature, Lond.* **178,** 1477–1478.
Waid, J. S. (1979). *In* "Soil Microbiology and Plant Nutrition" (W. J. Broughton, C. K. John, J. C. Rajarao and Beda Lim, Eds), pp. 33–39, University of Malaya, Kuala Lumpur.
Waid, J. S. and Woodman, M. J. (1957). *Pédobiologie* **7,** 155–158.
Waid, J. S., Preston, K. J. and Harris, P. J. (1971). *Soil Biol. Biochem.* **3,** 235–241.

Waid, J. S., Preston, K. J. and Harris, P. J. (1973). *In* "Modern Methods in the Study of Microbial Ecology" (T. Rosswall, Ed.), *Ecol. Res. Comm. Bull. (Stockholm)* **17,** 317–322.
Wainwright, M. (1978). *Plant Soil* **49,** 191–193.
Wainwright, M. and Killham, K. (1980). *Soil Biol. Biochem.* **12,** 555–558.
Warcup, J. H. (1950). *Nature, London* **166,** 117–118.
Warcup, J. H. (1951a). *Trans. Br. Mycol. Soc.* **34,** 376–399.
Warcup, J. H. (1951b). *Trans. Br. Mycol. Soc.* **34,** 515–518.
Warcup, J. H. (1955). *Nature, London* **175,** 953–954.
Warcup, J. H. (1957). *Trans. Br. Mycol. Soc.* **40,** 237–262.
Warcup, J. H. (1959). *Trans. Br. Mycol. Soc.* **42,** 45–52.
Warcup, J. H. (1967). *In* "Soil Biology" (A. Burges and F. Raw, Eds), pp. 51–110, Academic Press, London.
Warcup, J. H. and Baker, K. F. (1963). *Nature, London* **197,** 1317–1318.
Watling, R. (1971). *In* "Methods in Microbiology" (C. Booth, Ed.), Vol. 4, pp. 219–236, Academic Press, London.
Watling, R. (1981). "How to Identify Mushrooms to Genus. V: Cultural and Developmental Features", Mad River Press, Inc., Eureka, California.
Wells, J. M., Hughes, C. and Boddy, L. (1990). *New Phytol.* **114,** 595–606.
West, A. W. (1986). *J. Microbiol. Methods* **5,** 125–138.
West, A. W. (1988). *Biol. Fertil. Soils* **7,** 88–94.
West, A. W., Grant, W. D. and Sparling, G. P. (1987). *Soil Biol. Biochem.* **19,** 607–612.
Whipps, J. M. (1987). *Trans. Br. Mycol. Soc.* **89,** 199–203.
Whipps, J. M., Haselwandter, K., McGee, E. E. M. and Lewis, D. H. (1982). *Trans. Br. Mycol. Soc.* **79,** 385–400.
Wilcox, W. W. (1964). U.S. Dept. Agric. For. Serv. Res. Note FPL-056.
Wilcox, W. W. (1968). U.S. Dept. Agric. For. Serv. Res. Paper FPL-70.
Williams, E. N. D., Todd, N. K. and Rayner, A. D. M. (1984). *Trans. Br. Mycol. Soc.* **82,** 323–326.
Williams, S. T. and Parkinson, D. (1964). *J. Soil Sci.* **15,** 331–341.
Williams, S. T., Parkinson, D. and Burges, N. A. (1965). *Plant Soil* **22,** 167–186.
Wilson, J. M., Trinick, M. J. and Parker, C. A. (1983). *Soil Biol. Biochem.* **15,** 439–445.
Wissler, M. D., Tengerdy, R. P. and Murphy, V. G. (1983). *Dev. Ind. Microbiol.* **4,** 527–538.
Witkamp, M. (1974). *Soil Sci.* **118,** 150–155.
Wood, D. A. (1979). *Biotechnol. Lett.* **1,** 255–260.
Wood, D. A. and Goodenough, P. W. (1977). *Arch. Microbiol.* **114,** 161–165.
Wright, S. F. and Morton, J. B. (1989). *Appl. Environ. Microbiol.* **55,** 761–763.
Wright, S. F., Morton, J. B. and Sworobuk, J. E. (1987). *Appl. Environ. Microbiol.* **53,** 2222–2225.
Yang, C. S. and Wilcox, H. E. (1984). *Can. J. Bot.* **62,** 251–254.
Zak, B. (1976). *Can. J. Bot.* **54,** 1297–1305.
Zak, B. and Marx, D. H. (1964). *For. Sci.* **10,** 214–222.
Zak, J. C. (1988). *Proc. R. Soc. Edinb.* **94B,** 73–83.
Zelles, L. (1988). *Biol. Fertil. Soils* **6,** 125–130.
Zill, G., Engelhardt, G., Wallnöfer, P. R. (1988). *Z. Lebensm. Unters. Forsch.* **187,** 246–249.
Zollfrank, U., Sautter, C. and Hock, B. (1987). *Eur. J. For. Pathol.* **17,** 230–237.

# 12

# Methods for Studying the Ecology and Population Dynamics of Soil Myxomycetes

M. F. MADELIN
*Department of Botany, University of Bristol, Bristol BS8 1UG, UK*



## I. Introduction

It is only recently that attempts have been made to treat the ecology of myxomycetes (mycetozoa; acellular or plasmodial slime moulds) quantitatively. There is a great deal of information on the geographical and local distribution of these organisms, but there is little which reveals the nature and size of their ecological role. Judging from their neglect in most texts on microbial ecology the prevailing view appears to be that their role is unimportant. Recent work, however, has revealed their abundance in different kinds of soils from many regions worldwide and suggests that they are probably significant in the ecology of many terrestrial habitats (Feest and Madelin, 1985a, b, 1988a, b; Murray *et al.*, 1985; Feest, 1986; Feest and Campbell, 1986).

Myxomycetes have a curious life cycle which includes two very different trophic stages, one consisting of microscopic uninucleate amoeboid cells,

METHODS IN MICROBIOLOGY
VOLUME 22 ISBN 0-12-521522-3

with or without flagella, and the other consisting of the multinucleate plasmodium (Fig. 1). Because the plasmodium is often large and conspicuous whereas the uninucleate phases are rarely seen except in culture, it is sometimes thought that if myxomycetes have any ecological significance it is likely to be mainly through the feeding activities of their plasmodia. However Madelin (1984) suggested that in some species the plasmodium may represent an infrequent excursion from a predominantly unicellular life form and be primarily a precursor of sporulation.

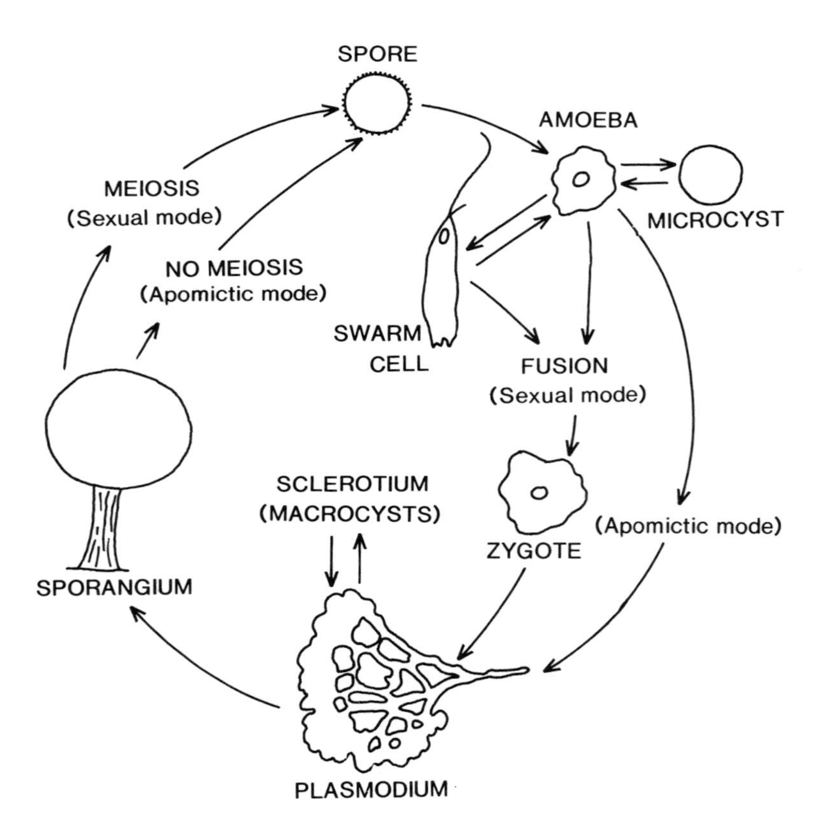


**Fig. 1.** Generalized myxomycete life cycle. In the sexual life cycle mode haploid and diploid states alternate, but in the apomictic mode the ploidy remains unchanged. Interconversion of life cycle modes has been reported.

Quantitative myxomycete ecology is still in its infancy. The principal need at the present time is objective measurement of the abundance of myxomycete organisms in their natural environments. In practice, this means being able to count the myxomycete population and if possible to characterize its state (e.g. whether active or resting). Methods for doing this will be described, and interpretation of the data discussed. Major sources of information on identifying, culturing and maintaining myxomycetes will be cited. Cellular slime moulds (e.g. dictyostelids) will not be considered; a recent text (Raper, 1984) is a comprehensive source of information on these quite different organisms.

## II. Enumeration and characterization of myxomycetes in soil

### A. The most probable number procedure

Feest and Madelin (1985a) have described a method for estimating numbers of soil myxomycetes which uses a suitably selective culture system and the most probable number (MPN) procedure. A fundamental problem is what should be, or indeed can be, counted when dealing with organisms with as great a variety of morphoses as myxomycetes possess (Fig. 1). The unit of presence enumerated by Feest and Madelin (1985a) was the 'plasmodium-forming unit' (PFU). This is any kind of myxomycetous entity capable of giving rise to a plasmodium in the Petri dish cultures used in the described system. The most probable number method involves using a series of known dilutions of a suspension of the population which is to be assayed to inoculate a series of replicate plates at each of several dilution steps, and recording the number of 'positives', i.e. plates in which the organism being assayed eventually appears. The amount of growth of the organism in question is unimportant; it is simply its presence or absence that matters. By choosing the presence or absence of plasmodia as the critical characteristic, Feest and Madelin (1985a) used an unambiguous indicator of the presence of myxomycetes. It is nevertheless likely that several different stages of the life cycle constitute 'plasmodium-forming units'. What these may be is discussed below. The method described by Feest and Madelin (1985a) was as follows.

The medium used is half-strength corn meal agar (CMA/2) with an overlay of washed yeast (*Saccharomyces cerevisiae*). CMA/2 is prepared by dissolving 8.5 g of Oxoid Corn Meal Agar (Oxoid CM 103) and 7.5 g of Oxoid Purified Agar (Oxoid L 28) in 1 l of distilled water, and is sterilized at 121°C for 20 min. *Saccharomyces cerevisiae* is grown in shaken Sabouraud's Dextrose Broth (Oxoid CM 147) for 4 days at 30°C, harvested by centrifugation, washed three times in sterile distilled water, and finally adjusted to a standard 1% packed-cell volume which can be stored at *ca* 5°C for several weeks. 0.5 $cm^3$ is pipetted into each 9-cm diameter Petri dish containing approximately 20 $cm^3$ of solidified CMA/2. This medium promotes the growth of plasmodia from a mixed inoculum, such as soil suspension. The use of *S. cerevisiae* restricts the growth of protozoa that would flourish if smaller microbial food were used.

The soil sample is best processed within a few hours of its collection. It is thoroughly mixed in a clean polythene bag by kneading and shaking, and a subsample of 50 $cm^3$, packed to approximately the same density as in its natural state, is weighed and added to 450 $cm^3$ of sterile 0.1% Brij 35 solution (a wetting agent; polyoxyethylene lauryl ether; BDH Ltd, Poole, UK). This is agitated on a wrist-action shaker for 20 min to yield a $10^{-1}$ dilution which is serially diluted in 0.1% Brij solution to give further tenfold dilutions. For each of three consecutive tenfold dilutions, 1 $cm^3$ is pipetted onto each of five plates of CMA/2 plus yeast. Usually dilutions of $10^{-1}$ to $10^{-3}$ are suitable, but where large numbers of myxomycetes are expected, dilutions of $10^{-2}$ to $10^{-4}$ are required. The 15 plates prepared from a single soil sample are placed in an unsealed polythene bag (to reduce drying) in a 20°C incubator for 1 week and then examined with a binocular dissecting microscope (×10 overall magnification) and a phase-contrast microscope (×10 eyepiece, ×20 objective). The former is usually adequate for observing phaneroplasmodia which often are large enough to be clearly visible to the naked eye (Fig. 2), but the phase-contrast microscope is necessary for observing aphanoplasmodia and protoplasmodia (see Gray and Alexopoulos, 1968, for their distinguishing features). After the first week, incubation is continued at room temperature (18–20°C) in daylight in a clear plastic seedling-propagating box (*ca* 56 × 28 × 23 cm high) to minimize drying and risk of contamination, and the plates examined at weekly intervals for the presence of plasmodia (Fig. 2) or

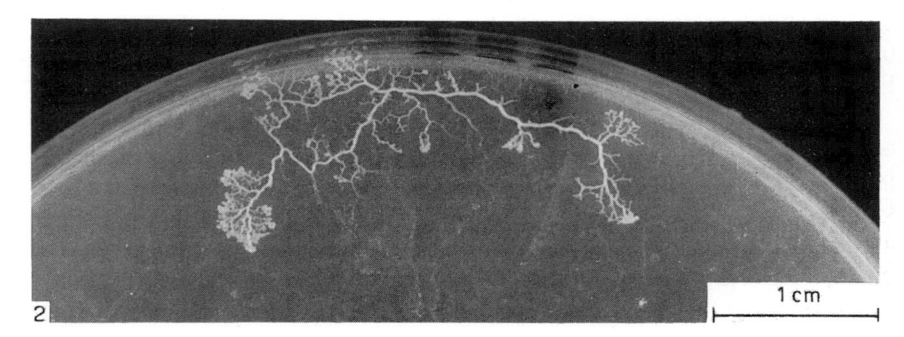


**Fig. 2.** A phaneroplasmodium from soil on a plate of CMA/2 plus yeast.

sporangia (Fig. 3). It is important that the surface of the agar should be kept suitably moist. If it loses its moisture film and appears dull, sterile distilled water is lightly sprayed onto it, sufficient to restore the film but not enough to flow freely across the surface.

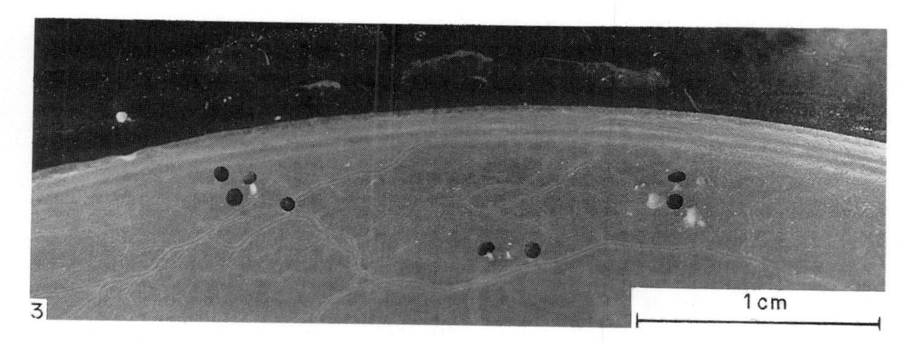


**Fig. 3.** Sporangia of a *Didymium* from soil, formed on a plate of CMA/2 plus yeast. Note the plasmodial tracks which serve to indicate that a plasmodium is, or has been, present.

The number of PFUs in the original soil sample is calculated from the number of positive plates (i.e. plates with plasmodia or sporangia or both) out of five at each tenfold dilution step with the aid of tables of most probable numbers (e.g. Taylor, 1962; Alexander, 1965; Anon., 1969), and expressed relative to soil volume, fresh weight or dry weight. The 95% confidence limits of an estimate based on five isolation plates at three tenfold dilution steps are provided by multiplying the estimate by 3.30 and its reciprocal (Cochran, 1950). Cochran (1950) lists the multipliers for 95% confidence limits if other numbers of plates and other dilution steps are employed, while Hurley and Roscoe (1983) present a BASIC computer program for calculating the most probable number and various statistics for MPN procedures involving any dilution series, any degree of replication at each dilution level, and any number of levels. The ratios of two separate estimates obtained by the above procedure which may be judged as significantly different at different critical levels of probability for two-tailed tests can be calculated from a formula given by Cochran (1950), and are 3.73 for $P = 0.1$, 4.80 for $P = 0.05$, 7.87 for $P = 0.01$, and 13.95 for $P = 0.001$ (Murray *et al.*, 1985).

Attempts to miniaturize the MPN assay procedure for PFUs have not been successful. A certain size of culture vessel appears to be necessary if the production of plasmodia on which the assay is based is to be reliably obtained. However if only small soil samples are available the initial preparation of inocula may be scaled down.

Occasionally the procedure yields improbable sequences of positives from the three dilution steps, i.e. the number of positives at the higher inoculum concentration is fewer than is to be expected on the basis of the numbers at lower concentrations, for example 3–3–0 for inoculum concentrations $10^{-1}$, $10^{-2}$, $10^{-3}$, respectively (Feest and Madelin, 1985a). The occurrence of unlikely most probable number results has been discussed by Taylor (1962). Feest and Madelin (1985a) suggested that the most likely explanation for such sequences is competition from other organisms introduced in the most concentrated inocula. Supporting evidence includes the absence or rarity of anomalous sequences in assays of artificial soil microcosms in which only myxomycetes, bacteria and angiosperm roots were present, and in assays of soil samples which had been frozen (see below) so that many competing organisms were destroyed.

The MPN method described allows concurrent enumeration of several other categories of soil organism such as naked soil amoebae, ciliates, dictyostelid cellular slime moulds (recognizable by their sorocarps), myxobacteria (recognizable by their characteristic fruit bodies) and nematodes (Feest and Madelin, 1985a, b, 1988a, b).

### B. The nature of 'plasmodium-forming units' (PFUs)

The described procedure estimates the number of entities in the soil suspension that are able to give rise to plasmodia in the MPN culture plates, that is, it estimates what are literally 'plasmodium-forming units'. Precisely what PFUs are is uncertain, but circumstantial evidence points to their being wholly or predominantly uninucleate phases of the myxomycete life cycle (see Fig. 1) rather than plasmodia, sclerotia or macrocysts (Feest and Madelin, 1985a, b). Firstly the rarity of plasmodia in isolation plates during the first week of incubation together with their speed of growth once they do appear suggests that they are not introduced in the inoculum, but arise amongst populations of myxamoebae that stem from it. Secondly, when plasmodia do appear they often do so in large numbers before somatic fusion usually reduces them to a single large one; this is unlikely to occur if the numerous plasmodia are independent introductions from the soil, but is likely if they are genetically related as they would be if arising in the culture plate amongst the descendants of a few uninucleate cells introduced in the inoculum. Thirdly, organisms with the same distinctive morphology and movement as myxomycete swarm cells are sometimes to be seen in plates within a few hours of inoculation, while plasmodia are not. Presumably they are derived directly from the soil or from swiftly excysting microcysts or spores. Fourthly, the numbers of PFUs recorded are sometimes so large that were they present in the inoculum as plasmodia they almost certainly would be noticed.

According to the equation of Chuang and Ko (1981) which relates propagule size to the maximum population density of the microorganisms in the soil, numbers of the order of a few thousand per gram of soil (not an uncommon abundance of PFUs) are compatible with the individuals being of the order of size of uninucleate myxomycete cells. Although in nature plasmodia are sometimes seen to have emerged from the soil, their existence there is probably a relatively brief one which immediately precedes sporulation. Nevertheless within restricted bodies of soil during limited periods of time they must constitute a very large microbial biomass. Our own data on numbers of PFUs in soil suggest that for most of the time the uninucleate phases are the predominant and most ecologically important ones. Populations of phagotrophic microorganisms in non-woodland soils are dominated numerically by naked amoebae (Feest and Madelin, 1988b). The latter category would include myxomycete amoebae since these are not readily distinguishable from many other kinds of naked amoebae. The numbers of PFUs in these soils throughout the year averaged 9.5% of the total naked amoebae and sometimes even exceeded 50%. It therefore seems probable that many published enumerations of amoebae in the soil have included a substantial proportion of unrecognized myxomycetes.

The described procedure possibly underestimates the numbers of viable myxomycete cells in soil because some myxomycetes may be unable to produce plasmodia under the conditions prevailing in the isolation plates. This kind of problem is not one peculiar to myxomycetes. However another reason for underestimation exists. If the species present are heterothallic, more than one viable cell of that species need be present in the inoculum for a plasmodium to form. Theoretically, if cells of the two mating types are present in equal proportion, on average three randomly chosen cells are required to give rise to a plasmodium, i.e. are equivalent to one PFU. If cells of one mating type are scarcer than those of the other, the number of cells to which a single PFU corresponds increases. This particular problem of underestimation could be overcome if instead of plasmodia one could recognize uninucleate myxomycete cells with confidence. This approach is considered in the next section.

### C. 'Myxoflagellate-forming units' (MFUs)

Feest and Madelin (1985a) use their MPN technique to enumerate 'presumptive myxoflagellates' in a soil by scoring the presence of microorganisms with the distinctive morphology and locomotory behaviour of myxomycete swarm cells (myxoflagellates). Feest and Campbell (1986) similarly scored these in wheat field soils as 'myxamoeba-forming units'. 'Myxoflagellate-forming unit' (MFU) is probably a preferable term since the method enumerates the entities in soil that are capable of yielding myxoflagellates in the enumeration plates. With experience one can recognize the presence of a population of myxoflagellates with a high degree of confidence even though individual cells may present difficulty. The myxoflagellates have a sharply conical flagellum-bearing anterior and are capable of rapid changes of shape and mode of movement when adequate moisture is present (Figs 4–6). They glide over surfaces with flagella sweeping from side to side, but when free in the water film they assume a comma shape for much

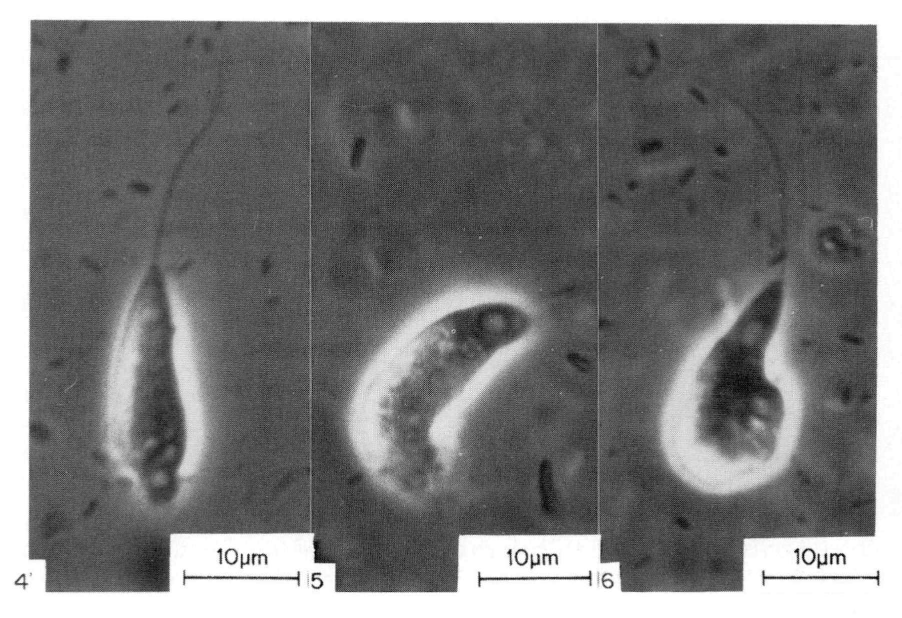


**Fig. 4.** A swarm cell (myxoflagellate) of *Didymium* creeping over an agar surface. Note the elongated (though changing) form and the rigid anterior cone bearing an apparently single flagellum.

**Fig. 5.** A swarm cell swimming with a vigorous jiggling movement in a water film over agar. The anterior cone is evident as is the characteristically curved shape.

**Fig. 6.** A swarm cell photographed within seconds of it contacting the agar surface and ceasing to swim; the jiggling movement stopped immediately, but the comma-like shape of the cell persisted for a short while.

of the time and display a vigorous jiggling movement that leads to erratic progress. Commonly only the longer flagellum is visible. In 10 replicate estimates of numbers in a single soil sample, Feest and Madelin (1985a) found MFUs exceeded PFUs sevenfold, and in soil samples from 38 different wheat fields, Feest and Campbell (1986) found that MFUs averaged twice the recorded numbers of PFUs. Enumeration of myxomycete populations in the soil in terms of PFUs may thus underestimate the actual numbers of discrete viable cells by an unknown amount, but because plasmodia are clearly recognizable and unambiguously myxomycetous it sets a clear lower limit on the numbers of myxomycetes present. On the other hand, it is probable that any organism that might be confused with a myxoflagellate when enumerating myxomycetes in terms of MFUs is a closely related organism anyway, such as a Protostelid (Spiegel, 1981).

### D. Trophic and dormant cells

It is desirable to be able to distinguish between vegetatively active (trophic) and dormant myxomycete cells in the soil. In practice, this usually means distinguishing between amoebae and myxoflagellates on the one hand and microcysts on the other. A method for partitioning active and inactive forms of dictyostelid cellular slime moulds developed by Kuserk *et al.* (1977) is based on their relative susceptibility to freezing at −16°C. Freezing for 24 h reduced recoveries of amoebae to zero, but spores and microcysts largely survived this treatment. A similar technique has been used with myxomycetes (Murray *et al.*, 1985; Feest and Campbell, 1986). Soil samples were frozen at −20°C for at least 24 h. However this treatment sometimes led to unexpected increases in the numbers of PFUs. These are believed to be attributable to the breaking of endogenous (constitutive) dormancy in a large population of resting cells (probably microcysts) that are undetectable in fresh soil. If enough such cells are present in a soil sample, the killing of active cells by freezing is more than offset by the breaking of dormancy in these freezing-resistant dormant cells.

It is not possible at present to partition the soil population between (a) active cells, (b) dormant but germinable cells (i.e. exogenously dormant), and (c) dormant cells that need to be frozen to become germinable (i.e. endogenously dormant). This is because the exogenously dormant population is a constituent not only of the population that is estimated in fresh soil but also of that recorded after freezing. For soil protozoa, other treatments besides freezing have been used to distinguish between active and resting populations, for example treatment overnight in 2%

hydrochloric acid (Singh, 1946) and treatment for a few minutes with 1% sodium dodecyl sulphate (Umeche, 1983). Their applicability to myxomycetes has not been studied.

### E. Soil-sampling procedures

To compare different soils or soil from one site at different times, it is necessary to adopt a suitably thorough sampling method, otherwise the variation that exists between individual small bodies of soil may obscure variation that is due to soil type or season. Feest and Madelin (1985a, 1988a, b) sampled within 1 $m^2$ quadrats divided into 100 10-cm squares. Sixty of the latter were selected by means of random numbers, and from each a cylindrical core of 1 $cm^2$ cross-sectional area was removed from the upper 4 cm of the soil profile. It was in this stratum that most PFUs were found. The 60 cores were thoroughly mixed in a polythene bag to yield the sample.

### F. Qualitative ecological study and identification of soil myxomycetes

The CMA/2 plus yeast medium for enumerating myxomycetes can also be used for isolating them from the soil, for fine-scale mapping of distribution, and for determining the species and mating-type composition of the myxomycete flora of soil. The soil sample should be dispersed dry into separate crumbs, one to several $mm^3$ in volume, in an empty, sterile Petri dish, and single crumbs gently spread over the surface of plates of CMA/2 plus yeast. If myxomycetes are present, plasmodia usually appear in 1 or 2 weeks and fructify a few days later.

Most of the identifiable myxomycetes recovered when using CMA/2 plus yeast in the described enumeration procedure have been species of *Didymium* (Feest and Madelin, 1985a, 1988a). This probably reflects their natural prevalence in the soil rather than selectivity of the isolation medium, because species of other genera have nevertheless sometimes been isolated, including *Arcyria, Clastoderma, Perichaena* and *Physarum*.

Major monographs for the identification of myxomycetes include those by Lister (1925), Martin (1949), Martin and Alexopoulos (1969) of which an abridged version for identification to the generic level is available (Martin *et al.*, 1983), Nannenga-Bremekamp (1974), Farr (1976) (for neotropical species), and two major works which refer specifically to India (Thind, 1977; Lakhampal and Mukerji, 1981). The myxomycetes of Japan have been comprehensively illustrated in colour (Emoto, 1977). We have found the monograph by Nannenga-Bremekamp (1974) to be particularly useful. Only the Dutch text is currently available but an English translation is in preparation.

## III. Culture and maintenance in the laboratory

Methods for isolating, culturing and maintaining myxomycetes are described by Gray and Alexopoulos (1968) and Carlile (1971), and more recent methods are reviewed by Aldrich (1982). Most species that have been cultured in the laboratory have been grown in crude culture on complex media such as oat or corn meal agar in the presence of associated microorganisms, usually bacteria. In only 19 of the approximately 500 known species has axenic culture of the plasmodial or myxamoebal stage or both been achieved (see Hu and Clark, 1986, for details). Axenic culture of myxamoebae has proven particularly difficult to accomplish. Axenic culture of plasmodia or myxamoebae on defined media is even rarer. However, for many ecological purposes including identification, crude culture on a weak medium such as CMA/2 often suffices as a first step.

## Acknowledgements

I thank Dr D.J. Patterson and Mr T. Colborn for photographic assistance.

## References

Aldrich, H. C. (1982). *In* "Cell Biology of Physarum and Didymium", Vol. 2 (H. C. Aldrich and J. W. Daniel, Eds), pp. 361–365, Academic Press, New York.

Alexander, M. (1965). *In* "Methods in Soil Analysis, Part 2. Chemical and Microbial Properties" (C. A. Black, Ed.), pp. 1467–1472, Am. Soc. Agronomy, Madison.

Anon. (1969). *In* "Ministry of Health; Ministry of Housing and Local Government: Reports on Public Health and Medical Subjects", No. 71, 4th Edn, HMSO, London.

Carlile, M. J. (1971). *In* "Methods in Microbiology", Vol. 4 (C. Booth, Ed.), pp. 237–265, Academic Press, London.

Chuang, T. Y. and Ko, W. H. (1981). *Soil Biol. Biochem.* **13,** 185–190.

Cochran, W. G. (1950). *Biometrics* **6,** 105–116.

Emoto, Y (1977). "The Myxomycetes of Japan", Sangyo Tosho Publishing Co., Ltd, Tokyo.

Farr, M. L. (1976). "Myxomycetes", Flora Neotropica Monograph No. 16, New York Botanical Garden, New York.

Feest, A. (1986). *Ekológia (CSSR)* **5,** 125–134.

Feest, A. and Campbell, R. (1986). *FEMS Microbiol. Ecol.* **38,** 99–111.

Feest, A. and Madelin, M. F. (1985a). *FEMS Microbiol. Ecol.* **31,** 103–109.

Feest, A. and Madelin, M. F. (1985b). *FEMS Microbiol. Ecol.* **31,** 353–360.

Feest, A. and Madelin, M. F. (1988a). *FEMS Microbiol. Ecol.* **53,** 133–140.

Feest, A. and Madelin, M. F. (1988b). *FEMS Microbiol. Ecol.* **53,** 141–152.
Gray, W. D. and Alexopoulos, C. J. (1968). "Biology of the Myxomycetes", The Ronald Press Co., New York.
Hu, F.-S. and Clark, J. (1986). *Mycologia* **78,** 478–482.
Hurley, M. A. and Roscoe, M. E. (1983). *J. Appl. Bacteriol.* **55,** 159–164.
Kuserk, F. T., Eisenberg, R. M. and Olsen, A. M. (1977). *J. Protozool.* **24,** 297–299.
Lakhampal, T. N. and Mukerji, K. G. (1981). "Taxonomy of the Indian Myxomycetes", J. Cramer, Vaduz.
Lister, A. L. (1925). "A Monograph of the Mycetozoa", 3rd Edn, Br. Mus. Nat. Hist., London.
Madelin, M. F. (1984). *Trans. Br. Mycol. Soc.* **83,** 1–19.
Martin, G. W. (1949). *N. Am. Flora* **1,** 1–190.
Martin, G. W. and Alexopoulos, C. J. (1969). "The Myxomycetes", University of Iowa Press, Iowa City.
Martin, G. W., Alexopoulos, C. J. and Farr, M. L. (1983) . "The Genera of Myxomycetes", University of Iowa Press, Iowa City.
Murray, P. M., Feest, A. and Madelin, M. F. (1985). *Bot. J. Linn. Soc.* **91,** 359–366.
Nannenga-Bremekamp, N. E. (1974). "De Nederlandse Myxomyceten", Koninklijke Nederlandse Natuurhistorische Vereniging, Zutphen.
Raper, K. B. (1984). "The Dictyostelids", Princeton University Press, Princeton.
Singh, B. N. (1946). *Ann. Appl. Biol.* **33,** 112–119.
Spiegel, F. W. (1981). *Biosystems* **14,** 491–499.
Taylor, J. (1962). *J. Appl. Bacteriol.* **25,** 54–61.
Thind, K. S. (1977). "The Myxomycetes of India", Indian Council of Agricultural Research, New Delhi.
Umeche, N. (1983). *Arch. Protistenk.* **127,** 127–130.

# 13

# Techniques for Studying the Microbial Ecology of Nitrification

SUSAN E. UNDERHILL

*Schools of Chemical Engineering, University of Bradford, West Yorkshire BD7 1DP, UK*



METHODS IN MICROBIOLOGY
VOLUME 22 ISBN 0-12-521522-3

## I. Introduction

Nitrification can be described as the biological oxidation of reduced forms of nitrogen, usually ammonia to nitrite followed by the oxidation of nitrite to nitrate. As such it is a term that collectively describes a conversion that is predominantly carried out by two distinct groups of bacteria, the chemoautotrophic ammonia- and nitrite-oxidizing bacteria. A number of prokaryotic and eukaryotic heterotrophs are also capable of nitrification (Focht and Verstraete, 1977). The rates of nitrification are far lower than those of autotrophs, and it has been reported that the nitrification products of at least one heterotroph, *Aspergillus flavus*, only appear after active growth has ceased (Killham, 1986). Kuenen and Robertson (1988) suggest that the nitrification rates of heterotrophs may be higher than measured in many studies due to simultaneous denitrification by these organisms which results in very little or no accumulation of nitrite. Heterotrophs may make a significant contribution to nitrification when conditions are unfavourable for growth of autotrophs.

This chapter will deal mainly with laboratory methods for the study of autotrophic nitrification. It must be remembered, however, that nitrification is an integral part of the whole nitrogen cycle (Fig. 1). In the natural environment outside the laboratory it is difficult to separate nitrification from the rest of the nitrogen cycle. For example, nitrification may depend on the rate of ammonification. When studying nitrification in the natural habitat, denitrification may have to be taken into account, particularly if activity of nitrifiers is assessed by nitrite and nitrate production. It is worth noting that nitrite and nitrate are not the only products of the autotrophic nitrifying bacteria which should be included in the nitrogen cycle. Some genera of ammonia oxidizers produce $N_2O$ at reduced oxygen concentrations from unstable intermediates and also by nitrite reduction (Ritchie and Nicholas, 1972; Goreau *et al.*, 1980). Poth (1986) showed that a *Nitrosomonas* sp. could reduce nitrite completely to dinitrogen, the final product of the denitrification process. Under anoxic conditions some strains of *Nitrobacter* can reduce nitrate to nitrite and gaseous products including $N_2O$ (Bock *et al.*, 1988).

## II. A little taxonomy and metabolic capabilities of nitrifiers

The family Nitrobacteraceae consists of the two groups of autotrophic ammonia- and nitrite-oxidizing bacteria. They are Gram-negative organisms which are categorized by their shape, size, arrangement of cytoplasmic membranes as well as by comparison of DNA base ratios and

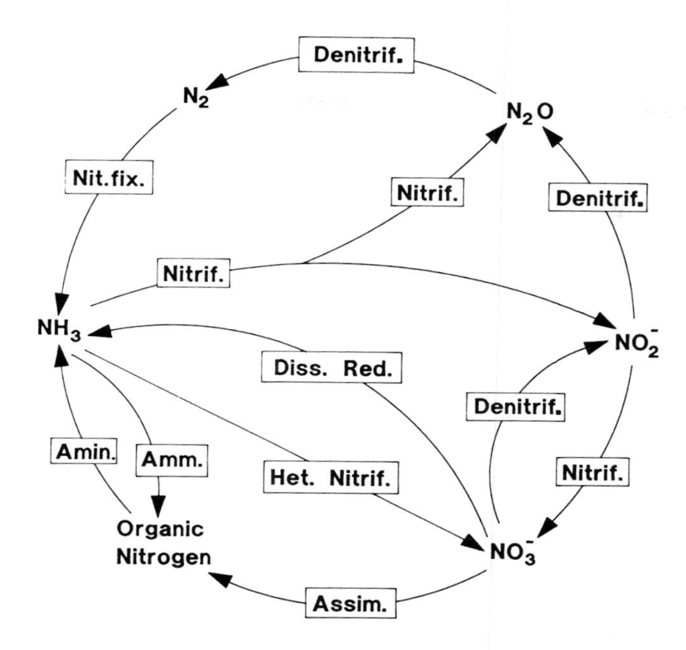


**Fig. 1.** The general nitrogen cycle. Amm = ammonification; Amin = amination; Denitrif = denitrification; Nit.fix = nitrogen fixation; Diss.red = dissimilatory nitrate reduction; Het.nitrif = heterotropic nitrification; Nitrif. = heterotropic and autotropic nitrification. Adapted from Kuenen and Robertson (1988) (with permission).

metabolic capabilities. A detailed account of their morphology is given by Walker (1975), Watson *et al.* (1981) and more recently by Bock *et al.* (1986).

### A. Ammonia oxidizers

There are five genera of ammonia oxidizers, *Nitrosomonas, Nitrosospira, Nitrosococcus, Nitrosolobus* and *Nitrosovibrio*. *Nitrosomonas europaea* is used extensively in pure culture experiments in the laboratory although it may not always be the predominant ammonia oxidizer in the natural habitat. MacDonald (1979) found *Nitrosolobus* to be dominant in a Rothamsted soil. *Nitrosomonas europaea* is the only species described in the genus *Nitrosomonas* (Winogradsky, 1892). Cells are short rods (0.8–1.0 × 1.0–2.0 μm) which may have 1–2 subpolar flagella. Marine strains tend to have an additional cell wall layer composed of subunits whose macromolecular arrangement may vary. Other physiological differences have been found which distinguish soil and freshwater isolates from sewage and marine isolates (Watson and Mandel, 1971). The presence of at least seven other species is indicated (Koops and Harms, 1985).

The genus *Nitrosococcus* includes at least three species, *N. nitrosus, N. oceanus* and *N. mobilis*, which range in size from 1.2 to 2.2 μm diameter. *N. oceanus* is an obligate halophile. Cells of *Nitrosospira* consist of tightly wound spirals with 3–20 turns. The genus includes at least five species (Bock *et al.*, 1986) but only one, *N. briensis*, has been described. *Nitrosolobus multiforms* has lobate, pleomorphic cells (1.0 × 1.5—1.0 × 2.5 μm), which are partially compartmentalized (Watson *et al.*, 1971). Again a second species of this genus has been detected (Koops and Harms, 1985). *Nitrosovibrio tenius* consists of curved rods (0.3–0.4 × 1.1–3.0 μm) isolated from soil.

All ammonia oxidizers use carbon dioxide as the main carbon source. There is only one report of heterotrophic growth (Pan and Umbreit, 1972) but Krümmel and Harms (1982) tested various organic compounds and found no evidence for heterotrophic growth without ammonia. Mixotrophic growth was observed and in some cases increased growth and cell yield was found in comparison to autotrophically grown cells. The ammonia oxidizers tested gave varied results however. Growth of *Nitrosospira* $Nsp_1$ was stimulated over 30% by addition of formate or acetate, but these substances had little effect on the growth of *Nitrosospira* $Nsp_5$ whilst pyruvate promoted ammonia oxidation by this strain and inhibited strain $Nsp_1$.

Recent investigations have expanded the list of metabolic capabilities of the ammonia oxidizers. The enzyme ammonia monooxygenase, which brings about the hydroxylation of ammonia, is also capable of using other organic substrates such as ethylene and CO (Hyman and Wood, 1984; Tsang and Suzuki, 1982). It also appears to be responsible for the oxidation of methane to methanol or $CO_2$ (Hyman and Wood, 1983; Jones and Morita, 1983). Due to the lack of specificity of ammonia monooxygenase methanol can be converted further to formaldehyde which inhibits nitrite production in growing cultures due to its reaction with hydroxylamine to produce formaldoxime (Voysey and Wood, 1987). Similarities between ammonia-oxidizing bacteria and methanotrophs have been noted. Methanotrophs can also oxidize ammonia to nitrite via the methane monooxygenase enzyme (Dalton, 1977); the maximum rate of oxidation by *Methylococcus thermophilus* is, however, only 0.2% of that of *Nitrosomonas*.

## B. Nitrite oxidizers

There are four genera of nitrite-oxidizing bacteria; *Nitrobacter, Nitrococcus, Nitrospina* and the most recently isolated *Nitrospira. Nitrobacter* is the most frequently isolated nitrite oxidizer. Two species are found in this genus, *Nitrobacter winogradski* and the more recently described *Nitrobacter hamburgensis* (Bock *et al.*, 1983). Both species are similar in shape, size and ultrastructure. Cells are rod-shaped (0.6–0.8 × 1.0–2.0 μm), may have one polar flagellum and are reported to reproduce by budding producing coccoid and pear-shaped forms. *Nitrobacter agilis* was reassigned to the species *Nitrobacter winogradski* (Watson and Mandel, 1971; Kalthoff *et al.*, 1979).

*Nitrobacter* was regarded as an obligate chemoautotroph for many years, however, mixotrophic and heterotrophic growth has been demonstrated (Smith and Hoare, 1968; Steinmuller and Bock, 1977; Bock, 1976). About half of the *Nitrobacter* strains studied by Watson (1971a) could be grown heterotrophically on acetate, the other half being reported as obligate chemoautotrophs. *Nitrobacter hamburgensis* (previously *Nitrobacter* strains X14 and Y) grows best mixotrophically, cell doubling time increasing when the bacteria are grown heterotrophically and being longest under chemoautotrophic conditions (Bock *et al.*, 1983). The intracytoplasmic membranes of *Nitrobacter* are more numerous and regularly arranged in auto- and mixotrophically grown cells than in heterotrophically grown cells. These membranes are the site of the nitrite-oxidizing system and the components of the reverse electron flow. Cytochrome content of membranes also depends on growth conditions and nitrite oxidase activity is repressed during heterotrophic growth with the absence of membrane particles associated with this activity (Sundermeyer and Bock, 1981; Milde and Bock, 1985). Further classification of the *Nitrobacter* genus may be necessary; the strains *Nitrobacter* 'Z' and 'Abl' may be assigned to a third, as yet unnamed species (Freitag *et al.*, 1987).

The other genera of nitrite oxidizers each contain only one species, *Nitrosococcus mobilis, Nitrospina gracilis* and *Nitrospira marina*, isolated from the marine environment (Watson and Waterbury, 1971; Watson *et al.*, 1986). Cells of *Nitrospira* are curved rods distinguishable by their wound spirals with up to 20 turns when grown autotrophically. Cells of *Nitrospina* are long slender rods (0.3–0.4 × 1.7–6.0 μm).

Bock *et al.* (1986) give full details of the morphology of the family Nitrobacteraceae and the reader is referred here for identification purposes. The need for further classification of the members of the family is evident and this may reflect the wide variations in growth kinetic parameters reported for particular species.

## III. Growth of nitrifiers

### A. Media and growth parameters

Nitrifying bacteria can be grown in the laboratory in axenic liquid culture using one of the inorganic media shown in Tables I and II. Because of the slow growth rate of nitrifiers in comparison to heterotrophs, strict aseptic technique must be employed. Production of extracellular organics by the nitrifiers allows growth of heterotrophic contaminants. Liquid media are more convenient for the routine maintenance of nitrifiers than solid media because of poor growth on the latter.

**TABLE I**

Media for the isolation and growth of terrestrial nitrifying bacteria. The trace salts are usually made up and supplied to the media as a separate stock solution chelated with $Na_2EDTA$, as well as a separate ion solution

| | Nitrite oxidizers | | Ammonia oxidizers | |
|---|---|---|---|---|
| | Smith and Hoare (1968) | Van Droogenbroek and Laudelout (1967) | Soranio and Walker (1968) | Watson *et al.* (1971) |
| $NaNO_2$ | 1.38 g | 1.38 g | | |
| $(NH_4)_2SO_4$ | | | 0.5 g | 2 g |
| $Na_2HPO_4.2H_2O$ | 2.55 g | 6.4 g | | |
| $KH_2PO_4$ | 0.27 g | 0.54 g | 0.2 g | |
| $K_2HPO_4$ | | | | 15.9 mg |
| $CaCl_2.2H_2O$ | 2.5 mg | | 40 mg | 20 mg |
| $MgSO_4.7H_2O$ | 20 mg | 10 mg | 40 mg | 200 mg |
| $FeSO_4.7H_2O$ | 10 mg | 5 mg | 0.5 mg | |
| chelated with $Na_2$-EDTA | 10 mg | 5 mg | 0.5 mg | |
| Chelated Fe[a] | | | | 1 mg |
| $H_3BO_3$ | 20 μg | | | |
| $CuSO_4.5H_2O$ | 100 μg | 20 μg | | 20 μg |
| $MnCl_2.4H_2O$ | | | | 200 μg |
| $MnSO_4.2H_2O$ | 20 μg | | | |
| $Na_2MoO_4.2H_2O$ | | 20 μg | | 100 μg |
| $(NH_4)_6Mo_7O_{24}.4H_2O$ | 20 μg | | | |
| $ZnSO_4.7H_2O$ | 150 μg | 20 μg | | 100 μg |
| $CoCl_2$ | 10 μg | | | |
| $CoCl_2.6H_2O$ | | | | 2 μg |
| Phenol red | | | 0.5 mg | |
| Distilled water | 1 l | 1 l | 1 l | 1 l |
| pH after autoclaving | 7.65 | 7.7 | ≅ 6.0[b] | |

[a] 13% Geigy chemical.
[b] pH adjusted to 7.5–8.0 by addition of sterile 5% $NaCO_3$ after autoclaving.

**TABLE II**
Media for the isolation of marine nitrifying bacteria. (A) Ammonia oxidizers; (B) nitrite oxidizers

| | (A) Watson (1965) | (B) Watson and Waterbury (1971) |
|---|---|---|
| $(NH_4)_2SO_4$ | 1.32 g | |
| $NaNO_2$ | | 69 mg |
| $K_2HPO_4$ | 114 mg | 1.74 mg |
| $CaCl_2.2H_2O$ | 20 mg | 6.0 mg |
| $MgSO_4.7H_2O$ | 200 mg | 100 mg |
| Chelated Fe[a] | 130 mg | 1.0 mg |
| $CuSO_4.5H_2O$ | 20 μg | 6.0 μg |
| $MnCl_2.4H_2O$ | 2 μg | 66 μg |
| $Na_2MoO_4.2H_2O$ | 1 μg | 30 μg |
| $ZnSO_4.7H_2O$ | 100 μg | 30 μg |
| $CoCl_2.6H_2O$ | 2 μg | 0.6 μg |
| Sea water | 1 l | 700 ml |
| Distilled water | | 300 ml |

[a]13% Geigy chemical.

The composition of the media used by different researchers varies. Generally media should include copper, sodium, calcium and magnesium. Loveless and Painter (1968) showed that these metals stimulated growth of pure cultures of *Nitrosomonas* and demonstrated effects of deficiencies. EDTA was inhibitory at low calcium concentrations only and was important in alleviating the inhibitory effect of copper. They demonstrated no stimulatory effect of the metal ions Zn, Co, Al, Sr, Pb or B (although inhibition was demonstrated above 0.08 mg $l^{-1}$ Zn and Co). Nitrifiers grown in media excluding the trace metals do have high specific growth rates and were used by some of the authors shown in Table III. The long-term effect of using such media for routine maintenance may be more important, as nitrifiers are prone to lose viability with continued culture over several years in such media (personal observation).

The energy source is usually supplied as ammonium sulphate for ammonia oxidizers and the sodium or potassium salt of nitrite for nitrite oxidizers. Substrate oxidation generally follows Michaelis–Menten type kinetics. Reported $K_m$ values vary widely for nitrifying bacteria ($K_m$ is the substrate concentration which gives half the maximum growth rate). Yoshioka *et al.* (1982) found that the specific growth rate of an ammonia oxidizer declined at more than 100 mM $NH_4^+$. Growth yield as well as specific growth rate declined at low $NH_4^+$ concentrations. For a nitrite-oxidizing bacterium growth yield decreased from $2 \times 10^3$ to $5 \times 10^3$ cells/nM $NO_2^-$ with increasing $NO_2^-$ concentration from 0.1 to 100 mM $NO_2^-$. Typical $K_m$ and growth yield values are shown in Table III.

## TABLE III

Typical growth constants of nitrifying bacteria (pure cultures)

| Reference | Bacteria | pH | Temperature (°C) | $\mu_{max}$ ($h^{-1}$) | $K_m$ (μg N $ml^{-1}$) | $Y_g$ (g biomass $g^{-1}$ N) | $m$ (g N g $biomass^{-1}$ $h^{-1}$) |
|---|---|---|---|---|---|---|---|
| Keen and[+] Prosser (1987) | *Nitrosomonas europaea* | 7.6 | 30 | 0.035 | 3.65 | 0.63 | 0.42 |
| | *Nitrobacter* | 8.0 | 30 | 0.039 | 2.8 | 0.056 | 0.70 |
| Yoshioka *et al.* (1982)[a] | Ammonia oxidizer | 7.6–7.8 | 20 | 0.024 | 3.6 | ($1.0 \times 10^4$ cells/nm $NH_4^+$) | ($1.1 \times 10^{-5}$ nm $NH_4^+$ $cell^{-1}$ $day^{-1}$) |
| | Nitrite oxidizer | 7.6–7.8 | 20 | 0.034 | 0.69 | | |
| Loveless and Painter (1968)[a] | *Nitrosomonas* (Jensen strain) | 8.0<br>7.6 | 25<br>25 | 0.057<br>0.036 | 1.0 | 0.1–0.03 | |
| Underhill and Prosser (1985)[a] | *Nitrosomonas europaea* | 7.6 | 27 | 0.030<br>0.038[b] | | | |
| | *Nitrobacter* sp. | 7.6 | 27 | 0.032<br>0.023[b] | | | |

[+]–values obtained by analysis of steady state data from continuous culture studies.
[a]Values obtained from liquid batch culture studies.
[b]Axenic soil incubation study.
$\mu_{max}$, maximum specific growth rate; $K_s$, substrate concentration giving half $\mu_{max}$; $Y_g$, true growth yield; $m$, maintenance coefficient.

It should be noted here that Suzuki *et al.* (1974) found that $K_m$ for $NH_4^+$ fell sharply with increasing pH. This was due to increasing concentrations of free $NH_3$ resulting from dissociation of $NH_4^+$ at the higher pH. $K_m$ values recalculated for $NH_3$ concentration showed little change with pH. This and other observations suggest that ammonia is the true substrate for ammonia monooxygenase and it is likely to penetrate the cells cytoplasmic membrane in its uncharged state. Most reported $K_m$ values and substrate concentrations in the literature are almost exclusively for $NH_4^+$. This obviously remains a grey area and so $NH_4^+$ concentrations are referred to in this article as appropriate. Kaplan (1983) shows $K_m$ values for undissociated $NH_3$ which were recalculated from $K_m$ values for $NH_4^+$ for marine strains at pH 8.3. The $K_m$ values were low enough for the calculated equilibrium concentration of $NH_3$ gas in the samples to be used as substrate!

## B. Culture conditions

### *1. Temperature, pH and oxygen*

In general the pH range for optimal growth of nitrifiers is between 7.0 and 8.0 with inhibition occurring below pH 6.0. Cultures are incubated aerobically between 25 and 30°C. Below 20°C and above 30°C growth rate falls off rapidly in pure cultures (Loveless and Painter, 1968); optimum temperature for *Nitrosomonas* appears to be 30°C. Flask batch cultures are usually incubated whilst shaking at 250 rpm to provide aeration. $K_m$ for oxygen is lower for *Nitrosomonas* than for *Nitrobacter*. Reported $K_m$ values for *Nitrosomonas* range from 0.15 to 0.3 mg $l^{-1}$ and 1 mg $l^{-1}$ for *Nitrobacter* (Focht and Verstraete, 1977; Goreau *et al.*, 1980).

It must be noted that culture conditions given here are for the optimum growth of nitrifiers. In many natural environments suboptimal conditions are usually found and in some laboratory studies of nitrification it may be desirable to devise media and conditions closer to those found in the natural habitat of interest.

Ammonia oxidizers produce acid during growth necessitating alkali addition. Growth will cease in liquid culture at approximately pH 6. Phenol red is often included in media as an indication of the need for neutralization. Sodium hydroxide or sodium carbonate is often used for pH adjustment. However, Loveless and Painter (1968) suggest caution in overaddition of the sodium ion which may inhibit growth of *Nitrosomonas* at 0.7% (w/v) in culture media. For routine maintenance of ammonia oxidizers in flask culture alkali can conveniently be included in a side arm and tipped into the culture as required reducing the risk of contamination (Fig. 2).

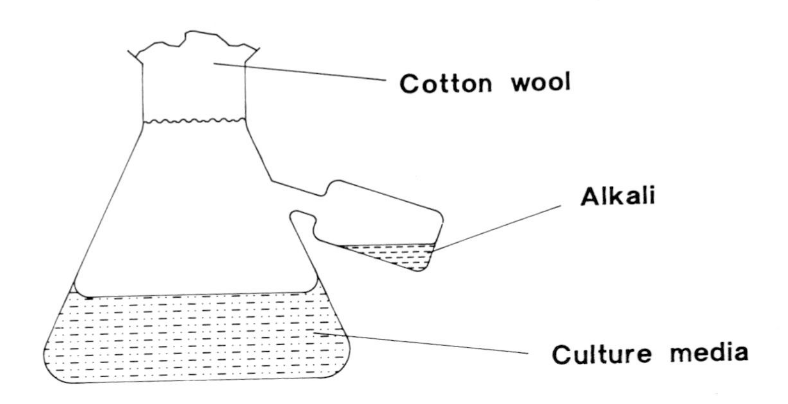


**Fig. 2.** Flask showing useful side arm adaptation for the neutralization of ammonia oxidizer cultures during routine growth.

*2. Special requirements of nitrifiers*

Nitrifying bacteria are extremely sensitive to carbon disulphide (Powlson and Jenkinson, 1971). Certain types of rubber release small quantities of $CS_2$. A red rubber bung or seal in a culture vessel will most likely completely inhibit growth. The use of silicone rubber is recommended.

*Nitrosomonas* is photosensitive (Hooper and Terry, 1974). Respiratory activity of 'resting cells' maintained under aerobic conditions without ammonia is fully inhibited within 10 min exposure to fluorescent and indirect room light! (Alleman *et al.*, 1987). Actively respiring cells or those in oxygen-free environments are not affected. Inhibited cells in this study regained 40% of their respiratory activity after 4 h of dark exposure. *Nitrobacter* is apparently more sensitive than *Nitrosomonas* (Bock, 1965; Olson, 1986).

Nitrification is usually absent in environments exposed to bright light and photoinhibition is due to the photo-oxidation of cytochrome c. In the laboratory *Nitrosomonas* should be grown in the dark especially if maximum growth rates are to be achieved. Glass vessels may be conveniently covered in aluminium foil for this purpose.

## IV. Enrichment and isolation of nitrifiers

Nitrifying bacteria can be isolated from a number of environments. In soil they are usually found in the upper 10 cm. Numbers are greater in alkaline and neutral soils and well-aerated soils are most suitable for nitrification. The surface layer of coastal and estuarine sediments is a favourable site for nitrifiers as concentrations of ammonium and oxygen are relatively high there. Nitrifiers are also found in the upper 200 m of the water column. Marine nitrification rates are lower at the water surface than in subsurface samples due to light inhibition, and maximum rates have been measured in the 50–100-m depth range (Olson, 1981). Distribution of nitrifying bacteria depends largely on the abundance of primary substrate, dissolved oxygen, light intensity and other environmental variables. In rivers and streams nitrifiers may be isolated from the sediment–water interface.

### A. Procedure for enrichment

A 1% inoculum should be used for enrichment cultures. For soil samples, suspensions can be made by mixing 10 g (field moist) soil with 95 ml of deionized water in a Waring blender for 1 min with the inclusion of 6 drops of Tween 80 (Difco) and 2 drops of antifoam (Belser and Schmidt, 1978a). A tenfold dilution of the suspension is made in sterile 1 mM phosphate buffer (pH 7.2). This method is given as a guide and sample preparations may vary. Techniques for studying the microbial ecology of soil are given in this volume.

Serial dilutions of the source material must be made as several ammonia or nitrite oxidizers may be present and dilution will allow isolation of both the dominant and the fastest growing organisms in the sample. The dominant organisms will be found in the higher dilutions and the fastest growing in the lower dilutions (Watson *et al.*, 1981). For example, in a most probable number (MPN) estimate (Section VII.A) of a suspension of silt loam soil, preliminary microscopic examination of all positive tubes showed only *Nitrosospira* in the $10^{-5}$ dilution. In lower dilutions ($10^{-2}$, $10^{-3}$) *Nitrosomonas* was the predominant genus (Belser and Schmidt, 1978).

#### *1. Ammonia oxidizers*

Media shown in Tables I and II for ammonia oxidizers can be used for enrichment cultures but Belser and Schmidt (1978a) showed that the media of Soranio and Walker (1968), Table I, with the addition of the trace elements given by Watson (1971b), required the shortest enumeration time for maximum counts in an MPN estimate for ammonia oxidizers. As the principle of dilution is the same as for enrichment it should also give the shortest time for isolation from the higher dilutions. The appropriate enrichment media can be inoculated with the serial dilutions of the source material or serial dilutions made in the enrichment medium. Different media may be selective for different genera of ammonia oxidizers. Test tubes are usually used for enrichment cultures to save space and these are incubated for up to 4 months at 25–30°C, testing periodically for nitrite production (see spot tests, Section V.A). If no nitrite is detected then the nitrate test should be done to establish if nitrite oxidizers are also present.

Growth of ammonia oxidizers is accompanied by a drop in pH which can conveniently be seen by the inclusion of phenol red in the medium which changes from red to yellow as pH drops. Further growth and ammonia oxidation will only take place if pH is readjusted. To obtain pure cultures, all dilutions showing nitrification must be again serially diluted and the procedure repeated using the highest dilution giving nitrification until pure cultures are obtained. Ammonia oxidizers may be isolated from turbid cultures by streaking onto agar plates, described below.

#### 2. *Nitrite oxidizers*

The same procedure should be adopted for the isolation of nitrite oxidizers using the appropriate media (Tables I and II). When no further nitrite is detected by spot tests, further serial dilutions may be undertaken or additional nitrite added to the media periodically to obtain turbid cultures from which pure cultures may be obtained by serial dilution or plating. Some nitrite oxidizers cannot tolerate more than 1 mM nitrite making it important to use both low and high concentrations of nitrite if these are to be isolated. Details of enrichment and isolation are given by Watson *et al.* (1981).

### B. Plating techniques

Plating techniques can be extremely tedious to use for nitrifying bacteria because of the small size of nitrifier colonies and extremely long incubation periods during which overgrowth by contaminants may occur. Inorganic medium containing either ammonium or nitrite is solidified with purified agar (1–1.5%) including 0.05 M HEPES for ammonia oxidizers and adjusted to pH 7.8–8.0. Plates should be incubated at 25–30°C in a moist chamber for 1–4 months after inoculation with a turbid enrichment culture.

Microscopic colonies may be detected in 3–4 days at 25°C and in 7–8 days colonies may be large enough to subculture. After 2–3 weeks colonies may reach 0.2 mm diameter (Soranio and Walker, 1968). Subculturing of the smaller colonies (20–30 μm diameter) requires a sterile Pasteur pipette drawn into a fine capillary tube. Soranio and Walker (1968) describe a micromanipulator for removing these small colonies. Contaminating heterotrophic colonies may appear early on in the incubation but their incidence should be reduced by the use of purified agar.

Microcolonies have no real distinguishing features and are difficult to recognize amongst contaminating colonies. Brown or orange colour may develop in the larger ammonia-oxidizer colonies. Numerous colonies should therefore be used to inoculate individual tubes containing liquid media which should be tested for nitrification and streaked out again on agar until pure cultures are obtained. (See also dilution plate method, Section VII.C.)

### C. Purity checks

Nitrifier cultures are commonly streaked onto nutrient agar to test for the presence of heterotrophic contamination. Plates are incubated at 25–30°C for up to 3 weeks. It is advisable, however, to use a number of organic media to test for heterotrophic contamination when first isolating nitrifiers and periodically during routine maintenance of pure cultures. Watson *et al.* (1981) suggest the use of 0.25 strength nutrient broth (Difco), 0.5 strength fluid thioglycolate and 0.25 strength trypticase soy broth (BBL). Enrichment cultures showing no growth on heterotrophic media should be examined microscopically for contamination and identification.

## V. Evaluation of growth and activity of nitrifiers by chemical analysis

Growth of nitrifying bacteria is often assessed by the oxidation of ammonium and nitrite or production of nitrite and nitrate by ammonia or nitrite oxidizers, respectively. Yoshioka *et al.* (1982) showed that for an ammonia oxidizer in pure culture, nitrite production showed the same tendancy as the increase in cell number measured by the fluorescent antibody method (Section VII.B). This also applied to the rate of nitrite consumption by a nitrite oxidizer. On some occasions, particularly in the presence of inhibitors of nitrification, substrate oxidation may not coincide with cell division. Underhill and Prosser (1987a) observed wide fluctuations in cell number in a potassium ethyl xanthate-inhibited, continuous culture of *Nitrobacter* whilst the concentration of $NO_2^-$ in the media remained steady.

## A. Spot tests for ammonium, nitrite and nitrate

The presence of ammonium in liquid media or samples may be detected by the addition of Nessler's reagent. A yellow-orange colour is produced which forms an orange-brown precipitate at higher $NH_4^+$ concentrations. This test is sensitive to concentrations below 1.0 μg $ml^{-1}$ $NH_4^+$. Nitrite may be detected by the addition of a drop of Greiss Ilosvay's reagent I (sulphanilic acid) followed by a drop of Greiss Ilosvay's reagent II (8-aminonaphthylene 2-sulphonic acid). In the presence of nitrite, diazonium compound formation results in the development of a pink or red colour. The intensity of the colour increases with increasing concentration of nitrite but above 50 μg $ml^{-1}$ $NO_2^-$ a yellow colour may be produced. The colour intensity may be used to quantify $NO_2^-$ in samples spectrophotometrically.

If no nitrite is detected by this test, the presence of nitrate may be determined by the addition of zinc dust to the sample.

$$NO_3 + 2Zn + H_2O \rightarrow NO_2 + 2ZnO + H_2$$

If a nitrite reaction (red colour) appears over the zinc dust then the presence of nitrate is confirmed.

An initial extraction must be undertaken for the analysis of soil samples. This may be conveniently done by shaking approximately 2.5 g of soil in 12.5 ml 1 M KCl for 25 min. The solution should then be filtered before analysis for nitrite, nitrate and ammonium.

## B. General methods

Bremner (1965a) describes methods for estimating ammonium using distillation methods or Nesslers reagent and nitrite using Greiss Ilosvay's reagents. Snell and Snell (1949) describe the phenoldisulphonic acid methods for determining nitrate. Nitrate concentration can also be calculated by subtracting initial nitrite concentration from the total nitrite after conversion of nitrate to nitrite using copper–cadmium granules. Again, Greiss Ilosvay's reagents are used. Further methods for the determination of ammonium nitrite and nitrate, and their adaptation for use in sea water, are given by D'Elia (1983).

Various autoanalyser systems are now available for the determination of ammonium, nitrite and nitrate and these are extremely useful when large numbers of samples have to be analysed. They also give good accuracy between the ranges of 0.1–50 μg $ml^{-1}$ nitrite and ammonium. Nitrate is reduced to nitrite by hydrazine sulphate under alkaline conditions using copper ions as catalysts and total nitrite measured, nitrate concentration being calculated again by subtraction of previously determined nitrite concentration.

### C. Electrode methods

Ammonia can be determined electrochemically using an ammonia gas-sensitive electrode which consists of a combined glass electrode and ammonia gas-permeable membrane. Equilibration of ammonia across the membrane causes a pH change in the internal electrolyte which is used to measure ammonia concentration in the range 0.1–100 μg $ml^{-1}$ $NH_4^+$. This method is unsatisfactory for on-line measurement of ammonia as the pH of the solution being tested must be raised above pH 11.0 to convert ammonium ions into volatile ammonia gas. The method is also greatly affected by impurities such as amines and precipitates of hydroxides of alkaline earth metals under the alkaline conditions employed.

It is appropriate to include here microbial electrodes in which the nitrifying bacteria themselves are used in the determination of ammonia and nitrite concentration. The ammonia sensor consists of an oxygen-sensitive electrode, a porous membrane and ammonia-oxidizing bacteria which are confined or immobilized by the membrane in direct or close contact with the diaphragm of the electrode (Hikuma *et al.*, 1981). An aerated aqueous sample is contacted with the sensor by pulse injection into a constant temperature flow cell (25–30°C) containing aerated buffer solution (pH 6–9) with a buffer flow rate of 3.9 ml $min^{-1}$. Ammonia gas, ammonium ions and oxygen can diffuse through a gas- or cation-permeable membrane but organic compounds cannot. The base current at a constant dissolved oxygen concentration is determined before sample injection and the reduction in current due to oxygen consumption during ammonia oxidation by the bacteria is detected.

A similar method for determination of sodium nitrite using immobilized *Nitrobacter* sp. is described by Karube *et al.* (1982). Nitrite ions are converted to nitrogen dioxide gas which diffuses through the electrode membrane. Here the gas is converted back to nitrite ions in pH 7.5 buffer where the bacteria metabolize these ions producing a consumption of oxygen detected by the oxygen electrode. The resultant current decrease has a linear response between 0.01 and 0.59 mM $NaNO_2$. These methods are being developed. The electrodes will require frequent renewal and preparation of immobilized cells (21 days or 400 assays in the case of the nitrite sensor).

## VI. Measurement of *in situ* nitrification rates

Measurement of nitrification rates in the environment presents many problems. Contributions to nitrite and nitrate production by other microorganisms makes direct estimation of autotrophic nitrification difficult. Many studies have involved the measurement of nitrate production under optimum conditions in laboratory-incubated samples taken from the environment (Chen *et al.*, 1972; Klingensmith and Alexander, 1983). The measurement of nitrification potentials provides little information however of *in situ* rates of nitrification. Removal of samples can affect concentration gradients that may have existed *in situ* and inactive cells, which did not contribute to nitrification in the environment, may be activated in laboratory incubations. Nitrifiers can survive periods of anaerobiosis. The problem has been addressed in several ways. Hall (1984) developed a method to study nitrification in intact sediment cores. The dilemma is discussed by Hall (1986) in relation to nitrification in lakes and sediments and many of the experimental problems are equally applicable to other environments such as soil.

### A. General methods

Isotopic methods are frequently used to measure nitrification rates in aquatic environments, the most commonly used being the $^{15}N$ tracer methods. In outline, a sample is collected and incubated after inoculation with substrate enriched in $^{15}N$. Nitrite is later extracted from the filtrate of the sample using an azo-dye formation reaction (see Schell, 1978; Olson, 1981). Isotopic tracer incorporation is measured by conversion to nitrogen gas and detection using mass spectrographic or radioisotopic analysis (Bremner, 1965b; Fiedler and Proksch, 1975). Harrison (1983) discusses the use of isotopes for studying nitrogen transformations in the marine environment. The method suffers from a number of drawbacks which may cause artefacts and these are discussed by Ward (1986).

The activity of nitrifiers in the environment has been assessed by the use of inhibitors of nitrification. Growth has been measured by $CO_2$ assimilation using $^{14}CO_2$ as a tracer. This should be a measure of nitrification rate if the relationship between $CO_2$ assimilation and nitrification can be determined. In the natural population $CO_2$ is assimilated by many other autotrophic microorganisms. The contribution of autotrophic nitrifiers is therefore assessed by measuring $^{14}CO_2$ assimilation in samples to which have been added specific inhibitors of nitrification. $^{14}CO_2$ assimilation is then compared in samples to which no inhibitor has been added. The difference represents the activity of the population of interest.

N-serve [2-chloro-6-(trichloromethyl) pyridine] is the most commonly used inhibitor (Billen, 1976; Somville, 1978). It is an inhibitor of ammonia oxidation.

Glover (1985) showed that for both ammonia and nitrite oxidizers the stoichiometric relationship between nitrification and organic carbon production was dependent on availability of inorganic nitrogen substrate. The organic carbon yield from $NH_4^+$ and $NO_2^-$ oxidation decreased with decreasing flow rates in chemostats. For ammonia oxidizers the ratio of carbon fixed to nitrogen oxidized declined with increasing age in batch culture. Gundersen *et al.* (1966) found that this ratio increased with decreasing partial pressures of oxygen. The situation must be clarified for a range of conditions and species if the measurement of $CO_2$ assimilation of nitrifiers is to be a useful indicator of nitrifier activity or nitrification. In addition, the inhibitor used must be specific for ammonium oxidation, i.e. it must not inhibit the activity of other microorganisms in the sample being tested.

Dark incorporation of $^{14}C$-bicarbonate in the presence and absence of N-serve has been used to estimate nitrification in the environment (Billen, 1976). Blockage of nitrite oxidation by chlorate and measurement of nitrite accumulation has been another approach used to estimate the *in situ* ammonia oxidation rate (Belser and Mays, 1980). Chlorate did not totally inhibit nitrite oxidation in samples with high initial activity and in addition to this nitrite oxidizers are able to reduce chlorate to chlorite, an effective inhibitor of ammonia oxidation. Jones *et al.* (1984) used carbon monoxide oxidation for estimation of *in situ* ammonia oxidation. CO acts as an alternative substrate for the ammonia monooxygenase system. Its oxidation should therefore act as a direct assay for ammonia monooxygenase activity and so for ammonia oxidation.

### B. CO oxidation for estimation of ammonia oxidation – a detailed example of *in situ* experimental techniques

The method is given here as a general example of the experimental protocol employed in isotope and inhibitor studies for assessing nitrification *in situ*.

It has many advantages such as small sample size, short incubation times and does not involve adding abnormally high ammonia concentrations to the samples as in many methods. The latter could affect the kinetics of the incubated population. There are old problems, such as specificity of the inhibitor. In addition, the possibility of competitive inhibition of ammonia oxidation by CO exists (Suzuki *et al.*, 1976) which may cause problems if the initial relative concentrations of both CO and $NH_4^+$ are not known.

The oxidation rates of each substrate may depend on their relative concentrations (Ward, 1986).

*1. Experimental procedure* (Jones *et al.*, 1984)

*(a) Sample preparation.* 1 ml of a pure culture is filtered onto a sterile membrane filter (0.45 μm pore size). The cells are washed with $NH_4^+$-free medium and the filter used in the assay. For soil samples 1.0 g wet weight is used. The filter or soil is placed in 60-ml serum bottles containing 25 ml of $NH_4^+$-free medium buffered with 0.2 M HEPES, pH 7.8. 25 ml of natural water samples may be used directly.

*(b) Total CO oxidation.* The serum bottles are gased with CO-free air for 30 s and sealed. 0.5 ml of $^{14}CO$ diluted in nitrogen is added to the head space (0.5 μCi $ml^{-1}$; specific activity 56 mCi $mmol^{-1}$; Amersham Corp.). Equilibrium concentration of CO in solution is determined by multiplying the mixing ratio of 10 by the Bunsen solubility coefficient and the reader is referred to Schmidt (1979) for details. The coefficient depends on temperature of incubation and salinity of samples. The relationship between CO concentration and CO oxidation in the studies of Jones *et al.* (1984) showed a doubling of rate with doubling of CO concentration at low CO concentration (<6.0 nM). Final range of CO in solution in this study was 2.0–2.23 nM CO and oxidation rates of samples from different sources were normalized to rates at 2.23 nM CO for comparative purposes.

Serum bottles prepared as above are incubated on a rotary shaker for 18 h at 25°C and 100 rpm. Note that temperature of incubation depends on the source of the sample and organism used. 1.0 ml of 0.5 N NaOH is injected into the bottles to terminate the reaction and bottles are shaken for 1 h at room temperature to absorb labelled $CO_2$ into the solution. Remaining labelled CO is removed by shaking unstoppered bottles in a fume hood for 30 min. Bottles are then closed with a serum bottle stopper fitted with a plastic rod and cap assembly containing a fluted strip of Whatman No. 1 chromatography paper (25 by 50 mm) onto which is placed 0.2 ml of B-phenylethylamine to trap the labelled $CO_2$. This is subsequently released by adding 2 ml of 10 N $H_2SO_4$ to the sample and shaking for 1 h at room temperature. The $CO_2$ trapping procedure and assay for radioactivity are described by Griffiths *et al.* (1977).

*(c) CO oxidation by autotrophic ammonia oxidizers.* N-serve is added to serum bottles to inhibit ammonia oxidation and the rate of CO oxidation measured in these samples is subtracted from the total CO oxidation.

Either 25 μl of 100 μg $\mu l^{-1}$ N-serve in DMSO or 25 μl DMSO (controls) is added to the serum bottles which are stoppered and preincubated for 1 h on a rotary incubator at 100 rpm, 25°C before addition of $^{14}CO$.

It must be noted that CO oxidation rate loses linearity after 3–6 h in the presence of N-serve. It is therefore advisable to incubate these samples for 6 h or less (Jones *et al.*, 1984). In the absence of inhibitor CO oxidation rate is linear for 48 h.

*(d) Complications.* A problem which is inherent in using inhibitors in the manner described above for analysis of samples taken from the field, is that the inhibitor must be totally specific for the activity of interest. The oxidation of CO by methane oxidizers is also completely inhibited by 100 mg $l^{-1}$ N-serve. It is therefore necessary to distinguish between inhibited CO oxidation of ammonia and methane oxidizer populations. Both methanotrophs and ammonia-oxidizing nitrifiers can oxidize methane (Jones and Morita, 1983) and CO. Jones *et al.* (1984) used the ratio of $CH_4$ to CO oxidation as a distinction between the two groups. For ammonia oxidizers the ratio varies between 0.0007 and 0.0428 whereas for methane oxidizers the ratio varies between 0.380 and 1.87. The ratios can be used to determine whether a system is methane or ammonia oxidizer dominated.

The same experimental procedure described for determining CO oxidation can be used to determine methane oxidation but with the inclusion of 1.0 ml $^{14}CH$ diluted in nitrogen 1.0 μCi $ml^{-1}$; specific activity 59 mCi $mmol^{-1}$; Amersham Corp) instead of $^{14}CO$.

*(e) Conversion to ammonia oxidation rates.* The *in situ* concentration of ammonium must be known. Jones *et al.* (1984) used previously determined $NH_4^+$ oxidation rates at 1.0 mg $l^{-1}$ $NH_4^+$ for $10^6$ cells $l^{-1}$ of three ammonia oxidizers and CO oxidation rates determined at 2.23 nM CO by *N. europaea* to calculate an average ratio of CO oxidized to $NH_4$ oxidized of $3.7 \times 10^{-4}$. Assuming first-order kinetics for ammonium concentrations <1.0 μg $l^{-1}$ $NH_4^+$ and knowing the *in situ* concentration of ammonium and the appropriate rate of CO oxidation, they calculated the ammonia oxidation rates for their samples.

## VII. Enumeration of nitrifying bacteria

### A. Most probable number method (MPN)

The MPN method is commonly used for the enumeration of nitrifying bacteria because of the difficulties experienced with plating techniques.

Alexander (1965) describes this technique for nitrifiers. Successive dilutions of the sample are made to reach an extinction point where no growth occurs. Three to ten replicates of each dilution are made. The MPN number of viable organisms present in the original sample is obtained by scoring the pattern of positive and negative tubes with respect to growth, and referring to a statistical table based on the Poisson distribution. The method suffers from a number of drawbacks with respect to media selectivity (Section IV.A) and long incubation times (up to 4 months at 25–30°C).

The main problem with MPN counts of samples taken from the natural environment is that they may greatly underestimate the nitrifier population if only one medium is used (Belser, 1979). All strains of ammonia oxidizers or nitrite oxidizers must be able to grow in this medium or they will not be detected.

Belser and Mays (1982) suggest the use of an index of counting efficiency in conjunction with MPN counting. This technique involves measuring maximum activities or nitrifying potentials of the sample. To measure ammonia oxidation, chlorate may be added to the sample to inhibit growth of nitrite oxidizers (10 mM). Nitrite production is then measured in incubated shaken soil or sediment slurries. The number of ammonia oxidizers that corresponds to the measured rate of nitrite accumulation is estimated from the equation $X = r/k$, where $X$ is the density of cells and $r$ is the rate of nitrite production. $k$ obeys Michaelis–Menten kinetics and so $k = K_0\ (NH_4^+)/K_m + (NH_4^+)$. The equation $X = r/k$ is modified by making $k$ equal to $K_0$ (its maximum value). A minimum estimate of cell density ($X_{min}$) is therefore obtained ($X_{min} = r/K_0$) since $k$ is always less than or equal to $K_0$ (Belser and Mays, 1982).

Unfortunately different genera of ammonia oxidizers have different $K_0$ values, and the dominant genus in the sample may not be known. Belser and Mays (1982) used a $K_0$ value of 0.023 pmol $NH_4^+$ oxidized $cell^{-1}\ h^{-1}$, on the basis of pure culture experiments of three major genera of ammonia oxidizers. Their counting efficiency index compares the cell density estimated from activity measurements with that obtained from MPN counts. The efficiency was so low in most samples analysed that they concluded that nitrifying potentials are a better quantitative estimate of nitrifier biomass than MPN counts in many circumstances.

### B. Fluorescent antibody techniques

The fluorescent antibody (FA) method has the potential for direct enumeration of nitrifying bacteria in samples, and also for the identification of different species which contribute to nitrification in a particular environment. It has limitations in that it relys on the preparation of FAs for all known strains of nitrifiers that may be expected in the sample, and serological diversity appears to be large. Full details of the preparation of FAs and cross-reaction testing are given by Belser and Schmidt (1978b), Schmidt *et al.* (1968), Lewis *et al.* (1964), Herbert *et al.* (1973) and Fliermans *et al.* (1974). The reader is also referred to the chapter of Chantler and McIllmurray (1987) in this series of volumes.

Belser (1979) characterized seven different serotypes for *Nitrobacter winogradski* and *Nitrobacter agilis*. Belser and Schmidt (1978b) found that serological cross reactions of their FAs occurred only for strains within the genera of three ammonia oxidizers tested. There was however considerable serological diversity, particularly within the *Nitrosospira* genus. They conclude that it would be desirable to produce a genus-specific FA reagent which is also strain comprehensive. This would obviously be convenient in sample testing.

The preparation of FA requires an initial pure culture of the bacteria which are used as the antigens and injected usually into rabbits. Problems in isolating all the nitrifier genera in an enrichment culture are outlined in Section IV. Indeed, new species of nitrifying bacteria have more recently been described and further characterization of the family appears to be necessary (Section II).

### C. Dilution plate method for ammonia oxidizers

Visualization of ammonia oxidizer colonies may be facilitated by a clear zone formed in solidified calcium carbonate mixture as a result of acid production by the colony. This enables counting of colonies and enumeration of bacteria by the usual method. This was the principle used by Winogradski and Winogradski (1933).

A similar method used by Soranio and Walker (1973) is given below.

A mixture containing N-sodium silicate and 5.5 N HCl in equal volumes is distributed in Petri dishes and allowed to set (24 h). The gel is dialysed in running tap water for 2–3 days and finally in distilled water and then pasteurized by immersion in hot water (80°C) for 2 min (purified silica gel may be obtained directly). 0.5 ml of a soil suspension is spread over the surface of the silica gel. 1.5 ml of an acid nutrient salt solution is then added containing 1.5 N $NH_3PO_4$, 50 ml; N HCl, 45 ml; N $H_2SO_4$, 5 ml; $(NH_4)_2.SO_4$, 10 g; $MgSO_4.7H_2O$, 2 g; 0.04% bromothymol solution, 12 ml; Fe/EDTA solution (0.2% Fe), 5 ml; and distilled water, 200 ml. A solution containing equal volumes of N sodium silicate and N potassium silicate is added to the portion of nutrient salts in the Petri dish to give a faintly alkaline reaction (0.5–0.6 ml); 3 drops of 0.04% bromomethyl blue solution and 0.25 g of sterile powdered $CaCO_3$ are also added and the mixture is left to set before incubating at 27°C for 3–4 weeks.

The method is far more tedious to use than the MPN method but, because the cells are spatially separated on the agar at lower sample dilutions, it should allow growth of several genera of ammonia oxidizers present in the original sample and colonies may also be identified microscopically.

## VIII. Laboratory systems for experimental and model studies of nitrification

Interpretation of experimental results from field studies is often difficult. The environment is uncontrolled and factors such as changing weather conditions and heterogeneity can influence results. The effect of other microorganisms on the population or species being studied cannot be accurately accounted for. Variability of environmental factors may make one field study quite unrelated to another. In the laboratory, microorganisms can be grown under well-defined conditions and physical parameters in particular, such as pH, temperature, aeration, moisture content, substrate concentration and composition of growth media, can be better controlled. These parameters can be varied in separate experiments in the laboratory. It is possible to relate these laboratory studies to those conducted in the field, and with caution to predict factors likely to influence the bacteria in their natural habitat.

Axenic laboratory culture experiments and model systems are therefore of greatest value when considered in conjunction with experiments carried out in the natural habitat and when they attempt to mimic particular features of the natural environment. Brock (1986) discusses the current state of microbial ecology and the problems associated with pure culture studies.

### A. Batch and continuous culture methods

In the laboratory the experimenter has to choose between batch and continuous systems. Both have features reflecting particular aspects of the natural environment. Continuous-flow systems allow the possibility of growing microorganisms at submaximal rates which, for most of the time, may reflect their activity *in situ*. Few natural habitats will maintain constant conditions for extended periods of time. Continuous-culture systems have been used to study the transient growth conditions. Steady-state and transient data have been obtained in a variety of systems ranging from glass bead columns (Prosser and Gray, 1977; Cox *et al.*, 1980; Bazin *et al.*, 1982) to free cell culture in a chemostat (Bazin and Saunders, 1973; Keen and Prosser, 1987a), Some experimental systems are used specifically to test theoretical models. Bazin and Menell describe methods for modelling microbial systems in this volume. Gottschal describes ecological studies using the chemostat and turbidostat, also in this volume.

Soil batch incubation studies have been used traditionally for studying nitrification, and these represent a closed system. They are useful in that replication of experiments is more easily achieved because of the simplicity of the technique. Underhill and Prosser (1985) compared substrate oxidation rates of nitrifiers and their sensitivity to an inhibitor in soil and liquid batch culture flask experiments. Soil was sterilized, pH and moisture adjusted and additional ammonia or nitrite added before inoculation with axenic nitrifier cultures and incubation at optimal temperatures. Such experiments give little informatlon about actual nitrification rates in the environment, but are designed to test a specific interaction of nitrifiers with one facet of the natural environment, in this case interaction with surfaces.

In this study a reduced sensitivity of nitrifiers to inhibitor in soil by comparison to liquid culture was investigated further by studying the effect of adhesion to a range of surfaces considered to be analogous to highly charged and inert surfaces found in soil (ion exchange resins and glass). In batch and continuous-culture immobilized cell studies, a mechanism for protection of bacteria from inhibitory effects by interaction with charged surfaces emerged (Underhill, 1982). Anion and cation exchange resins were used in batch culture to show that the site of substrate adsorption is a major feature in the initial colonization of surfaces by *Nitrosomonas* and *Nitrobacter* (Underhill and Prosser, 1987b).

Any detailed study on the effect of surface attachment on growth and activity of bacteria in soil is complicated by the soil's heterogeneous and poorly defined nature. Model surfaces reduce the number of experimental variables allowing the elucidation of causative factors. Such studies can help us to understand and control nitrification in the field. Methods for studying adhesion and attachment to surfaces are given by Fletcher in this volume.

### B. Soil columns as batch and continuous-culture methods

The soil reperfusion system of Lees and Quastel (1946) has been used extensively for the study of nitrification (Fig. 3). In its simplest form it is a closed system where the perfusate is recycled through the column until ammonium is fully utilized. The perfusate can be directly analysed for ammonium, nitrite and nitrate without an initial soil extraction and without disturbing the soil, unlike flask incubation studies. Soil is also aerated by the application of positive pressure to the top of the soil column.

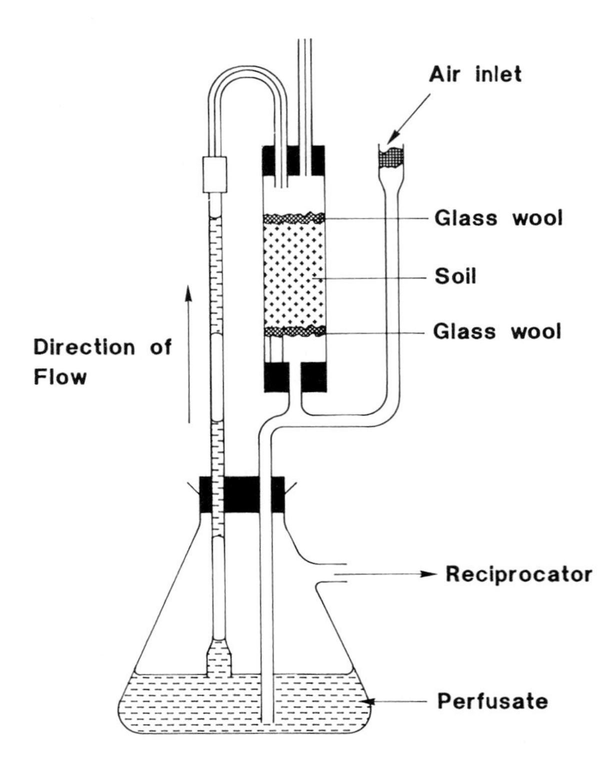


**Fig. 3.** Soil perfusion apparatus. Adapted from Lees and Quastel (1946). The action of the reciprocator causes air to be slowly forced out of and then rapidly sucked into the reservoir. Pressure in the reservoir flask forces perfusate up the lift tube. Suction of air out of the reservoir causes air to be drawn in from outside, bubbling through the perfusate, which is therefore mixed and aerated.

Continuous-flow columns have also been used by many researchers (Schloesing and Muntz, 1877; Macura and Kunc, 1965; Ardakani *et al.*, 1973; Elrick *et al.*, 1966). A constant supply of substrate is added to the top of the column and collected from the base without reperfusion. Continuous soil columns allow study of growth at submaximal rates and under substrate limitation unlike the reperfusion columns.

All of these experimental systems, their uses in nitrification studies and theoretical modelling, are detailed by Prosser (1982, 1986).

Laboratory systems of the type described here have been used extensively to study the effect of inhibitors on nitrification and these studies are described by Powell (1986). Continuous free cell culture studies have demonstrated surprising results such as stimulation of ammonia and nitrite oxidation at low concentration of a potential agricultural inhibitor of nitrification, potassium ethyl xanthate (Underhill and Prosser, 1987a). The stimulation appeared to be a feature of substrate-limited cells and was not readily demonstrated in batch culture experiments which are typically used to test the efficiency of inhibitors of nitrification.

### C. Air-lift columns

Even further removed from the natural environment than packed soil and glass bead columns are studies which have used air-lift columns for the growth of nitrifiers on particulate surfaces (Fig. 4). These were used by Keen and Prosser (1987b), Underhill (1982) and Powell (1985). These systems are used solely to allow a more precise study of the effects of attachment on cell growth, activity and inhibition. In packed columns, gradients of biomass, substrate and product concentrations make interpretation of results and mathematical modelling difficult. Circulation of beads in the air flow of these columns causes sufficient mechanical attrition to allow a steady-state biofilm level to be achieved where growth and detachment are balanced in a homogeneous system.

Criticism has been directed at the use of such systems by traditional microbial ecologists. It is important to appreciate that many model systems aim to mimic a particular feature of the more complex natural environment and to clarify observations of *in situ* experiments which could not possibly be tackled *in situ*.

## IX. Summary

It is hoped that a representative sample of the common techniques used in the study of the microbial ecology of nitrification is given in this article. Laboratory techniques for the isolation and assessment of growth in pure culture are given in some detail. These are a necessary requirement for workers who are field based and those using laboratory model systems. An appreciation of the general features of the family Nitrobacteraceae and their recently discovered metabolic capabilities is also important. *In situ* experiments themselves and the concepts that must be considered are an enormous area and the reader is given only an introduction which should demonstrate the principles involved and allow the development of

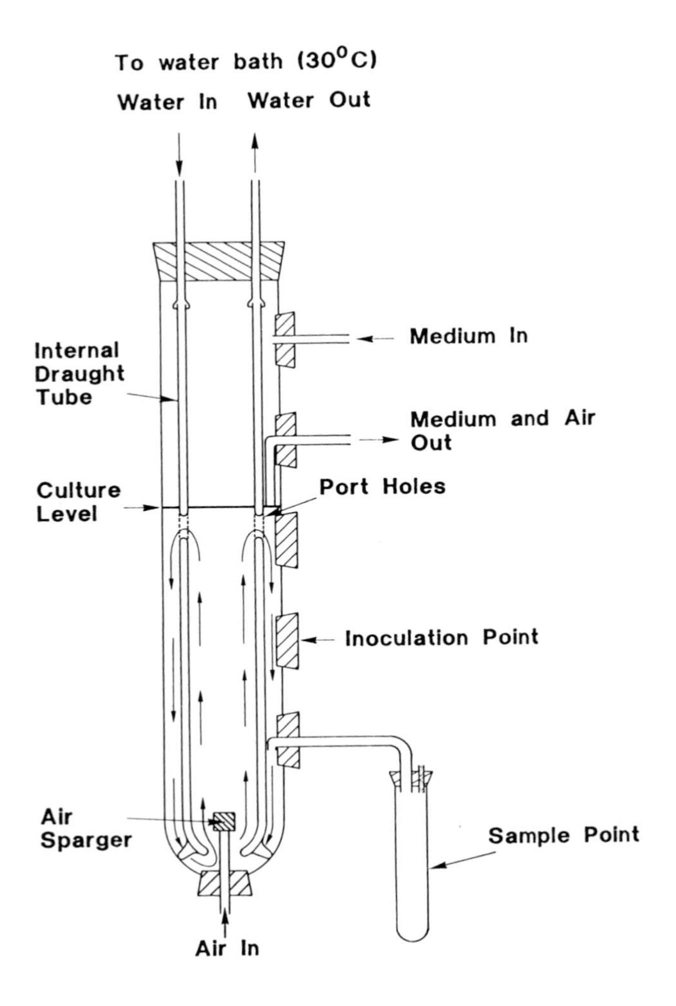


**Fig. 4.** Air-lift column apparatus. Air is introduced at the base of the column. Particles in the liquid medium are lifted in the air flow inside the internal draught tube, pass through ports in the draught tube near the liquid surface and fall back to the base of the column where they are again drawn back into circulation by the airflow. Bacteria immobilized on ion exchange resins or other carrier particles are circulated in this fashion.

independent techniques. The use of model systems has been common and sometimes controversial in studies of nitrification and their general use is discussed elsewhere in this book.

## References


Alexander, M. (1965). *In* "Methods of Soil Analysis, Part 2, Chemical Microbiological Properties" (C. A. Black, Ed.), pp. 1467–1472, American Society of Agronomy, Madison, Wisconsin.
Alleman, J. E., Keramida, V. and Pantea-Kiser, L. (1987). *Water Res.* **21,** 499–501.
Ardakani, M. S., Rehbock, J. T. and McLaren, A. D. (1973). *Proc. Soil Sci. Soc. Am.* **37,** 53–56.
Bazin, M. J. and Saunders, P. T. (1973). *Soil Biol. Biochem.* **5,** 531–543.
Bazin, M. J., Cox, D. J. and Scott, R. I. (1982). *Soil Biol. Biochem.* **14,** 477–487.
Belser, L. W. (1979). *Ann. Rev. Microbiol.* **33,** 309–333.
Belser, L. W. and Mays, E. L. (1980). *Appl. Environ. Microbiol.* **39,** 505–510.
Belser, L. W. and Mays, E. L. (1982). *Appl. Environ. Microbiol.* **43,** 945–948.
Belser, L. W. and Schmidt, E. L. (1978a). *Appl. Environ. Microbiol.* **36,** 584–588.
Belser, L. W. and Schmidt, E. L. (1978b). *Appl. Environ. Microbiol.* **36,** 589–593.
Billen, G. (1976). *Water Res.* **10,** 51–52.
Bock, E. (1965). *Arch. Microbiol.* **51,** 18–41.
Bock, E. (1976). *Arch. Microbiol.* **108,** 305–312.
Bock, E., Sundermeyer-Klinger, H. and Stackebrandt, E. (1983). *Arch. Microbiol.* **136,** 281–284.
Bock, E., Koops, H. and Harms, H. (1986). *In* "Nitrification: Special Publications of the Society for General Microbiology, Vol. 20" (J. I. Prosser, Ed.), pp. 17–38, IRL Press, Oxford.
Bock, E., Wilderer, P. A. and Frietag, A. (1988). *Water Res.* **22,** 245–250.
Bremner, J. M. (1965a). *In* "Methods of Soil Analysis" (C. A. Black, Ed.), pp. 1179–1232, American Society of Agronomy, Madison, Wisconsin.
Bremner, J. M. (1965b). *Agronomy* **9,** 1257–1286.
Brock, T. D. (1986). *In* "Ecology of Microbial Communities" (M. Fletcher, T. R. G. Gray and J. G. Jones, Eds), Symposium of the Society for General Microbiology, Vol. 41, pp. 1–17, Cambridge University Press.
Chantler, S. and McIllmurray, M. B. (1987). *In* "Methods in Microbiology, Vol. 19" (R. R. Colwell and R. Grigorova, Eds), pp. 273–332, Academic Press, London.
Chen, R. L., Keeney, D. R. and Konrad, S. G. (1972). *J. Environ. Qual.* **1,** 151–154.
Cox, D. J., Bazin, M. T. and Gull, K. (1980). *Soil Biol. Biochem.* **12,** 241–246.
Dalton, H. (1977). *Arch. Microbiol.* **114,** 273–279.
D'Elia, C. F. (1983). *In* "Nitrogen in the Marine Environment" (E. J. Carpenter and D. G. Capone, Eds), pp. 731–762, Academic Press, New York.
Elrick, D. E., Erh, H. T. and Krupp, H. K. (1966). *Water Resour. Res.* **2,** 717–727.
Fiedler, R. and Proksch, G. (1975). *Anal. Chim. Acta.* **78,** 1–62.
Fliermans, C. B., Bohlool, B. B. and Schmidt, E. L. (1974). *Appl. Microbiol.* **27,** 124–129.
Focht, D. D. and Verstraete, W. (1977). *Adv. Microb. Ecol.* **1,** 135–214.
Freitag, A., Rudert, M. and Bock, E. (1987). *FEMS Microbiol. Lett.* **48,** 105–109.
Glover, H. (1985). *Arch. Microbiol.* **142,** 45–50.
Goreau, T. J., Kaplan, W. A., Wofsy, S. C., McElroy, M. B., Valois, F. W. and Watson, S. W. (1980). *Appl. Environ. Microbiol.* **40,** 526–532.
Griffiths, R. P., Hayasaka, S. S., MaNamara, J. M. and Morita, R. Y. (1977). *Appl. Environ. Microbiol.* **34,** 801–805.

Gunderson, K., Carlucci, A. F. and Böstrom, F. (1966). *Experientia* **22,** 229–230.
Hall, G. H. (1984). *Microb. Ecol.* **10,** 25–36.
Hall, G. H. (1986). *In* "Nitrification: Special Publications of the Society for General Microbiology, Vol. 20" (J. I. Prosser, Ed.), pp. 127–156, IRL Press, Oxford.
Harrison, W. G. (1983). *In* "Nitrogen in the Marine Environment" (E. J. Carpenter and D. G. Capone, Eds), pp. 763–807, Academic Press, New York.
Herbert, G., Pelham, P. L. and Pitman, B. (1973). *Appl. Microbiol.* **25,** 26–36.
Hikuma, M., Kubo, T., Yasuda, T., Karube, I. and Suzuki, S. (1981). United States Patent, 4297173A, October 27th, 11 pp.
Hooper, A. B. and Terry, K. R. (1974). *J. Bacteriol.* **119,** 899–906.
Hyman, M. R. and Wood, P. M. (1983). *Biochem. J.* **212,** 31–37.
Hyman, M. R. and Wood, P. M. (1984). *Arch. Microbiol.* **137,** 155–158.
Jones, R. D. and Morita, R. Y. (1983). *Appl. Environ. Microbiol.* **45,** 401–410.
Jones, R. D., Morita, R. Y. and Griffiths, R. P. (1984). *Mar. Ecol. Prog. Ser.* **17,** 259–269.
Kalthoff, H., Fehr, S., Sundermeyher, H., Renwrantz, L. and Bock, E. (1979). *Current Microbiol.* **2,** 375–380.
Kaplan, W. A. (1983). *In* "Nitrogen in the Marine Environment" (J. E. Carpenter and D. G. Capone, Eds), pp. 139–190, Academic Press, New York.
Karube, I., Okada, T., Suzuki, S., Suzuki, H., Hikuma, M. and Yasuda, T. (1982). *Eur. J. Appl. Microbiol. Biotechnol.* **15,** 127–132.
Keen, G. A. and Prosser, J. I. (1987a). *Arch. Microbiol.* **147,** 73–79.
Keen, G. A. and Prosser, J. I. (1987b). *Soil Biol. Biochem.* **19,** 665–670.
Killham, K. (1986). *In* "Nitrification: Special Publications of the Society for General Microbiology, Vol. 20" (J. I. Prosser, Ed.), pp. 117–126, IRL Press, Oxford.
Klingensmith, K. M. and Alexander, V. (1983). *Appl. Environ. Microbiol.* **46,** 1084–1092.
Koops, H-P. and Harms, H. (1985). *Arch. Microbiol.* **141,** 214–218.
Krümmel, A. and Harms, H. (1982). *Arch. Microbiol.* **133,** 50–54.
Kuenen, J. G. and Robertson, L. A. (1988). *In* "The Nitrogen and Sulphur Cycles" (J. A. Cole and S. J. Ferguson, Eds), pp. 161–218, Symposium 42, The Society for General Microbiology, Cambridge University Press.
Lees, H. and Quastel, J. H. (1946). *Biochem. J.* **40,** 803–912.
Lewis, V. J., Jones, W. L., Brooks, B. and Cherry, W. B. (1964). *Appl. Microbiol.* **12,** 343–348.
Loveless, J. E. and Painter, H. A. (1968). *J. Gen. Microbiol.* **52,** 1–14.
MacDonald, R. M. (1979). *J. Appl. Ecol.* **16,** 529–535.
Macura, J. and Kunc, F. (1965). *Folia Microbiol.* **10,** 125–135.
Milde, K. and Bock, E. (1985). *FEMS Microbiol. Lett.* **26,** 135–139.
Olson, R. J. (1981). *J. Mar. Res.* **39,** 227–238.
Pan, P. and Umbreit, W. W. (1972). *J. Bacteriol.* **109,** 1149–1155.
Poth, M. (1986). *Appl. Environ Microbiol.* **52,** 957–959.
Powell, S. J. (1985). Ph.D. Thesis, University of Aberdeen, Scotland.
Powell, S. J. (1986). *In* "Nitrification: Special Publications of the Society for General Microbiology, Vol. 20" (J. I. Prosser, Ed.), pp. 79–97, IRL Press, Oxford.
Powlson, D. S. and Jenkinson, D. S. (1971). *Soil Biol. Biochem.* **3,** 267–269.
Prosser, J. I. (1982). *In* "Experimental Microbial Ecology" (R. G. Burns and J. H. Slater, Eds), pp. 178–193, Blackwell Scientific Publications, Oxford.

Prosser, J. I. (1986). *In* "Nitrification: Special Publications of the Society for General Microbiology, Vol. 20" (J. I. Prosser, Ed.), pp. 63–78, IRL Press, Oxford.
Prosser, J. I. and Gray, T. R. G. (1977). *J. Gen. Microbiol.* **102,** 111–117.
Ritchie, G. A. F. and Nicholas, D. J. D. (1972). *Biochem. J.* **126,** 1181–1191.
Schell, D. M. (1978). *In* "Microbiology" (D. Schlessinger, Ed.), pp. 292–295, American Society of Microbiology, Washington.
Schloesing, T. and Muntz, A. (1877). *C.R. Acad. Sci. (Paris)* **85,** 1018–1020.
Schmidt, U. (1979). *Tellus* **31,** 68–79.
Schmidt, E. L., Bankole, R. O. and Bohlool, B. B. (1968). *J. Bacteriol.* **95,** 1987–1997.
Smith, A. J. and Hoare, D. S. (1968). *J. Bacteriol.* **95,** 844–855.
Snell, F. D. and Snell, C. T. (1949). "Colorimetric Methods of Analysis" 3rd Edn, Vol. II, Van Nostrand, Princeton.
Somville, M. (1978). *Water Res.* **12,** 843–848.
Soranio, S. and Walker, N. (1968). *J. Appl. Bacteriol.* **31,** 493–497.
Soranio, S. and Walker, N. (1973). *J. Appl. Bacteriol.* **36,** 523–529.
Steinmuller, W. and Bock, E. (1977). *Arch. Microbiol.* **115,** 51–54.
Sundermeyer, H. and Bock, E. (1981). *In* "Biology of Inorganic Nitrogen and Sulfur" (H. Bothe and A. Trebst, Eds), pp. 317–324, Springer-Verlag, Berlin.
Suzuki, I., Dular, U. and Kwok, S.-C. (1974). *J. Bacteriol.* **120,** 556–558.
Suzuki, I., Kwok, S.-C. and Dular, U. (1976). *FEBS Lett.* **72,** 117–120.
Tsang, D. C. Y. and Suzuki, I. (1982). *Can. J. Biochem.* **60,** 1018–1024.
Underhill, S. E. (1982). Ph.D. Thesis, University of Aberdeen, Scotland.
Underhill, S. E. and Prosser, J. I. (1985). *Soil Biol. Biochem.* **17,** 229–233.
Underhill, S. E. and Prosser, J. I. (1987a). *J. Gen. Microbiol.* **133,** 3237–3245.
Underhill, S. E. and Prosser, J. I. (1987b). *Microb. Ecol.* **14,** 129–139.
Van Droogenbroek, R. and Landelout, H. (1967). *Ant. Leeuw. J.* **33,** 287–296.
Voysey, P. A. and Wood, P. M. (1987). *J. Gen. Microbiol.* **33,** 283–290.
Walker, N. (1975). *In* "Soil Microbiology" (N. Walker, Ed.), pp. 133–146, Butterworths, London.
Ward, B. B. (1986). *In* "Nitrification: Special Publications of the Society for General Microbiology, Vol. 20" (J. I. Prosser, Ed.), pp. 157–184, IRL Press, Oxford.
Watson, S. W. (1965). *Limn. Oceanog.* **10,** R274–R289.
Watson, S. W. (1971a). *Int. J. Syst. Bacteriol.* **21,** 254–270.
Watson, S. W. (1971b). *Arch. Mikrobiol.* **75,** 179–188.
Watson, S. W. and Mandel, M. (1971). *J. Bacteriol.* **107,** 563–569.
Watson, S. W. and Waterbury, J. B. (1971). *Arch. Mikrobiol.* **77,** 203–230.
Watson, S. W., Graham, L. B., Remsen, C. C. and Valois, F. W. (1971). *Arch. Mikrobiol.* **76,** 183–203.
Watson, S. W., Valois, F. W. and Waterbury, J. B. (1981). *In* "The Prokaryotes" (M. P. Starr, H. Stolp, H. G. Trüper, A. Balows and H. G. Schlegel, Eds), Vol. 1, pp. 1005–1022, Springer Verlag, Berlin.
Watson, S. W., Bock, E., Valois, F. W., Waterbury, J. B. and Schlosser, U. (1986). *Arch. Microbiol.* **114,** 1–7.
Winogradsky, S. (1892). *Arch. Sci. Biol.* **1,** 86–137.
Winogradsky, S. and Winogradsky, H. (1933). *Ann. Inst. Pasteur.* **50,** 291–343.
Yoshioka, T., Terai, H. and Saijo, Y. (1982). *J. Gen. Appl. Microbiol.* **28,** 169–180.

# 14

# Methods for Studying the Microbial Ecology of the Rhizosphere

R. CAMPBELL AND M. P. GREAVES

*Department of Botany, University of Bristol, Bristol, BS8 1UG, UK*

and

*Department of Agricultural Sciences, University of Bristol, AFRC Institute of Arable Crops Research, Long Ashton Research Station, Bristol, BS18 9AF, UK*



## I. Introduction

Studies of the rhizosphere use the methods of general soil microbiology with additions and modifications to meet the needs of studying the wide range of microorganisms near, on or just within the plant root. There have

METHODS IN MICROBIOLOGY
VOLUME 22 ISBN 0-12-521522-3

been several recent reviews of rhizosphere microbiology in general (Bowen and Coleman, 1984; Curl and Truelove, 1986; Foster, 1988; Vancura and Kunc, 1988; Lynch, 1990) which include information on methods. Conversely, there have been reviews of microbiological methods which have relevance to rhizosphere studies (Burns and Slater, 1982; Poindexter and Leadbetter, 1986; Gray this volume, Chapter 10). The aim of this chapter is not, therefore, to reiterate details of standard methods, but rather to stress the advantages, limitations and uses of particular methods which might assist workers in this area of study to select appropriate techniques. Because of the difficulties inherent in rhizosphere studies it is often necessary to use several different, but complimentary, methods simultaneously.

One basic methodological problem in rhizosphere studies is that of definition. The endorhizosphere is the root cortex and epidermis which is colonized by microbes, usually saprotrophs. The root surface, if it can be recognized, is the rhizoplane, but it rapidly becomes discontinuous as the epidermal and cortical cells die. In older roots (more than a few weeks) it may be difficult to identify the rhizoplane. The ectorhizosphere extends out into the soil for an indeterminant distance which depends on the plant species, age, soil conditions, etc.

In some vegetation types (e.g. old grass swards) and some experimental conditions (e.g. closely spaced plants in pots) roots may be so densely packed in the soil that it is all effectively rhizosphere which may then be sampled using a corer or auger. None of these methods is strictly quantitative and the amount of soil sampled and its distance from the root depends not only on the root density and the care taken in handling the plants, but also on soil texture, compaction, water availability, etc.

Some grasses, cereals and desert plants have a more clearly defined rhizosphere in which a covering of soil, the so-called rhizosheath, adheres firmly to the roots bound by mucilage and root hairs (Duell and Peacock, 1985; Vermeer and McCully, 1982).

Once having obtained a sample of soil from close to the root, or root samples with most of the general soil removed (see Section IV.B below) the whole range of techniques to study microbial ecology is potentially available. There is, however, the complication that it is often necessary, but very difficult, to distinguish between, or separate, the roots and microorganisms. This is a special problem with any attempt to measure metabolic activity or the concentrations of metabolites or enzymes common to both plant and microbe. It complicates direct examination methods, but is less of a problem with cultural studies where media and conditions select for microbes rather than higher plant cells.

## II. Root structure and exudation

Understanding the rhizosphere, and selecting suitable methods to study it, depends on a knowledge of basic root structure and physiology. This information is conveyed in many biology tests. The knowledge of structure is mostly based on light microscopy and there is still much to be gained from its skilful use on both fresh and prepared specimens in the study of the rhizosphere. Fresh whole mounts of freehand sections, in particular, may show structure and orientation before artefacts are introduced by fixation, embedding and staining (McCully, 1987). There is a tendency to get overwhelmed with high technology and to ignore basic methods. Techniques for light microscopy are described by O'Brian and McCully (1981) from the setting up of the microscope for optimum performance, through specimen preparation, staining, sectioning, etc., the use of epi-fluorescence, polarizing, phase and Nomarksi interference microscopes, photography for microscopy and the formulation of stains, fixatives, etc.

Roots produce a variety of soluble exudates, colloidal gels, live and dead cells which are the basis of the rhizosphere effect (Curl and Truelove, 1986; Vancura and Kunc, 1988; Lynch, 1990). Usually more microorganisms grow, and are more active, near roots than in the bulk soil where they are limited by the supply of carbon, nitrogen and other mineral nutrients. Actually determining what a root normally produces is very difficult. Micro-organisms can affect the quantity and quality of soluble exudents (Curl and Truelove, 1986) and may themselves produce exudates, metabolize insoluble materials and act as sinks to establish diffusion gradients. The usual way of determining what the plant produces is to grow a gnotobiotic plant (without other organisms or with known combinations of organisms) by surface sterilizing seed (with alcohol, sodium hypochlorite, silver nitrate or other sterilants) and growing the resulting seedling in sterile nutrient solution, sand or $\gamma$-irradiated soil. A great variety of different shaped vessels, perfusion equipment, etc., have been described (Dhingra and Sinclair, 1985) and Curl and Truelove (1986) list 40 studies with different species of plant and different collection and analysis methods. All this work is done on seedlings or very young plants, usually in sterile conditions. There are no methods for old plants or perennials, let alone trees, which are growing in anything approximating to natural conditions. Analysis of bulk soil, with and without roots, can give the net effect of plant plus microorganisms on soil chemistry for plants of different ages (Bokhari *et al.*, 1979) and does show differences between axenic plants and those growing under more natural conditions, and between older plants and seedlings.

The only answer to this problem has been to study root activity in terms of carbon and nitrogen balance, with and without microorganisms, by the use of isotopic tracers. They may be continuously supplied, so the plant is uniformly labelled (Whipps and Lynch, 1983), or pulses are supplied to unlabelled plants (Warembourg *et al.*, 1982). Again, there are a variety of controlled environment cabinets and plant containers to hold plants during exposure, usually to $^{14}CO_2$ or $^{15}N$ in the form of dinitrogen gas or a soluble salt. The ingoing and outgoing gases can be monitored, exudates collected and finally the whole plant assayed for isotopic label. Cabinets, for continuously labelled plants especially, may be elaborate and expensive (Whipps and Lynch, 1983) since they not only have to contain and monitor the radioactivity but also allow plant growth over long periods under controlled and possibly sterile conditions. It is also desirable to be able to separately monitor the products (such as $CO_2$) deriving from roots and leaves. No information on the quantities of individual chemicals produced or used at given times is obtained, only overall budgets for the plants and/or microorganisms (see Section V for the time period between sampling). By comparison of different treatments and controls it is therefore possible to determine the microbial contribution to the carbon and nitrogen balances, assuming exudation is not greatly affected by the presence of the microorganisms. Partition of available carbon between different symbionts and saprotrophs in the rhizosphere has also been studied (Paul and Kucey, 1981).

## III. Direct examination of the rhizosphere

### A. General considerations

Direct observation of microorganisms in the root region allows estimation of the positions of microbes in relation to each other, to microhabitats and to the root, a relationship which is destroyed, or not addressed, by culturing (Section IV) or activity measurements (Section V). It is also possible to see organisms which will not grow in culture or which are dormant. The sole aim may be to observe microbial communities in order to stimulate the development of hypotheses about their interactions and interrelationships. Alternatively the aim may be to enumerate organisms, and possibly measure sizes for biomass estimates.

There are problems with direct examination. Firstly the organisms must be present in reasonable numbers. At microscope magnifications needed to see microorganisms, the area and depth of the field of view is necessarily small so unless quite high numbers are present, there will be few or no organisms visible in most fields. This is not usually a problem with bacteria and fungi in general, but if only one species or strain is being observed (as in some fluorescent antibody techniques, Section III.C), then this can be a problem. Similarly for protozoa, only one light microscope field in 10 is likely to contain an organism (Clarholm, 1981) so observation is possible, but not enumeration.

Secondly, the opacity of the soil matrix and the large size of the roots (in relation to the microorganisms) means that microbes may often be 'hidden'. There is no satisfactory solution to this problem, although various methods of dispersing particles and spreading them on agar-coated slides, membrane filters, etc., have been devised (Poindexter and Leadbetter, 1986) and fluorescence microscopy (Section III.C) is also useful.

### B. Visible light microscopy

O'Brien and McCully (1981) give an extensive summary of methods for light microscopy, especially of roots. The material may need to be cleared (made transparent) before observation and this is usually done with lactophenol, often combined with a stain such as aniline blue. More drastic methods such as boiling chloral hydrate or autoclaving in KOH may also be used, especially for the observation of fungi within roots (see Section VII for mycorrhiza). Phase-contrast and interference-contrast microscopy may be useful for studies of root and nearby soil, but in general the confusion from thick specimens such as whole roots, curved epidermal cell surfaces and extraneous material is difficult to interpret. Bright-field microscopy with staining may be preferable and cotton blue, aniline blue and carbol erythrosin have been used successfully (Rovira *et al.*, 1974; Campbell, 1982). Different staining times are needed for different species or ages of root in order to get stained organisms against an almost unstained background. These techniques are simple and quick, do not require specialized microscopes and the stain does not fade or quench (see below for some fluorochromes which do fade).

There remains, however, the problem of distinguishing bacteria, especially, from small pieces of organic debris which may also stain. Fungi may be counted, but being filamentous the 'number' has little meaning and what is usually needed is the length of hyphae which may be determined by direct measurement or more quickly by the line intersect method (Rovira *et al.*, 1974; Campbell, 1982).

Organisms growing near roots may be observed by placing microscope slides or other surfaces in the soil and observing them after colonization (Dhingra and Sinclair, 1985), although this may tell more about the colonization of slides than about the true soil flora and fauna.

There have been many attempts to observe roots *in situ* by the use of various observation chambers and specially designed boxes to make roots grow against glass or clear plastic panels so that they can be observed without disturbance (Dhingra and Sinclair, 1985). The glass may be a microscope cover glass to enable close observation (Polonenko and Mayfield, 1979). There are limitations on the size of plant which can be used and there is often limited optical resolution because of the physical constraints on the system. This is even more true of field studies using buried tubes or channels and fibre optic light systems which allow observations from the almost undisturbed soil (Rush *et al.*, 1984; Taylor, 1987).

Apart from using direct examination just to count microbial populations on roots and in nearby soil, their distribution pattern(s) can be determined and examined statistically. General observation will give information on whether bacteria are present individually or in microcolonies, although it may be difficult to detect pattern amongst the random variability by simple inspection. Specially designed sampling systems are needed to detect small changes in the size or intensity of groupings with distance up the root, or with age of the plant, etc. Direct observations are usually done by superimposing a grid of squares on the field of view and counting organisms within a known area on the grid (the grid may be an eyepiece graticule for light microscopy or an overlay for electron micrographs: see Campbell, 1982, for example). By using single or contiguous squares, groups of squares, lines of groups across a root, clusters of lines, different roots and so on, the variances are determined for the different sizes of sampling unit. These may then be compared to detect clusters of organisms or patterns of colonies which represent an uneven, but not random, distribution of the organisms (Newman and Bowen, 1974; Polonenko *et al.*, 1978; Polonenko and Mayfield, 1979). Such pattern analysis techniques can quantify purely descriptive information on microbial distribution which is obtained by direct examination with both light and electron microscopes.

## C. Ultraviolet light microscopy (epifluorescence microscopy)

These methods attempt to overcome some of the problems of opacity of soil particles mentioned above, and to give direct microscopy a diagnostic value by differentiating between live and dead organisms and by making some of the stains species or even strain specific. A special adaption to the standard microscope is needed (O'Brien and McCully, 1981) but these are readily available for modern research microscopes. Stains (fluorochromes) which absorb ultraviolet light and emit in the visible spectrum are available to stain different cell components (e.g. nucleic acids, proteins, cell wall carbohydrates, etc.) and the ones in use for rhizosphere studies are ethidium bromide (Roser, 1980), acridine orange (van Vuurde and Elenbaas, 1978), fluorescein diacetate (FDA; Soderström, 1977; Schnurer *et al.*, 1985), europium chelate and fluorescent brightner (Fig. 1) (Anderson and Slinger, 1975; Johnen and Drew, 1978; Johnen, 1978) and 8-anilino-l-naphthalene sulphonic acid (Polonenko *et al.*, 1978; Polonenko and Mayfield, 1979). However, many of these are rather general in their action and give much non-specific staining of soil debris and roots which makes interpretation difficult. There are possible ways to avoid this; with acridine orange and europium chelate for example various counter stains and blocking procedures have been proposed (van Vuurde and Elenbaas, 1978; Schans *et al.*, 1982; Johnen, 1978). Some fluorochromes fade (quench) especially under ultraviolet illumination, so that the image disappears. However, this is not generally serious with fluorochromes in common use which maintain their colour for long enough (minutes or hours) for the material to be counted or photographed. There are also additives which slow the rate of fading and may increase the colour contrast between different sorts of organism or between the organism and the background (Wynn-Williams, 1985).

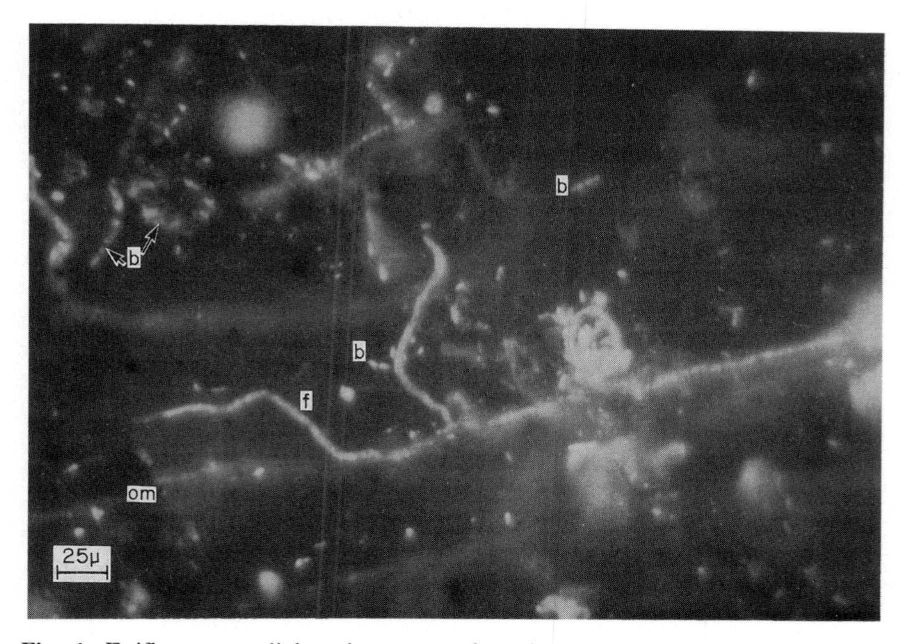


**Fig. 1.** Epifluorescent light microscopy of a wheat root stained with europium chelate and counterstained with 'Calcafluor' (Johnen and Drew, 1978). Fungal hyphae (f) and bacteria (b) grow down boundaries between epidermal cells as well as in chains and groups on the cell surface. Organic matter (om) remains unstained. (Bar = 25 μ.)

Fluorescence microscopy can be made much more specific by combining it with immunological methods. These have been extensively reviewed and described, so will only be outlined here (Bohlool and Schmidt, 1980; Chantler and McIllmurray, 1987). The object is to identify species or strains of an organism, distinguishing them from the general population. A culture of the organism is required so all the problems with culturing soil organisms are involved (Section IV.A). Antibodies to fungi have proved difficult to make specific, but strain specificity is possible for bacteria and the situation should improve as monoclonal antibodies are used rather than polyclonal (Chantler and McIllmurray, 1987). Purified antibody should attach specifically to the organisms of the same type as were used in the initial cultures, although there is usually some cross-reaction with related organisms or strains. The antibodies are visualized by chemically bonding them to a fluorochrome (usually fluoroscein isothiocyanate) and using ultraviolet fluorescence microscopy. Other markers may be used such as ferritin (Jacobs *et al.*, 1985) or colloidal gold for electron microscopy or an enzyme which can produce a coloured product for visible light microscopy. It may be easier to use a two-stage process in which the specific antibody (produced in rabbit) is marked with commercially available, ready-labelled goat-antirabbit antibodies. These processes are time consuming and often need considerable skill and immunological knowledge. There may be unacceptable levels of background fluorescence and non-specific staining in soil and on roots so that the tests are in fact done on cultures from the root (Section IV) and then each isolate is tested to identify the required strain. Obviously the latter must be present in considerable numbers to make this worthwhile or to see it on direct examination (see above, Section III.A). So far fluorescent antibody techniques have only been used in rhizosphere studies for identifying *Rhizobium* strains (which are naturally concentrated in nodules) and to follow bacteria inoculated, for plant disease control or plant growth promotion, in very high numbers (Schank *et al.*, 1979). Unless present in excess of $10^6$ $g^{-1}$ soil or root they are unlikely to be detected by direct microscopy without enrichment cultures. Many experimental studies have been done on many different bacteria from soil and from roots and the technique is 'promising' or 'may be useful if only . . .' this or that problem may be overcome (Bohlool and Schmidt, 1980), but much work is being done to improve and simplify the technique because of the need to track genetically engineered and other released organisms in the environment.

### D. Transmission electron microscopy

We are not concerned here with the methods for the study of the plant root itself, but only with those which include some rhizosphere soil and its associated microorganisms. The fixation and embedding techniques are, with minor variations, standard glutaraldehyde-osmium followed by dehydration and embedding in the standard resins (e.g. Faull and Campbell, 1979; Foster and Rovira, 1976; Foster, 1981, 1982, 1986, 1988; Greaves and Darbyshire, 1972; Greaves and Sargent, 1985; Kilbertus, 1980). There may be further staining or the use of cross-linking stabilizers (Foster, 1986) to retain and stain mucilage and other polysaccharides, with ruthenium red (Greaves and Darbyshire, 1972; Foster, 1981), lanthanum hydroxide (Foster 1981, 1982), periodic acid with silver methionine or Thiery's silver proteinate (Foster, 1981). These special stains may require some modification of the fixation schedules (Foster, 1988).

There is a major problem with sectioning such material containing soil: embedding tends to be uneven because of the different hardness of the mixed materials and it is impossible to cut sand grains. Soils or parts of the root without sand grains must be selected. Diamond knives help in obtaining good sections but are damaged by contact with even small sand grains.

All sectioned material is necessarily dehydrated and this can alter the apparent structure and the dimensions of mucilage. In most transmission electron micrographs (and most dried scanning electron micrographs, see below) the mucilage appears to be fibrous, in sheets, or lamellae. In normal soils this mucilage will be a hydrated and expanded gel. Care should therefore be exercised in measuring sizes and relative positions in dehydrated materials.

There are various other analytical procedures and instruments which may be used in association with transmission electron microscopy (Bisdom, 1983; Foster, 1986). Electron microprobe analysis can give information on the composition of soil particles for elements above calcium in the periodic table, but is not useful for most organic materials. Bisdom (1983) describes various techniques for soil examination and analysis such as laser microprobe mass analysis which have not been used on roots or rhizosphere soils.

### E. Scanning electron microscopy

Scanning electron microscopy has been used to study the rhizosphere using standard preparatory techniques, usually freeze-drying or critical point drying (Robards, 1978; Giddings and Wray, 1986; Cole, 1986). Fresh or air-dried material is often so distorted as to be useless.

Freeze drying involves rapidly freezing a small specimen (such as a part of a root up to 1 or 2 mm in diameter, or a freehand section) in isopentane or freon held at their melting points by immersion in liquid nitrogen. The ice is then slowly sublimed from the specimen under vacuum either in a special apparatus where the temperature can be controlled or more simply by placing them in good thermal contact with a metal block cooled in liquid nitrogen which gradually warms up in the vacuum. This drying may take some hours or even days if a low temperature is maintained: it is commonly done overnight so that after gold coating, the specimens are ready next day. Frozen material or dried specimens may be broken to reveal the internal structure of the root and any microorganisms within the endorhizosphere. There is general shrinkage of the root and mucilage (see below) but freeze drying is an easy method of preparation, resulting in visually attractive material, without any very expensive apparatus.

Critical point drying attempts to overcome the main drying problem in a different way. The main distortion to specimens results from surface tension effects as they dry. Freeze drying gets over this by subliming from solid ice, in critical point drying the liquid goes to vapour at the same density without the formation of a gas–liquid interface, and hence no surface tension effects. This is possible at the critical temperature and pressure for the liquid, which for the most frequently used one (carbon dioxide) is 31.1°C and 73 atm ($7.4 \times 10^3$ kN $m^{-2}$). The specimen is fixed, usually in glutaraldehyde, dehydrated in a water–ethanol series and transferred to acetone or absolute alcohol in the critical point drying apparatus. The ethanol is replaced with liquid carbon dioxide under pressure and below 31.1°C, after which the temperature of the sealed vessel is raised. Again dried specimens may be fractured to reveal internal structure. There is some shrinkage of roots under this system but delicate structures such as root hairs are generally less distorted than in freeze drying. The method does, however, require special apparatus and, more seriously, involves fixative, water washing and dehydration in organic solvents which may remove some soluble components of the specimen and/or disturb spatial arrangements when the root and soil is immersed in liquid.

Both freeze drying and critical point drying involve drying the specimen, however carefully, before viewing in the scanning electron microscope. The hydrated mucilage which covers the root is invariably reduced from a layer several tens of micrometers thick to a thin dried sheet (Campbell and Rovira, 1973; Rovira and Campbell, 1974; Old and Nicolson, 1978; Foster *et al.*, 1983). Some of the mucilage is also washed off. This enables the bacteria, fungi, etc. contained within the mucilage that remains to be seen as an aggregation of cells dried down onto the root surface. While visually attractive, and bearing some relation to the original arrangement of the root, the mucilage and the microorganisms, there has been considerable distortion and simplification (Campbell and Porter, 1982).

Some of these problems resulting from the dehydration of mucilage can be overcome by the use of low-temperature scanning electron microscopy with fully or partly hydrated specimens (Beckett and Read, 1986). The specimen is rapidly frozen in slushed nitrogen and thereafter kept in a dry argon atmosphere below −130°C to prevent condensation and ice recrystallization. The specimen may be observed as it is fractured, etched to remove surface water, and/or gold coated before examination in the scanning microscope with a cold stage maintained below −175°C. Fully hydrated specimens have all the water surfaces and interfaces around the root intact and the mucilage is hydrated as a gel stretching out into the surrounding soil (Campbell and Porter, 1982) (Fig. 2). It is possible to dissect away parts of the specimen and then re-examine, so that a root may just be observed covered in soil and then observations continued as soil is removed to get nearer to the root (Campbell and Porter, 1982). Relationships between microorganisms and surfaces, including liquid surfaces, can be seen (Campbell, 1983), but most of the microorganisms are within the mucilage and so are observed only on fracturing or etching. The interpretation of these specimens is more complex than dried material, but the structure observed is much closer to the natural situation. Specimen preparation can be very quick (a few minutes) but specialized apparatus and additional equipment on the microscope are needed.

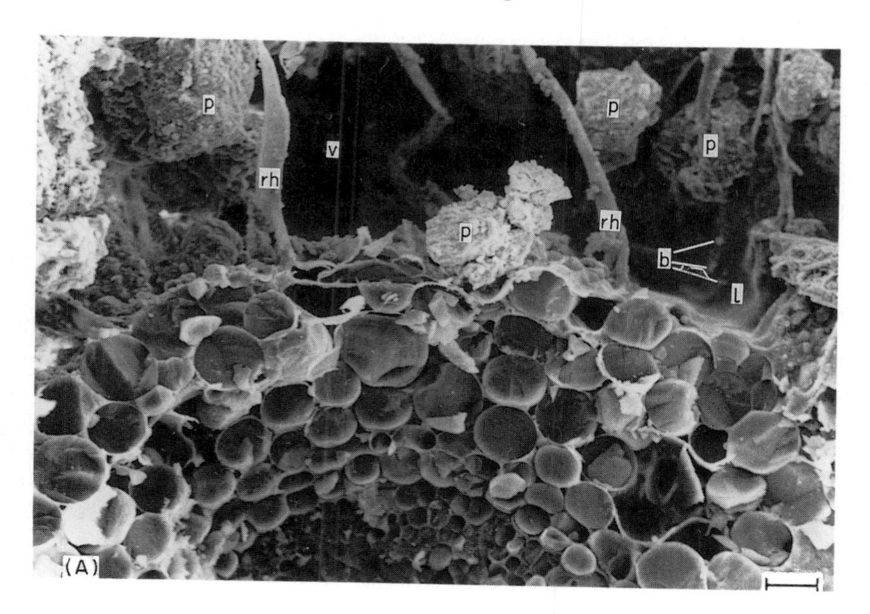


**Fig. 2A.** See caption on next page.

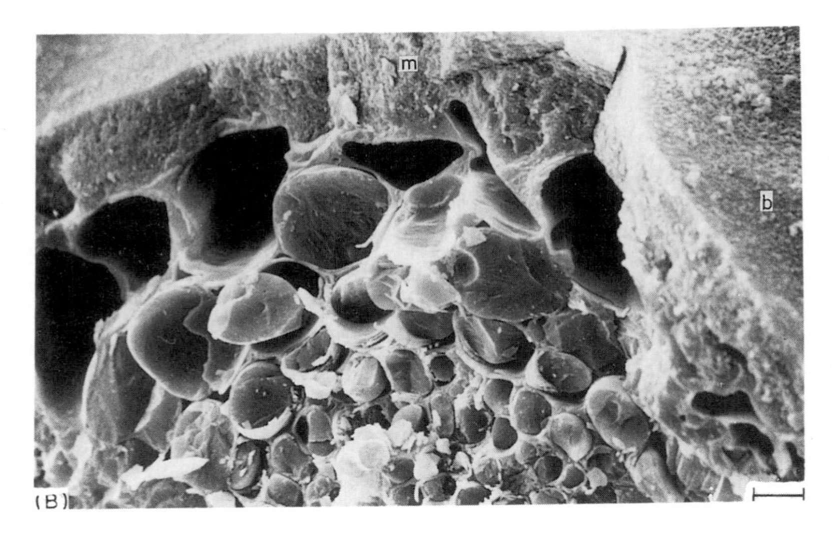


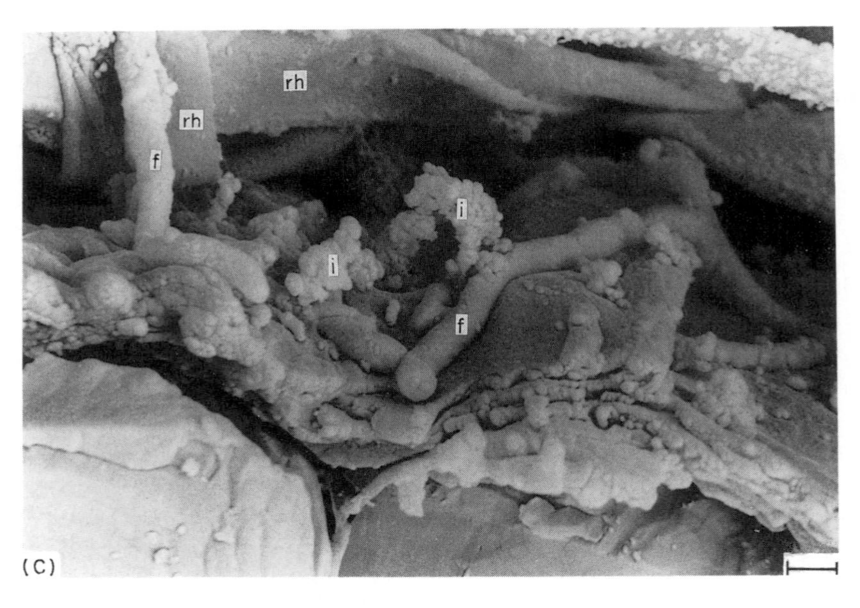


**Fig. 2.** Frozen hydrated specimens of wheat roots viewed in the scanning electron microscope showing: (A) root hairs (rh), a gas void (v), soil particles (p) and liquid interfaces (l) in which bacteria (b) are growing (bar = 25 μm); (B) the mucigel layer (m). Bacteria (b) grow both at the surface of this layer and within its matrix (bar = 10 μm); (C) fungal hyphae (f) growing on the root surface amongst root hairs (rh). The irregular, smooth particles (i) are ice (bar = 4 μm).

Scanning electron microscopy can give much useful information on the structure of the rhizosphere and its microbial communities and interactions. There is a temptation to use it for quick illustrations of some point derived by other methods, but in practice the use requires much time and patience to get the correct interpretation. Along with the transmission electron microscope, the scanning electron microscope has the advantage of high magnification to give a picture of even the smallest microorganism, but the converse is that the field of view will be correspondingly small, so sampling frequency is likely to be very low in relation to the whole root system of even a single plant.

## IV. Culturing microorganisms from the rhizosphere

### A. General considerations

It is generally accepted that, while studies of microorganisms in cultures, as opposed to *in vivo*, offer practical advantages of relative ease of handling, there are considerable drawbacks. In particular, isolated organisms may be stimulated to artificially high metabolic rates by growth in laboratory media, which are usually nutrient rich, although repeated subculture can result in almost total loss of some physiological characteristics. Many organisms, adapted to low-nutrient conditions in soil, will not grow on normal laboratory media (see Section IV.D; Olsen and Bakken, 1987). Furthermore, such organisms are removed from their normal ecological associations which may exert significant regulatory influences. Thus, while interesting data can be obtained using cultural techniques, their value is limited. At best, the data can indicate the potential for certain organisms to express defined characteristics in their natural habitat. They can never prove that the expression does occur and should never be used to extrapolate to the situation in the rhizosphere.

This broad criticism applies principally to studies of pure cultures in relation to eliciting ecological information. There is, of course, a range of situations in which pure culture studies are essential. Not only are they a key to the identification of microorganisms but, with the current emphasis on agricultural biotechnology, they are the principal means of isolating, identifying and improving microorganisms with potential for exploitation.

### B. Sampling the rhizosphere

As discussed earlier (Section I), the rhizosphere includes microorganisms colonizing root cortex and epidermis, occupying the spaces both within and between cells, the outer surfaces of the root and the soil immediately adjacent to the root. Further, the nature of the rhizosphere flora will vary according to the region of the root supporting it (and therefore, the root age) and with the precise characteristics of the soil.

Clearly, it is necessary to use a variety of methods to separate these different parts of the rhizosphere. These methods will be selected to suit the characteristics of the root being studied and the soil in which it is grown.

In principle, the root system must be removed from the soil with as little damage as possible. In pot-grown plants this may be relatively simple but for field studies there can be considerable problems, especially with deep-rooted species. In practice, roots are excavated with a fork or spade and the amount of root recovered for each plant is a function of root morphology and strength, soil structure and the time available. In the majority of cases extraction of roots in the field produces a mixed sample of roots from several more or less adjacent plants. Should roots from only single plants in a crop be required, the plants must be grown so that the roots are prevented from intermingling. Steen (1991) has used fine mesh cylinders sunk into the soil before sowing for this purpose. Metal or plastic cylinders inserted in the soil can serve a similar purpose (Christie *et al.*, 1978), localized sections of roots can be selectively sampled when plants are grown in root boxes or lysimeters with detachable panels. Roots can be observed during growth in such equipment and sampled at appropriate times.

Once removed from the soil, the roots will carry a soil load which is determined by root morphology and mass as well as by soil characteristics. Rhizosphere soil is, in practical terms, that soil which is tightly held on the roots, forming a thin 'cylinder' around the roots. An extreme example of this is the 'rhizosheath' referred to earlier (Section II). With plants grown in sieved soils in pots or in light soils in the field, it is usually sufficient to shake the roots gently in air to remove loose 'non-rhizosphere' soil. In many instances, especially on clays or highly organic soils, stones and large aggregates of soil or organic matter will be quite firmly held within the root system. If not removed, these can significantly dilute the rhizosphere sample. Normally soil aggregates can be removed by lightly crushing between sheets of greaseproof paper or by using forceps. Whichever method is used care must be taken to minimize damage to the roots in the process.

### C. Separation of rhizosphere components

Separation of rhizosphere soil from roots can be achieved by a variety of methods. The method chosen will depend on the plant species and the soil it is grown in. At the simplest level small aggregates of soil can be removed from the root using forceps. More commonly shaking the roots in water (Harley and Waid, 1955), sometimes with glass beads (Rovira *et al.*, 1974), is enough. In some instances, the microbial cells may be released more effectively by using an electrolyte solution (e.g. 0.1 M NaCl) in place of the water. Isotonic solutions also may prevent some potential loss of microorganisms through the osmotic shock effect of dispersal in water.

Often the suspension of rhizosphere soil can be cultured directly but in some cases it may prove necessary to release organisms from the forces which bond them to the soil particles and then to extract them from the suspension. Macdonald (1986a,b,c) has described a process which achieves this. The soil suspension is shaken gently with the $Na^+$ form of Dowex A1 ion exchange resin beads for 2 h at 4°C. Addition of a non-denaturing anionic detergent, sodium cholate, increases the yield of microorganisms. Non-filamentous microorganisms are separated from the suspension by sieving through meshes of decreasing porosity. The final mesh size of $<30$ μm excludes vegetative fungi, actinomycetes and filamentous algae from the final suspension. This contains silt and clays as well as microorganisms and these are separated by density gradient centrifugation. Before that it is necessary to remove some of the mineral material by elutriation, a process exploiting the differential sedimentation rates of organisms and mineral particles. This dilutes the sample so excess water is removed by hollow fibre ultrafiltration. Organisms in the concentrated suspension are then extracted by density gradient centrifugation. Elutriation and sieving can be adjusted so that any major size class of microorganism can be separated. Macdonald (1986c) describes this procedure as applied to bulk soil, but there is no reason why it cannot be applied to rhizosphere soil, although the small amounts of rhizosphere soil usually obtained from single plants, especially young ones, may necessitate some adjustments in the method.

Removal of the rhizosphere soil, even by gentle shaking in water, will also remove some microorganisms from the root itself. However, most organisms will remain firmly attached to the root surfaces or embedded in the mucigel. The nature of the root surfaces will vary according to its age and will only rarely be an intact layer of relatively smooth epidermis cells. Most usually the surface comprises a complex of broken cells, cell fragments, soil debris and mucigel. In either case considerable force will be required to remove the microorganisms. The technique used most frequently involves vigorously shaking the soil-free roots in water for long periods with glass beads (Louw and Webley, 1959; Greaves and Webley, 1965; Rovira *et al.*, 1974). The collisions between beads and roots effectively remove most of the organisms on the root tissue, but will also remove some microbial cells from within the crushed tissue. Gentler techniques which remove mucigel, such as shaking roots in 1 M $NH_4Cl$ solution (Jenny and Grossenbacher, 1963) or the resin technique of Macdonald (1986a) have not been applied to studies of root microorganisms but might offer some advantages.

Microorganisms colonizing the root internal tissues, the endorhizosphere, can be released by a variety of techniques. For small samples all-glass tissue macerators can be very effective. Larger samples can be disrupted in Waring blenders or similar apparatus or, more efficiently and conveniently, in a 'Stomacher' (Colworth Laboratories) as described by Greaves *et al.* (1978).

As well as applying the above techniques to bulk samples of roots they can be used, perhaps with modification to their scale, to treat quite small selected samples of root. It is well known that the density and composition of the root microflora varies along the length of the root. Dissection can be used to separate different zones of the root such as root cap, meristem, elongation zone, root hair zone and progressively older segments back along the root behind the root hair zone. These latter will include root segments from which lateral roots have emerged. Much useful information on microbial distribution on roots can be obtained in this way. There are also methods available to dissect soil away from roots at known distances (Papavizas and Davey, 1961), but they rarely yield enough material for detailed analysis. Dissection methods in general are tedious and time consuming to do with precision. Inevitably, therefore, sample sizes obtained tend to be small and the subsequent separation of rhizosphere components requires the techniques to be scaled down appropriately. Such techniques are, however, the only ones that allow precise localization of samples in relation to the root (though see Section III.E).

It cannot be stressed too much that the methods used to separate roots from bulk soil or rhizosphere soil, and to extract microorganisms from the separated components, will require to be selected on the basis of the nature of the roots being studied, the characteristics of the soils in which they are grown and the precise objectives of the research. In addition, each selected method will inevitably require some calibration. For example, the time of shaking roots to remove rhizosphere soil, or of stomaching to release endorhizosphere microorganisms, will vary according to the plant species under study. It is essential that preliminary studies are made to establish optimum conditions for each project.

### D. Isolation and culture of rhizosphere microorganisms

Many rhizosphere studies require microorganisms to be isolated and put into culture in some way. The rhizosphere is characterized by high numbers of a diverse range of microorganisms, therefore the methods used for isolation and culture must also be diverse.

Detection of some specific microorganisms or groups of microorganisms can be achieved using relatively simple selective techniques. Plating of small amounts of rhizosphere soil or pieces of washed root on the surface of solid highly selective media can be used effectively (Harley and Waid, 1955). Root pieces can be surface sterilized to obtain cultures of organisms which colonize the internal root tissues. This technique is commonly used to isolate pathogens from infected plant material and reduces interference from saphrophytes (Dhingra and Sinclair, 1985).

Selective isolation can also be achieved by inoculation of rhizosphere soil, soil suspensions, root macerates, etc. into selective liquid media. Subsequently, the cultures are inoculated into fresh medium and the process repeated until a stable pure or mixed culture is obtained. Similar results can be obtained using perfusion columns or sophisticated fermentation equipment. These approaches are commonly used to isolate organisms with specific nutritional requirements or with specific metabolic activities. For example, degraders of xenobiotics are readily obtained in this way (Lappin, 1984).

For the enumeration of microorganisms in the rhizosphere, the soil or root suspensions invariably require some degree of dilution in order to produce densities of viable cells which can be counted. The wide range of organisms in soil is present in equally wide ranges of numbers. This presents several problems. Ideally, counting techniques should be as non-selective as possible, in order to include the majority of organisms present. In practice, there is no such thing as a non-selective medium and the physical procedures used invariably impose some selection. Dilution of microbial suspensions, for example, will either be so high as to eliminate those organisms with low population density from the count or, if they are to be included, will be so low that antagonism, antibiosis, lytic activity and competition will all affect the final population of colonies developing on dilution plates and so give a false picture of the density and composition of the microbial community being analysed. One way to overcome this problem, or at least to reduce its effects, is to count different components of the population separately by using selective media (Figs 3 and 4). It is not surprising that the total microbial numbers in soils or the rhizosphere has never been fully assessed, even after more than 100 years of research.

Cultural methods of enumeration involve encouraging individual microbial cells to grow to the point where they can be seen as colonies or turbidity in solid or liquid media. The traditional pour-plate method is used most commonly though better results are often obtained with the spread

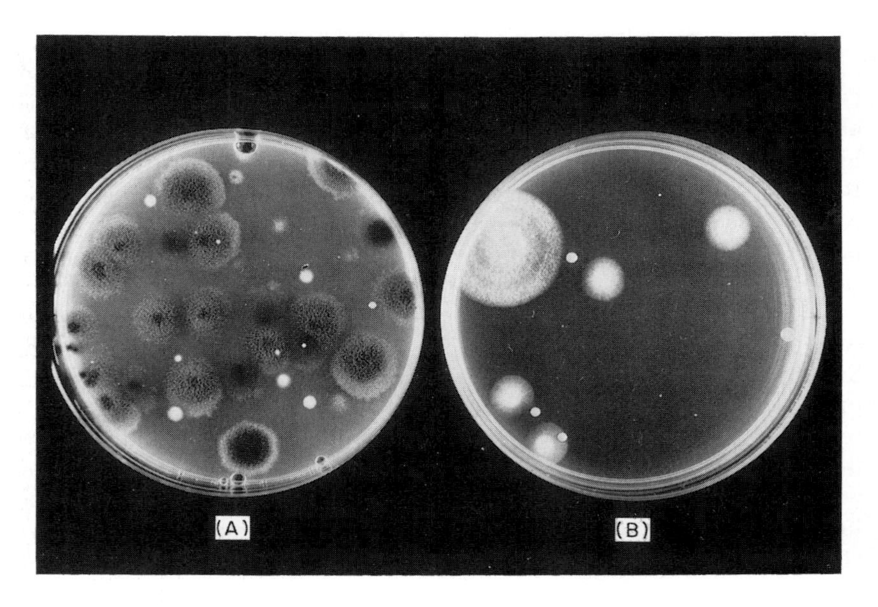


**Fig. 3.** Traditional pour plates of soil fungi showing the effect of medium composition on the composition of the isolated population. (A) Potato dextrose agar, pH 5.5. (B) Soil extract agar, pH 5.5.

plate method or a most probably number method (e.g. Darbyshire *et al.*, 1974). More sophisticated spiral plates coupled with instrumental analysis of colony density are available and counting of colonies on plates can also be made quicker and more accurate with modern image analysers. With both these instrumental methods problems can be encountered with thin spreading colonies which are translucent.

Methods exploiting statistical means of calculating most probable numbers (MPN) are commonly used. These may use microtitre plates where dilution and growth are achieved in the same plate and most stages in preparation can be done automatically (Darbyshire *et al.*, 1974). Alternatively, simpler approaches with test tubes are effective. Where specialized organisms are the subjects of study, highly selective media can be used. The basis of this method is that aliquots of serial dilutions of the original soil or root suspensions are inoculated into replicate samples of liquid medium. The number of samples then showing growth after appropriate incubation can be used to derive a 'most probable number' of cells in the original sample. In theory a single microbial cell can give rise to growth and the statistical probability of a single cell arising from a population by a

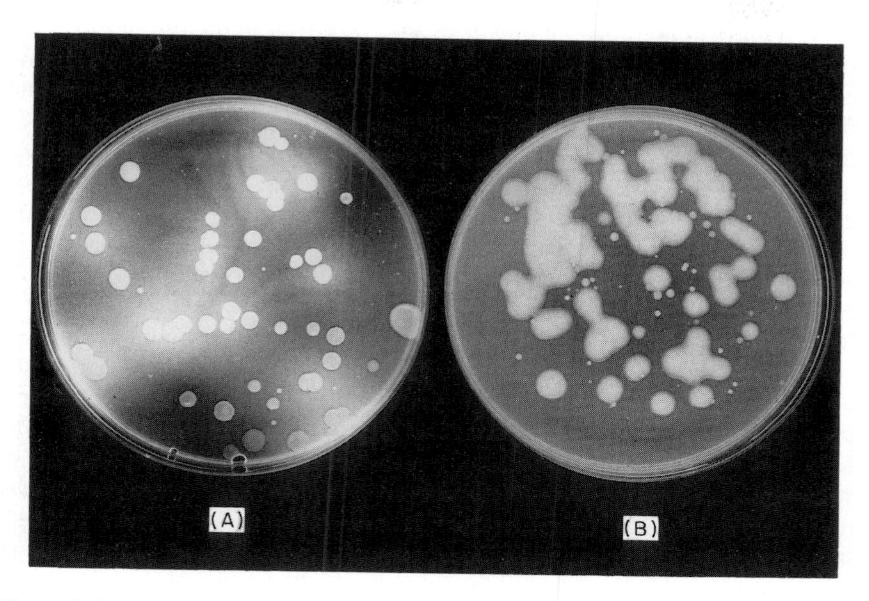


**Fig. 4.** Selective isolation of fluorescent *Pseudomonas* spp. from root-free soil and the rhizosphere of wheat using the pour plate technique and Martin's medium (Martin, 1975). (A) Rhizosphere. (B) Root-free soil.

number of serial dilutions can be defined precisely, so the efficiency of the technique should be high. In practice the technique usually, but not always, produces higher counts than traditional plate counts (Darbyshire *et al.*, 1974; Olsen and Bakken, 1987).

A specialized variation of the MPN technique has been developed for counting organisms such as root symbionts and soil-borne pathogens. Rather than inoculating rhizosphere suspensions into liquid media, soil is diluted serially with sterile inert material, such as sand, and specific host plants are then grown in replicate samples of each dilution. The numbers of plants forming symbiotic associations or developing disease symptoms are then used to determine the most probable number of propagules in the original soil.

Organisms with specific metabolic abilities can be enumerated using a similar technique. Soils are again diluted with sterile inert material and pellets or otherwise localized samples of the substrate placed in or on replicate samples of the soil dilutions. After incubation, metabolites may be detected by colour reactions either directly or after addition of appropriate reagents.

All these cultural methods suffer from the problem that they permit the growth and enumeration of only a small proportion of the organisms in the rhizosphere. Apart from the factors mentioned earlier in this section, many cells fail to grow in laboratory media because of dormancy, nutritional fastidiousness and physiochemical changes in environment between their natural habitat and the laboratory medium. In addition to using selective media, medium pH can be altered, as can incubation temperature, aeration and $CO_2$ level. Various antibiotics can be incorporated into the media to repress the growth of selected microbial groups. Obviously the combinations of media components and incubation conditions which are possible to selectively grow the range of organisms in the rhizosphere is infinite. It would be fruitless to list even a selection of them within the constraints of this chapter. Suffice it to say that the researcher must exercise his or her ingenuity to select or develop media which give the best possible results in the circumstances of the specific research project, whilst keeping within what is logistically possible in the laboratory. It goes without saying that a major consideration must be the nature of the organisms to be included in the study. Bacteria, actinomycetes, fungi, protozoa and algae are all present in the rhizosphere and all demand very different approaches and modifications to the basic techniques described above. A sound knowledge of the basic biology of each group is an essential prerequisite to allow judicious selection of techniques. It is, also, particularly important to be aware of, and to allow for, the limitations that are always inherent in the techniques.

Once cultured, it is possible to locate particular strains of bacteria and fungi by comparing total cell protein profiles after polyacrylamide gel electrophoresis (Lambert *et al.*, 1987). This may allow the localization of particular strains of interest or may be used to study the variation in number of different types of organisms and time, using either the protein profiles as such or by linking these profiles to those of known organisms.

### E. Microbial interactions

Clearly a habitat such as the rhizosphere, which is densely colonized by microorganisms, will contain many different species. These will often grow as monospecific microcolonies, but also as random distributions of individual cells or small groups of cells. Some organisms will be surrounded by slime or will be embedded within the root mucigel; others will be more exposed. Whatever the growth form many organisms will interact with other neighbouring species in a wide variety of ways, ranging from commensalism through competition to parasitism and predation (Stotzky, 1974; Campbell, 1985). These interactions will be profoundly affected by spatial distribution of the organisms and a wide variety of environmental factors such as pH and clay minerals. As with virtually all aspects of rhizosphere microbiology, little success has been achieved in developing methods for *in situ* use, although electron microscopy can be used to examine interactions where some visible distortion of growth, lysis or other damage results (see Section III.D and E; Campbell, 1983; Faull and Campbell, 1979; Homma, 1984; Homma *et al.*, 1979). Most information available has been derived from simple studies of the interactions of organisms isolated from the rhizosphere and grown in laboratory systems such as agar plates or liquid cultures which may not reflect growth and interactions in soil.

Bennett and Lynch (1981) and Turner and Newman (1984) have used gnotobiotic plants whose roots were inoculated with two readily distinguishable bacteria from the rhizosphere to study interactions; while this type of approach is much preferable to simplistic studies in synthetic laboratory media, it still suffers from the isolation of the organisms from the influence of the many other organisms in the normal rhizosphere population. Thus, any interaction found can only be described as a potential rather than an actual interaction which does occur in the natural system. Another approach is to modify the normal flora and fauna, by chemical means or by $\gamma$-irradiation, and then control conditions or add inoculum of particular organisms or microbial groups (bacteria but not fungi for example). It is then possible to establish quite stable populations that are different from the original ones and whose interactions and effects in the plant are also different (Newman *et al.*, 1977). The advent of genetic engineering, which enables the researcher to specifically and uniquely mark organisms and, thus, to allow them to be distinguished from all others in the ecosystem, offers exciting opportunities to study microbial interactions and behaviour on the rhizosphere (Ford and Olson, 1988).

## V. Microbial activity in the rhizosphere

A major requirement of studies of the ecology, population dynamics and function of microorganisms in the rhizosphere is to measure growth rates and metabolic activities of the species in the population or of the population itself. These measurements are difficult to perform in bulk soil and few successful techniques have been published. Stotzky (1974) has cited requirements for successful techniques as: permitting recovery of marker species over a given time span with minimum disturbance of the soil to prevent movement of cells, substrates and soil particles; simulating conditions that exist in soil; providing precise information on rates of growth and microgeographic distribution of individual species even in a mixed population; being rapid and reproducible and enabling both alteration of the environment and statistical analysis of the results. No single method meets all these requirements.

In view of the difficulties of achieving these requirements in bulk soils it is hardly surprising that there has been little or no progress in measuring microbial growth or activity directly in the rhizosphere, especially the endorhizosphere where microorganisms are totally surrounded by both dead and living plant cells.

In theory, it should be possible to analyse at least some metabolic activities in rhizosphere soil, which can be removed from the root and incubated with specific substrates. However, removal of soil will also remove some root cells, especially root hairs and cap cells, thus potentially adding to microbial activities. Further, the quantities of soil obtainable from the rhizosphere are generally very small, tedious to obtain and vary according to factors such as root morphology and soil moisture at time of sampling. In consequence, any method of measuring activity, be it of a specific enzyme or of a physiological process such as respiration, must be capable of precision and sensitivity with only very small amounts of soil. This aspect is further complicated as different parts of the root will probably support microbial populations with different activity levels. If such variations are to be quantified, the amounts of rhizosphere soil available for analysis will be smaller than ever.

An indirect approach has been adopted frequently where rhizosphere microorganisms are counted by dilution plate methods and representative samples of the organisms on the dilution plates isolated for study in pure culture. Such approaches generally indicate that not only are the total numbers of organisms present in the rhizosphere larger than those in the root-free soil, but that the proportion of the isolated population having a particular activity (e.g. producing non-specific phosphatases) is also higher. That is to say, there is preferential selection of certain microbial types in the rhizosphere. Indeed, it is suggested that individual species may have higher activities when growing in the rhizosphere than in the absence of root. For this sort of study the very wide range of methods of measuring microbial activity which is available from medical, industrial and food microbiology, is appropriate (Skerman, 1969).

Direct measurements of activities in the rhizosphere await development of appropriate methods and we can do no more than suggest some approaches which may prove to be of value.

Respiration is used frequently as an indication of overall microbial activity in soils and many methods of determining it have been described. These measure oxygen uptake, using for example Gilson or Warbourg respirometers or $O_2$-electrodes. These techniques require relatively large soil samples and the latter has the disadvantage that the electrodes may consume large quantities of oxygen, and so affect microbial respiration. Measurement of $CO_2$ output is generally too insensitive to apply to small soil samples, although addition of a substrate such as glucose may give useful increase in output. Where measurement is confined to the first 24 h after glucose addition, to non-rhizosphere soil, the results may be assumed to indicate the activity of the standing crop of organisms present without undue interference from proliferation of the microflora (Malkomes, 1987). It has not yet been proved that this assumption is always valid, especially for rhizosphere soils where microorganisms can proliferate very rapidly.

Martin and Foster (1985) have described a model system for studying the biochemistry and biology of the root–soil interface. They used this model in which plants are grown in attapulgite columns, to determine the production of $CO_2$ from the rhizosphere of plants grown with different microbial loadings. While the system is not claimed to reproduce natural growth conditions, it does appear to give data which provide useful insight into the activity of root-associated microorganisms.

A promising approach to characterizing the respiratory activity of the rhizosphere microflora, while still present around the root, was described by Waremboug (1977). This approach has not received the attention that it deserves. Essentially it depends on providing plants with a short pulse of $^{14}CO_2$ and then making a kinetic analysis of the liberation of labelled $CO_2$ in the rooted soil. Two peaks of production are invariably found; an initial peak ascribed to root respiration and a second, later peak arising from microbial respiration using labelled root exudates as substrates (Warembourg and Billes, 1979). This method has been applied to pot-grown plants and to natural vegetation. In the latter case, soil cores were taken from plant communities in the field and transported to the laboratory for study.

Dobbins and Pfaender (1988) have pointed out that measurement of biodegradation in non-homogeneous environments, using respiration activity alone, results in inadequate quantification of turnover rates. They stress the need for simultaneous measurement of respiration, cellular uptake and sorption of the substrate and its metabolites. The development of techniques for effectively separating cells from particulate matter allowed them to develop a methodology using radiolabelled substrates, appropriate for assessing uptake into cells and sorption as well as respiration by soil microbial communities. Thus, a comprehensive mass balance can be derived. As with all other methods for studying metabolic activity of soil microbial populations, that of Dobbins and Pfaender (1988) requires relatively large quantities (85 g) of soil and, thus, is not readily applicable to samples of rhizosphere soil from individual root systems or, especially, from selected regions of the root. Nonetheless, it could find useful application to comparisons of unplanted (non-rhizosphere) soil with densely planted soil, which may be assumed to give a bulk representation of the effect of root function on the soil microflora. Furthermore, the technique may offer scope for scaling down to a point where it is more suitable for use with the small amounts of rhizosphere soil which can be readily obtained from plant roots. It must be remembered, however, that the problems associated with the presence of detached root cells in the soil sample will still be present.

The approach to measuring microbial activity in the rhizosphere by comparison of the metabolism of selected substrates in densely planted and unplanted soils is essentially limited to substrates known not to be taken up by plant roots or broken down by root extracellular enzymes. It has found some application in the study of microbial degradation of certain herbicides and pesticides (Mudd *et al.*, 1983, 1985).

Some promising developments have been made in the use of specific substrates as a means of measuring overall microbial activity. In some cases the techniques are sufficiently sensitive to allow the use of only small samples of soil and, thus, lend themselves to application in studies of rhizosphere soils. A notable example of such a technique is that described by Schnurer and Rosswall (1982). This utilizes fluorescein diacetate, which is hydrolysed by a range of enzymes (proteases, lipases and esterases) to release fluorescein. Soil samples as small as 1 g gave easily measurable hydrolysis. As microbial activity in the rhizosphere is often considerably greater than in root-free soil, much smaller amounts of rhizosphere soil may give satisfactory data.

Substrate metabolism, or at least the presence of microbial intracellular enzymes required for metabolism of specific substrates, is frequently studied using appropriate histochemical techniques and the electron microscope (Sexton and Hall, 1978). Despite uncertainties about the precise location of enzymes detected in this way, and lack of information on the levels of activity, histochemical analysis can shed light on activity potentials on the rhizosphere microbial community.

Primary requirements for measurements of total microbial activity are that they should be non-specific, sensitive and require only short incubation periods. Although enzyme determinations match many of these requirements, a major drawback is that non-specificity means they will measure also activity from roots. Therefore, they are suited principally to root-free systems. As pointed out earlier, removal of rhizosphere soil from roots, however gently it is achieved, also removes root cells. Thus activities measured in such samples will be overestimates of microbial activity in rhizosphere soil. Equally, much of the total rhizosphere activity is associated with microorganisms living in the endorhizosphere and this will not be accounted for in soil detached from roots.

Despite these difficulties, enzyme analysis may provide useful evidence of activity in the microflora associated with roots. In particular, such an approach may be useful as a means of detecting the effects of factors such as xenobiotics, changes in soil conditions and changes in plant physiological activity which affect root exudation. It is beyond the scope of this chapter to give detail of the multitude of enzyme assays which are available. Suffice it to say that most, if not all, of those published will need modification before they can be applied validly to the rhizosphere soil.

This is a general conclusion to the vast majority of techniques which might be applied to studies of metabolic activity in the rhizosphere, regardless of their source. There are numerous opportunities to adapt techniques developed for studies of metabolism in all areas of microbiology. For example, microcalorimetry (Sparling, 1981) and measurements of adenylate energy change (Brookes *et al.*, 1983, 1987) are two approaches which have considerable potential. Marriage of these two techniques for separating microbial cells from soil particulate matter (Dobbins and Pfaender, 1988), and, especially, from plant cells will give useful benefit.

It should be possible to measure growth rates of microorganisms by incorporation of tritiated thymidine, but again it is impossible to separate activity of a particular organism from the root and rhizosphere in general. However, this approach has been used in simplified model systems to study growth rates of particular species of bacteria on gnotobiotic plants (Christensen *et al.*, 1989).

## VI. Colonization studies

The methods used are basically those already described to enumerate microorganisms on roots, but modified or expanded to allow expression of the numbers of microorganisms in relation to root age or position on the root. Thus, organisms introduced into gnotobiotic systems may be counted by direct examination (Bennett and Lynch, 1981), isolated onto standard media in the absence of other bacteria from which the introduced organisms would have to be distinguished (Turner and Newman, 1984). In more natural environments, individual introduced strains or species can be followed by selecting antibiotic-resistant mutants and then plating the total soil suspension on media with the antibiotics so that only selected strains will grow (Kloepper, *et al.*, 1980; Loper *et al.*, 1984, 1985; Suslow and Schroth, 1982; Weller, 1983, 1984).

Sequences of colonization require that root age is known. This may be done by using root observation boxes (Section III.A) and tracing roots during their development. Small parts of the root(s), each of known age, are then excised and microbes cultured or counted by direct examination (van Vuurde and Schippers, 1980). With grasses and cereals, in which adventitious roots are produced from the crown or nodes, they can be marked with the date of origin as they are produced so that the overall root age is known even if a precise age cannot be assigned to each part which might be sampled.

## VII. Mycorrhiza

The amount of mycorrhizae can be assessed by basic methods of microscopy or even direct visual examination. Vesicular arbuscular mycorrhizae are not easily cultured so there is no other method but direct examination, usually after clearing the root and staining with trypan blue (Gionvannetti and Mosse, 1980). Ectomycorrhizae have a characteristic morphology and can be visually assessed. They are also subjected to a wide range of light and electron microscope techniques (Section III). Even though the fungi will grow in culture, this is not generally suitable as a means of assessment because it cannot be quantified.

The interaction between the host and the symbiont usually involves studying movement of minerals, such as phosphorus, with isotopes ($^{32}P$ or $^{15}N$) or the assessment of the response of the plant (Abbott and Robson, 1984).

## VIII. Data handling

It should be stressed that data on rhizosphere populations are particularly in need of careful analysis and rigorous assessment because of variability. Standard statistical techniques can be used, but care is needed that the assumptions implicit in the use of such tests are valid (Gilligan, 1986). Specialized ecological techniques, originally devised for higher plant and animals studies (Moore and Chapman, 1986) such as population biology and vegetation analysis may also be useful. However, many of the methods, for similarity indices between populations for example, depend on having accurate identification, which is rarely available for microbial populations on roots and anyway depends on cultures as representative of the population. The use of diversity indices in microbial ecology has been much criticized (Brock, 1987).

More complex analysis of data is also possible using techniques such as principle components and factor analysis (Rosswall and Kvillner, 1978) but these have rarely been used in rhizosphere studies (though see Skyring and Quadling, 1969). They are powerful analyses to show relationships between groups of microorganisms, but they do require cultures with all that this implies about isolation techniques, etc. (Section IV).

Attempts to make mathematical models (see Bazin and Menell, Chapter 4) of rhizosphere populations and interactions are rare (Newman and Watson, 1977). Components such as colonization studies of surfaces in general (Caldwell *et al.*, 1981) and roots in particular (Gilligan, 1983), or nutrient fluxes around roots (Cushman, 1982, 1984) have been studied. However, information is at present so fragmentary that it is difficult to devise meaningful models for lack of basic data: the main purpose of models at this stage is to indicate which are the most important areas for study in order to improve our understanding of the rhizosphere.

## IX. Conclusions

The rhizosphere is a very difficult environment to sample for microorganisms, especially if quantitative data are required, because of its complexity and the steep environmental gradients which exist. All methods have their disadvantages, but the situation is compounded in the rhizosphere because of the presence of plant roots which impose additional problems in extracting microbes and in metabolic measurements. Information on the location of microbes and how they interact requires direct examination, taxonomic identification requires that they are cultured and, in order to characterize their function, metabolic and activity measurements are needed. Comparison of general methods has been made (Domsch *et al.*, 1979), but not all of these are applicable to the rhizosphere; direct observation and counting had large errors due to inability to see all organisms and non-uniform distribution of microbes. Conversions of most measurements to biomass involved gross oversimplification and unfounded assumptions. Plate counts were subject to great variation and were not reproducible. There may be poor correlation between estimates based on different methods. There is, therefore, a need for caution in comparing data derived from different methods or even from different laboratories with apparently the same method.

There is an urgent need for much further work on rhizosphere methodology. Research on the release of genetically engineered organisms, for example, is limited by the fact that it is not possible to locate small numbers of particular bacteria or to get reliable estimates of survival and growth rates. The methods for studying protozoa on roots are particularly poor and insensitive, yet what evidence there is suggests that they may be important in biomass terms and in their effect on nutrient turnover.

Brock (1987) reviewed many aspects of methodology which are appropriate for, or at least used in rhizosphere studies. He is particularly critical, with justification, of cultural procedures though these still have a place in the study of particular organisms (e.g. introduced bacteria) which can be recovered quantitatively from the soil. The importance of direct examination is stressed by Brock (1987), if only to see the microhabitat where the microorganisms live, but when combined with autoradiography and microchemical tests it is an effective method. There are problems with this in the complex root environment rather than the more simple aquatic environments for which it has been most used, but these methods are worthy of more serious investigation. It is to be hoped that such investigation in rhizosphere microbiology will be undertaken rapidly and comprehensively in the near future. There is no doubt that the rhizosphere holds the key to effective, environmentally benign advances in crop production and crop protection.

Economic and commerical pressures, as well as environmental protection agencies, are demanding the development of better, more rigorous methods and rhizosphere microbiology must respond to this if the potential it offers is to be exploited to the full.

## References

Abbott, L. K. and Robson, A. D. (1984). *In* "VA Mycorrhiza" (C. L. P. Powell and D. J. Bagyaraj, Eds), pp. 113–130, CRC Press, Boca Raton, Florida.

Anderson, J. R. and Slinger, J. M. (1975). *Soil Biol. Biochem.* **7,** 205–209.

Beckett, A. and Read, N. D. (1986). *In* "Ultrastructure Techniques for Microorganisms" (H. C. Aldrich and W. J. Todd, Eds), pp. 45–86, Plenum Press, New York.

Bennett, R. A. and Lynch, J. M. (1981). *J. Gen. Microbiol.* **125,** 95–102.

Bisdom, E. B. A. (1983). *Adv. Agronomy* **36**, 55–96.

Bohlool, B. B. and Schmidt, E. L. (1980). *Adv. Microb. Ecol.* **4**, 203–241.

Bokhari, U. G., Coleman, D. C. and Rubink, A. (1979). *Can. J. Bot.* **57**, 1473–1477.

Bowen, G. D. and Coleman, D. C. (1984). *In* "Current Perspective in Microbial Ecology" (M. J. Klug and C. A. Reddy, Eds), pp. 237–238, American Society for Microbiology, Washington DC.

Brock, T. D. (1987). *In* "Ecology of Microbial Communities" (M. Fletcher, T. R. G. Gray and J. G. Jones, Eds), pp. 1–17, Society for General Microbiology, Sym. 41, Cambridge University Press, Cambridge.

Brookes, P. C., Tate, K. R. and Jenkinson, D. S. (1983). *Soil Biol. Biochem.* **15**, 9–16.

Brookes, P. C., Newcombe, A. P. and Jenkinson, D. S. (1987). *Soil Biol. Biochem.* **19**, 211–217.
Burns, R. G. and Slater J. H. (1982). "Experimental Microbial Ecology", Blackwell, Oxford.
Caldwell, D. E., Brannan, D. K., Morris, M. E. and Betlach, M. R. (1981). *Microb. Ecol.* **7**, 1–11.
Campbell, R. (1982). *In* "Sourcebook of Experiments for the Teaching of Microbiology" (S. B. Primrose and A. C. Wardlaw, Eds), pp. 652–659, Academic Press, London.
Campbell, R. (1983). *Can. J. Microbiol.* **29**, 39–45.
Campbell, R. (1985). "Plant Microbiology", Edward Arnold, London.
Campbell, R. and Porter, R. (1982). *Soil Biol. Biochem.* **14**, 241–245.
Campbell, R. and Rovira, A. D. (1973). *Soil Biol. Biochem.* **5**, 747–752.
Chantler, S. M. and McIllmurray, M. B. (1987). *In* "Methods in Microbiology" (R. R. Colwell and R. Grigorova, Eds), pp. 273–332, Academic Press, London.
Christensen, H., Funck-Jensen, D. and Kjøller, A. (1989). *Soil Biol. Biochem.* **21**, 113–117.
Christie, P., Newman, E. I. and Campbell, R. (1978). *Soil Biol. Biochem.* **10**, 521–527.
Clarholm, M. (1981). *Soil Biol. Biochem.* **7**, 343–350.
Cole, G. T. (1986). *In* "Ultrastructure Techniques for Microorganisms" (H. C. Aldrich and W. J. Todd, Eds), pp. 1–44, Plenum Press, New York.
Curl, E. A. and Truelove, B. (1986). "The Rhizosphere", Springer-Verlag, Berlin.
Cushman, J. H. (1982). *Soil Sci. Soc. Am. J.* **46**, 704–709.
Cushman, J. H. (1984). *Plant Soil* **79**, 122–141.
Darbyshire, J. F., Wheatley, R. E., Greaves, M. P. and Inkson, R. H. E. (1974). *Rev. Ecol. Biol. Sol.* **11**, 465–475.
Dhingra, O. D. and Sinclair, J. B. (1985). "Basic Plant Pathology Methods", CRC Press, Boca Raton, Fl.
Dobbins, D. C. and Pfaender, F. K. (1988). *Microb. Ecol.* **15**, 257–273.
Domsch, K. H., Beck, T., Anderson, J. E., Soderstrom, B., Parkinson, D. and Trolldenier, G. (1979). *Z. Pflanzenernaehr. Bodenkd.* **142**, 520–533.
Duell, R. N. and Peacock, G. R. (1985). *Crop. Sci.* **25**, 880–883.
Faull, J. L. and Campbell, R. (1979). *Can. J. Bot.* **57**, 1800–1808.
Ford, S. and Olson, B. H. (1988). *Adv. Microb. Ecol.* **10**, 45–79.
Foster, R. C. (1981). *New Phytol.* **89**, 263–273.
Foster, R. C. (1982). *New Phytol.* **91**, 727–740.
Foster, R. C. (1986). *A. Rev. Phytopath.* **24**, 211–234.
Foster, R. C. (1988). *Biol. Fertil. Soil* **76**, 189–203.
Foster, R. C. and Rovira, A. D. (1976). *New Phytol.* **6**, 343–352.
Foster, R. C., Rovira, A. D. and Cock, T. W. (1983). "Ultrastructure of the Root–Soil Interface", Am. Phyt. Soc., St Paul, Minnesota.
Giddings, T. H. and Wray, G. P. (1986). *In* "Ultrastructure Techniques for Microorganisms" (H. C. Aldrich and W. J. Todd, Eds), pp. 241–265, Plenum Press, New York.
Gilligan, C. A. (1983). *Ann. Rev. Phytopath.* **21**, 45–64.
Gilligan, C. A. (1986). *Adv. Plant Pathol.* **5**, 225–261.
Giovannetti, M. and Mosse, B. (1980). *New Phytol.* **84**, 489–500.
Greaves, M. P. and Darbyshire, J. F. (1972). *Soil Biol. Biochem.* **4**, 443–449.
Greaves, M. P. and Sargent, J. A. (1985). *Weed Sci.* **34**, (Suppl. 1), 50–53.
Greaves, M. P. and Webley, D. M. (1965) *J. Appl. Bact.* **28**, 454–465.

Greaves, M. P., Cooper, S. L., Davies, H. A., Marsha, J. A. P. and Wingfield, G. I. (1978). Tech. Rep. Ag. Res. Council, Weed Res. Org. **45**.
Harley, J. L. and Waid, J. S. (1955). *Trans. Br. Mycol. Soc.* **38**, 104–108.
Homma, Y. (1984). *Phytopath.* **74**, 1234–1239.
Homma, Y., Sitton, J. W., Cook, R. J. and Old, K. M. (1979). *Phytopathology* **69**, 1118–1122.
Jacobs, M. J., Bugbee, W. M. and Gabrielson, D. A. (1985). *Can. J. Bot.* **63**, 1262–1265.
Jenny, H. and Grossenbacher, K. (1963). *Proc. Soil Sci. Soc. Am.* **27**, 273–277.
Johnen, B. G. (1978). *Soil Biol. Biochem.* **10**, 495–502.
Johnen, B. G. and Drew, E. A. (1978). *Soil Biol. Biochem.* **10**, 487–494.
Kilbertus, G. (1980). *Rev. Ecol. Biol. Sol.* **17**, 543–557.
Kloepper J. W., Schroth, M. N. and Miller, T. D. (1980). *Phytopathology* **70**, 1018–1024.
Lambert, B., Leyns, F., van Rooyen, L., Gosselé, F., Papon, Y. and Swings, J. (1987). *Appl. Environ. Microbiol.* **53**, 1866–1871.
Lappin, H. M. (1984). Ph.D. Thesis, Univ. Warwick.
Loper, J. E., Suslow, T. V. and Schroth, M. N. (1984). *Phytopathology* **74**, 1454–1460.
Loper, J. E., Haack, C. and Schroth, M. N. (1985). *Appl. Environ. Microbiol.* **49**, 416–422.
Louw, H. A. and Webley, D. M. (1959). *J. Appl. Bacteriol.* **22**, 216–226.
Lynch, J. M. (Ed) (1990) *The Rhizosphere*, John Wiley, Chichester.
Macdonald, R. M. (1986a). *Soil Biol. Biochem.* **18**, 399–406.
Macdonald, R. M. (1986b). *Soil Biol. Biochem.* **18**, 407–410.
Macdonald, R. M. (1986c). *Soil Biol. Biochem.* **18**, 411–416.
Malkomes, H.-P. (1987). *In* "Pesticide Effects on Soil Microflora" (L. Somerville and M. P. Greaves, Eds), pp. 81–96, Taylor and Francis, London.
Martin, J. K. (1975). *Soil Biol. Biochem.* **7**, 401–402.
Martin, J. K. and Foster, R. C. (1985). *Soil Biol. Biochem.* **17**, 261–269.
McCully, M. E. (1987). *In* "Root Development and Function" (P. J. Gregory, J. V. Lake and D. A. Rose, Eds), pp. 53–70, Cambridge University Press, Cambridge.
Moore, P. D. and Chapman, S. B. (Eds) (1986). "Methods in Plant Ecology", 2nd Edn, Blackwell Scientific Publications, Oxford.
Mudd, P. J., Hance, R. J. and Wright, S. J. L. (1983). *Weed Res.* **23**, 239–246.
Mudd, P. J., Greaves, M. P. and Wright, S. J. L. (1985). *Weed Res.* **25**, 423–432.
Newman, E. I. and Bowen, H. J. (1974). *Soil Biol. Biochem.* **6**, 205–209.
Newman, E. I. and Watson, A. (1977). *Plant Soil.* **48**, 17–56.
Newman, E. I., Campbell, R. and Rovira, A. D. (1977). *New Phytol.* **79**, 107–118.
O'Brien, T. P. and McCully, M. E. (1981). "The Study of Plant Structure: Principles and Selected Methods", Termarcarphi Pty Ltd, Melbourne.
Old, K. M. and Nicholson, T. H. (1978). *In* "Microbial Ecology" (M. W. Loutit and J. A. R. Miles, Eds), pp. 291–294, Springer-Verlag, Berlin.
Olsen, R. A. S. and Bakken, L. R. (1987). *Microb. Ecol.* **13**, 59–74.
Papavizas, G. C. and Davey, C. B. (1961). *Plant Soil* **14**, 215–236.
Paul, E. A. and Kucey, R. M. N. (1981). *Science* **213**, 473–474.
Poindexter, J. S. and Leadbetter, E. R. (1986). Bacteria in Nature, Vol. 2. "Methods and Special Applications in Bacterial Ecology", Plenum Press, New York.
Polonenko, D. R. and Mayfield, C. I. (1979). *Plant Soil* **51**, 405–420.

Polonenko, D. R., Pike, D. J. and Mayfield, C. I. (1978). *Can. J. Microbiol.* **24**, 1262–1271.
Robards, A. W. (1978). *In* "Electron Microscopy and Cytochemistry of Plant Cells (J. L. Hall, Ed.), pp. 343–415, Elsevier/North-Holland Biomedical Press, Amsterdam.
Roser, D. J. (1980). *Soil Biol. Biochem.* **12**, 329–336.
Rosswall, T. and Kvillner, E. (1978). *Adv. Microb. Ecol.* **2**, 1–45.
Rovira, A. D. and Campbell, R. (1974). *Microbiol. Ecol.* **1**, 15–23.
Rovira, A. D., Newman, E. I., Bowen, H. J. and Campbell, R. (1974). *Soil Biol. Biochem.* **6**, 211–216.
Rush, C. M., Upchurch, D. R. and Gerik, T. J. (1984). *Phytopathology* **74**, 104–105.
Schank, S. C., Smith, R. L., Wesser, G. C., Zuberer, D. A., Bonton, J. H., Quesenberry, K. H., Tyler, M. E., Milam, J. R. and Littel, R. C. (1979). *Soil Biol. Biochem.* **11**, 287–295.
Schans, J., Mills, J. T. and van Caeseele, L. (1982). *Phytopathology* **72**, 1582–1586.
Schnurer, J. and Rosswall, T. (1982). *Appl. Environ. Microbiol.* **43**, 1256–1261.
Schnurer, J., Clarholm, M. and Rosswall, T. (1985). *Soil Biol. Biochem.* **17**, 611–618.
Sexton, T. and Hall, J. L. (1978). *In* "Electron Microscopy and Cytochemistry of Plant Cells" (J. L. Hall, Ed.), pp. 63–147, Elsevier, Amsterdam.
Skerman, V. B. D. (1969). "Abstracts of Microbiological Methods" (V. B. D. Skerman, Ed.), Wiley-Interscience, London.
Skyring, G. W. and Quadling, C. (1969). *Can. J. Microbiol.* **15**, 473–488.
Soderstrøm, B. (1977). *Soil Biol. Biochem.* **9**, 59–63.
Sparling, G. P. (1981). *Soil Biol. Biochem.* **13**, 93–98.
Steen, E. (1991). *In* "Plant Root Systems: Their Effect on Ecosystem Composition and Structure" (D. Atkinson, D. Robinson and I. Alexander, Eds), Blackwell Scientific Publications, Oxford (in press).
Stotzky, G. (1974). *In* "Microbial Ecology" (A. Laskin and H. Lechevalier, Eds), pp. 57–135, CRC Press, Cleveland.
Suslow, T. W. and Schroth, M. N. (1982). *Phytopathology* **72**, 199–206.
Taylor, H. M. (Ed.) (1987). "Minirhizotron Observation Tubes: Methods and Applications for Measuring Rhizosphere Dynamics", American Society of Agronomy, Special publication 50, Madison, Wisconsin.
Turner, S. M. and Newman, E. I. (1984). *J. Gen. Microbiol.* **130**, 505–512.
Vancura, V. and Kunc, F. (Eds) (1988). "Soil Microbial Associations: Control of Structures and Functions", Developments in Agricultural and Managed-Forest Ecology, 17, Elsevier, Amsterdam.
Vermeer, J. and McCully, M. E. (1982). *Planta* **156**, 45–61.
van Vuurde, J. W. L. and Elenbaas, P. F. M. (1978). *Can. J. Microbiol.* **24**, 1272–1275.
van Vuurde, J. W. L. and Schippers, B. (1980). *Soil Biol. Biochem.* **12**, 559–565.
Warembourg, F. (1977). Thesis, Université des Sciences et Techniques due Languedoc.
Warembourg, F. R. and Billes, G. (1979). *In* "The Soil–Root Interface" (J. L. Harley and R. Scott Russel, Eds), pp. 183–196, Academic Press, London.
Warembourg, F. R., Montage, D. and Bardin, R. (1982). *Physiol. Plant* **56**, 46–55.
Weller, D. M. (1983). *Phytopathology* **73**, 1548–1553.
Weller, D. M. (1984). *Appl. Environ. Microbiol.* **48**, 897–899.
Whipps, J. M. and Lynch, J. M. (1983). *New Phytol.* **95**, 605–623.
Wynn-Williams, D. D. (1985). *Soil Biol. Biochem.* **17**, 739–746.

# 15

# A Review of Methods Used to Study the Microbial Ecology of Timber and Forest Products

D. J. DICKINSON AND J. F. LEVY

*Department of Pure and Applied Biology, Imperial College, London SW7 2BB, UK*



## I. Introduction

Wood as a substrate and the microbial ecology of timber in service, with its subsequent development, have probably received more attention than all other natural materials. As a consequence, a whole range of methods have evolved, which, in reality, tend to be variations on the same theme. That theme is the removal of wood samples from the timber in question and the isolation, culture and identification of the microorganisms present therein. All of these variations have their relative merits and describing each in detail seems rather pointless in such a review, without discussing the development of our understanding of this complex subject and the way in which the various methods have enabled this to be done. The important point is the significance of the results of the various microbial studies carried out on timber, which have primarily been directed to try to understand and control the performance of timber in service.

This review, therefore, has been limited to pinpointing the most significant developments in the study of the microbial ecology of wood.

METHODS IN MICROBIOLOGY
VOLUME 22 ISBN 0-12-521522-3

The specific details of each of the experimental procedures are best determined from examination of the cited literature.

The work discussed has fallen into three main phases.

(a) Identifying and understanding the organisms involved in the colonization and decay of timber.

(b) Understanding the relationships between the organisms themselves and between the organisms and the substrate.

(c) Extension of these studies on natural wood to determine the strengths and weaknesses of wood chemically treated to impart durability, i.e. the performance of preservative-treated wood and the action mechanisms of the introduced toxins.

Timber and other forest products derived from wood have a well-defined structure composed of the wood cell walls. The pattern of these structures will be characteristic for any one species of tree, but different from all others, often markedly so. This gives timber as a material a series of anisotropic structures with well-defined water-conducting channels, which play a very important role in the movement of microorganisms, liquids and gases into and through wood.

Levy (1987) summarizes the situation in the following paragraph:

> "To the succession of bacteria and fungi colonising it, wood consists of a series of conveniently orientated holes surrounded by food. The degradation of wood leading to the utilisation of that food supply takes place in a number of clearly defined stages, each of which results from the activity of a particular group of organisms. The success of the entire process is governed by the moisture content of the wood. This has to be high enough to promote a flow path for the products of enzyme action yet low enough to prevent water-logging, which would produce an anaerobic habitat hostile to the normal wood decay organisms".

There is, however, a serious problem for anyone attempting to determine exactly what is happening inside a piece of wood. There is, as yet, no method of making direct observations of the presence, movement, or interaction between organisms at the time they occur. What organisms are present and where? What is the state of activity of each organism at any moment? When and by what route did each arrive in the wood? Butcher (1971) set this out clearly and since then no new technique has emerged which enables us to make visual observations of the situation that develops at the time it happens during the process of decay.

Much of the early work on fungal decay was concerned with the isolation of climax organisms from heavily decayed wood and the resulting lists of fungal isolates that could be identified were substantially Basidiomycete species which gave rise to two types of decay, the brown rots (which destroyed the holocellulose and left the lignin largely unaltered) and the white rots (which destroyed all three cell wall polymers, cellulose, hemicelluloses and lignin). It was not until 1950 that Findlay and Savory presented evidence to the Seventh International Botanical Congress that a type of degrade that had hitherto been regarded as due to chemical action, was in fact caused by a fungus (Findlay and Savory, 1950). Savory's classic papers followed (1954a, b, 1955), defining this type of decay as 'soft rot' with a characteristic type of destruction of the middle layer of the wood cell wall. This opened up a new field of work and many researchers across the world started to look again at decayed wood to examine the distribution and economic importance of the soft rot fungi and the range and variety of the organisms that caused it. The literature was reviewed by Duncan (1960), Knudsen (1963) and Levy (1965, 1967a). Nevertheless, Findlay (1966) was still able to comment that the ecology of these fungi and other organisms associated with wood had received surprisingly little attention.

## II. Sampling methods

The early work on the ecology of the fungi associated with wood was hampered by lack of suitable techniques. Corbett (1963) and Corbett and Levy (1963a, b) examining a field trial of preservative-treated fence posts, observed that the untreated controls could be divided into a number of ecological zones, below ground, at the ground line, above ground and at the top of the post. A series of posts was set up at the field site and sampled over a period of 18 months.

Three sampling techniques were used. (a) Small blocks, 1.5 × 1.0 × 0.7 cm were cut from the posts in the field, taken to the laboratory, cut to expose an uncontaminated face from which small slivers of wood were taken and transferred to sterile nutrient agar plates. The resulting isolates were subcultured and identified. (b) Similar sample blocks were taken to the laboratory and sectioned under aseptic conditions. Adjacent sections were used in two ways, one being transferred to a sterile nutrient agar plate and the fungal isolates examined, whilst the other section was mounted on a slide for microscopic examination. (c) The pressler or increment borer (a standard forestry implement for taking cores from growing trees, particularly to assess their growth rate), was used to extract a core sample from one side of the post to the other. This core was split into 0.5-cm sections and isolations were made from each, with an adjacent core being examined microscopically.

These three sampling techniques provided interesting information and formed the basis for further development. The next problem to be faced was the enormous number of isolates that had to be handled and identified. Greaves and Savory (1965), reported the isolation from four different species of wood of 241 fungal isolates, comprising some 184 species with 57 cultures not fully identified. Shigo (1966) made more than 70 000 isolations from 1000 trees in New Hampshire, USA in 6 years. Okigbo (1966) reviewed the published techniques and devised a series of methods for making isolations in depth from wood. He achieved the staggering total of 7400 isolates from a single fence post. These numbers, suggested that while it might be theoretically possible to use them to plot a three-dimensional diagram of the distribution of each fungal species within the wood, they did create a frightening problem with regard to their culture and identification.

Corbett (1963) and Corbett and Levy (1963b) noted that there were few references in the literature concerning the early colonization or succession of fungi where they were known to invade wood, one exception being the work of Jones (1963), using test blocks submerged in the sea. Corbett concluded that there were some indications of a fungal succession in the fence posts they examined and made the interesting observation that Basidiomycetes were only isolated from the outermost layers of the post during the first 3 months of exposure and were not isolated from the centre of the post until some 18 months had elapsed.

Greaves (1966a, b), Greaves and Levy (1965) and Levy and Greaves (1966) introduced another factor into the picture by claiming that bacteria had a part to play in the process of wood decay. Levy (1967) elaborated these observations and postulated a possible explanation. All these observations indicated the extreme complexity of the situation and pointed to the urgency of developing adequate techniques for obtaining a clear and accurate picture of what was happening inside wood and what interactions, if any, were occurring between organisms. Greaves (1966a) and Okigbo (1966), working to some extent independently, reviewed all the published techniques for isolating fungi from wood. By modifying existing techniques and devising new ones, they were able to put forward a series of methods for isolating a wide range of organisms from any depth inside wood as well as from the surface (Okigbo *et al.*, 1966). It appeared to be a useful basis for beginning a critical examination of the ecology of fungi in wood. The problem is basically four-fold: (a) to isolate the organisms at depth; (b) to use a range of suitable media for growing all the organisms present and bring them all into pure culture; (c) to identify and classify the isolates; (d) to determine whether the isolates are simply wood-inhabiting organisms or can cause active decay.

Greaves (1966a) compared four methods of isolation and 11 types of culture medium (Greaves and Savory, 1965). He found that a simple method, using a small saw blade to produce fine sawdust directly onto the culture medium, gave the best results on both sugar-rich and cellulose-rich culture media. He screened all his isolates for their wood-rotting ability by a variety of means, including biochemical, and concluded that the most reliable and quickest results were obtained by exposing sterile orientated 1-cm cubes of beech, birch and Scots pine sapwood to cultures of each isolate, from which sections could be cut for microscopic observation.

Okigbo (1966) used Greaves 'saw cut' technique and devised a modification which involved using a metal drill bit instead of a saw. He also devised two methods for sampling the surface: (a) a type of adhesive tape that could be sterilized by heat treatment. This was cut into strips, sterilized and pressed against the wood surface, removed and put onto culture media and the isolates subcultured as they started to grow. (b) A modified drill bit, flattened and slightly toothed at the end to grind the wood surface into fine sawdust which was used as the inoculum for fungal isolations.

A later modification was used by Asmah (1967) whereby the method for sampling in depth was to use an 0.25-inch drill bit to reach the required depth followed by an 0.125-inch drill bit to take the sample for isolation of organisms. He also commented on the difficult problem of identifying and classifying the organisms isolated and posed the question 'Does it matter that we do not know what it is called, providing we know what it does in wood?'.

Further work on fungal succession in wood was described by Merrill and French (1966), who found that moulds were the first colonizers followed by soft rot fungi. Kaarik (1967, 1968) found that in 6 months pine and spruce poles were colonized by Ascomycetes and Fungi Imperfecti with Basidiomycetes in localized patches, but after 18 months Basidiomycetes occurred in high frequency. Butcher (1968a, b) reported that the groundline zone of pine stakes was colonized first by moulds, followed by soft rot fungi and then Basidiomycetes and secondary moulds. Once the Basidiomycetes became established the activity of the soft rot fungi declined. Grant and Savory (1969) summarized the methods described to that date.

Banerjee (1970) and Banerjee and Levy (1970, 1971) made further modifications to the techniques used by Okigbo (1966) and Asmah (1967). This consisted of three special drill bits, a surface grinder, a twist drill calibrated along its length for boring holes to a known depth and a grinder small enough to be inserted into the bored hole which was also calibrated along its length. Both grinders had a serrated end which would collect small particles of wood which could then be transferred to suitable culture media. The method described was to use the surface grinder first. This tool fitted into a standard carpenter's brace and had a 22-mm-diameter working face. The grinder was sterilized by dipping in ethanol and flaming lightly and then placed against the wood surface and rotated slowly. The particles of wood which collected on the serrated end were picked off using sterile forceps and transferred to the variety of culture media used for isolation of organisms. The surface sampled was cleaned thoroughly with sterilized cotton wool swabs following which a hole was drilled through it to a depth of 5 mm using a sterilized twist drill of 15 mm diameter. The hole was thoroughly cleaned using sterilized cotton wool swabs and the smaller grinder (13 mm diameter), inserted to full 5-mm depth and then rotated slowly under pressure after which it was withdrawn and the fine wood particles transferred to suitable culture media. The same procedure was used to extract samples from 25-mm and 45-mm depths below the wood surface. The transfer of the inocula onto the plates of culture medium was carried out in the field and the local airspora was monitored by exposing a series of culture plates for 15 min each for the whole sampling period. This enabled the progressive movement of certain groups of fungi into and through the stakes to be monitored.

## III. Growth media

The selection of suitable media for the isolation of organisms was another major problem to be overcome. Corbett and Levy (1963a) observed that some isolates grew quickly and very rapidly swamped the slower developers which made it necessary to provide alternative culture media to slow the growth of some isolates and encourage the growth of others. Greaves (1966a) and Greaves and Savory (1965) used 11 types of culture media, whilst Banerjee and Levy (1970) used seven. At this stage the bacteria were often regarded as unimportant in the colonization sequence and specific culture media were introduced to eliminate them. The use of various selective media has always been a mainstay in microbial ecology. Probably one of the most important developments in this area was the use of selective fungicides. This includes the use of benomyl to suppress the growth of microfungi and select for Basidiomycetes, whilst the use of toxins such as copper encouraged the growth of soft rot fungi from treated wood (Clubbe, 1978, 1980a, b; Clubbe and Levy, 1982; Carey, 1980, 1983).

As Butcher (1971) has pointed out, however many varieties of culture medium are used, there is no way that one can be sure that every microorganism present in the wood at the time of sampling is represented amongst the isolates. Neither can one know the state of each of the fungi isolated when they were in the wood. They may have been virile actively growing young hyphae. They may have been moribund and almost dead, but have revived when supplied with suitable nutrient and shown active growth in culture. They may even have been spores or resting propagules which will germinate when given a suitable growth medium. What one needs to know is the activity that is occurring at any one moment and the great problem that has to be overcome is the development of a method to enable this to be done. There is no sure means of finding out, but visual observation by some form of microscopy can sometimes be very helpful and the advent of the scanning electron microscope gave rise to parallel investigations being made to attempt to find out what, if any, visible changes were being made by each individual group of organisms to the wood cell wall structure, including the pit membranes. This was named by Corbett (1965) as the 'micromorphology' of wood decay.

Whilst techniques had been developed to obtain a more accurate picture of the organisms inhabiting wood, it had often been done to the exclusion of any attempt at microscopic analysis, so that the species list gave little or no indication of their role in the colonization and decay process. In order to fully understand the decay process it is essential to know what effects the microorganisms have on the wood as well as identifying the component members of the microflora as soon as they first appear. Butcher (1971) not only outlined the problems of the analysis of wood-inhabiting fungi with great clarity, but went on to attempt to quantify the observations by using frequency of isolation as a measure of abundance. Dwyer and Levy (1976) considered this work to be the first successful attempt with this complex problem to make an objective analysis of objectively collected data, most previous work being subjective analyses of objectively collected data. Butcher's analysis was based on methods commonly applied to the ecology of higher plant communities. Dwyer and Levy (1976) describe Dwyer's extension of Butcher's concept, which involved the development of a veneer-quadrat sampling technique from a model system for studying wood decay. This method does, however, make the basic assumption that the isolation and identification of organisms is fully representative of the components of the flora within the wood. This, as has already been discussed, may not be true and any statistical analysis of the data must take due regard of this or else a totally incorrect picture may be drawn.

In an attempt to overcome this source of error, Dwyer devised a technique for making direct quantitative assessment of the type and presence of the organisms in the different cell tissues. Statistical analysis of this type of data would be valid and would relate the presence of organisms to the incidence of wood decay, although it would not necessarily identify the individual components of the flora present. The method was based on the observations made by previous researchers that the anatomy of wood provided the pathways for both fluids and organisms. Corbett (1963) had shown that soft rot fungi penetrated more rapidly through 1-cm cubes of sapwood from transverse face to transverse face than from tangential face to tangential face and very much faster than from radial face to radial face. Greaves (1966a) and Greaves and Levy (1965) had shown that bacteria, staining fungi, soft rot fungi and the two groups of Basidiomycetes, the brown rots and the white rots, all colonized the ray parenchyma of the sapwood of both hardwoods and softwoods before moving into the vessels, tracheids or fibres. Dwyer, therefore, made his model system of small stakes of sapwood of birch and Scots pine, 20 × 40 × 200 mm, cut so that the orientation of grain of the wood was parallel to the longer dimension and the 20 × 200 mm sides were in the tangential plane. One tangential face was left unsealed whilst the other three lateral sides were sealed to render them impervious to either liquids or organisms. After being sterilized the stakes were buried to half their length in soil and were sampled by removing three replicates at regular intervals. These replicates were each cut across at the groundline to form a block 20 × 20 × 40 mm with the 20 × 20 face being in the tangential plane. These blocks were sampled by being cut into veneers 1 mm thick; each veneer was then cut into three strips and each strip into eight inocula which were plated out onto suitable culture media and the resulting organisms subcultured and identified. A similar number of replicate stakes were sampled by being cut into a similar groundline block, but these were sectioned parallel to and at known distances from the tangential face and were 25 μm thick. The sections so cut constituted the sample to be analysed and the random quadrats were delineated by the field of view at ×50 magnification when photographed under the light microscope on 35 mm film. Each photograph was projected and an analysis made by scoring the presence and type of fungal structure present in tracheids, vessels, fibres, ray cells and resin canals, with estimates of the degree and type of wall degradation, if any. The data were processed by computer to determine any correlations. Dwyer and Levy (1976) concluded that the technique had its limitations. On the positive side the maps plotted from the veneer isolates did give a picture of the distribution of organisms across the block and some impression of the movement of organisms into the block from the exposed tangential face. However, the data recorded depend once again on the experimenter's ability to isolate each species of organism present in the wood and culture it well enough to make a positive identification. Statistical correlations would, therefore, have little meaning on such imprecise data. The direct observation technique cannot be criticized on these grounds; the statistical analysis being based on precise data has much greater intrinsic value. In spite of this the method had still not been able to distinguish between the individual members of the microflora inside the wood.

## IV. The colonization of wood

The production of lists of fungi colonizing timber and the establishment of their natural succession is only part of the story. As far as wood is concerned, the most significant feature will be the modification of the substrate by the activities of the organisms, which leads to a change in its structure and properties and consequently to its performance in service. Many of the early colonizers, such as moulds and staining fungi, have only the ability to utilize readily available nutrients in wood, such as cell contents, particularly stored food reserves. From the point of view of timber in service, it is the degradation of the structural components that has received most attention.

Closely associated studies of the micromorphology of the fungal colonizers were carried out to determine whether the isolates were affecting the cell wall structure in any way. Staining fungi had long been recognized by the way in which they penetrated laterally through cell walls by the development of a transpressorium when the hypha makes contact with the wall surface, followed by lateral penetration through the wall by a fine constriction of the hypha (Liese and Schmid, 1964; Karkanis, 1966). The anisotropic nature of wood made it possible to cut small cubes of wood, sized between 0.75 mm and 1.0 mm, with opposite sides in the transverse, tangential and radial planes with respect to the 'grain' of the wood. These could be exposed to fungal cultures in Petri dishes, and replicate samples removed from time to time and sectioned sequentially from each of the three surface planes for microscopic examination to determine the mode of penetration of the organisms into and through wood. Corbett (1963), Greaves (1966a), Okigbo (1966), Asmah (1967) and Banerjee (1970) all used this technique to screen their isolates. Olofinboba (1967) and Levy and Olofinboba (1967) used the technique to determine the mode of entry of the staining fungus *Lasiodiplodia theobromae* into two African hardwoods. By sampling daily for the first week and weekly thereafter they were able to show that the fungus penetrated from a cut transverse face to colonize the vessels within 24 h and then spread rapidly from there to the adjacent parenchyma and then into the rays. From the practical point of view this indicated that the incidence of blue stain in the two timbers had to be tackled immediately the tree had been felled, otherwise only a day or two on the forest floor would be too late.

The colonization sequence, as Butcher (1971) pointed out, is a purely ecological one. Wood, being no longer a living material, makes no pathological response to the invading organisms. It can only be changed by some action of an invading organism or by water in movement (Baines, 1983; Baines and Levy, 1979) or the movement of nutrients in the moving water (Uju, 1979; Uju *et al.*, 1980; King *et al.*, 1976). The invading organisms may be involved with one or more of a number of activities. These include: destruction of the pit membranes; scavenging for nitrogen; nitrogen fixation; synergism between organisms; antagonism between organisms; fine penetration through cell walls; cavity formation in cell walls; destruction of cell walls; competition for nutrients.

Levy (1969) postulated a spectrum of interaction between the colonizing fungal species and the wood, with different species having a different role to play. There are a number of ecological niches to be filled and the exact species which fills a particular niche will depend on the particular circumstances that pertain at the time, such as the wood species, pH, temperature and moisture content. Levy (1975a) asserted that what was important was not the exact species, since these may differ with time and place, but the particular physiological action it may have on its substratum and habitat.

The destruction by bacteria of the pit membranes of both bordered and crossfield pits of the sapwood of timber that had been stored in water or transported by floating down rivers had been reported by many authors (Harmsen and Nissen, 1965; Shigo, 1966; McQuire, 1970; Liese, 1970; Boutelje and Bravery, 1968; Rossell *et al.*, 1973; Banks and Dearling, 1973; Dunleavy *et al.*, 1973). Banks and Dearling (1973) had used both light and scanning electron microscopy to show bacteria in the act of breaking down the pit membranes, whilst Dunleavy *et al.* (1973) and Levy (1975b) described the work of Rossell who used orientated cubes of pine sapwood in soil and demonstrated the progressive breakdown of the pit membranes as the bacteria penetrated deeper into the wood. She also pointed out that bacteria could be either synergistic or antagonistic to other organisms, including fungi. Nitrogen is in short supply in wood and Cowling (1961), amongst others, pointed out that invading organisms were likely to follow nitrogen pathways or gradients. Banerjee (1970), using orientated cubes was able to show the hypha of one fungus penetrating into the hypha of another and put forward the concept that early colonizing organisms could scavenge for nitrogen and then be parasitized or saprophytized by the later colonizing species. Baines and Millbank (1976) and Baines *et al.* (1977) were able to demonstrate nitrogen fixation by bacteria in wood using a larger wood block, also orientated with respect to the grain, replicates of which were sampled in two ways: (a) by being split into small slivers which were examined in the mass spectrometer for associated nitrogen, and (b) sections taken for microscopic examination to ensure that bacteria were present. Baines was able to demonstrate the progressive increase of nitrogen along the rays and along the grain by this method, which he suggested could be important in localized pockets in wood, but, if distributed evenly throughout the wood, was likely to be insufficient to make a marked difference.

Thin sections of wood examined by light microscopy were useful to identify staining fungi and soft rot by virtue of the effect on the wood cell wall. Findlay (1970) was one of the first to use the scanning electron microscope to examine wood and show the potential of a three-dimensional view of a very small orientated cube of a range of wood species. Findlay and Levy (1969) illustrated its use as an aid in the study of wood anatomy and decay. Findlay (1970), went on to use the transmission electron microscope to examine the cavities made by the soft rot fungi in the middle layer of the secondary cell wall and produced what are still some of the best micrographs of the subject. Bravery (1971, 1972, 1975), using the scanning electron microscope, made some striking observations of the micro-morphology of white rot and brown rot attack in wood, which Nasroun (1971) confirmed on other fungi belonging to each group. As a result, both types of Basidiomycete decay could now be recognized by microscopic examination. This was an important discovery and gave rise to much speculation as to the precise mechanism of action by which these two important wood-rotting groups of fungi destroyed the wood cell wall. Bravery used small orientated cubes of wood, some of which had been treated with a wood preservative, and examined the effect of wood preservatives on the colonizing organisms. He used both fungal isolation and microscopical examination to sample the colonized wood blocks.

## V. Chemically treated wood

Some of the most interesting and commercially important microbial ecological studies of wood have been the investigations into the performance of chemically treated wood in the natural environment. The earlier studies of the ecology of decay had provided the background and the choice of organisms for controlled laboratory assessments of preservative-treated wood, many of which have become standard practice. But to attempt to understand the performance of treated wood in service, an ecological approach is essential. From a microbial ecological point of view, this simply represents a modified substrate, where the toxins form an additional abiotic factor in the environment.

In an attempt to understand the interactions of different wood species and a commercial wood preservative (copper–chrome–arsenate), Sorkhoh

(1976), Sorkhoh and Dickinson (1976), Dickinson and Sorkhoh (1976) and Dickinson *et al.* (1976) used the small orientated cubes of beech, birch and Scots pine, to demonstrate the effect of wood preservatives and their microdistribution on the colonization of wood by microorganisms. The replicate samples of the various treatments used were measured for weight loss, to demonstrate gross loss of structure due to decay, and then sampled in different ways for examination by light, scanning electron and transmission electron microscopy, including microanalytical techniques. These authors were able to demonstrate that the location of the toxin within the cell types and within the layers of the cell wall varied with the wood species concerned, with a consequence that the micromorphology of the fungal colonizers also varied. Some wood species were clearly more difficult to protect from decay than others.

This was further demonstrated by Clubbe (1980a, b) and Clubbe and Levy (1982), who used the basic concept outlined by Dwyer and Levy (1976) but developed the technique still further. Carefully orientated stakes of sapwood of Scots pine and birch which had been sealed on three sides, leaving the outer tangential face unsealed, were partly buried in soil as a field trial. Half the number of stakes of each species were treated with a subtoxic dose of a wood preservative, whilst the remainder were untreated. Three replicate stakes of each of the types were sampled at 2, 4, 8, 16, 32, 96, 184, 265 and 363 days after burial in the soil, taken to the laboratory and analysed in two ways. The stakes were cut so that a 10-mm-deep block was cut transversely across the stake at the groundline. This block was cut radially into two equal halves. One half was used for the isolation of organisms whilst the other was used for microscopic examination. The samples for isolation were cut parallel to the outer tangential face at three defined depths and a thick tangential section cut at each point. Each section was divided into 32 inocula and 4 plated out on to each Petri dish of selective media (Clubbe, 1978; Clubbe and Levy, 1977). The samples for microscopic examination were cut radially into two equal portions and thin radial sections cut from each half. These were examined in the light microscope at ×50 magnification in 10 microquadrats at each of the same three defined distances from the outer tangential face from which the inocula sections had been taken. The microscopic observations involved the quantification of the number of cells of each type colonized by a defined range of hyphal types and quantification of the cells of each type which showed a defined range of degrade of the cell wall. Clubbe classified the organisms into six groups; bacteria, primary moulds, staining fungi, soft rots, wood-rotting Basidiomycetes (white rots and brown rots), and secondary moulds.

The results of Clubbe's work were by nature of a breakthrough. He showed that the bacteria had colonized to a depth of 3 mm from the exposed surface by the second day and were the first colonizers in both treated and untreated wood and immediately began the destruction of the pit membranes. They were followed by the fourth day by the primary moulds and staining fungi. The soft rot fungi did not appear for about 3 months but then began to degrade the cell wall of the untreated wood. They were followed by the wood-rotting Basidiomycetes (white rots and brown rots), which suppressed the soft rots and rapidly became the dominant climax microflora causing decay of the untreated wood. They were followed by the secondary moulds—mainly cellulolytic organisms, incapable of destroying wood by themselves but able to utilize the cellulose of the wood cell walls when this had been made available by the opening up of the wall structure by the wood-rotting species in the course of their destruction of the wall. Some secondary moulds could utilize the breakdown products of cell wall degradation. In the stakes treated with what had been considered a subtoxic treatment with a wood preservative, however, whilst bacteria remained the first colonizers, the primary moulds and staining fungi showed greatly reduced numbers and the wood-rotting Basidiomycetes were totally eliminated. The soft rot fungi, however, took over and became the climax microflora followed by the secondary moulds where decay of the cell wall occurred. In this particular experiment, treated pine stakes showed no sign of decay after 18 months exposure, whereas the fibres in the treated birch were heavily decayed. The significance of soft rot fungi becoming the climax microflora has proved to be the most challenging problem which has had to be faced by the wood preservation industry in recent years.

Another superb example of the use of techniques developed in studies of the microbial ecology of wood, as well as being a superb example of microbial ecology in its own right, are the studies of the decay of window joinery by the use of simulated joinery components. Carey (1980, 1983) developed a test system, which consisted of two pieces of wood joined together at right angles by a simple mitre joint to represent the bottom corner of a window frame. These samples became known as L-joints and were sealed at the top of the vertical component (the mortise) and the distal end of the horizontal component (the tenon) with a waterproof seal. The samples were carefully painted with a normal three-coat system. Some samples were preservative treated before being painted, whilst others were not. To accelerate the potential entry of decay organisms, the joint was opened after the paint was completely dry and then put together again. The L-joints were exposed on racks out of ground contact as a field trial and sampled with a similar frequency to Clubbe's stakes. The moisture content was monitored regularly with a special moisture meter and at the same time specimens were removed for sampling, they were X-rayed using soft X-rays which showed up the presence of excess water. Destructive sampling again involved both isolation of organisms on to selective media and microscopic examination, both of which were carried out at pre-determined distances from the joint along the tenon. In addition, tests were carried out to monitor changes in the porosity of the wood caused by the early microbial colonizers. The white rot fungi invariably proved to be the climax microflora. The use of benomyl in a selective medium was shown to be invaluable. The unexpected result of this work was to show that the ecological sequence of organisms colonizing the wood was very similar to that determined by Clubbe from his work with wood in ground contact. In hindsight it is perhaps less surprising since the common factor in both cases is the wood itself and clearly the ecological sequence is necessary to provide the right conditions for active decay to occur, which must mean a progressive alteration to the habitat until the conditions are right. Mendes (1982) and Le Poidevin (1986) used this technique very effectively to monitor a range of surface finishes for wood and to establish criteria for the protection of window joinery. The technique has since evolved into a standard test procedure for assessing the performance of treated wood in this out-of-ground contact environment. The standard method uses the isolation of the climax white rot fungi and modification of the substrate, i.e. changes in the porosity of the wood, as the criteria on which this assessment is judged.

Levy and Dickinson (1980) reviewed the importance of the work of both Clubbe and Carey and the implications their results might have on other related areas of investigation. Murphy and Dickinson (1981) and Murphy (1982) extended the isolation of organisms from preservative-treated wood in ground contact to an investigation of the soil microflora in contact with preservative-treated posts and at 1 cm and 10 cm from them. They showed that there was a change in the composition of the microflora in the soil in immediate contact with the post, but little or none 10 cm away. A similar effect was shown to occur with untreated naturally durable posts. Gray and Dickinson (1982, 1983, 1989) and Gray (1983), working on the deposition of toxic elements in the wood cell wall, developed a technique which involved measuring the deflection of the ministakes at regular intervals as a non-destructive measure of the onset and development of decay. It was Morris and Dickinson (1981a, b), Morris *et al.* (1984) and Morris (1980, 1983) who benefitted most from the knowledge and understanding of the microbial ecology of wood decay in understanding the performance of treated wood. Morris was investigating methods to control the internal decay of inadequately treated electricity poles and the possibility of using antagonistic organisms to provide a biological control system. He used the pattern of the organisms present at the time of an examination, particularly the secondary moulds, to determine the past history of the infection and subsequent decay. He was also faced with the need to take repeated isolations from individual poles in order to monitor the changes which occurred with time. Although the selective media techniques were well developed, it was impossible to remove repeated samples from one pole. To overcome this he used sterile wooden dowels which could be inserted into predrilled holes in the post, left for a period then removed directly on to selective media, to monitor fungal growth within the pole. Dowels infected with decay organisms were also used to monitor the effect of biological control organisms previously introduced into the pole. Morris used a drill for making a specific pattern of borings into which he could insert a dowel of suitable length. Similar work was carried out into the decay of railway sleepers.

In treated softwood under certain situations of exposure, bacteria have been shown to be early colonizers, detoxifiers and decay organisms. Wyles (1987) and Wyles and Dickinson (1987) have modified a technique first described by Nilsson (1984) for the examination of this type of decay. This involves the exposure of obliquely transverse sections of suitably treated wood to wet soil. The sections are put into small envelopes made from fine plastic mesh and buried in the soil and incubated for the required time, after which they are removed from the soil, taken out of the envelopes and mounted on a slide for examination under the light microscope. This technique has also been used to investigate the effectiveness of possible preservative systems that might be used to control this type of decay.

All the methods described in this article have been used successfully to unravel part of the complex of organisms in wood. This knowledge and the techniques have proved invaluable in further studying the performance of treated wood. This in turn has led to the need to understand how the organisms involved overcome the problem of the presence of toxins. This has led to a major interest in the study of detoxification mechanisms, i.e. the removal of toxins from the substrate and thus making it susceptible again to wood rotting organisms (Daniel and Nilsson, 1989).

Further work on the microbial ecology of wood decay must be directed into developing methods to determine the long-term predictive performance of wood in service. Such an approach will almost certainly call upon the wealth of data and experience obtained from past fundamental studies in this subject.

## References

Asmah, J. (1967). M.Sc. thesis, Univ. of London.

Banerjee, A. K. (1970). Ph.D. thesis, Univ. of London.

Banerjee, A. K. and Levy, J. F. (1970). *Int. Biodetn. Bull.* **6,** 37–41.

Banerjee, A. K. and Levy, J. F. (1971). *Mat. Org.* **6,** 1–25.
Baines, E. F. (1983). *In* "5th Int. Biodetn. Symposium" (T. A. Oxley and S. Barry, Eds), pp. 26–37, John Wiley & Sons, London.
Baines, E. F. and Levy, J. F. (1979). *J. Inst. Wood Sci.* **8,** 109–113.
Baines, E. F. and Millbank, J. W. (1976). *Mat. Org.*, Heft **3,** 167–173.
Baines, E. F., Dickinson, D. J., Levy, J. F. and Millbank, J. W. (1977). *Rec. 1977 Ann. Conv. Br. Wood Pres. Assoc.,* 33–45.
Banks, W. B. and Dearling, T. B. (1973). *Mat. Org.* **8,** 39–49.
Boutelje, J. B. and Bravery, A. F. (1968). *J. Inst. Wood Sci.* **4,** 47–57.
Bravery, A. F. (1971). *J. Inst. Wood Sci.* **5,** 13–19.
Bravery, A. F. (1972). Ph.D. thesis, Univ. of London.
Bravery, A. F. (1975). *In* "Biological Transformation of Wood by Micro-organisms" (W. Liese, Ed.), pp. 129–142, Springer-Verlag, Berlin.
Butcher, J. A. (1968a). *In* "Biodetn. of Materials" (A. H. Walters and J. J. Elphick, Eds), pp. 444–459, Elsevier Publ. Co., London.
Butcher, J. A. (1968b). *Can. J. Bot.* **46,** 1577–1589.
Butcher, J. A. (1971). *Mat. Org.* **6,** 209–232.
Carey, J. K. (1980). PhD. thesis, Univ. of London.
Carey, J. K. (1983). *In* "5th Int. Biodetn. Symposium" (T. A. Oxley and S. Berry, Eds), pp. 13–25, John Wiley and Sons Ltd., London.
Clubbe, C. P. (1978). Int. Res. Group on Wood Pres., Document No. IRG/WP/186.
Clubbe, C. P. (1980a). Int. Res. Group on Wood Pres. Document No. IRG/WP/1107.
Clubbe, C. P. (1980b). Ph.D. thesis, Univ. of London.
Clubbe, C. P. and Levy, J. F. (1977). Int. Res. Group on Wood Pres., Document No. IRG/WP/159.
Clubbe, C. P. and Levy, J. F. (1982). *Mat. Org.* **17,** 21–34.
Corbett, N. H. (1963). Ph.D. thesis, Univ. of London.
Corbett, N. H. (1965). *J. Inst. Wood Sci.* **14,** 18–29.
Corbett, N. H. and Levy, J. F. (1963a). *Br. Wood Pres. Assoc. Newsletter* **27,** 1–3.
Corbett, N. H. and Levy, J. F. (1963b). *Br. Wood Pres. Assoc. Newsletter* **28,** 1–10.
Cowling, E. J. (1961). U.S. Dept. Agric. For. Serv. Tech. Bull. No. 1258.
Daniel, G. F. and Nilsson, T. (1989). *J. Inst. Wood Sci.* **11,** 162–171.
Dickinson, D. J. and Sorkhoh, N. A. A. H. (1976). SEM/76 **2,** 549–554.
Dickinson, D. J., Sorkhoh, N. A. A. H. and Levy, J. F. (1976). *Rec. 1976 Ann. Conv. Br. Wood Pres. Assoc.,* 25–40.
Duncan, C. G. (1960). U.S. For. Prod. Lab., Report No. 2173.
Dunleavy, J. A., Moroney, S. and Rossell, S. E. (1973). *Rec. 1973 Ann. Conv. Br. Wood Pres.,* 127–148.
Dwyer, G. and Levy, J. F. (1976). *Mat. Org.*, Heft **3,** 13–20.
Findlay, G. W. D. (1970). Ph.D. thesis, Univ. of London.
Findlay, G. W. D. and Levy, J. F. (1969). *J. Inst. Wood Sci.* **4,** 57–63.
Findlay, W. P. K. (1966). *Mat. Org.*, Heft **1,** 199–211.
Findlay, W. P. K. and Savory, J. G. (1950). *Proc. VIIth. Int. Bot. Congr.* **7,** 315.
Grant, C. and Savory, J. G. (1969). *Int. Biodetn. Bull.* **5,** 77–94.
Gray, S. M. (1983). Ph.D. thesis, Univ. of London.
Gray, S. M. and Dickinson, D. J. (1982). Int. Res. Group on Wood Pres., Document No. IRG/WP/3201, 13 pp.
Gray, S. M. and Dickinson, D. J. (1983). Int. Res. Group on Wood Pres., Document No. IRG/WP/3244, 20 pp.

Gray, S. M. and Dickinson, D. J. (1989). *Rec. 1989 Ann. Conv. Br. Wood Pres. Assoc.*, 65–77.
Greaves, H. (1966a). Ph.D. thesis, Univ. of London.
Greaves, H. (1966b). *Met. Org.*, Heft **1**, 61–68.
Greaves, H. and Levy, J. F. (1965). *J. Inst. Wood Sci.* **3**, 55–63.
Greaves, H. and Savory, J. G. (1965). *J. Inst. Wood Sci.* **3**, 45–50.
Harmsen, L. and Nissen, T. V. (1965). *Nature* **206**, 319.
Jones, E. B. G. (1963). *J. Inst. Wood Sci.* **3**, 14–23.
Kaarik, A. (1967). *Mat. Org.* **2**, 97–108.
Kaarik, A. (1968). *Mat. Org.* **3**, 185–198.
Karkanis, A. G. (1966). M.Sc. thesis, Univ. of London.
King, B., Oxley, T. A. and Long, K. D. (1976). *Mat. Org.* **1**, 264–276.
Knudsen, M. V. (1963). *Trae. Ind.* **13**, 76–77.
Le Poidevin, J. (1986). M.Phil. thesis, Univ. of London.
Levy, J. F. (1965). *Adv. Bot. Res.* **2**, 323–357.
Levy, J. F. (1967). *Rec. 1967 Ann. Conv. Br. Wood Pres. Assoc.*, 147–161.
Levy, J. F. (1969). *Rec. 1969 Ann. Conv. Br. Wood Pres. Assoc.*, 71–78.
Levy, J. F. (1975a). *In* "Biological Transformation of Wood by Microorganisms" (W. Liese, Ed.), pp. 16–24, Springer-Verlag, Berlin.
Levy, J. F. (1975b). *In* "Biological Transformation of Wood by Microorganisms" (W. Liese, Ed.), pp. 64–73, Springer-Varlag, Berlin.
Levy, J. F. (1987). *Phil. Trans. R. Soc. Lond.* **A321**, 423–433.
Levy, J. F. and Dickinson, J. D. (1980). *Rec. 1980 Ann. Conv. Br. Wood Pres. Assoc.*, 67–73.
Levy, J. F. and Greaves, H. (1966). *Br. Wood Pres. Assoc. News Sheet* **68**, 1–6.
Levy, J. F. and Olofinboba, M. (1967). *Br. Wood Pres. Assoc. News Sheet* **80**, 1–2.
Liese, W. (1970). *Rec. 1970 Ann. Conv. Br. Wood Pres.*, 81–97.
Liese, W. and Schmid, R. (1964). *Phytopath. Zeit.* **51**, 385–393.
McQuire, A. J. (1970). Ph.D. thesis, Univ. of Leeds.
Mendes, F. (1982). Ph.D. thesis, Univ. of London.
Merrill, W. and French, D. W. (1966). *Phytopathology* **56**, 301–303.
Morris, P. I. (1980). "Abstracts, 2nd Int. Symp. of Microbial Ecology", p. 216, Academic Press, London.
Morris, P. I. (1983). Ph.D. thesis, Univ. of London.
Morris, P. I. and Dickinson, D. J. (1981a). Int. Res. Group on Wood Pres., Document No. IRG/WP/1130.
Morris, P. I. and Dickinson, D. J. (1981b). *Int. Biodetn. Bull.* **17**, 95–100.
Morris, P. I., Dickinson, D. J. and Levy, J. F. (1984). *Rec. 1984 Ann. Conv. Br. Wood Pres. Assoc.*, 42–55.
Murphy, R. J. (1982). Ph.D. thesis, Univ. of London.
Murphy, R. J. and Dickinson, D. J. (1981). Int. Res. Group on Wood Pres., Document No. IRG/WP/1131.
Nasroun, T. A. H. (1971). Diploma thesis, Imperial College, London.
Nilsson, T. (1984). Int. Res. Group on Wood Pres., Document No. IRG/WP/1234.
Okigbo, L. C. (1966). M.Sc. thesis, Univ. of London.
Okigbo, L. C., Greaves, H. and Levy, J. F. (1966). *Br. Wood Pres. Assoc. News Sheet* **65**, 1–3.
Olofinboba, M. (1967). Ph.D. thesis, Univ. of Ibadan, Nigeria.
Rossell, S. E., Abbott, E. and Levy, J. F. (1973). *J. Inst. Wood Sci.* **6**, 28–35.
Savory, J. G. (1954a). *J. Appl. Bacteriol.* **17**, 213.
Savory, J. G. (1954b). *Ann. Appl. Biol.* **41**, 336–347.

Savory, J. G. (1955). *Rec. 1955 Ann. Conv. Br. Wood Pres. Assoc.*, 3–35.
Shigo, A. L. (1966). *Mat. Org.* **1,** 309–324.
Sorkhoh, N. A. A. H. (1976). Ph.D. thesis, Univ. of London.
Sorkhoh, N. A. A. H. and Dickinson, D. J. (1976). *Mat. Org.*, Heft **3,** 287–293.
Uju, G. C. (1979). Ph.D. thesis, Univ. of London.
Uju, G. C., Baines, E. F. and Levy, J. F. (1980). *J. Inst. Wood Sci.* **9,** 23–26.
Wyles, A. M. (1987). Ph.D. thesis, Univ. of London.
Wyles, A. M. and Dickinson, D. J. (1987). Int. Res. Group on Wood Pres., Document No. IRG/WP/2286.

# 16

# Methods for Detecting Microbial Pathogens in Food and Water

CHARLES W. KASPAR* AND CARMEN TARTERA**

*University of Maryland, Department of Microbiology, College Park, MD, USA*



## I. Introduction

Newly developed methods for the detection of bacteria and viruses have provided microbiologists with the means to rapidly identify and monitor specific microorganisms in food and water. Traditional methods of testing involve culture techniques to increase numbers of the organism to a detectable level; followed by isolation and biochemical identification. Methods of propagating some enteropathogenic viruses (i.e., Norwalk) are lacking and reports of viable but non-culturable bacteria (Colwell *et al.*, 1985) suggest direct methods of detection are warranted.

* *Present address: San Labs, 405 Eighth Avenue S.E., Cedar Rapids, Iowa 52401, USA;* ** *Present address: Department of Microbiology, Faculty of Biology, University of Barcelona, 08071-Barcelona, Spain.*

METHODS IN MICROBIOLOGY
VOLUME 22 ISBN 0-12-521522-3

The literature is replete with culture methodologies for the isolation and identification of food- and waterborne pathogens. Emphasis here will be on methodologies to detect pathogens and indicator organisms; however, the methods described are applicable to most bacteria. This chapter will focus on the use of nucleic acid and antibody probes which have the potential to circumvent the need to culture the organism prior to identification. It is beyond the scope of this chapter to cover all of the rapid methods and culture media described for the isolation and detection of specific bacteria, or groups of bacteria. For reviews of rapid methods, see Tilton (1981), Pierson and Stern (1986) and Lányi (1987).

## II. Food- and waterborne pathogens

### A. Bacterial pathogens

As detection and isolation methods have improved, a growing number of pathogens have been identified as important food- and waterborne pathogens (Table I). Several of the bacterial pathogens are widely distributed in soil, marine and estuarine waters, the intestinal tract of warm-blooded animals, or water contaminated with faecal matter. The challenge to microbiologists has been to develop effective monitoring procedures and control measures for a variety of samples and pathogens.

Both water and food have served as vehicles for bacterial pathogens. In the United States from 1946 to 1980, 672 waterborne outbreaks were reported (Lippy and Waltrip, 1984). Although the causative agent for one-half of the outbreaks was never established, *Salmonella* was identified as the aetiological agent in 75 of the outbreaks, *Shigella* in 61, *Escherichia coli* in 5 and *Campylobacter* in 2 (Lippy and Waltrip, 1984). A variety of foods have served as vehicles for bacterial pathogens (Table I). Bryan (1988) analysed 1586 cases of foodborne illness reported from 1977 to 1984 in the United States. The most frequently identified vehicles were seafoods (24.8%), meats (23.2%), poultry (9.8%), salads (8.8%), and others (<5%). Other foods implicated included: raw clams, fried rice, and Mexican-style foods. *Salmonella* has been the most common aetiological agent identified in foodborne cases, followed by *Campylobacter jejuni, Yersinia enterocolitica, Clostridium perfringens, Staphylococcus aureus* and *Bacillus cereus* (Beckers, 1988; Todd, 1988). A number of studies of foodborne outbreaks and the vehicles and pathogens involved have been reported by others (Blake *et al.*, 1980; Hauschild and Bryan, 1980; Sours and Smith, 1980; Remis *et al.*, 1984; Archer, 1988; Archer and Young, 1988; Beckers, 1988; Bryan, 1988; Hackney and Dicharry, 1988).

**TABLE I**
Habitat, pathogenic characteristics, and vehicles of major foodborne bacterial pathogens

| Organism | Characteristics | Foods |
|---|---|---|
| *Bacillus cereus* | Common to soil, vegetation and water; produce one of two enterotoxin types, causing either diarrhoea or a vomiting response; infective dose $10^6$–$10^8$ cells $g^{-1}$ of food | Diarrhoeal-type; cereals, potatoes, vegetables, meat products, and puddings Vomiting-type: fried and boiled rice |
| *Campylobacter jejuni* | Habitat is the intestinal tract of animals, widely distributed in nature; infective dose $10^2$ cells | Meats, poultry, poultry products, unpasteurized dairy products, mushrooms |
| *Clostridium botulinum* | Indigenous to soil, widely distributed; produces potent exotoxin | Meats, poultry, fish, home-canned vegetables and fruits |
| *Clostridium perfringens* | Indigenous to soil, widely distributed; produces exotoxin; infective dose $>10^6$ cells | Meats, poultry, fish; some frozen foods, fruits, vegetables |
| *Escherichia coli* | Habitat is the intestinal tract of animals and man, widely disseminated in nature; enteropathogenic, enterotoxigenic, enteroinvasive and haemorrhagic strains | Meats, fish, poultry, milk, dairy products, vegetables, rice |
| *Listeria monocytogenes* | Widely distributed in soil, vegetation, water and the intestinal tract of animals; haemolytic strains are pathogenic | Vegetables, dairy products, raw and pasteurized milk, soft cheeses, poultry, meats |
| *Salmonella* | Habitat is the intestinal tract of animals and man; widely disseminated in nature | Eggs, egg products, unpasteurized milk, poultry, meat and meat products |
| *Shigella* | Habitat is the intestinal tract of man and primates; invasive, shiga toxin may be produced; infective dose $10^1$–$10^2$ cells | Salads, seafoods, Mexican foods |

**TABLE I** – *continued*

| Organism | Characteristics | Foods |
|---|---|---|
| *Staphylococcus aureus* | Associated with the nasal cavities and skin of man; produces exotoxin; $10^6$–$10^8$ cells $g^{-1}$ of food necessary for sufficient toxin production | Meats, poultry, fish, shrimp, dairy products, vegetables |
| *Vibrio parahaemolyticus* | Widely distributed in marine and coastal waters; virulent strains generally Kanagawa positive, produce heat-stable haemolysin | Seafoods; oysters, shrimp, crabs, lobster, clams |
| *Vibrio cholerae* | Widely distributed in bay and estuarine waters; disease caused primarily by 01 serotype, cholera toxin produced by some | Raw and undercooked oysters and clams |
| *Yersinia enterocolitica* | Widely distributed in nature, man, and animals; pathogenicity associated with calcium dependency, autoagglutination, and binding of Congo Red dye | Meats, water, tofu, milk |

## B. Viral pathogens

Over 100 types of viruses are disseminated from human faeces and urine into the environment. The source and physical characteristics of enteric viruses are shown in Table II. Enteric viruses are not normal inhabitants of the gastrointestinal and respiratory tracts of man and may be classified as pathogens although several produce asymptomatic infections. These viruses may multiply in the host reaching numbers between $10^6$ and $10^{10}$ $g^{-1}$ of faeces (Sabin, 1955). Symptomatology of enteric virus infections has been reported by Melnick (1984).

Despite the fact that, in most countries, sewage is treated prior to discharge into the environment, enteric viruses have been found in river water (Simkova and Wallnerova, 1973; Block, 1983), lakes (Vaughn and Landrey, 1977), groundwater (Wellings *et al.*, 1974, 1975; Vaughn *et al.*, 1978), bathing and coastal waters (Metcalf and Stiles, 1967; Goyal *et al.*, 1978) and drinking water (Coin *et al.*, 1965; Mack, 1973; Deetz *et al.*, 1984). The presence of these viruses in environmental waters indicates that

**TABLE II**
The characteristics, source, and immunoassays for potentially pathogenic viruses excreted into the environment

| Virus | Origin | Virus size/genome[a] | Immunoassay[b] | Growth in cell lines |
|---|---|---|---|---|
| Enteroviruses | Faeces | 28 nm; ss-RNA | | Yes |
| Poliovirus | | | FIA, EIA | |
| Echovirus | | | FIA | |
| Coxsackievirus A | | | | |
| Coxsackievirus B | | | FIA, EIA | |
| New enteroviruses (type 68–71) | | | | |
| Hepatitis A (type 72) | | | EIA, RIA | Slow |
| Rotavirus | Faeces | 70–80 nm; ds-RNA | FIA, EIA | Slow |
| Reovirus | Faeces | 70–80 nm; ds-RNA | FIA | Slow |
| Adenovirus | Faeces, Urine | 80 nm; ds-DNA | | Yes |
| Norwalk virus | Faeces | 27 nm; incomplete | | No |
| Calicivirus | Faeces | 35–40 nm; ss-RNA | | No |
| Astrovirus | Faeces | 28 nm; incomplete | | No |
| Coronavirus | Faeces | 80–180 nm; ss-RNA | | No |
| Snow mountain | Faeces | 27 nm; incomplete | | No |
| Non A–non B hepatitis | Faeces | 27–30 nm; incomplete | | No |

[a] Nucleic acid type; ss = single stranded; ds = double stranded.
[b] RIA = radioimmunoassay; EIA = enzyme immunoassay; FIA = fluorescent immunoassay.

current disinfection treatments are not thoroughly effective for viruses. Enteric viruses have been found to be more resistant to chlorine treatment and to survive longer in the environment than bacteria, and these properties are accentuated in the presence of particulate material (Berg *et al.*, 1978). It is not surprising that enteric viruses have been found in waters which lack bacterial indicators and are considered microbiologically safe (Craun, 1978; Rose *et al.*, 1987). Recently, indicators more closely related to enteric viruses in structure, morphology, chemical composition, and size, like coliphage, F-male phages and *Bacteriodes fragilis* bacteriophages, have been used as indicators of enteric viruses. Bacterial and viral indicators are discussed below.

Although it is accepted that viruses are transmitted via water, few outbreaks have identified a virus as the causative agent. Outbreaks have been reported for hepatitis A (Poskanzer and Beadenkopf, 1961; Bryan *et al.*, 1974; Craun, 1981; Hejkal *et al.*, 1982), Norwalk virus (Grohmann *et al.*, 1980; Gerba *et al.*, 1985), rotavirus (Gerba *et al.*, 1985), poliovirus (Mosley,

1967), hepatitis non-A, non-B (Khuroo, 1980; Wong *et al.*, 1980), and adenovirus (contracted in swimming pools and involved in eye and respiratory infections; (Foy *et al.*, 1968; D'Angelo *et al.*, 1979; and gastroenteritis, Uhnoo *et al.*, 1984).

The dissemination of viruses by food is not as well documented as by water; however, interest has increased with the development of methods that detect low numbers of virus. Poliovirus has been implicated in outbreaks involving milk (Dingman, 1916; Aycock, 1927; Lipari, 1951) and has been isolated from raw ground beef (Sullivan *et al.*, 1970). Echovirus 4 found in coleslaw was the aetiological agent involved in an outbreak of meningitis (US Department of Health, Education, and Welfare, 1976). While, hepatitis A has been transmitted to food through food handlers (Cliver, 1971), other enteric viruses have been isolated from raw ground beef (Sullivan *et al.*, 1970), and fruits and vegetables irrigated with sewage (Bagdasar'yan, 1964; Larkin *et al.*, 1976; Katzenelson and Mills, 1984). Shellfish are perhaps the most important food vehicle for viruses because as filter feeders they concentrate viruses from surrounding waters and are frequently consumed raw or undercooked. A number of outbreaks of hepatitis A (Mason and McLean, 1961; Stille *et al.*, 1972; O'Mahony *et al.*, 1983; Gerba *et al.*, 1985; Richards, 1985) and Norwalk virus (Murphy *et al.*, 1979; Gunn *et al.*, 1982; Guzewich and Morse, 1986) have been traced to the consumption of raw shellfish. Recently, Snow Mountain agent, astrovirus and calicivirus have been incriminated in shellfish-associated gastroenteritis (Dolin *et al.*, 1987; Kurtz and Lee, 1987). Enteroviruses and hepatitis non-A, non-B viruses have been isolated from shellfish (Gerba and Goyal, 1978; Goyal *et al.*, 1979; Ellender *et al.*, 1980; Caredda *et al.*, 1981; Alter *et al.*, 1982), but have not been incriminated in any outbreaks.

## III. Indicator organisms

Because it is not feasible to test food and water for all pathogens, indicator organisms are used to signal the presence of faecal contamination and the possible presence of intestinal pathogens. Conceptually, indicators are found in samples at higher numbers and are more easily identified than pathogens. Standard indicators include coliforms, faecal coliforms, *E. coli*, and faecal streptococci. The following have been proposed as indicators: *Bacteriodes fragilis* (Fiksdal *et al.*, 1985), *Clostridium perfringens* (Bonde, 1966), *Bifidobacterium* sp. (Evison and James, 1975), and *Rhodococcus coprophilus* (Mara and Oragui, 1981). Because bacterial indicators fail to correlate with the presence of viral pathogens, coliphages (Guelin, 1948), F-male phages (Havelaar *et al.*, 1985), and phages against *B. fragilis* (Jofre *et al.*, 1986; Tartera and Jofre, 1987), which have survival properties similar to those of viral pathogens, have been proposed as indicators of viral contamination. The use, shortcomings, advantages, and detection methods for indicator organisms have been reviewed extensively (Hoadley and Dutka, 1977; Mossel, 1978, 1982; Matches and Abeyta, 1983; Reinhold, 1983; Splittstoesser, 1983; Tompkin, 1983; Hartman *et al.*, 1986).

## IV. DNA probes

Nucleic acid probes have become a valuable diagnostic reagent in the identification of human and animal pathogens and made possible the identification of viruses and bacteria which are difficult, if not impossible, to cultivate. DNA probes have also proved to be a useful tool for identifying and monitoring organisms in food and the environment (Moseley *et al.*, 1982; Fitts *et al.*, 1983; Hill *et al.*, 1983a,b; Pace *et al.*, 1985; Holben *et al.*, 1988; Stahl *et al.*, 1988). Nucleic acid probes to a number of important pathogens found in food and water have been generated (Table III). DNA probes to genes encoding for toxin (Moseley *et al.*, 1982; Kaper *et al.*, 1981, 1982) or hemolysin (Datta *et al.*, 1987; Morris *et al.*, 1987) have been used to identify virulent members among an inocuous population and to study the epidemiology of the pathogen (Moseley *et al.*, 1982; Kaper *et al.*, 1981, 1982). Other organisms have been identified using probes to characteristic plasmids (Totten *et al.*, 1983; Hill *et al.*, 1983b), chromosomal DNA (Grimont *et al.*, 1985; Fitts *et al.*, 1983), whole and fragments of viral genomes (Berninger *et al.*, 1982) and 16S rRNA (Stahl *et al.*, 1988).

The major limitations of nucleic acid probes have been the limited shelf life of radiolabelled probes and the time and problems associated with cultivating specific organisms to a detectable level. However, progress with non-radioactive probes and direct detection methods will generate wider use of nucleic acid probes in the detection of microorganisms in food and water in the near future.

### A. Methods

The use of nucleic acid probes in the analysis of food and water has provided an alternative to conventional biochemical identification of pathogens and indicator organisms. Nucleic acid probes are more specific, can detect pathogenic members of a population, and results are generally obtained faster than when using standard identification methods. When testing food and water, pathogen-specific probes have been typically used in modifications of the colony hybridization procedure of Grunstein and Hogness (1975). Colony hybridization has been used to detect virulent

**TABLE III**
Nucleic acid probes to important food- and waterborne pathogens

| Organism | References |
|---|---|
| *Campylobacter* spp. | Stolzenbach *et al.*, 1988 |
| *Campylobacter jejuni* | Stollar and Rashtchian, 1987 |
| | Korolik *et al.*, 1988 |
| Enteroinvasive *E. coli* | Boileau *et al.*, 1984 |
| | Pal *et al.*, 1985 |
| | Sethabutr *et al.*, 1985 |
| | Venkatesan *et al.*, 1988 |
| Enterotoxigenic *E. coli* | Moseley *et al.*, 1980, 1982 |
| | Hill *et al.*, 1983a |
| Haemolytic *Listeria monocytogenes* | Datta *et al.*, 1987 |
| *Listeria* spp. | Klinger and Johnson, 1988 |
| *Salmonella* spp. | Fitts *et al.*, 1983 |
| | Flowers *et al.*, 1987 |
| Invasive *Shigella* | Boileau *et al.*, 1984 |
| | Sethabutr *et al.*, 1985 |
| | Venkatesan *et al.*, 1988 |
| Toxigenic *Vibrio cholerae* | Kaper *et al.*, 1981 |
| *Vibrio parahaemolyticus* | Nishibuchi *et al.*, 1986 |
| *Vibrio vulnificus* | Morris *et al.*, 1987 |
| *Yersinia enterocolitica* | Hill *et al.*, 1983b |
| | Jagow and Hill, 1986 |
| Enteric Adenoviruses | Takiff *et al.*, 1985 |
| Enteroviruses | Hyypiä *et al.*, 1984 |
| Hepatitis A | Berninger *et al.*, 1982 |
| | Scotto *et al.*, 1983 |
| | Jiang *et al.*, 1986, 1987 |
| Rotavirus | Dimitrov *et al.*, 1985 |

*Yersinia enterocolitica* (Hill *et al.*, 1983b; Jagow and Hill, 1986) and toxigenic *E. coli* (Hill *et al.*, 1983a) in artifically contamined food, and toxigenic *E. coli* (Echeverria *et al.*, 1982; Moseley *et al.*, 1982) and *Vibrio cholerae* (Kaper *et al.*, 1981, 1982) in water.

### *1. Colony hybridization*

1. Food samples are homogenized if necessary and 0.1 ml of appropriate dilutions are spread onto sterile nitrocellulose filters previously placed on a suitable agar medium.
2. Plates are incubated for 24 h at 37°C (or other appropriate temperature).
3. Colonies are lysed on the nitrocellulose filters by placing the filters colony side upwards on filter paper saturated with 0.5 M NaOH for 10 min followed by three successive transfers to filter paper saturated with 1.0 M ammonium acetate and 0.2 N NaOH for 1 min each. The filters are then

transferred to a fourth ammonium acetate–NaOH-saturated paper for 10 min.
4. The filters are then air dried and baked in a vacuum oven at 80°C for 2 h and stored until probed.
5. Prior to hybridization, the filters are incubated for 3 h at 37°C in hybridization solution which consists of: 50% formamide, 5× SSC (1×SSC = 0.15 M sodium chloride and 0.015 M sodium citrate), 0.1% sodium dodecyl sulphate, 1 mM EDTA, and Denhardt's solution [0.02% Ficoll (molecular weight 400 000), 0.02% polyvinyl pyrrolidone (molecular weight 360 000), and 0.02% bovine serum albumin].
6. The filters are then transferred to fresh hybridization solution containing approximately $10^5$ cpm of heat-denatured probe DNA $ml^{-1}$ and 75 μg of heat-denatured calf thymus DNA $ml^{-1}$ and incubated overnight at 37°C.
7. The filters are then washed in 5× SSC with 0.1% sodium dodecyl sulphate for 45 min at 65°C, rinsed briefly with 2× SSC at room temperature, and air dried.
8. The membrane is then attached to a piece of Whatman 3MM paper, covered with plastic wrap, placed on X-ray film with a single intensification screen, and held at −70°C for 24 h.
9. The X-ray film is then developed as specified by the manufacturer.

In a similar procedure, *Salmonella* have been detected in foods following growth in pre-enrichment broth (Fitts *et al.*, 1983). Detection of salmonellae by DNA hybridization was faster than standard biochemical identification and serological confirmation, which required an additional 2–3 days to complete. Flowers *et al.* (1987) compared the standard culture method with the DNA hybridization assay; hybridization was conducted following both pre-enrichment and selective enrichment of samples. Results from the testing of 1600 samples of food showed that the DNA hybridization method was as effective as the standard culture procedure and was significantly better with some foods. For additional details and protocols on colony hybridization, see Grunstein and Hogness (1975), Moseley *et al.* (1980) and Maniatis *et al.* (1982).

The above procedures do not eliminate the need to culture the organism; however, rapid methods (Palva, 1983; Miller *et al.*, 1988) and advances in nucleic acid labelling and extraction will undoubtedly lead to quicker and more sensitive detection of methods.

### B. Non-radioactive labels

The most common means of labelling nucleic acid probes is with $^{32}P$-tagged nucleotides which are incorporated into the probe using the nick translation procedure of Rigby *et al.* (1977). Although radiolabelled probes are used in research laboratories with few difficulties, their application to large-scale testing of food and water samples is undesirable because of the short-shelf life of $^{32}$P-labelled probes, high cost, hazards, disposal problems associated with radioactive waste, and public acceptance.

The biotin–avidin system has been the most common non-radioactive means of labelling probes. Like radioactive labelling, nick translation is used to incorporate biotinylated analogues of nucleotides, rather than $^{32}$P-labelled nucleotides, into the DNA probe. Avidin, which has a strong affinity for biotin, is usually tagged with an enzyme (Leary *et al.*, 1983; Sethabutr *et al.*; 1985, Yokota *et al.*, 1986; Bialkowska-Hobrzanska, 1987) and used to detect hybridized, biotinylated probe. Enzyme-labelled antibodies to biotin, in place of avidin, have also been used (Langer-Safer *et al.*, 1982). Hybridized probe is detected following the addition of enzyme substrate which when cleaved results in the production of a visible end product. Biotin-labelled probes have been used in the detection of enterotoxigenic *E. coli* (Bialkowska-Hobrzanska, 1987), *Shigella* and enteroinvasive *E. coli* (Sethabutr *et al.*, 1985), and hepatitis B virus (Yokota *et al.*, 1986). For a thorough review of the applications of the biotin–avidin system, see Wilchek and Bayer (1988).

Other non-radioactive markers include haptens, such as dinitrophenol and 2-acetylaminofluorene (Vincent *et al.*, 1982; Landegent *et al.*, 1985), enzymes cross-linked to single-stranded DNA (Renz and Kurz, 1984; Seriwatana *et al.*, 1987), and antibodies specific for RNA–DNA hybrids (Rudkin and Stollar, 1977; Boguslawski *et al.*, 1986). Miller *et al.* (1988) utilized alkaline phosphatase-labelled anti-DNA–RNA antibodies to detect hybrids formed between latex immobilized DNA (probe) and samples containing rRNA complementary to the DNA probe. Hybridization was rapid, complete within 15 min, and the procedure detected as few as 500 cells. Alternatively, the anti-RNA–DNA antibodies can be immobilized on polystyrene and used to capture RNA–DNA hybrids from solution (Stollar and Rashtchian, 1987). This method produces little background and has been used to detect *Campylobacter jejuni* rRNA.

With refinement of current systems and the development of more efficient detection systems and labels, the use of nucleic acid probes will no longer be restricted because of limitations imposed by radioactive labels.

### C. Gene amplification and direct detection

Another impediment to the widespread use of nucleic acid probes has been the need to propagate the organism to a level sufficient for detection by hybridization. Saiki *et al.* (1988) have developed a gene amplification method (the polymerase chain reaction) whereby specific DNA sequences are selectively amplified by a factor of $10^5$–$10^6$. The procedure involves two primers which hybridize to opposite strands of DNA on each end of target DNA. The primers are oriented in opposite directions so DNA synthesis proceeds across the target DNA and between the two primers. Because the primers also bind to each of the newly synthesized strands, repeated cycles of heat denaturation, annealing, and DNA synthesis doubles the amount of target from the previous cycle. The technique has been improved by replacing the Klenow fragment of *E. coli* DNA polymerase I with heat stable *Thermus aquaticus* DNA polymerase (Taq), which eliminates the need to add fresh DNA polymerase after each denaturation, and by automation with a DNA Thermal Cycler (Perkin-Elmer Cetus Corporation, Norwalk, CT). Steffan and Atlas (1988) used the polymerase chain reaction to amplify a specific region of DNA from *Pseudomonas cepacia* which was then identified by dot-blot hybridization. Initial quantities of 0.3 pg of target DNA could be detected following amplification, an increase in sensitivity of $10^3$ over non-amplified samples. Following extraction of DNA from sediment, $10^2$ *P. cepacia* cells 100 per g of sediment (or 1 cell $g^{-1}$) could be detected despite the presence of $10^{11}$ other organisms.

Alternatively, methods which enhance the sensitivity of probes might also avoid culturing the organism prior to hybridization. Polymerization-enhanced and single-stranded probes have both been reported to increase the sensitivity of radiolabelled probes (Holben *et al.*, 1988; Somerville *et al.*, 1988a). Polymerization enhancement increased the sensitivity by at least two orders of magnitude. In this procedure, an oligomer probe served as a primer for DNA synthesis. Specificity and DNA synthesis were contingent upon the probe (primer) binding to the target DNA while sensitivity was based on DNA synthesis and the incorporation of labelled nucleotides downstream from the primer (Somerville *et al.*, 1988a). Holben *et al.* (1988) were capable of detecting $4 \times 10^4$ bacteria $g^{-1}$ of soil or 0.01–0.02 pg of DNA using $\alpha$-$^{32}$P-labelled, single-stranded DNA probes. The single-stranded probe was generated from M13 containing the sequence of interest (Holben *et al.*, 1988). Advantages of the M13-generated single-stranded probe include a high specific activity, lower background, and the elimination of probe–probe hybridization in the reaction mixture.

Microbes have been detected directly, without cultivation, by extraction and isolation of nucleic acids from environmental samples, followed by hybridization with specific probes (Ogram and Sayler, 1988; Holben *et al.*, 1988; Stahl *et al.*, 1988; Jiang *et al.*, 1986). By extracting nucleic acids directly from samples, lengthy incubation times and problems associated with difficult to grow and non-culturable organisms are eliminated. Stahl *et al.* (1988) were able to monitor *Bacteriodes succinogenes* and *Lachnospira multiparus* in the bovine rumen without culturing. Detection was accomplished by extraction of total nucleic acids followed by hybridization with oligonucleotides to species-specific 16S rRNA segments. Because many ribosomes are present ($10^4$ in actively growing cells) fewer numbers of bacteria are needed for a positive signal when 16S rRNA segments are used as targets. Giovannoni *et al.* (1988) have extended this procedure to the detection of a single cell by combining microautoradiography with microscopy.

Direct detection methods require an efficient means of harvesting cells and extracting nucleic acids from an environmental sample. Methods for extraction of DNA from soils (Holben *et al.*, 1988), sediments (Ogram *et al.*, 1987), and water (Fuhrman *et al.*, 1988; Sommerville *et al.*, 1988b) have been reported. Extraction from aquatic environments requires that large sample volumes be taken to accrue enough DNA for testing. Somerville *et al.* (1988b) described an inexpensive method for concentrating microorganisms from litres of water on a single cylindrical filter membrane. Cell lysis and proteolysis were executed in the filter housing yielding high molecular weight DNA/RNA solutions which could be tested immediately, concentrated, or purified. Hepatitis A has been detected in concentrated estuarine water samples (Jiang *et al.*, 1986). The RNA was extracted from the sample, purified, blotted onto nitrocellulose, and hybridized with a hepatitis A probe. This method, and others which bypass cultivation of the organism, save time and can be used to account for members of the population which resist cultivation (i.e. Norwalk virus). Additional information on the use of gene probes to study microbial communities can be found in a review by Ogram and Sayler (1988).

Although these procedures have not been applied to the detection of microbes in foods, the methods are applicable to most nucleic acid probes and samples. Certainly, these methods will be of great value in the detection of pathogens and studies on the ecology of food and water.

## V. Immunoassays

The basis for serological identification (i.e. agglutination, precipitation, etc.) of viral and bacterial pathogens is the presence of a pathogen-specific antigenic determinant(s). Because viral and bacterial pathogens are found in low numbers in food and water, immunological detection requires enrichment of the organism to obtain a sufficient number of cells for detection. The use of polyclonal antisera sometimes necessitates selective media to prevent growth of cross-reacting organisms. The specificity and sensitivity of immunological assays have been enhanced with monoclonal antibodies and new combinations of enzyme labels and substrate. These advancements have facilitated the development of methods to detect microorganisms directly in samples without prior culture enrichment. As stated above, direct detection methods are important in light of recent reports that several bacterial pathogens are not culturable using standard culture methods yet are still metabolically active (Colwell *et al.*, 1985; Roszak and Colwell, 1987). Likewise, several groups among the enteric viruses cannot be propagated using standard tissue culture techniques, but are still of public health concern (e.g. Norwalk virus). Because it is not possible to cover all of the immunological methods used to detect pathogens, emphasis will be on the more sensitive techniques used in the detection and identification of important food- and waterborne pathogens.

### A. Fluorescent immunoassay

In fluorescent immunoassays (FIA), fluorochrome molecules are used to label immunoglobulins. The fluorochrome absorbs short-wavelength light and then emits light at a higher wavelength which can be detected using fluorescent microscopy. Fluorescein isothiocyanate and rhodamine isothiocyanate–bovine serum albumin are the most common fluorochromes used to tag antibodies and counterstain samples, respectively. These fluorochromes emit light of different wavelengths permitting their use in the same assay.

The FIA, initially developed by Coons *et al.* (1941), has been applied to the detection and identification of microorganisms because of the specificity, sensitivity, and rapid nature of the procedure. The direct and indirect procedures are the most commonly used to test food and water samples. In both methods the antigen or organism is immobilized and either (a) stained directly with an organism-specific, fluorescein-tagged antibody, or (b) stained indirectly, by first reacting the sample with an organism-specific antibody and then a fluorescein-labelled anti-immunoglobulin species antibody.

Food samples tested by FIA are typically from enrichment cultures because the number of bacteria in the original sample is insufficient to be detected directly and the food particulates, which are diluted in the medium, can produce background fluorescence. Water samples have been analysed directly by concentrating bacteria using membrane filtration. Polycarbonate filters, previously stained with Irgalan Black to reduce background and improve contrast, are commonly used in this procedure (Hobbie *et al.*, 1977).

A number of bacterial pathogens have been detected in food and water (Table IV) since Thomason *et al.* (1957) first applied FIA to the detection

**TABLE IV**
Microbial pathogens detected by immunofluorescence in food and water samples

| Organism | Source | Reference |
|---|---|---|
| *Listeria* | Food | Donnelly and Baigent, 1986 |
| *Salmonella* | Food | Insalata *et al.*, 1972 |
| | Food/water | Cherry *et al.*, 1975 |
| | Water | Thomason *et al.*, 1975 |
| | Food | Barrell and Paton, 1979 |
| *Vibrio cholerae* 01 | Water | Brayton *et al.*, 1987 |
| | Water | Brayton and Colwell, 1987 |
| Poliovirus | Water | Katzenelson, 1976 |
| | Water | Guttman-Bass *et al.*, 1981 |
| Coxsackievirus B5 | Water | Guttman-Bass *et al.*, 1981 |
| Echovirus 7 | Water | Guttman-Bass *et al.*, 1981 |
| Reovirus | Water | Ridinger *et al.*, 1982 |
| Rotavirus | Water | Hejkal *et al.*, 1982, 1984 |
| | Water | Smith and Gerba, 1982 |
| | Water | Bates *et al.*, 1984 |
| | Water | Deetz *et al.*, 1984 |
| | Water | Metcalf *et al.*, 1984 |
| | Water | Guttman-Bass *et al.*, 1987 |
| | Water | Bosch *et al.*, 1988 |

of *Salmonella* in foods. A shortcoming of this technique is the presence of cross-reacting antibodies which cannot be removed from antisera without a subsequent drop in titre to the organism of interest. Monoclonal antibodies may provide a solution to this problem.

Monoclonal antibodies specific for *Vibrio cholerae* 01 (Brayton *et al.*, 1987) have been used to detect this pathogen in water by FIA. Another shortcoming of FIA is the inability to differentiate viable from non-viable cells; both appear fluorescent. Brayton and Colwell *et al.* (1987) described an FIA procedure for the enumeration of viable *V. cholerae* 01 in environmental samples. Samples were incubated with a small quantity of yeast extract to supply nutrients and nalidixic acid which prevents replication by inhibition of DNA gyrase. Metabolically active cells elongate because they are unable to divide in the presence of nalidixic acid. Viable (elongated) cells are then distinguished from non-viable (unelongated) cells, microscopically. When combined with plate counts, this method provided the means to determine the numbers of viable (culturable), dead, and viable but non-culturable cells present within a sample (Colwell *et al.*, 1985; Brayton and Colwell, 1987). The significance of these viable but non-culturable cells to public health has yet to be fully elucidated.

The fluorescent antibody staining technique has been described in a number of papers (Gray and Kreger, 1985; Brayton and Colwell, 1987), and the method is essentially as follows.

*Procedure*

1. The sample or a dilution of the sample is spread on a glass slide, allowed to air dry, and then fixed with either 95% ethanol or heat.
2. A drop of rhodamine isothiocyanate–bovine serum albumin (RITC–BSA), diluted approximately 1 : 20, is placed on the sample. A coverslip is placed on top to distribute the RITC–BSA evenly over the sample.
3. Incubate at 37°C for 30 min in a dark moist chamber.
4. Rinse the slide three times in phosphate-buffered saline (PBS; per litre: NaCl, 8.5 g; $Na_2HPO_4$, 9.1 g; $KH_2PO_4$, 1.5 g; pH 7.3) and air dry.
5. A drop of appropriately diluted antiserum (titred prior to assay) is placed on the sample and covered with a coverslip. If the direct antibody-staining method is being used, the FITC-pathogen-specific antibody is added and the procedure continued at step 9.
6. Incubate at 37°C for 30 min in a dark moist chamber.
7. Rinse the sample three times in PBS and air dry.
8. Place a drop of FITC-anti-immunoglobulin (Ig of animal species from which pathogen-specific antibody was derived) on the sample and cover with a coverslip.
9. Incubate at 37°C for 30 min in a dark moist chamber.
10. Rinse three times in PBS and air dry.
11. Place a small drop of mounting fluid (pH 9) on the sample and place a coverslip on top.
12. Examine using an epifluorescent microscope and a 450–490 nm band pass filter. The antigen (or pathogen) if present will exhibit a green fluorescence and can be graded according to intensity.

FIA has also been applied to the detection of enteric viruses present in water. Detection involves the concentration of virus from the sample, inoculation into a suitable cell line, incubation to allow for adsorption, followed by the addition of new media and incubation for 20–24 h. The cells are then dried, fixed, and stained with a fluorescein-labelled virus-specific antibody. Fluorescent cells are then observed and quantified using fluorescent microscopy. A number of enteric viruses have been isolated from waters and enumerated using this technique (Tables II and IV). The FIA requires only 6–9 h or less to complete after infection, and is equal to or more sensitive than the plaque assay (Kedmi and Katzenelson, 1978; Ridinger *et al.*, 1982). The technique can also detect viruses that multiply poorly in cell lines (i.e. human rotavirus; Smith and Gerba, 1982).

Alternative fluorescent labels to those mentioned above, although not commonly used in microbiology, include lanthanide elements like europium and terbium chelates (Soini and Hemmilä, 1979; Soini and Kojola, 1983). The advantages of using these compounds are the high quantum yield and narrow emission peaks produced upon excitation. These labels have a short lifespan, in the microsecond range, that allows measurements to be made after the background emission has decayed. The procedure, called timed-resolved fluoroimmunoassay, couples the rare-earth elements to the antibody molecule in a non-fluorescent state. Following antibody attachment, the lanthanide element is released and it then combines with a chelating agent and emits an intense fluorescence that can be quantitated using a fluorometer. This technique has been used to detect antigens of hepatitis B, rotavirus and adenovirus in clinical samples (Siitar *et al.*, 1983; Soini, 1985), but has not been applied to the detection of food- and waterborne pathogens.

The major shortcomings of FIA are the limited availability of pathogen-specific antibody, the inability to differentiate viable from non-viable pathogens, the tedious nature of the procedure, and the expense of automation.

Flow cytometry has been applied to the detection of *Listeria* in milk (Donnelly and Baigent, 1986). In this technique, *Listeria* identification was based upon morphology, nucleic acid content, and surface antigens, detected with fluorescein-labelled antibodies. The different parameters were analysed simultaneously with a laser cytofluorograf. Flow cytometry has not been used extensively in the analysis of food and water because of the high cost of the equipment involved.

### B. Enzyme immunoassay

In enzyme immunoassays (EIA), enzyme–antibody conjugates are used to identify and quantitate antibody–antigen complexes. The enzyme catalyses the conversion of substrate into a quantifiable end product, amplifying the 'signal' over time, whereas, with labels such as fluorescein and $^{125}I$, the quantity of label or 'signal' is fixed. Thus, the enzyme tag grants immunoassays a sensitive means of detecting small quantities of antigen (Engvall and Perlmann, 1972; Yolken, 1980). The assay, developed in 1971 by Engvall and Perlmann, has been extensively used because of its specificity, sensitivity, use of non-hazardous reagents, the stability of the reagents, and low cost. Preparation of enzyme–antibody conjugates is typically carried out as described by Engvall and Perlmann (1972).

A variety of EIAs have been developed and are shown in Table V. Details and descriptions of the assays can be found in the references provided. EIAs can be divided into two types; homogeneous and heterogeneous (solid phase). In homogeneous EIAs, neither antibody nor antigen is immobilized to a solid matrix eliminating the need to remove unbound reactants with repetitive washing steps. Homogeneous EIAs are normally used to quantitate small molecules but have been used in the detection of macromolecules (Tan *et al.*, 1981). Heterogeneous EIAs involve a solid matrix to which one of the immunoreactants (antibody or antigen) is immobilized. Unbound or weakly bound material must be removed between steps.

**TABLE V**
The various classes of enzyme immunoassay

| | | | | |
|---|---|---|---|---|
| EIA[a] | Homogeneous | Competitive | | Rubenstein *et al.*, 1972<br>Carrico *et al.*, 1976<br>Schroeder *et al.*, 1976<br>Burd *et al.*, 1977<br>Wei and Riebe, 1977<br>Gibbons *et al.*, 1980<br>Ngo *et al.*, 1981<br>Bacquet and Twumasi, 1984 |
| | | Non-competitive | | Ngo and Lenhoff, 1981 |
| | Heterogeneous | Competitive | Ag binded | Voller *et al.*, 1979<br>Tijssen and Kurstak, 1981<br>Friguet *et al.*, 1983 |
| | | | Ab binded | Engvall and Perlmann, 1971<br>Van Weemen and Schuurs, 1971<br>Belanger *et al.*, 1976<br>Van Weemen *et al.*, 1978 |
| | | Non-competitive | Ag binded | Guesdon *et al.*, 1979<br>Butler *et al.*, 1980<br>Guesdon and Avrameas, 1980<br>Yolken and Leister, 1981<br>Madri and Barwick, 1983 |
| | | | Ab binded | Crook and Payne, 1980<br>Barbara and Clark, 1982<br>Koening and Paul, 1982<br>De Jong, 1983 |

[a] Enzyme immunoassay.

Homogeneous and heterogeneous EIAs can be subclassified into non-competitive and competitive assays. In non-competitive EIAs, the antigen (microbe or substance of interest) within a sample is detected using complementary enzyme-labelled antibody. The antigen concentration is directly proportional to the amount of enzyme-labelled antibody bound and end product formed. In competitive assays, the antigen to be detected within a sample is usually purified and labelled with an enzyme. The enzyme-labelled antigen is added to the sample and incubated with an immobilized antigen-specific antibody. Absence of the antigen will result in the production of end product due to the binding of the enzyme-labelled antigen to the antigen-specific antibody. The presence of antigen within the sample (unlabelled) reduces the amount of end product formed because of the competition with the enzyme-labelled antigen for antibody binding sites. Thus, in competitive EIAs, the quantity of antigen within a sample is inversely proportional to the amount of end product formed. Competitive assays are faster than non-competitive EIAs, but require purification and labelling of the antigen of interest. Although non-competitive EIAs are slower, requiring several incubation and washing steps to avoid non-specific binding and background, it is more commonly used for the detection of microbial antigens.

Although several protocols have been described (Polin and Kennett, 1980; Robison *et al.*, 1983; Berdal *et al.*, 1981; Minnich *et al.*, 1982), a typical non-competitive EIA would be conducted as follows.

*Procedure*

1. Enrichment broth is inoculated with the food or water sample and incubated. The growth is centrifuged and resuspended in coating buffer (per litre: $NaCO_3$, 1.7 g; $Na_2CO_3$, 1.7 g; $NaHCO_3$, 2.9 g; pH 9.6). One hundred μl are added to each of duplicate wells of a microtitration plate and incubated overnight at 4°C.
2. The wells are then washed three times with washing buffer (per litre: NaCl, 7.6 g; $Na_2HPO_4$, 0.7 g; $KH_2PO_4$, 0.2 g; Tween 20, 0.5 ml; pH 7.4).
3. The wells are blocked with 250 ml of a 2% solution of bovine serum albumin (BSA) in washing buffer and incubated for 1 h at 37°C to block non specific binding sites.
4. Wash the wells three times with washing buffer.
5. One hundred μl of pathogen-specific antibody diluted appropriately in washing buffer containing 2% BSA is added per well and incubated at 37°C for 1 h. In a direct EIA, an enzyme-labelled antibody would be added above and the procedure continued at step 8.
6. Wash the wells three times with washing buffer.

7. Enzyme-labelled anti-immunoglobulin (Ig of animal species from which pathogen-specific antibody was derived), diluted in washing buffer plus 2% BSA is added to each well (100 μl $well^{-1}$) and incubated at 37°C for 1 h.
8. Wash the wells three times with washing buffer.
9. Enzyme substrate in an appropriate buffer is added (100 μl $well^{-1}$), and after incubating for a specified period of time (usually 30 min), the reaction is stopped.

For alkaline phosphatase-conjugated antibodies, *p*-nitrophenylphosphate at 1 mg $ml^{-1}$ of diethanolamine buffer (97 ml diethanolamine, 100 mg $MgCl.6H_2O$, 800 ml distilled water; adjusted to pH 9.8 and the final volume brought to 1 l with distilled water) is used as substrate. After incubation at 37°C for 30 min, the reaction is stopped with 30 μl of 3 N NaOH and the colour intensity quantified spectrophotometrically. With horseradish peroxidase-conjugated antibodies, 2,2′-azino-bis(3-ethylbenzthiazoline-6-sulphonic acid), 5 mg, in 5 ml of 0.05 M citric acid and 15 μl of hydrogen peroxide (prepared immediately before use) is used as substrate. After incubation at 37°C for 30 min the reaction is stopped and the colour intensity quantified spectrophotometrically.

EIAs have been used extensively in the detection of foodborne pathogens and their toxins (Table VI). Early difficulties with cross-reactions between closely related pathogens have been somewhat alleviated with the advent of monoclonal antibodies, but development of specific monoclonal antibodies which react with members of a particular genus or species can be challenging (Kaspar and Hartman, 1987). Another impediment to testing food and water for pathogens using EIA is the requirement for a minimum of $10^4$–$10^5$ organisms $ml^{-1}$. Rarely do pathogens reach this number in food and water. In most cases, the organism to be detected must first be grown in enrichment culture, to increase numbers, and then identified using EIA. Membrane filtration coupled with EIA has been used to detect a number of enterotoxigenic bacteria (Table VI). Following filtration of the sample onto a nitrocellulose filter and incubation, the bacterial colonies formed on the surface of the filter are screened by EIA for a particular pathogen or toxin. Using an assay similar to the immunofluorescence assay, virus-infected cells have been detected using a peroxidase-labelled antibody. If bound, peroxidase catalyses the production of a stain which enables foci to be quantitated microscopically.

### 1. *Enzyme labels*

Several enzymes have been used as labels in immunoassays. Important characteristics of an enzyme used to tag immunoglobulins include: (a)

**TABLE VI**
Microbial pathogens and toxins detected by enzyme immunoassay in food and water samples

| Organism | Source | Reference |
|---|---|---|
| Toxigenic *Clostridium perfringens* | Food | Stelma *et al.*, 1985 |
| *Listeria* | Food | Farber and Speirs, 1987 |
| | Food | Butman *et al.*, 1988 |
| | Food | Mattingly *et al.*, 1988 |
| *Salmonella* | Food | Minnich *et al.*, 1982 |
| | Food | Robison *et al.*, 1983 |
| | Food | Smith and Jones, 1983 |
| | Food | Aleixo *et al.*, 1984 |
| | Food | Mattingly, 1984 |
| | Food/water | Rigby, 1984 |
| | Food | Anderson and Hartman, 1985 |
| | Food | Farber *et al.*, 1985 |
| | Food | Mattingly *et al.*, 1985 |
| | Food | Cerqueira-Campos *et al.*, 1986 |
| | Food | Ibrahim and Lyons, 1987 |
| Staphylococcal Enterotoxin | Food | Saunders and Bartlett, 1977 |
| | Food | Morita and Woodburn, 1978 |
| | Food | Stiffler-Rosenberg and Fey, 1978 |
| | Food | Berdal *et al.*, 1981 |
| | Food | Freed *et al.*, 1982 |
| | Food | Notermans *et al.*, 1983 |
| | Food | Fey *et al.*, 1984 |
| | Food | Peterkin and Sharpe, 1984 |
| | Food | Thompson *et al.*, 1986 |
| | Food | Ocasio and Martin, 1988 |
| *Vibrio cholerae* 01 | Water | Tamplin *et al.*, 1987 |
| Enteric viruses | Water | Payment and Trudel, 1985, 1987 |
| Coxsackievirus B5 | Food | Loh *et al.*, 1985 |
| Hepatitis A | Water | Nasser and Metcalf, 1987 |
| Poliovirus | Food | Loh *et al.*, 1985 |
| Rotavirus | Water | Steinmann, 1981 |
| | Water | Raphael *et al.*, 1985 |
| | Water | Guttman-Bass *et al.*, 1987 |

attachment should produce a minimal effect on the binding properties of the antibody (Takashi and Kayoto, 1977; Chandler *et al.*, 1982; (b) a high specific activity; (c) a low molecular weight to maximize the quantity of enzyme per immunoglobulin molecule; (d) the enzyme should be stable;

and (e) a substrate which yields a quantifiable product should be available. Enzymes commonly used in the detection of foodborne pathogens are shown in Table VII.

The avidin–biotin system is an important tool for EIAs. Avidin, a glycoprotein, has a very high affinity for biotin, a vitamin (Bayer and Wilcher, 1980; Guesdon *et al.*, 1979). Biotin is generally used to tag immunoglobulins, and an avidin–enzyme conjugate used to detect bound biotin–labelled antibody. The biotin–avidin system acquires its high degree of sensitivity from the avidin molecule that has two pairs of binding sites acting as a bridge between biotinylated molecules.

Alternative assays include thermometric (Mattiason *et al.*, 1977) and cyclic EIAs (Harper and Orengo, 1981). In thermometric EIAs, the heat generated by an enzyme, such as catalase, is measured rather than a visible end product. In cycling EIAs, a portion of the substrate which has been converted into end product is immediately regenerated into substrate (Harper and Orengo, 1981). Neither of these assays has been applied to the detection of foodborne pathogens.

### 2. *Enzyme substrates*

The most commonly used enzyme substrates (chromogenic) in EIAs release a visible end product that can be measured spectrophotometrically. New methods (e.g. avidin–biotin system) and new enzyme–substrate combinations have enhanced the sensitivity of EIA. For example, enzyme substrates which release a fluorogenic portion when cleaved are available for a number of enzymes (Table VII; Guilbault *et al.*, 1968). The sensitivity with fluorogenic substrates has been reported to be 10–100-fold greater than chromogenic substrates (Yolken and Leister, 1982; Swaminathan *et al.*, 1985).

EIAs which utilize radioactive substrates are highly sensitive, but find limited use in the food industry due to the hazards and disposal problems associated with the use of radioactive materials. Radioactive substrates have been used with several enzymes to detect toxins and viral antigens (Harris *et al.*, 1979); however, the radioactive end product must be isolated from the reaction mixture, generally by ion-exchange methods, prior to measurement (Yolken, 1980). Fields *et al.* (1981) elminated the need for separation steps by using glutamate decarboxylase which cleaves $^{14}CO_2$ from L-$^{14}$C-glutamic acid; the $^{14}CO_2$ is captured from the atmosphere above the reaction mixture and measured. The sensitivity of this assay was reported to be 100 times higher than radioimmunoassay.

Chemiluminescent substrates, like luminol, 3-aminophthalhydrazide (Puget *et al.*, 1977) and its derivatives (Cheng *et al.*, 1982) generate light

**TABLE VII**
Enzymes and substrates commonly used in enzyme immunoassay

| Enzyme | Substrate | | | |
|---|---|---|---|---|
| | Chromogenic | Fluorogenic | Radioactive | Luminescent |
| Alkaline phosphatase | *p*-Nitrophenyl phosphate<br>Disodium phenyl phosphate + aminoantipyrene | 4-Methylumbelliferyl phosphate<br>Fluorescein methyl phosphate<br>NADH<br>3-(*p*-Hydroxyphenyl) propionic acid | $^{3}$H-adenosin monophosphate<br>$^{3}$H-nitrophenyl phosphate<br>$^{14}$C-nitrophenyl phosphate | |
| Peroxidase | 5-Aminosalicylic acid + $H_2O_2$<br>*o*-Dianisidine + $H_2O_2$<br>2,2-Azino-di(3-ethyl benzothiazolin sulfone-6)<br>3,3′-Diaminobenzidine + $H_2O_2$<br>*o*-Toluidine + $H_2O_2$<br>*o*-Phenylendiamine + $H_2O_2$ | | | Isoluminol + $H_2O_2$<br>D-Luciferin + luminol + $H_2O_2$<br>D-Luciferin + 7-dimethylamino-naphthalene 1,2 dicarbonic acid hydrazide + $H_2O_2$<br>Pyrogallol + $H_2O_2$ |
| β-Galactosidase | *o*-Nitrophenyl β-D-galactopyranoside | 4-Methylumbelliferyl-β-D-galactopyranoside | $^{3}$H-β-galactose phosphate | |

**TABLE VII** – *continued*

| | | | | |
|---|---|---|---|---|
| Glucose oxidase | Glu[a] + 5-aminosalicyclic acid<br>Glu + p-nitro blue tetrazolium chloride<br>Glu + thiazolyl<br>Sodium, 3,5-dichloro-2 hydrobenzene sulfonate<br>4-Aminoantipyrine glucose peroxidase<br>*o*-Dianisidine + horseradish peroxidase | NADH<br>β-Hydroxyphenylacetic acid | $^{3}H$-Glucose | Glu + peroxidase + luminol |
| Urease | Bromocresol purple + urea | | | |
| β-Lactamase | Starch + iodine + penicillin G | | | |
| Catalase | $H_2O_2$ | | | |
| Glutamate decarboxylase | | | $^{14}C$-glutamate | |
| Glucose 6-phosphate dehydrogenase | Glu 6-phosphate+ $NAD^+$ | | | Glu 6-phosphate + NADP + luciferase |

[a] Glucose.

when enzymatically degraded. The reaction is quantitated by measuring the emitted light. The sensitivity of EIAs using chemiluminescent substrates is not as high as with other assays. Similarly, bioluminescent assays utilize enzyme–substrate combinations which shuttle electrons to luciferase and result in the production of bioluminescence. Bioluminescent EIAs have sensitivities comparable to radioimmunoassays and require shorter reaction times than assays using chromogenic substrates. A number of excellent reviews of enzyme substrates have been published (Ishikawa *et al.*, 1983; Yolken, 1984; Swaminathan and Konger, 1986).

## C. Radioimmunoassay

Radioimmunoassay (RIA), described by Yalow and Berson (1959), can be used in all of the variations of EIA described above. RIA combines the specificity of immunoassays with the sensitivity of radioisotopic methods, detecting nanogram to picogram quantities of antigen (Yung *et al.*, 1977; Kalmakoff *et al.*, 1977). Although $^{3}H$, $^{14}C$ and $^{131}I$ have been utilized as labels, $^{125}I$ is more commonly used because of its high specific activity and short half-life. Several methods are available to prepare $^{125}I$-RIA reagents (Miles and Hales, 1968; Hunter, 1978; Marchalonis, 1969).

Several variations of radioimmunoassay have been used to test food and water samples (Robern *et al.*, 1975, 1978; Miller *et al.*, 1978). The method outlined below was described by Pierce and Klinman (1976).

*Procedure*

1. Following inoculation and incubation, cells from enrichment broth (for the organism of interest) are pelleted by centrifugation and washed in PBS (per litre: NaCl, 7.6 g; $Na_2HPO_4$, 0.7 g; $KH_2PO_4$, 0.2 g; pH 7.4).
2. To each of duplicate wells of a 96-well polyvinyl plate, 100 μl of the cell suspension is added and incubated at 4°C (usually overnight).
3. The wells are then washed three times with RIA buffer (PBS containing 0.1% BSA).
4. To minimize non-specific binding, 250 μl of PBS containing 2% BSA is added to each well and incubated for 1 h at 37°C.
5. Wash the wells three times with RIA buffer.
6. The organism-specific antibody is diluted appropriately in RIA buffer, and 100 μl is added to each well and incubated for 1 h at 37°C.
7. The wells are then washed three times with RIA buffer.
8. Between 50 and 100 μl of $^{125}I$ anti-immunoglobulin (10 μCi $\mu g^{-1}$) is added to each well (10 000 cpm $well^{-1}$) and incubated at 37°C for 1 h.
9. Unbound $^{125}I$ anti-immunoglobulin is then removed by washing the wells three times.

10. The wells are then separated and counted in a gamma scintillation counter. The background values for controls should not exceed 100–200 cpm.

Enterotoxigenic bacteria have been detected within natural populations (Shah *et al.*, 1982) by blotting colonies onto polyvinyl membranes and probing with specific radiolabelled antibody. A radioimmunofocus assay (Lemon *et al.*, 1983) has been used to detect viruses following multiplication in suitable cell lines. Infected cells are detected and enumerated using $^{125}$I-labelled antibodies and autoradiography. In Table VIII are listed pathogens which have been detected in food and water by RIA.

**TABLE VIII**
Pathogens and toxins detected in food and water using radioimmunoassay

| Pathogen/toxin | Source | Reference |
|---|---|---|
| Staphylococcal | Food | Collins *et al.*, 1973 |
| Enterotoxin | Food | Robern *et al.*, 1975 |
| | Food | Orth, 1977 |
| | Food | Pober and Silverman, 1977 |
| | Food | Miller *et al.*, 1978 |
| | Food | Robern *et al.*, 1978 |
| *Cl. perfringens* Enterotoxin | Food | Stelma *et al.*, 1983 |
| Hepatitis A | Water | Hejkal *et al.*, 1982 |

Despite the high sensitivity of the assay, RIA has not been used extensively in the food industry because of the problems associated with the use and disposal of radioactive materials.

## VI. Enzyme substrates

The ability to detect specific enzymes rapidly using chromogenic and fluorogenic substrates has led to the development of a number of rapid methods for the identification of several bacteria (Kilian and Bülow, 1976; Godsey *et al.*, 1981; Facklam *et al.*, 1982; Trepeta and Edberg, 1984; O'Brien and Colwell, 1985; Feng and Hartman, 1982; Littel and Hartman, 1983; Petzel and Hartman, 1985; Freier and Hartman, 1987). Feng and Hartman (1982) tested a variety of food, water, and milk samples using lauryl tryptose broth containing 4-methylumbelliferyl-β-D-glucuronide (MUG). After incubation for 24 h, *E. coli* was presumptively identified and confirmed in most-probable-number tubes using lactose fermentation (gas production) and MUG (fluorescence, detected under long-wave ultraviolet light), respectively. The lauryl tryptose broth–MUG–MPN was sensitive, produced few false-positive reactions, detected anaerogenic strains of *E. coli*, and yielded faecal coliforms from 90% of the tubes which were gas and fluorescence positive. The MUG test has compared favourably with standard MPN methods for *E. coli* (Alvarez, 1984; Robison, 1984; Koburger and Miller, 1985; Peterson *et al.*, 1987). The basis for the MUG test is the presence of β-glucuronidase which cleaves MUG to release the fluorescent 4-methylumbelliferyl portion. β-Glucuronidase is found in a number of bacterial genera; however, among the *Enterobacteriaceae* it is restricted to *E. coli* (96%), a few *Salmonella*, and *Shigella* (50%) (Kilian and Bülow, 1976; LeMinor, 1979).

The incorporation of chromogenic and/or fluorogenic substrates into a selective medium can eliminate the need for subculture and subsequent biochemical testing, saving time, supplies and money. Several substrates have been incorporated into media or detection schemes to aid in the identification of coliforms, faecal coliforms (Warren *et al.*, 1983), enterococci (Bosley *et al.*, 1983), faecal streptococci (Littel and Hartman, 1983), *E. coli* (Feng and Hartman, 1982; Ley *et al.*, 1988; Watkins *et al.*, 1988) and *Vibrio cholerae* (O'Brien and Colwell, 1985). Although not feasible at this time, these substrates may be of use in direct detection procedures in the future.

## VII. Concluding remarks

Molecular biology techniques have made it possible to develop highly specific nucleic acid and antibody probes to an organism or group of organisms. The application of these probes and methods to the detection of important food and waterborne microorganisms has increased in recent years. New gene and signal amplification techniques, as well as the use of rRNA genes, could circumvent the need to propagate these organisms prior to hybridization, saving time and allowing the detection of nonculturable organisms. Likewise, with new labels and/or enzyme substrates, immunological methods may be of even greater value in direct detection procedures in the near future.

## References

Aleixo, J. A. G., Swaminathan, B. and Minnich, S. A. (1984). *J. Microbiol. Meth.* **2,** 135–145.

Alter, M. J., Gerety, R. J., Smallwood, L. A., Sampliner, R. E., Tabor, E.,

Deinhardt, F., Frosner, G. and Matanoski, G. M. (1982). *J. Infect. Dis.* **145,** 886–893.

Alvarez, R. J. (1984). *J. Food Sci.* **49,** 1186–1187.

Anderson, J. M. and Hartman, P. A. (1985). *Appl. Environ. Microbiol.* **49,** 1124–1127.

Archer, D. L. (1988). *Food Technol.* **42,** 53–58.

Archer, D. L. and Young, F. E. (1988). *Clin. Microbiol. Rev.* **1,** 377–398.

Aycock, W. (1927). *Am. J. Hyg.* **7,** 791–803.

Bacquet, C. and Twumasi, D. (1984). *Anal. Biochem.* **136,** 487–490.

Bagdasar'yan, G. A. (1964). *J. Hyg. Epidemiol.* **8,** 497–505.

Barbara, D. J. and Clark, M. F. (1982). *J. Gen. Virol.* **58,** 315–322.

Barrell, R. A. E. and Paton, A. M. (1979). *J. Appl. Bacteriol.* **46,** 153–159.

Bates, J., Goddard, M. R. and Butler, M. (1984). *J. Hyg. Camb.* **93,** 639–643.

Bayer, E. A. and Wilcher, M. (1980). *Meth. Biochem. Anal.* **26,** 1–43.

Beckers, H. J. (1988). *J. Food Protect.* **51,** 327–334.

Belanger, L., Hamel, D., Dufour, D. and Pouliot, M. (1976). *Clin. Chem.* **22,** 198–204.

Berdal, B. P., Olsvik, O. and Omland, T. (1981). *Acta Path. Microbiol. Scand. Sect. B.* **89,** 411–415.

Berg, G., Dahling, D. R., Brown, G. A. and Berman, D. (1978). *Appl. Environ. Microbiol.* **36,** 880–884.

Berninger, M., Hammer, M., Hoyer, B. and Gerin, J. L. (1982). *J. Med. Virol.* **9,** 57–68.

Bialkowska-Hobrzanska, H. (1987). *J. Clin. Microbiol.* **25,** 338–343.

Blake, P. A., Weaver, R. E. and Hollis, D. G. (1980). *Ann. Rev. Microbiol.* **34,** 341–367.

Block, J. C. (1983). *In* "Viral Pollution of the Environment" (G. Berg, Ed.), pp. 117–145, CRC Press, West Palm Beach, FL.

Boguslawski, S. J., Smith, D. E., Michalak, M. A., Michelson, K. E., Yehle, C. O., Patterson, W. L. and Carrico, R. J. (1986). *J. Immunol. Meth.* **89,** 123–130.

Boileau, C. R., d'Hauteville, H. M. and Sansonetti, P. J. (1984). *J. Clin. Microbiol.* **20,** 959–961

Bonde, G. J. (1966). *Health Lab. Sci.* **3,** 124–128.

Bosch, A., Pinto, R. M., Blanch, A. R. and Jofre, J. T. (1988). *Water Res.* **22,** 343–348.

Boseley, G. S., Facklam, R. R. and Grossman, D. (1983). *J. Clin. Microbiol.* **18,** 1225–1277.

Brayton, P. R. and Colwell, R. R. (1987). *J. Microbiol. Meth.* **6,** 309–314.

Brayton, P. R., Tamplin, M. L., Huq, S. A. and Colwell, R. R. (1987). *Appl. Environ. Microbiol.* **53,** 2862–2865.

Bryan, F. L. (1988). *J. Food. Protect.* **51,** 498–508.

Bryan, J. A., Lehmann, J. D., Setrady, I. F. and Hatch, M. H. (1974). *Am. J. Epidemiol.* **99,** 145–154

Burd, J. F., Carrico, R. J., Fetter, M. C., Buckler, R. T., Johnson, R. D., Boguslaski, R. C. and Christner, J. E. (1977). *Anal. Biochem.* **77,** 56–67.

Butler, J. E., Cantarero, L. A., Swanson, P. and McGivern, P. L. (1980). *In* "Enzyme Immunoassay" (E. G. Maggio, Ed.), CRC Press, West Palm Beach, FL.

Butman, B. T., Plank, M. C., Durham, R. J. and Mattingly, J. A. (1988). Abstract P-5, 88th Ann. Meeting Am. Soc. Microbiol., Miami, FL.

Caredda, F., D'Aminio Monforte, A., Rossi, E., Lopez, S. and Moroni, M. (1981). *Lancet* **2,** 48.
Carrico, R. J., Christner, J. E., Boguslaski, R. C. and Yeung, K. K. (1976). *Anal. Biochem.* **72,** 271–282.
Cerqueira-Campos, M. L., Peterkin, P. I. and Sharpe, A. N. (1986). *Appl. Environ. Microbiol.* **52,** 124–127.
Chandler, H. M., Cox, J. C., Healey, K., MacGregor, A., Premier, R. R. and Hurrell, J. G. R. (1982). *J. Immunol. Meth.* **53,** 187–194.
Cheng, P., Hemmilä, I. and Lövgren, T. (1982). *J. Immunol. Meth.* **48,** 159–168.
Cherry, W. B., Thomason, B. M., Gladden, J. B., Holsing, N. and Murlin, A. M. (1975). *Ann. N.Y. Acad. Sci.* **254,** 350–368.
Cliver, D. O. (1971). *Crit. Rev. Environ. Control* **1,** 551–579.
Coin, L., Menetrier, M. L., Labonde, J. and Hanoun, M. C. (1965). *In* "Advances in Water Pollution Research" (O. Jagg, Ed.), pp. 1–10, Pergamon Press, UK.
Collins, W. S., Johnson, A. D., Metzger, J. F. and Bennett, R. W. (1973). *Appl. Microbiol.* **25,** 774–777.
Colwell, R. R., Brayton, P. R., Grimes, D. J., Roszak, D. B., Huq, S. A. and Palmer, L. M. (1985). *Biotechnol.* **3,** 817–820.
Coons, A. H., Creech, H. J. and Jones, R. N. (1941). *Proc. Soc. Exp. Biol. Med. (New York)* **47,** 200–202.
Craun, G. F. (1978). *In* "Evaluation of the Microbiology Standards for Drinking Water" (C. W. Hendricks, Ed.), pp. 21–35, U.S. Environmental Protection Agency, Washington, D.C.
Craun, G. F. (1981). *J. Am. Water. Works. Assoc.* **73,** 360–369.
Crook, N. E. and Payne, C. C. (1980). *J. Gen. Virol.* **46,** 29–37.
D'Angelo, L. J., Hierholzer, J. C., Keenlyside, R. A., Anderson, L. J. and Martone, W. J. (1979). *J. Infect. Dis.* **140,** 42–47.
Datta, A. R., Wentz, B. A. and Hill, W. E. (1987). *Appl. Environ. Microbiol.* **53,** 2256–2259.
De Jong, P. J. (1983). *J. Clin. Microbiol.* **17,** 928–930.
Deetz, T. R., Smith, E. M., Goyal, S. M., Gerba, C. P., Vollet III, J. J., Tsai, L., DuPont, H. L. and Keswick, B. H. (1984). *Water Res.* **18,** 567–571.
Dimitrov, D. H., Graham, D. Y. and Estes, M. K. (1985). *J. Infect. Dis.* **152,** 293–300.
Dingman, C. (1916). *N.Y. State J. Med.* **16,** 589.
Dolin, R., Treanor, J. J. and Madore, P. (1987). *J. Infect. Dis.* **155,** 365–376.
Donnelly, C. W. and Baigent, G. J. (1986). *Appl. Environ. Microbiol.* **52,** 689–695.
Echeverria, P., Seriwatana, J., Chityothin, O., Chaicumpa, W. and Tirapat, C. (1982). *J. Clin. Microbiol.* **16,** 1086–1090.
Ellender, R. D., Mapp, J. B., Middlebrooks, B. L., Cook, D. W. and Cake, E. W. (1980). *J. Food. Protect.* **43,** 105–110.
Engvall, E. and Perlmann, P. (1971). *Immunochem.* **8,** 871–874.
Engvall, E. and Perlmann, P. (1972). *J. Immunol.* **109,** 129–135.
Evison, L. M. and James, A. (1975). *Prog. Water Technol.* **7,** 57.
Facklam, R. R., Thacker, L. G., Fox, B. and Eriquez, L. (1982). *J. Clin. Microbiol.* **15,** 987–990.
Farber, J. M. and Speirs, J. I. (1987). *J. Food Protect.* **50,** 479–484.
Farber, J. M., Peterkin, P. I., Sharpe, A. N. and D'Aoust, J. Y. (1985). *J. Food Prot.* **48,** 790–793.
Feng, P. C. S. and Hartman, P. A. (1982). *Appl. Environ. Microbiol.* **43,** 1320–1329.

Fey, H., Pfister, H. and Rüegg, O. (1984). *J. Clin. Microbiol.* **19,** 34–38.
Fields, H. A., Davis, C. L., Dreesman, G. R., Bradley, D. W. and Maynard, J. E. (1981). *J. Immunol. Meth.* **47,** 145–159.
Fiksdal, L., Maki, J. S., Lacroix, S. J. and Staley, J. T. (1985). *Appl. Environ. Microbiol.* **49,** 148–150.
Fitts, R., Diamond, M., Hamilton, C. and Neri, M. (1983). *Appl. Environ. Microbiol.* **46,** 1146–1151.
Flowers, R. S., Mozola, M. A., Curiale, M. S., Gabis, D. A. and Silliker, J. H. (1987). *J. Food Sci.* **52,** 781–785.
Foy, H. M., Cooney, M. K. and Hatlen, J. B. (1968). *Arch. Environ. Health* **17,** 795–802.
Freed, R. C., Evenson, M. L., Reiser, R. F. and Bergdoll, M. S. (1982). *Appl. Environ. Microbiol.* **44,** 1349–1355.
Freier, T. A. and Hartman, P. A. (1987). *Appl. Environ. Microbiol.* **53,** 1246–1250.
Friguet, B., Djavadi-Ohaniance, L., Pages, J., Bussard, A. and Goldberg, M. (1983). *J. Immunol. Meth.* **60,** 351–358.
Fuhrman, J. A., Comeau, D. E., Hagstrom, A. and Chan, A. M. (1988). *Appl. Environ. Microbiol.* **54,** 1426–1429.
Gerba, C. P. and Goyal, S. M. (1978). *J. Food Prot.* **41,** 743–754.
Gerba, C. P., Singh, S. N. and Rose, J. B. (1985). *CRC Crit. Rev. Environ. Control* **15,** 213–236.
Gibbons, I., Skold, C., Rowley, G. L. and Ulman, E. F. (1980). *Anal. Biochem.* **102,** 167–170.
Giovannoni, S. J., DeLong, E. F., Olsen, G. J. and Pace, N. R. (1988). *J. Bacteriol.* **170,** 720–726.
Godsey, J. H., Matteo, M. R., Shen, D., Tolman, G. and Gohlke, J. R. (1981). *J. Clin. Microbiol.* **13,** 483–490.
Goyal, S. M., Gerba, C. P. and Melnick, J. L. (1978). *J. Water Poll. Control Fed.* **50,** 2247–2256.
Goyal, S. M., Gerba, C. P. and Melnick, J. L. (1979). *Appl. Environ. Microbiol.* **37,** 572–581.
Gray, L. D. and Kreger, A. S. (1985). *Diagn. Microbiol. Infect. Dis.* **3,** 461–468.
Grimont, P. A. D., Grimont, F., Desplaces, N. and Tchen, P. (1985). *J. Clin. Microbiol.* **21,** 431–437.
Grohmann, G. S., Greengerg, H. B., Welch, B. M. and Murphy, A. M. (1980). *J. Med. Virol.* **6,** 11–19.
Grunstein, M. and Hogness, D. S. (1975). *Proc. Natl. Acad. Sci. U.S.A.* **72,** 3961–3965.
Guelin, A. (1948). *Ann. Inst. Pasteur* **75,** 485–496.
Guesdon, J. L. and Avrameas, S. (1980). *J. Immunol. Meth.* **39,** 1–13.
Guesdon, J. L., Ternynck, T. and Avrameas, S. (1979). *J. Histochem. Cytochem.* **27,** 1131–1139.
Guilbault, G. G., Brignac, P. Jr. and Juneau, M. (1968). *Anal. Chem.* **40,** 1256–1263.
Gunn, R. A., Janowski, H. T., Lieb, S., Prather, E. C. and Greenberg, H. (1982). *Am. J. Epidemiol.* **115,** 348–351.
Guttman-Bass, N., Tchorsh, Y., Nasser, A. and Katzenelson, E. (1981). *In* "Viruses and Wastewater Treatment" (M. Goddard and M. Butler, Eds), pp. 247–251, Pergamon, Oxford, UK.
Guttman-Bass, N., Tchorsh, Y. and Marva, E. (1987). *Appl. Environ. Microbiol.* **53,** 761–767.

Guzewich, J. J. and Morse, D. L. (1986). *J. Food Prot.* **49,** 389–394.
Hackney, C. R. and Dicharry, A. (1988). *Food Technol.* **42,** 104–109.
Harper, J. R. and Orengo, A. (1981). *Anal. Biochem.* **113,** 51–57.
Harris, C. C., Yolken, R. H., Krokan, H. and Hsu, I. C. (1979). *Proc. Natl. Acad. Sci. U.S.A.* **76,** 5336–5339.
Hartman, P. A., Petzel, J. P. and Kaspar, C. W. (1986). *In* "Foodborne Microorganisms and their Toxins: Developing Methodology" (M. D. Pierson and N. J. Stern, Eds), pp. 175–217, Marcel Dekker, NY.
Hauschild, A. H. W. and Bryan, F. L. (1980). *J. Food Prot.* **43,** 435–440.
Havelaar, A. H., Hogeboom, W. M. and Pot, R. (1985). *Wat. Sci. Technol.* **17,** 645–655.
Hejkal, T. W., Keswick, B., Labelle, R. L., Gerba, C. P., Sanchez, Y., Dreesman, G., Hafkin, B. and Melnick, J. L. (1982). *J. Am. Wat. Works Assoc.* **74,** 318–321.
Hejkal, T. W., Smith, E. M. and Gerba, C. P. (1984). *Appl. Environ. Microbiol.* **47,** 588–590.
Hill, W. E., Madden, J. M., McCardell, B. A., Shah, D. B., Jagow, J. A., Payne, W. L. and Boutin, B. K. (1983a). *Appl. Environ. Microbiol.* **45,** 1324–1330.
Hill, W. E., Payne, W. L. and Aulisio, C. G. G. (1983b). *Appl. Environ. Microbiol.* **46,** 636–641.
Hoadley, A. W. and Dutka, B. J. (Eds). (1977). "Bacterial Indicators/Health Hazards Associated with Water", American Society for Testing and Materials, Philadelphia.
Hobbie, J. E., Daley, R. J. and Jasper, S. (1977). *Appl. Environ. Microbiol.* **33,** 1225–1228.
Holben, W. E., Jansson, J. K., Chelm, B. K. and Tiedje, J. M. (1988). *Appl. Environ. Microbiol.* **54,** 703–711.
Hunter, W. M. (1978). *In* "Handbook of Experimental Immunology" (D. M. Weir, Ed.), Vol. 1, Blackwell Scientific Publications, Oxford, UK.
Hyypiä, T., Stålhandske, P., Vainionpää, R. and Pettersson, U. (1984). *J. Clin. Microbiol.* **19,** 436–438.
Ibrahim, G. F. and Lyons, M. J. (1987). *J. Food Prot.* **1,** 59–61.
Insalata, N. F., Mahnke, C. W. and Dunlap, W. G. (1972). *Appl. Microbiol.* **24,** 645–649.
Ishikawa, E., Imagawa, M. and Nashida, S. (1983). *In* "Immunoenzymatic Techniques" (S. Avrameas *et al.*, Eds), pp. 219–232, Elsevier, New York.
Jagow, J. and Hill, W. E. (1986). *Appl. Environ. Microbiol.* **51,** 441–443.
Jiang, X., Estes, M. K., Metcalf, T. G. and Melnick, J. L. (1986). *Appl. Environ. Microbiol.* **52,** 711–717.
Jiang, W., Estes, M. K. and Metcalf, T. G. (1987). *Appl. Environ. Microbiol.* **53,** 2487–2495.
Jofre, J., Bosch, A., Lucena, F., Girones, R. and Tartera, C. (1986). *Wat. Sci. Technol.* **18,** 167–173.
Kalmakoff, J., Parkinson, A. J., Crawford, A. M. and Williams, B. R. G. (1977). *J. Immunol. Meth.* **14,** 73–84.
Kaper, J. B., Moseley, S. L. and Falkow, S. (1981). *Infect. Immun.* **32,** 661–667.
Kaper, J. B., Bradford, H. B., Roberts, N. C. and Falkow, S. (1982). *J. Clin. Microbiol.* **16,** 129–134.
Kaspar, C. W. and Hartman, P. A. (1987). *J. Appl. Bacteriol.* **63,** 335–341.
Katzenelson, E. (1976). *Archv. Virol.* **50,** 197–206.
Katzenelson, E. and Mills, D. (1984). *In* "Monographs in Virology" (J. L. Melnick, Ed.), pp. 216–220, Karger, Basel.
Kedmi, S. and Katzenelson, E. (1978). *Archv. Virol.* **56,** 337–340.

Khuroo, M. S. (1980). *Am. J. Med.* **68,** 818–824.
Kilian, M. and Bülow, P. (1976). *Acta. Pathol. Microbiol. Scand. Sect. B* **84,** 245–251.
Klinger, J. D. and Johnson, A. R. (1988). *Food Technol.* **42,** 66–70.
Koburger, J. A. and Miller, M. L. (1985). *J. Food Protect.* **48,** 244–245.
Koening, R. and Paul, H. L. (1982). *J. Virol. Meth.* **5,** 113–125.
Korolik, V., Coloe, P. J. and Krishnapillai, V. (1988). *J. Gen. Microbiol.* **134,** 521–529.
Kurtz, J. B. and Lee, T. W. (1987). *In* "Novel Diarrhoea Viruses", John Wiley, UK.
Landegent, J. E., Jansen in de Wal, N., Van Ommen, G.-J. B., Bass, F., de Vijlder, J. J. M., Van Duijn, P. and van der Ploeg, M. (1985). *Nature (London)* **317,** 175–177.
Langer-Safer, P. R., Levine, M. and Ward, D. C. (1982). *Proc. Natl. Acad. Sci. U.S.A.* **79,** 4381–4385.
Lányi, B. (1987). *In* "Methods in Microbiology" (R. R. Colwell and R. Grigorova, Eds), pp. 1–67, Academic Press, London.
Larkin, E. P., Tierney, J. T. and Sullivan, R. (1976). *J. Environ. Engin. Div. Am. Soc. Civ. Engrs.* **102/EE1,** 29–35.
Le Minor, L. (1979). *Zentralbl Bakteriol. Parasitenkd. Infekleonskr. Hyg. Abt.* I, **A243,** 321–325.
Leary, J. J., Brigati, D. J. and Ward, D. C. (1983). *Proc. Natl. Acad. Sci. U.S.A.* **80,** 4045–4049.
Lemon, S. M., Binn, L. N. and Marchwicki, R. H. (1983). *J. Clin. Microbiol.* **17,** 834–839.
Ley, A. N., Bowers, R. J. and Wolfe, S. (1988). Abstract Q-35. 88th Ann. Meeting Am. Soc. Microbiol., Miami, FL.
Lipari, M. (1951). *N.Y. State J. Med.* **51,** 362.
Lippy, E. C. and Waltrip, S. C. (1984). *Am. Water Works. Assoc. J.* **76,** 60–67.
Littel, K. J. and Hartman, P. A. (1983). *Appl. Environ. Microbiol.* **45,** 622–627.
Loh, P. C., Dow, M. A. and Fujioka, R. S. (1985). *J. Virol. Meth.* **12,** 225–234.
Mack, W. N. (1973). *J. Am. Water Works. Assoc.* **65,** 345–348.
Madri, J. A. and Barwick, K. W. (1983). *Lab. Invest.* **48**, 98–107.
Maniatis, T., Fritsch, E. F. and Sambrook, J. (1982). "Molecular Cloning: a Laboratory Manual", Cold Spring Harbor, New York.
Mara, D. D. and Oragui, J. I. (1981). *Appl. Environ. Microbiol.* **42,** 1037–1042.
Marchalonis, J. J. (1969). *Biochem. J.* **113,** 299–305.
Mason, J. O. and McLean, W. R. (1961). *Am. J. Hyg.* **75,** 90–111.
Matches, J. R. and Abeyta, C. (1983). *Food Technol.* **37,** 114–117.
Mattiason, B., Borrebaeck, C., Sanfridson, B. and Mosbach, K. (1977). *Biochem. Biophys. Acta* **483,** 221–227.
Mattingly, J. A. (1984). *J. Immunol. Meth.* **73,** 147–156.
Mattingly, J. A. and Gehle, W. D. (1984). *J. Food. Sci.* **49,** 807–809.
Mattingly, J. A., Butman, B. T., Plank, M. C., Durham, R. J. and Robison, B. J. (1988). *J. Assoc. Off. Anal. Chem.* **71,** 679–681.
Melnick, J. L. (1984). *In* "Monographs in Virology" (J. L. Melnick, Ed.), Karger, New York.
Metcalf, T. G. and Stiles, W. C. (1967). *In* "Transmission of Viruses by the Water Route" (G. Berg, Ed.), pp. 439–447, Wiley-InterScience, NY.
Metcalf, T. G., Rao, V. C. and Melnick, J. L. (1984). *Monogr. Virol.* **15,** 97–110.
Miles, L. E. M. and Hales, C. N. (1968). *Nature (London)* **219,** 186–189.
Miller, B. A., Reiser, R. F. and Bergdoll, M. S. (1978). *Appl. Environ. Microbiol.* **36,** 421–426.

Miller, C. A., Patterson, W. L., Johnson, P. K., Swartzell, C. T., Wogoman, F., Albarella, J. P. and Carrico, R. J. (1988). *J. Clin. Microbiol.* **26,** 1271–1276.
Minnich, S. A., Hartman, P. A. and Heimsch, R. C. (1982). *Appl. Environ. Microbiol.* **43,** 877–883.
Morita, T. N. and Woodburn, M. J. (1978). *Infect. Immun.* **21,** 666–668.
Morris, Jr., J. G., Wright, A. C., Roberts, D. M., Wood, P. K., Simpson, L. M. and Oliver, J. D. (1987). *Appl. Environ. Microbiol.* **53,** 193–195.
Moseley, S. L., Huq, I., Alim, A. R. M. A., So, M., Samadpour-Motalebi, M. and Falkow, S. (1980). *J. Infect. Dis.* **142,** 892–898.
Moseley, S. L., Echeverria, P., Seriwatana, J., Tirapat, C., Chaicumpa, W., Sakuldaipeara, T. and Falkow, S. (1982). *J. Infect. Dis.* **145,** 863–869.
Moseley, J. W. (1967). *In* "Transmission of Viruses by the Water Route" (G. Berg, Ed.), pp. 5–23, InterScience, New York.
Mossel, D. A. A. (1978). *Food Technol. Aust.* **30,** 212–219.
Mossel, D. A. A. (1982). *Antonie van Leeuwenhoek J. Microbiol.* **48,** 609–611.
Murphy, A. M., Grohmann, G. S., Christopher, P. J., Lopez, W. A., Davey, G. R. and Millsom, R. H. (1979). *Med. J. Austr.* **2,** 329–333.
Nasser, A. M. and Metcalf, T. G. (1987). *Appl. Environ. Microbiol.* **53,** 1192–1195.
Ngo, T. T. and Lenhoff, H. M. (1981). *Biochem. Biophys. Res. Commun.* **99,** 496–503.
Ngo, T. T., Carrico, R. J., Boguslaski, R. C. and Burd, J. F. (1981). *J. Immunol. Meth.* **42,** 93–103.
Nishibuchi, M., Hill, W. E., Zon, G., Payne, W. L. and Kaper, J. B. (1986). *J. Clin. Microbiol.* **23,** 1091–1095.
Notermans, S., Boot, R., Tips, P. D. and Nooij, M. P. (1983). *J. Food Prot.* **46,** 238–241.
O'Brien, M. and Colwell, R. R. (1985). *J. Clin. Microbiol.* **22,** 1011–1013.
Ocasio, W. and Martin, S. E. (1988). Abstract P-49, 88th Ann. Meeting Am. Soc. Microbiol., Miami, Fla.
O'Mahony, M. C., Gooch, C. D., Smyth, D. A., Thrussel, A. J., Bartlett, C. L. R. and Noah, N. D. (1983). *Lancet* **1,** 518–520.
Ogram, A. V. and Sayler, G. S. (1988). *J. Indust. Microbiol.* **3,** 281–292.
Ogram, A. V., Sayler. G. S. and Barkay, T. (1987). *J. Microbiol. Meth.* **7,** 57–66.
Orth, D. S. (1977). *Appl. Environ. Microbiol.* **34,** 710–714.
Pace, N. R., Stahl, D. A., Lane, D. J. and Olsen, G. J. (1985). *Am. Soc. Microbiol. News* **51,** 4–12.
Pal, T., Echeverria, P., Taylor, D. N., Sethabutr, O. and Hanchalay, S. (1985). *Lancet* **2,** 785.
Palva, A. M. (1983). *J. Clin. Microbiol.* **18,** 92–100.
Payment, P. and Trudel, M. (1985). *Appl. Environ. Microbiol.* **50,** 1308–1310.
Payment, P. and Trudel, M. (1987). *Can. J. Microbiol.* **33,** 568–569.
Peterkin, P. I. and Sharpe, A. N. (1984). *Appl. Environ. Microbiol.* **47,** 225–234.
Peterson, E. H., Nierman, M. L., Rude, R. A. and Peeler, J. T. (1987). *J. Food Sci.* **52,** 409–410.
Petzel, J. P. and Hartman, P. A. (1985). *Appl. Environ. Microbiol.* **49,** 925–933.
Pierce, S. K. and Klinman, N. R. (1976). *J. Exp. Med.* **144,** 1254–1262.
Pierson, M. D. and Stern, N. J. (Eds) (1986). "Foodborne Microorganisms and their Toxins: Developing Methodology", Marcel Dekker, New York.
Pober, Z. and Silverman, G. J. (1977). *Appl. Environ. Microbiol.* **33,** 620–625.
Polin, R. A. and Kennett, R. (1980). *J. Clin. Microbiol.* **11,** 332–336.

Poskanzer, D. C. and Beadenkopf, W. G. (1961). *Publ. Hlth. Rep.* **76,** 745–751.
Puget, K., Michelson, A. M. and Avrameas, S. (1977). *Anal. Biochem.* **79,** 447–456.
Raphael, R. A., Sattar, S. A. and Springthorpe, V. S. (1985). *J. Virol. Meth.* **11,** 131–140.
Reinbold, G. W. (1983). *Food Technol.* **37,** 111–113.
Remis, S. R., MacDonald, I. L., Riley, L. W., Puhr, N. D., Wells, J. G., Davis, B. R., Blake, P. A. and Cohen, M. L. (1984). *Ann. Intern. Med.* **101,** 624–626.
Renz, M. and Kurz, C. (1984). *Nucleic Acid Res.* **12,** 3434–3444.
Richards, G. P. (1985). *J. Food Prot.* **48,** 815–823.
Ridinger, D. N., Spendlove, R. S., Barnett, B. B., George, D. B. and Roth, J. C. (1982). *Appl. Environ. Microbiol.* **43,** 740–746.
Rigby, C. E. (1984). *Appl. Environ. Microbiol.* **47,** 1327–1330.
Rigby, P. W., Diekman, M., Rhodes, C. and Berg, P. (1977). *J. Mol. Biol.* **113,** 237–251.
Robern, H., Dighton, M., Yano, Y. and Dickie, N. (1975). *Appl. Microbiol.* **30,** 525–529.
Robern, H., Gleeson, T. M. and Szabo, R. A. (1978). *Can. J. Microbiol.* **24,** 436–439.
Robison, B. J. (1984). *Appl. Environ. Microbiol.* **48,** 285–288.
Robison, B. J., Pretzman, C. I. and Mattingly, J. A. (1983). *Appl. Environ. Microbiol.* **45,** 1816–1821.
Rose, J. B., Mullinax, R. L., Singh, S. N., Yates, M. V. and Gerba, C. P. (1987). *Water Res.* **21,** 1375–1381.
Roszak, D. B. and Colwell, R. R. (1987). *Appl. Environ. Microbiol.* **53,** 2889–2983.
Rubenstein, K. E., Schneider, R. S. and Ullman, E. F. (1972). *Biochem. Biophys. Res. Commun.* **47,** 846–851.
Rudkin, G. T. and Stollar, B. D. (1977). *Nature (London)* **265,** 472–473.
Sabin, A. B. (1955). *Am. J. Med. Sci.* **230,** 1–8.
Saiki, R. K., Gelfand, D. H., Stoffel, B., Scharf, S. J., Higuchi, R., Horn, G. T., Mullis, K. B. and Erlich, H. A. (1988). *Science* **239,** 487–491.
Saunders, G. C. and Bartlett, M. L. (1977). *Appl. Environ. Microbiol.* **34,** 518–522.
Schroeder, H. R., Vogelhut, P. O., Carrico, R. J., Boguslaski, R. C. and Bucker, R. T. (1976). *Anal. Chem.* **48,** 1933–1937.
Scotto, J., Hadchouel, M., Henry, C., Alvarez, F., Yuart, J., Tiollais, P., Bernard, O. and Brechot, C. (1983). *Gut* **24,** 618–624.
Seriwatana, J., Echeverria, P., Taylor, D. N., Sakuldaipeara, T., Changchawalit, S. and Chivoratanond, O. (1987). *J. Clin. Microbiol.* **25,** 1438–1441.
Sethabutr, O., Echeverria, P., Hanchalay, S., Taylor, D. N. and Leksombon, U. (1985). *Lancet* **2,** 1095–1097.
Shah, D. B., Kauffman, P. E., Boutin, B. K. and Johnson, C. H. (1982). *J. Clin. Microbiol.* **16,** 504–508.
Siitar, H., Hemmilä, I., Soini, E., Lovgren, T. and Koistinen, V. (1983). *Nature (London)* **301,** 258–260.
Simkova, A. and Wallnerova, Z. (1973). *Acta Virol. (Praha)* **17,** 363.
Smith, A. M. and Jones, C. (1983). *Appl. Environ. Microbiol.* **46,** 826–831.
Smith, E. M. and Gerba, C. P. (1982). *Appl. Environ. Microbiol.* **43,** 1440–1450.
Soini, E. (1985). *In* "Rapid Methods and Automation in Microbiology and Immunology" (K. O. Hebermehl, Ed.), pp. 414–421, Springer-Verlag, New York.

Soini, E. and Hemmilä, I. (1979). *Clin. Chem.* **25,** 353–361.
Soini, E. and Kojola, H. (1983). *Clin. Chem.* **29,** 65–68.
Sommerville, C. C., Knight, I. T., Straube, W. L. and Colwell, R. R. (1988a). Abstract #70, First Int. Conf. Release Genet. Eng. Micro. Cardiff, Wales, UK.
Sommerville, C. C., Knight, I. T., Straube, W. L. and Colwell, R. R. (1988b). Abstract #40, First Int. Conf. Release Genet. Eng. Micro. Cardiff, Wales, UK.
Sours, H. E. and Smith, D. G. (1980). *J. Infect. Dis.* **142,** 122–125.
Splittstoesser, D. F. (1983). *Food Technol.* **37,** 105–106.
Stahl, D. A., Flesher, B. A., Mansfield, H. R. and Montgomery, L. (1988). *Appl. Environ. Microbiol.* **54,** 1079–1084.
Steffan, R. J. and Atlas, R. N. (1988). *Appl. Environ. Microbiol.* **54,** 2185–2191.
Steinmann, J. (1981). *Appl. Environ. Microbiol.* **41,** 1043–1045.
Stelma, G. N. Jr., Wimsatt, J. C., Kauffman, P. E. and Shah, D. B. (1983). *J. Food Prot.* **46,** 1069–1073.
Stelma, G. N., Johnson, C. H. and Shah, D. B. (1985). *J. Food Prot.* **48,** 227–231.
Stiffler-Rosenberg, G. and Fey, H. (1978). *J. Clin. Microbiol.* **8,** 473–479.
Stille, W., Kunkel, B. and Nerger, K. (1972). *Dtsch. Med. Wochenschr.* **97,** 145–147.
Stollar, B. D. and Rashtchian, A. (1987). *Anal. Biochem.* **161,** 387–394.
Stolzenbach, F., Phillips, L. A. P., Yang, Y. Y., Enns, R. K. and You, M. S. (1988). Abstract P-53, 88th Ann. Meeting Am. Soc. Microbiol., Miami, Fla.
Sullivan, R., Fassolitis, A. C. and Read, R. B. Jr. (1970). *J. Food Sci.* **35,** 624–626.
Swaminathan, B. and Konger, R. L. (1986). *In* "Foodborne Microorganisms and their Toxins: Developing Methodology" (M. D. Pierson and N. J. Stern, Eds), pp. 253–281, Marcel Dekker Inc., New York.
Swaminathan, B., Aleixo, J. A. G. and Minnich, S. A. (1985). *Food Technol.* **39,** 83–89.
Takashi, K. and Kayoto, S. (1977). *J. Gen. Virol.* **36,** 345–349.
Takiff, H. E., Seidlin, M., Krause, P., Rooney, J., Brandt, C., Rodriguez, W., Yolken, R. and Straus, S. E. (1985). *J. Med. Virol.* **16,** 107–118.
Tamplin, M. L., Brayton, P. R., Jalali, R. and Colwell, R. R. (1987). Abstract Q-86, 87th Ann. Meeting Am. Soc. Microbiol., Atlanta, GA.
Tan, C. T., Chan, S. W. and Hsia, J. C. (1981). *Meth. Enzymol.* **74,** 152–161.
Tartera, C. and Jofre, J. (1987). *Appl. Environ. Microbiol.* **53,** 1632–1637.
Thomason, B. M., Cherry, W. B. and Moody, M. D. (1957). *J. Bacteriol.* **74,** 525–532.
Thomason, B. M., Biddle, J. W. and Cherry, W. B. (1975). *Appl. Microbiol.* **30,** 764–767.
Thompson, N. E., Razdan, M., Kuntsmann, G., Aschenbach, J. M., Evenson, M. L. and Bergdoll, M. S. (1986). *Appl. Environ. Microbiol.* **51,** 885–890.
Tijssen, P. and Kurstak, E. (1981). *J. Virol.* **37,** 17–23.
Tilton, R. C. (1981). *In* "Rapid Methods and Automation in Microbiology" (R. C. Tilton, Ed.), pp. 125–129, American Society for Microbiology, Washington, D. C.
Todd, E. C. D. (1988). *J. Food Prot.* **51,** 56–65.
Tompkin, R. B. (1983). *Food Technol.* **37,** 107–110.
Totten, P. A., Holmes, K. K., Hansfield, H. H., Knapp, P. L., Perine, P. L. and Falkow, S. (1983). *J. Infect. Dis.* **148,** 462–471.
Trepeta, R. W. and Edberg, S. C. (1984). *J. Clin. Microbiol.* **19,** 172–174.
US Department of Health, Education, and Welfare (1976). p. 11. Center for Disease Control, Atlanta, GA.
Uhnoo, I., Wadell, G., Svensson, L. and Johannson, M. E. (1984). *J. Clin. Microbiol.* **20,** 365–372.

Van Weemen, B. K. and Schuurs, A. H. W. M. (1971). *FEBS Lett.* **15,** 232–236.
Van Weemen, B. K., Bosch, A. M. G., Dawson, E. C., Van Hell, H. and Schuurs, A. H. W. M. (1978). *Scand. J. Immunol.* **8,** 73.
Vaughn, J. M. and Landry, E. F. (1977). *In* "Virus Study Interim Report Series", **6.** Nassau-Suffolk Regional Planning Board, Hauppauge, NY.
Vaughn, J. M., Landry, E. F., Raranosky, L. J., Beckwith, C. A., Dahl, M. and Delihas, N. C. (1978). *Appl. Environ. Microbiol.* **36,** 47–51.
Venkatesan, M., Buysse, J. M., Vandendries, E. and Kopecko, D. J. (1988). *J. Clin. Microbiol.* **26,** 261–266.
Vincent, C., Tchen, P., Cohen-Solal, M. and Kourilsky, P. (1982). *Nucleic Acid Res.* **10,** 6787–6796.
Voller, A., Bidwell, D. and Bartlett, A. (1979). Nuffield Laboratories of Comparative Medicine, The Zoological Society of London, London, UK.
Warren, L. S., Benoit, R. E. and Jessee, J. A. (1978). *Appl. Environ. Microbiol.* **35,** 136–141.
Watkins, W. D., Rippey, S. R., Clavet, C. R., Kelley-Reitz, D. J. and Burkhardt, W. (1988) *Appl. Environ. Microbiol.* **54,** 1874–1875.
Wei, R. and Riebe, S. (1977). *Clin. Chem.* **23,** 1386–1388.
Wellings, F. M., Lewis, A. L. and Mountain, C. W. (1974). *In* "Virus Survival in Water and Wastewater Systems" (J. F. Malina and B. P. Sagik, Eds), pp. 253–260, Univ. of Texas Press, Austin, TX.
Wellings, F. M., Lewis, A. L., Mountain, C. W. and Pierce, L. V. (1975). *Appl. Microbiol.* **29,** 751–757.
Wilchek, M. and Bayer, E. A. (1988). *Anal. Biochem.* **171,** 1–32.
Wong, D. C., Purcell, R. H., Sreenivasan, M. A., Prasad, S. R. and Pavri, K. M. (1980). *Lancet* **2,** 876–879.
Yalow, R. S. and Berson, S. A. (1959). *Nature (London)* **184,** 1648–1649.
Yokota, H., Yokoo, K. and Nogata, Y. (1986). *Biochem. Biophys. Acta* **868,** 45–50.
Yolken, R. H. (1980). *Yale J. Biol. Med.* **53,** 85–92.
Yolken, R. H. (1984). *In* "Rapid Methods and Automation in Microbiology and Immunology" (K. O. Habermehl, Ed.), pp. 401–407, Springer-Verlag, New York.
Yolken R. H. and Leister, F. J. (1981). *J. Immunol. Meth.* **43,** 209–218.
Yolken, R. H. and Leister, F. J. (1982). *J. Clin. Microbiol.* **15,** 757–760.
Yung, L. L., Loh, W. and TerMeulen, V. (1977). *Med. Microbiol. Immunol.* **163,** 111–123.

# 17

# Methods for Studying the Ecology of Actinomycetes

ALAN J. McCARTHY AND STANLEY T. WILLIAMS
*Department of Genetics and Microbiology, University of Liverpool, Liverpool L69 3BX, UK*



## I. Introduction

Actinomycetes have long been recognized as a group of microbes which are distinct from other bacteria and fungi, and which play important roles in many natural environments. Nevertheless, the criteria used to define the actinomycetes (order Actinomycetales) have varied in the past, and are still developing. A 'classical' definition is that they are Gram-positive bacteria which form branching hyphae at some stage of their development and may produce a spore-bearing mycelium. This definition has its

METHODS IN MICROBIOLOGY
VOLUME 22 ISBN 0-12-521522-3

problems. Those with transient mycelium may be difficult to distinguish from other Gram-positive bacteria. Also, application of chemical and molecular techniques suggests that the composition and boundaries of the order do not always correlate with morphological characteristics (see Goodfellow, 1989a).

Similar problems are faced when defining genera within the actinomycetes. Over 45 genera are now validly described, with an increasing tendency to define them on the basis of characters such as wall chemotype, menaquinone composition, nucleic acid pairing and RNA oligonucleotide sequencing. Thus it is not uncommon to encounter morphologically similar isolates which fall into different genera when such criteria are applied.

These techniques, together with the development of specific molecular probes (see Section V) will undoubtedly have a marked effect on our understanding of actinomycete ecology. However, one must recognize that many ecologists studying or merely encountering actinomycetes in the environment may not have the time, necessity or facilities to apply molecular and biochemical techniques. As actinomycetes are widespread in the environment and metabolically diverse (see Section II), the possible objectives of ecological studies are legion. Methods applied obviously reflect the aims, but the unique growth form, reproduction, survival and dispersal mechanisms of actinomycetes should not be overlooked. Thus the 'classical' definition of these microbes still has ecological validity, and most of the genera discussed in this chapter conform to it.

A similarly pragmatic choice has been made in our consideration of habitats. Most genera are strict saprophytes, occurring in a wide variety of natural and man-made habitats, and some form symbiotic associations with plants. These have been included. Human pathogens, such as *Mycobacterium leprae* and *Corynebacterium diptheriae*, together with plant pathogens such as certain *Corynebacterium* spp., have been omitted.

Therefore our main aim is to assess the principles and problems of studying the occurrence and distribution of saprophytic actinomycetes in a variety of natural habitats. Particular attention will be paid to methods of isolation, enumeration and identification, from which much ecological data are still derived, but determination of activity *in vivo* and *in vitro* will also be considered.

## II. Occurrence and distribution

The ubiquity of saprophytic actinomycetes in the natural environment (Table I) is due to two main factors—metabolic diversity and the evolution of specific mechanisms for dispersal. Colonization of solid substrates by actinomycete hyphae is followed by sporulation or fragmentation. Many actinomycetes rely on air dispersal and form an aerial mycelium bearing hydrophobic spores. The remainder appear to be adapted for water-mediated dispersal and range from the actinoplanetes which produce motile spores within dessication-resistant sporangia to *Nocardia* and related actinomycetes whose primary vegetative mycelium simply undergoes fragmentation into coccoid and bacillary elements. Micromonosporas are closely related to actinoplanetes but lack an aerial mycelium and produce single non-motile spores. These actinomycetes compensate for this apparent lack of specialization by producing very large numbers of hydrophilic spores, and this undoubtedly contributes to the prevalence of micromonosporas in freshwater environments. This raises the more general question of whether detection indicates colonization within a particular environment or simply the presence of dormant spores or hyphal fragments. The fact that actinomycetes are so well adapted for growth on solid substrates suggests that their occurrence in aquatic environments is merely the result of 'wash in', but for some taxa at least there is evidence of growth on suspended or sedimented material (Cross, 1981). A more specific ecosystem where detection does not necessarily imply activity is the intestine of invertebrates, where many but by no means all actinomycetes recovered are symbionts (Williams *et al.*, 1984).

**TABLE I**
The major habitats of some actinomycete taxa

| Taxon | Habitats |
|---|---|
| *Actinoplanes* | Fresh water, plant litter, soil |
| *Frankia* | Non-legume root nodules |
| *Micromonospora* | Fresh water, sediments, wet soil |
| *Nocardia amarae* | Activated sewage sludge |
| *Rhodococcus coprophilus* | Animal dung, water, soil |
| *Saccharopolyspora hirsuta* | Self-heated bagasse |
| *Saccharopolyspora rectivirgula* | Mouldy hay |
| *Streptomyces* | Soil, plant litter, water |
| *Thermoactinomyces* | Composts, mouldy fodders and other self-heated materials |

Saprophytic actinomycetes are important primary colonizers of soil organic material, the bulk of which is in the form of insoluble polymers. Actinomycetes' ability to penetrate and solubilize these polymers, whether of plant (lignocellulose) or animal (chitin) origin, allows them to persist in the microbial succession beyond the initial phase of rapid bacterial growth. Actinomycetes, like other soil microbes, are unevenly distributed according to the occurrence of microenvironments providing sites of activity governed by indeterminate factors. Consequently summary statements on the effects of environmental parameters on occurrence and distribution are of limited value. The general preference of saprophytic actinomycetes for neutral to alkaline pH and absolute requirement for aerobic conditions are however established, and clearly important in actinomycete ecology. The recovery of acidophilic actinomycetes from soils of low pH (Khan and Williams, 1975; Hagedorn, 1976) and identification of a novel actinomycete population colonizing a salt marsh (Hunter *et al.*, 1981) are good examples of the ability of actinomycetes to adapt to a range of environmental factors.

Temperature is always an important ecological factor, but in addition to obvious seasonal effects on population sizes there is a diverse group of thermophilic actinomycetes associated with overheated accumulations of organic material, usually in the form of composts or mouldy fodders. In fact, most thermophilic actinomycetes have growth temperature optima in the range 50–60°C and are unable to grow above 65°C. Thermoactinomycetes and streptomycetes generally predominate but a number of substrates have profound and inexplicable effects on population profiles. Good examples are the high numbers of *Saccharopolyspora rectivirgula* in mouldy hay, the predominance of thermomonosporas in mushroom compost and the occurrence of *Thermoactinomyces sacchari* almost exclusively in mouldy bagasse (Lacey, 1973, 1978). The association of individual actinomycete taxa with particular environments is not confined to thermophiles. In addition to the obvious associations between actinomycete parasites and symbionts and their plant or animal hosts/partners, there are apparent relationships between substrates and saprophytic actinomycetes. Examples include the development of large populations of *Nocardia amarae* in activated sludge plants (Lechevalier and Lechevalier, 1974) and animal dung as the source of *Rhodococcus coprophilus* isolates from both terrestrial and freshwater environments (Rowbotham and Cross, 1977).

Soil is clearly the origin of most saprophytic actinomycetes recovered from rivers and lakes but the additional factors present in the marine environment (i.e. salinity and, at depth, very low temperatures and high pressures) suggest that distinct forms could have developed. The differentiation of survivors, adapted terrigenous actinomycetes and possible indigenous forms is discussed and reviewed by Weyland and Helmke (1988).

## III. Isolation and enumeration

### A. General principles

There have been several detailed reviews of isolation and the problems involved (Goodfellow and Williams, 1986; Nolan and Cross, 1988; Goodfellow and O'Donnell, 1989; Williams and Vickers, 1988). Actinomycetes are usually both isolated and counted using standard dilution plate procedures, but the two objectives are not always compatible. As the aims of workers attempting to detect actinomycetes vary considerably, so do the appropriate isolation procedures. Thus an ecologist is often interested in both the qualitative and quantitative nature of a mixed population, or sometimes a single genus or species. A taxonomist may be happy to obtain a single novel isolate – indeed new taxa are still erected on such evidence! However, by far the most actinomycetes are isolated by industrial microbiologists who screen many thousands of strains from an ever increasing range of habitats every week. Yet, ironically, this mass isolation provides little or no ecological data, due to secrecy or disinterest. A typical example is the isolation of *Kibdelosporangium philippinense* sp. *nov.*, which was isolated 'from soil collected in the Philippines by using selective isolation procedures' (Mertz and Yao, 1988).

Whatever the objectives, the selectivity of isolation–enumeration procedures may be influenced deliberately or accidentally at a number of sequential stages.

#### *1. Selection of samples*

The number of geographical, climatic and vegetational zones is inexhaustible, as is particularly well illustrated by the flood of new taxa isolated in China over the past few years. However, there is little apparent correlation between such factors and the distribution of most taxa. Associations are more likely to be found between highly adapted taxa growing on specific substrates (e.g. *Saccharopolyspora hirsuta* on sugar cane bagasse) or in specific niches (e.g. *Frankia* in root nodules).

Collection of samples involves standard aseptic procedures and storage at 0–4°C if not analysed within a few hours. Special procedures may be needed to obtain samples from some environments and substrates, such as concentration of the populations in sea water by filtration and centrifugation (Okazaki and Okami, 1972) and concentration of spores in the air (see Sections III.C and D).

### 2. *Pretreatment of samples*

Once samples are obtained, a variety of pretreatments may be applied to them prior to plating. The common aim is to increase the chances of one or more actinomycete taxa developing on isolation plates. Hence they are designed to inactivate other microbes as far as possible. As most fungi can be selectively inhibited by incorporation of antifungal antibiotics in isolation media (see below), pretreatments are designed to reduce the proportion of other bacteria in the sample. Many treatments involve heating or drying of samples and their effectiveness is presumably due to the greater tolerance of many actinomycete propagules compared with that of the cells of non-sporing bacteria. Two examples are the heating of air-dried soil samples at 120°C for 1 h to isolate *Streptosporangium* and *Microbispora* strains (Nonomura, 1989a,b), and keeping soil at 100°C for 15 min to isolate *Actinomadura* and other genera (Athalye *et al.*, 1981). Other examples are given below.

### 3. *Selective media*

Many media formulations have been proposed for the isolation of a broad range of actinomycetes or selected taxa. However, due to our lack of knowledge of the specific nutrient requirements of this metabolically diverse group of chemo-organotrophs, most formulae are not based on precise nutritional or physiological criteria. It is clearly impossible to devise a general isolation medium for actinomycetes but several, such as starch–casein (Küster and Williams, 1964), colloidal chitin (Hsu and Lockwood, 1975) and M3 medium (Rowbotham and Cross, 1977), have been widely used with some success. However, evaluation of the efficiency of an isolation medium is a complex process. Thus, for example, it is possible to get 'cleaner' plates of streptomycetes using chitin medium or increased NaCl concentrations (MacKay, 1977), but the majority of *Streptomyces* spp. do not hydrolyse chitin or tolerate raised salt concentrations (Williams *et al.*, 1983). A more rational choice of media constituents for isolation of specific taxa can be achieved by analysis and selection of data bases obtained from numerical taxonomic studies (Williams and Vickers, 1986, 1988; Goodfellow and O'Donnell, 1989).

Antimicrobial agents, particularly antibiotics, have been effective in improving the selectivity of isolation media. It is standard practice to reduce fungal contamination by supplementing media with antifungal antibiotics such as cycloheximide or nystatin which have no effect on actinomycetes. Selection of antibacterial antibiotics must be more targeted as actinomycetes as a group are likely to be susceptible to these inhibitors.

Some examples of the use of antibacterial antibiotics in the selective isolation of actinomycete genera are given in Table II; others are mentioned below and a comprehensive review was given by Nolan and Cross (1988).

**TABLE II**
Some examples of antibacterial antibiotics used for the selective isolation of actinomycetes

| Antibiotic | Actinomycetes selected | References |
|---|---|---|
| Bruneomycin | *Actinomadura* | Preobrazhenskaya *et al.* (1975) |
| Novobiocin and streptomycin | *Glycomyces* | Labeda (1986) |
| Oxytetracycline | *Streptoverticillium* | Hanka *et al.* (1985) |
| Penicillin and nalidixic acid | *Saccharothrix* | Labeda (1986) |
| Tunicamycin | *Micromonospora* | Wakisaka *et al.* (1989) |
| Vancomycin | *Amycolatopsis* | Lechavalier *et al.* (1986) |

### 4. *Incubation*

The temperature and length of incubation are the major variables. Most actinomycetes are mesophilic and grow well between 25 and 30°C. Thermophilic species are best grown at 50°C (see Section III.C). Members of commonly isolated genera, such as *Micromonospora* and *Streptomyces*, can be detected after 7–14 days, while thermophiles like *Saccharopolyspora rectivirgula* appear after 2–3 days' incubation. However, the rarer, slower growing strains in both groups may be overlooked unless incubation periods are extended. Nonomura and Ohara (1971) isolated several new taxa when they incubated plates at 30°C and 40°C for up to 1 month. Prolonged incubation is also essential for the isolation of *Frankia* symbionts (see Section III.E).

### 5. *Colony selection*

This is the most time-consuming and subjective process in the isolation procedure. Depending on the aims of the study, colonies may be selected randomly or with some degree of choice. The former can be particularly tedious if performed by hand, but pharmaceutical companies operating large throughput screens rely increasingly on automated colony selection systems backed up by image analysis. Meanwhile, most ecologists transfer their colonies by hand; a sterile wooden cocktail stick is a convenient tool.

In all cases, direct microscopic examination is useful for identifying actinomycete colonies by their growth form (using a stereomicroscope) or preliminary assignment to taxa based on morphological observations under high power (using a long working distance objective lens). This is particularly important in commercial screening procedures where minimization of duplication is almost essential (see also Section IV).

## B. Soil

Actinomycetes constitute a significant proportion of the microbial population in most soils, their viable count often exceeding 1 million per gram. The soil is also the most prolific source of isolates, many of which produce antibiotics and other useful metabolites *in vitro*. Representatives of over 90% of actinomycete genera have been isolated from soil, and some examples are given in Table III. There is no doubt that members of the genus *Streptomyces* are the most widespread and hence the most studied. Nevertheless, the frequency and importance of other genera may be underestimated as they may not develop or be overlooked when using so-called general isolation procedures. With the increasing use of more selective methods, the frequency of isolation of long-established genera increases and there is also a steady stream of new genera and species. Actinomycetes, particularly streptomycetes, have roles in many soil processes although more definitive information on their growth and activities in this habitat is still needed. Reviews on actinomycete ecology in soil and methods of study were given by Williams and Wellington (1982), Williams *et al.* (1984), and Goodfellow and Williams (1983).

Methods for the isolation and enumeration of actinomycetes in soil are based on the same principles and face the same problems as those outlined in the previous section, reflecting the predominance of ecological studies on this environment. Therefore, consideration will be restricted to some methods and problems of particular relevance to soil.

Samples of soil taken randomly from different geographical or climatic regions are of limited value not only to ecologists, but also to industrial microbiologists seeking to devise a more logical strategy for the isolation of strains producing useful metabolites. Soil is a complex habitat, so precise details of the sampling site, sample location within the soil and its physical and chemical properties are desirable. These should include information on vegetational cover, soil type, profile, depth of sample and at least the physical nature, organic content, moisture content and reaction of soil in the sample. For example, the streptomycete populations of a highly organic, acidic horizon and a predominantly mineral, neutral one in the same soil can be totally different (Williams *et al.*, 1971).

**TABLE III**
Occurrence and isolation of some genera in soil

| Genus | Comments | Reference for isolation procedure[a] |
|---|---|---|
| *Actinomadura* | Rare, but possibly underestimated | Meyer (1989) |
| *Actinoplanes* | Occasionally detected in wide range of soils | Palleroni (1989) |
| *Dactylosporangium* | Rare, but detected in various soil types worldwide | Vobis (1989) |
| *Frankia* | Limited evidence for status outside nodules | Lechevalier and Lechevalier (1989) |
| *Geodermatophilus* | Rare but habitats range from deserts to Mt Everest | Luedemann and Fonesca (1989) |
| *Glycomyces* | Increasingly isolated, colonies small and slow-growing | Labeda (1989) |
| *Kibdelosporangium* | Readily isolated but colonies resemble *Streptomyces* and *Nocardia* | Shearer *et al.* (1989) |
| *Kitasatosporia* | Readily isolated but colonies resemble *Streptomyces* | Omura *et al.* (1989) |
| *Microbispora* | Quite common, most readily isolated from dried, treated samples | Nonomura (1989a) |
| *Micromonospora* | Common in many soils, particularly if the moisture content is high | Kawamoto (1989) |
| *Nocardia* | Quite common, frequency up to $10^4$ $g^{-1}$ using selective methods | Goodfellow and Lechevalier (1989) |
| *Rhodococcus* | Some species common and readily isolated | Goodfellow (1989b) |
| *Streptomyces* | The dominant genus in almost all soils. Methods to isolate selected species increasing | Williams *et al.* (1989) |
| *Streptosporangium* | Quite common, up to $10^4$ $g^{-1}$ from dried, heated soil | Nonomura (1989b) |
| *Streptoverticillium* | Isolated from a wide range of soils, but seldom in high numbers | Locci and Schofield (1989) |

[a]The references, which give information on ecology and isolation methods for these and other genera, are from *Bergey's Manual of Systematic Bacteriology*, 9th Edition, Volume 4 (1989).

A wide range of pretreatments of samples prior to plating have been applied to soils. As many of these are designed to reduce the numbers of other bacteria, most involve drying and/or heating of the sample. The spores of most actinomycetes withstand desiccation and show a slightly higher resistance to dry or wet heat than vegetative bacterial cells. It is generally useful to air-dry samples for 2–7 days, the period depending on the moisture content and moisture-holding capacity of the soil. It is also essential to air-dry before applying dry heating methods (e.g. 100°C for isolation of *Microbispora* and other genera) to avoid wet heat damage. Lower temperatures (55–60°C) may be applied to soil suspensions to isolate, for example, *Micromonospora* spp. (Goodfellow and Haynes, 1984) and *Rhodococcus coprophilus* (Rowbotham and Cross, 1977). These mild heat treatments are primarily designed to kill competing Gram-negative soil bacteria. Soil suspensions are usually prepared by standard procedures, but detachment and dispersal of cells may be improved by high-speed homogenization (Baecker and Ryan, 1987).

A large and increasing range of media and selective inhibitors is used to isolate and enumerate soil actinomycetes, including those claimed to be general and those designed to detect particular genera (see III.A). Space does not permit us to give details here, but information on all genera is given in the 9th Edition of *Bergey's Manual of Systematic Bacteriology*, Volume 4 (1989). As previously mentioned, the composition of media and choice of inhibitors, and indeed their evaluation, have often been somewhat subjective. However, it is becoming increasingly realized that the large data banks accumulated in numerical taxonomic studies can provide a sound basis for the formulation of selective media. These data provide a numerical assessment of character states for all taxa studied, including nutritional, physiological and inhibitor resistance profiles which can be used to devise targeted isolation media. Their potential selective value for taxa included in the matrix can be assessed using the DIACHAR program (Sneath, 1980), and those with highest scores used to formulate the isolation medium. This approach has been used with some success to isolate selected *Streptomyces* spp. from soil (Vickers *et al.*, 1984; Williams and Vickers, 1986) and is being applied to other genera.

Whatever the isolation medium, several other factors can influence the efficiency of the isolation procedure at this stage. Plates are usually inoculated with a sample from an appropriate dilution of a soil suspension. This may be incorporated into molten media or spread onto solidified media. The former reduces spread of motile bacteria but hinders the distinction between colonies of actinomycetes and other bacteria. The latter is satisfactory if the media surface is thoroughly dried both before and after inoculation (e.g. Vickers and Williams, 1987). Another technique is to pass the soil suspension through a membrane filter ($<0.5$ μm pore size) and place it face downwards on the medium (Trolldenier, 1966). Actinomycetes are more able to grow through the filter than other bacteria. A simple method, purely for isolation, is to pick up a small crumb of soil with a glass spreader and inoculate up to six dried plates. This provides a crude dilution effect, is rapid and can be used in the field.

Isolation and counting of developing colonies is a difficult exercise, but is facilitated by experience. Surface-inoculated plates encourage the typical powdery, mycelial form of many actinomycetes. Pigmentation of the surface and reverse of colonies is also a useful diagnostic criterion, but some isolation media (e.g. colloidal chitin) do not encourage pigment production. Streptomycetes usually predominate on soil isolation plates, but the colonies of some other genera (e.g. *Actinomadura, Glycomyces, Kitoa satosporia*) which resemble streptomycetes macroscopically, grow more slowly and may be out-competed. This problem can be ameliorated by adding high titres of polyvalent streptomycete phage to the soil suspension immediately before plating (Williams and Vickers, 1988; Kurtboke, unpublished data).

## C. Composts and mouldy fodders

These habitats contain large diverse populations of actinomycetes whose proliferation has been encouraged by moist, aerobic and neutral to alkaline pH conditions. Microbial activity in general is stimulated by the high organic nutrient content, and self-heating results in a classical succession from a mesophilic to thermophilic microflora. The predominance of actinomycetes is well-established, and can even be directly observed as a surface bloom, termed 'fire fang', in composts.

Mixing to maintain aerobic conditions is an important part of the composting process and has implications for sampling since the actinomycete population should therefore be evenly distributed. In fodders, which include substrates with very different properties, e.g. hay and grain, localised areas of actinomycete activity are more likely to have developed, with the possibility of anaerobic, and consequently acidic, regions which should be avoided during sampling. Enrichment procedures to increase actinomycete numbers are generally unnecessary and since bacilli dominate the remainder of the bacterial population, heat treatments to reduce their numbers are ineffective. For air dispersal techniques, samples must be completely dried, while for suspension in liquid diluent, some form of comminution is required. An alternative to the latter is to sample compost or fodder dust fractions.

Recovery of actinomycetes from fodders by air dispersal derives from the importance of airborne actinomycete spores as respiratory allergens, and was first described by Gregory and Lacey (1963). These workers used a wind tunnel to generate a spore cloud which could then be sampled with an Andersen sampler. A more convenient alternative is the sedimentation chamber (Lacey and Dutkiewicz, 1976a) in which a spore cloud is produced by simply shaking samples within the chamber and sampling after 1–2 h sedimentation. These techniques are effective because bacilli adhere to the dried substrate and more often occur in clumps so that fewer are suspended in the air. The surface of the isolation plates must be dry to ensure that any colonies of competing bacilli remain discrete and therefore do not interfere with the development of actinomycete colonies.

Although originally used for mouldy fodders, the sedimentation chamber/Andersen sampler technique has been successfully applied to dried samples of compost and other materials (McCarthy and Broda, 1984). The reduction in numbers of bacilli recovered on isolation plates enables the use of relatively rich isolation media and this results in a greater diversity of actinomycete isolates. The main disadvantage of air dispersal techniques is the difficulty in relating colony counts to cfu per g substrate, however dilution plating as a method of estimating actinomycete population sizes is not without its own inherent sources of error. Air dispersal of actinomycete spores as an isolation technique may not necessarily require specialized equipment. Treuhaft and Arden-Jones (1982) have demonstrated that a range of thermophilic actinomycetes can be isolated from mouldy fodder dust by simply shaking samples in a small plastic bag followed by puffing the spore cloud onto dry nutrient agar plates.

Suspension of spores and hyphal fragments in liquid diluent can be achieved by simple shaking or homogenization. Saline or buffer solutions serve as diluent, and low concentrations ($\leqslant 0.01\%$) of peptone may be incorporated to aid recovery of sublethally impaired propagules. While addition of detergents has improved recovery of bacteria from soil (Olsen and Bakken,1987) and has been applied to the isolation of actinomycetes from compost (Millner, 1982), it is one of a number of modifications to the dilution procedure which may be of little significance (Amner *et al.*, 1988).

The composition of the isolation medium is the most important aspect of actinomycete recovery by both dilution plating and spore cloud sampling. Tryptone soy and nutrient agars made up at half strength are used routinely subsequent to the work of Lacey and Dutkiewicz (1976b) on recovery of actinomycetes from mouldy hay. Further reducing the nutrient concentration is an approach commonly used to isolate actinomycetes by virtue of the limiting effect on the growth of competing bacteria. However, this has not been a useful strategy for recovering thermophilic actinomycetes from composts and fodders, resulting in low numbers and limited diversity, i.e. predominantly thermotolerant streptomycetes. In fact a number of thermophilic actinomycete species require especially rich media for their isolation, e.g. yeast–malt agar for *Thermoactinomyces sacchari* (Lacey, 1974) and peptone–corn starch agar for *Tha.dichotomicus* (Agre, 1964). Furthermore, unexpected selective effects can result from the use of nutrient-rich media as exemplified by the improved recovery of *Saccharomonospora viridis* from compost on a medium designed for regeneration of streptomycete protoplasts (Amner *et al.*, 1989). Addition of antibiotics or other compounds to render media selective for species or groups is a common approach (Table II) also used in the isolation of thermophilic actinomycetes from composts and fodders. Novobiocin (50 $\mu g\ ml^{-1}$) for thermoactinomycetes (Cross, 1968) and kanamycin (25 $\mu g\ ml^{-1}$) for *Thermomonospora chromogena* (McCarthy and Cross, 1981) are well established examples. In the case of the farmer's lung organism, *Saccharopolyspora rectivirgula*, addition of NaCl (7%) to the medium is equivocal since it only improves recovery if the substrate contains low numbers of this species.

For the recovery of thermophilic actinomycetes, incubation at 50°C is to be preferred since a number of species grow poorly at ⩾55°C. It is important that spreading growth of bacilli is limited by ensuring that the plates are free of surface water, but at the same time taking steps to prevent desiccation of the agar since some species may require 5–7 days to form colonies upon isolation. Finally, composting and the moulding of fodders are initiated by the growth of mesophilic microorganisms and so composts also serve as excellent sources of mesophilic actinomycetes (Lacey,1980; McCarthy and Broda,1984).

### D. Air

The Andersen sampler is used almost universally for the recovery of actinomycete spores from the aerial environment. Early studies on spore release from hay also used a cascade impactor and liquid impinger (Gregory *et al.*, 1963), and where a microscopic assessment of spore loads in air is also required, the former is still used (Lacey, 1977). Obtaining quantitative data on airborne actinomycete spores outdoors requires particular attention to sampling regimes. For example, Lacey (1975) located an Andersen sampler within a mobile customized spore trap to assess airborne fungal and actinomycete spores in pastures, and Millner *et al.* (1980) used an elaborate network of carefully chosen sample stations to study the release of spores from a sewage sludge compost site. Sampling indoor environments is more straightforward, but in agricultural buildings containing mouldy fodders, Andersen sampler plates may become overloaded. This problem can be overcome by plating out a dilution series prepared from spores washed off the agar surface of the fifth stage of the Andersen sampler (Lacey and Lacey, 1964).

The ability of actinomycete spores to act as respiratory allergens has provided the main impetus for studies on the recovery of actinomycetes from air. Their further implication in some cases of humidifier fever or ventilation pneumonitis has extended the scope of this work beyond the traditional agricultural or industrial (e.g. tobacco factories and cotton mills) environments. Some qualitative information can be obtained using simple sedimentation plates where low numbers are present, but passage of a known amount of air through an Andersen sampler loaded with plates of a broad spectrum actinomycete isolation medium such as half-strength nutrient agar, is more appropriate. This relationship between actinomycete spores and respiratory disease has generated a substantial amount of information on numbers and species of actinomycetes found in air and the reader is referred to Lacey (1988) for a thorough review of this subject.

## E. Roots

Actinomycetes occur in the rhizosphere and rhizoplane of many plants, and *Frankia* strains are nitrogen-fixing endophytes in root nodules of various non-leguminous shrubs and trees. Actinomycetes can be isolated and enumerated in rhizosphere soil using standard procedures applied to other soil samples. There have been surprisingly few quantitative studies, but R:S ratios based on viable counts are often low in fertile soils, while those in sand dunes ranged from 16 to 50 (Watson and Williams, 1974). Few clear and consistent qualitative differences between rhizosphere and general soil populations have yet been established. Isolations from the rhizoplane may be achieved by plating serially washed root segments on suitable media, after drying them thoroughly on sterile blotting paper to reduce bacterial spread. However, contamination by bacteria and fungi can still be troublesome from these point inoculations. Maceration of washed roots, followed by dilution of the suspension can facilitate the counting and isolation of actinomycetes. Studies of the rhizoplane have been very limited, but there seems to be no reason why qualitative differences between root surface and soil populations should not occur as they do for other soil bacteria and fungi. Most studies on these habitats have involved screening of isolates for inhibition of root pathogens (particularly fungi), with a view to their use in biological control (see Williams and Vickers, 1986).

The endophyte observed in the root nodules of non-leguminous plants was long suspected to be an actinomycete, which was confirmed by its initial isolation from *Comptonia* (Callaham *et al.*, 1978). It has now been isolated from 14 genera, including *Alnus, Casuarina* and *Myrica*, but it remains to be extracted from several other nodule-bearing genera (Lechevalier and Lechevalier, 1989). The evidence for saprophytic activity of *Frankia* in soil is still equivocal, but Baker (1988) suggested that it is quite likely to be suited for life as a soil saprophyte.

A review of the various isolation procedures for *Frankia* was given by Lechevalier and Lechevalier (1989), but some general points are worthy of note. Nodules are best plated out while fresh, or after storage at −20°C. Surface sterilization can be achieved by various agents, e.g. mercuric chloride (Callaham *et al.*, 1978), sodium hypochlorite (Baker *et al.*, 1979) and osmium tetroxide (Labonde *et al.*, 1981). Isolation procedures include dilution of crushed nodule suspensions, microdissection to isolate vesicle clusters and sucrose density gradient centrifugation (Baker and Torrey, 1979). A variety of isolation media have been used. While many strains grow on relatively simple media such as 0.5% yeast extract (Callaham *et al.*, 1978), others require complex growth factors such as 'special' fatty acids (Quispel *et al.*, 1983). After incubation at 25–33°C, colonies may appear within 3–10 days, but usually 4–8 weeks is required, and in a few cases up to 1 year! Colonies are small and sometimes visible only under the microscope. Another problem in ecological studies of *Frankia* is that the spores do not germinate consistently, thus *Frankia* can be quantified in the natural environment only by measuring numbers of infection particles, using the most probable number method with host plant seedlings (Baker, 1988).

Frankiae fall into at least two physiological groups (Lechevalier *et al.*, 1983). Group A strains are morphologically and chemically diverse. They are aerobic, easily maintained and relatively fast-growing. The strains are also serologically and genetically diverse, and usually not infective or effective for their host root. Group B is more homogenous, containing microaerophilic strains which are more difficult to maintain in culture. Most induce effective nodules in their host plant. Thus reinoculation of roots presents some problems and there is also no clear general pattern of endophyte–host specificity. Thus, for example, one Group B strain induced effective nodules in *Alnus, Comptonia* and *Myrica* (Baker and Torrey, 1980), while another was confined to the *Elaeagnaceae* (Baker *et al.*, 1980).

## F. The aquatic environment

Actinomycetes are not adapted for growth in aquatic habitats but can nevertheless be recovered readily from fresh water, sea water and sediment samples. Taxonomic diversity comparable to that in soil is usually encountered and this leads to the suggestion that isolates are merely the result of 'wash-in'. In fact, opinions on the origin and role of actinomycetes in the aquatic environment are rife, and it is a subject which has been critically reviewed by Cross (1981).

There are no sampling requirements peculiar to the actinomycetes. Obtaining water and particularly sediment samples from the sea present special problems which are overcome using a range of grabs, dredges and coring devices. Some examples of specifications for such equipment are provided by Okami (1972). Particular attention to sampling locations is required when actinomycete species are being used as indicators of terrestrial input. Monitoring *Rhodococcus coprophilus* in a stream system (Rowbotham and Cross, 1977) and thermoactinomycetes in a complex estuarine system (Attwell and Colwell, 1984) are good examples. Recovery of actinomycetes from water may also be maximized by sampling surface foam rather than the water itself. It has been demonstrated for river water at least that a range of actinomycete species along with other bacteria can be concentrated 100–1000-fold in river foam (Al-Diwany and Cross, 1978).

Standard dilution plating procedures are used to recover actinomycetes from aquatic habitats on the usual range of selective and non-selective media (see Section III.A). Sea water usually replaces distilled water in such media when the marine environment is being sampled, although this does not select against terrestrial actinomycetes (Okami and Okazaki, 1972; Okazaki and Okami, 1972). More recently, Barcina *et al.* (1987) conducted a more systematic study of methods for isolating and enumerating actinomycetes from sea water and sediment. Of the nine isolation media used, glycerol–asparagine agar (Pridham and Lyons, 1961) proved the most consistently effective. Interestingly, these workers found that heat treating samples improved actinomycete recovery from sediments but not from sea water. This is in direct contrast to the dramatic increase in actinomycete numbers routinely recovered from freshwater samples heat treated, usually 6 min at 55°C, prior to dilution plating (Cross, 1982). Other pretreatments which may find general application include the homogenization or sonication of sediment samples. Johnston and Cross (1976) used this technique to detect an increase in the numbers of *Micromonospora* colonies recovered from lake sediment, and therefore imply their presence as active hyphae in addition to dormant spores. Thermophilic thermoactinomycetes are present in very low numbers in aquatic habitats and this, together with the unequivocal fact that they do not grow in such habitats, has led to their use as indicator organisms. Large samples of water can be membrane-filtered and the membranes incubated at 50°C for 24 h on Czapek-Dox yeast extract casamino acids (CYC) agar made selective for thermoactinomycetes by incorporation of 25 μg $ml^{-1}$ novobiocin (Al-Diwany *et al.*, 1978). These workers also found that thermoactinomycete recovery was significantly affected by the type or commercial source of membrane used. *Rhodococcus coprophilus* is used as an indicator of farm effluent contamination of rivers and lakes, since this species is present on herbivore dung in high numbers. The medium designed specifically for the preferential recovery of this species from fresh water, M3 agar (Rowbotham and Cross, 1977), is actually a useful isolation medium for actinomycetes in general.

Actinomycetes which produce motile spores within a dessication-resistant sporangium (actinoplanetes) have a strong association with the freshwater environment. They colonize leaves and vegetation which are periodically submerged in water, and a range of special isolation techniques have been developed for them. Methods are considered in detail by Makkar and Cross (1982) and include dilution plating zoospore-containing supernatant from settled samples of rehydrated organic matter, and isolating actinoplanetes from colonized floating hair and pollen baits. Another method based on the chemotactic response of actinoplanete zoospores to potassium chloride held within a floating 1-μm capillary has been described by Palleroni (1980).

### G. Miscellaneous habitats

Recovery of actinomycetes from a variety of environments beyond those discussed above can usually be achieved by simple adaptation of routine procedures. There are a number of situations however where the nature of the habitat has necessitated the development of specialized techniques. Isolation of actinomycetes from the intestines of arthropods generally involves surface sterilization and aseptic dissection prior to dilution plating on nutrient media. Inclusion of an enrichment step in cellulose broth led to the successful isolation of cellulolytic actinomycetes from the hindguts of four different termites (Pasti and Belli, 1985). Standard dilution plating techniques were also used by Dzingov *et al.* (1982) to recover actinomycetes from the gut contents and excrement of millipedes. *Promicromonospora*-like actinomycetes were identified as predominant in contrast to the microflora on the surface of millipede legs, which was dominated by streptomycetes. The natural movement of the millipedes responsible for acquiring a surface population of actinomycetes in the soil, was used to sample the microflora by repeatedly placing millipedes on plates containing a range of isolation agars.

Actinomycetes, particularly nocardioforms, are often associated with the degradation of hydrocarbons in the environment. Hydrocarbon-baiting techniques for recovering nocardiae from soil are long-established (McClung, 1960) and hydrocarbons are used as carbon sources in isolation media for nocardioforms which contribute to scum and foam problems in sewage treatment plants (Lechevalier and Lechevalier, 1974; Lemmer and Kroppenstedt, 1984). In other environmental situations, enrichment techniques can be used to good effect. A hydrocarbon-utilizing *Nocardia* sp. was isolated from a beach subsequent to an oil pollution incident, by first incubating pebbles in minimal medium containing the pollutant. The strain was isolated from this enrichment by plating samples on a minimal agar, with hydrocarbon-impregnated filter paper on the Petri dish lid (Mulkins-Phillips and Stewart, 1974). A similar enrichment approach has been used to obtain rhodococci from oilfield groundwater by incubating samples in a mineral salts solution in a propane- or butane-filled atmosphere (Ivshina *et al.*, 1981). This ability to utilize hydrocarbons has implications for the biodeterioration of rubber rings and seals in water pipes. Hutchinson *et al.* (1975) sonicated samples of rubber to dislodge actinomycetes and bacteria from the surface prior to dilution plating. They found that isolation plates from decayed rubber were rich in streptomycetes and nocardiae, while those from intact rings were dominated by unicellular bacteria.

Actinomycetes have been isolated from decayed wood using a variety of techniques. Haraguchi (1975) preincubated powdered decayed wood with Rose bengal, presumably to suppress fungi, prior to the successful recovery of nocardioforms by dilution plating on glucose asparagine agar. A number of workers have also incorporated heat treatments, 1 h at 60–65°C, in their isolation procedures for recovery of actinomycetes from decayed or merely discoloured wood (Harmsen and Nissen, 1965; Blanchette *et al.*, 1981). More recently, Baecker *et al.* (1988) have carried out a systematic study on quantitative isolation of streptomycetes from wood. These workers vacuum impregnated wood blocks with *Streptomyces albus* spore suspensions and recovered 2–3% of the inocula by dilution plating a homogenized suspension of millings from hammer-milled blocks. This apparently represented a 1000-fold improvement compared to conventional recovery by directly sprinkling wood block millings on the surface of isolation plates.

Within the Actinomycetales, there is only one species which fits the definition of an extremophile. *Actinopolyspora halophila* is a supposed contaminant of solar salt and the isolation procedure suggested (Gochnauer *et al.*, 1989) aims to eliminate osmotically sensitive halobacteria by washing with distilled water prior to collecting cells on membrane filters which are then incubated on an agar medium containing 20% NaCl. The species has yet to be isolated from natural high-salt habitats which are its most likely origin.

## IV. Sorting and identification of isolates

It is becoming increasingly difficult to identify actinomycetes to the genus or species level without access to relatively specialized techniques or expertise. As Lechevalier (1989) stated in his *Practical Guide to Generic Identification of Actinomycetes*, 'although nothing beats morphology for simplicity, even with experience, generic identification based on morphology alone is rarely secure'. The well-established use of cell wall types and whole cell sugar patterns has now been joined by determinations of phospholipids, mycolic acids, menaquinones and, increasingly, nucleic acid sequencing and homology data. These techniques have had a major impact on classification, particularly when applied to the so-called nocardioform genera which were difficult to distinguish by more traditional methods. Species identification, especially in large genera like *Streptomyces*, is also a difficult and time-consuming process and identification problems are compounded by the fundamental necessity in many ecological investigations to study large and diverse populations. Further details of genus and species characteristics are given in *Bergey's Manual of Systematic Bacteriology*, 9th Edition, Volume 4. Here we will suggest a few practical guidelines for the non-specialist.

It is rarely possible to identify colonies on isolation plates unless a particular taxon is expected to be present and has a characteristic appearance, such as *Saccharomonospora viridis* from composts and related substrates (see Section III.C). However, it is often possible to group colonies with similar, distinctive growth form and pigmentation. If colonies are isolated at random, they should be transferred to slopes or plates, preferably using two media (e.g. oatmeal agar, yeast–malt extract agar) which encourage growth and pigmentation of many genera and lead to more accurate grouping of isolates. When soil streptomycete isolates were grouped on the basis of spore colour, reverse colour and diffusible pigments, most members of each group identified to the same species when a wide range of characters were determined (Williams and Vickers, 1988), supporting the validity of this approach.

A key for tentative identification of genera based on colony features and microscopic characters is presented in Table IV. For determination of characters, isolates should be inoculated in a crosshatch pattern onto several nutrient media. Microscopic examination is facilitated if some inoculum is also placed onto the upper surface of a sterile coverslip inserted into the medium at an angle of about 45°. Growth occurs on the coverslip which can be removed and examined microscopically with the growth surface uppermost, preferably as unstained preparations. Alternatively, colonies can be observed directly using a microscope fitted with a long working distance objective lens. The distinction between substrate and aerial growth is important. Aerial growth appears dry, powdery or cottony to the naked eye, while substrate growth is usually moist and smooth. Unstained aerial growth appears darker than the substrate mycelium, under the microscope. Other useful advice and methods for examination of actinomycetes have been given by Cross (1989).

**TABLE IV**

A working key for tentative identification of some actinomycete genera

| | |  |  |
|---|---|---|---|
| A. | | No aerial mycelium; substrate mycelium soon fragmenting into various sized rod-coccoid elements. Colony texture soft, bacterial-like | |
| | (a) | Colonies often pink–orange | *Rhodococcus* |
| | (b) | Fragmentation elements motile | *Oerskovia* |
| | (c) | Hyphae dividing in *all* planes to form motile cocci; colony black | *Geodermatophilus* |
| B. | | No aerial mycelium; substrate mycelium not fragmenting. Colony texture tough | |
| | (a) | Single spores, often in clusters. Colony often yellow–orange–red, surface darkening on sporing | *Micromonospora* |
| | (b) | Globose to irregular vesicles (sporangia), 5–22 μm diameter with many spores, formed on colony surface; spores motile. Colony often red–orange | *Actinoplanes* |
| | (c) | Clusters of finger-like vesicles (sporangia) (4–6 × 1–1.2 μm) with row of 3–4 motile spores on colony surface | *Dactylosporangium* |
| C. | | Dry, powdery–cottony aerial mycelium formed; substrate mycelium fragmenting into various sized rod–coccoid elements. Colony texture moderately soft | |
| | (a) | Aerial mycelium white–pink, sometimes sparse; may be stable, fragmenting, or have short spore chains. Substrate mycelium often yellow–orange | *Nocardia* |
| | (b) | Aerial mycelium and substrate mycelium both fragment into rod–coccoid elements | *Nocardioides* |
| D. | | Dry, powdery–cottony aerial mycelium formed; substrate mycelium not fragmenting. Colony texture tough | |
| | (a) | Single, heat-resistant spores on aerial and substrate mycelium. Colonies white–buff–yellow. Mostly thermophilic | *Thermoactinomyces* |
| | (b) | Single, heat-susceptible spores on aerial mycelium only or also on substrate mycelium. Colonies white–brown. Mostly thermophilic | *Thermomonospora* |
| | (c) | Single, heat susceptible spores on aerial mycelium only. Colonies green–grey above, dark green beneath. Thermophilic | *Saccharomonospora* |
| | (d) | Paired spores on aerial mycelium. Some thermophilic. | *Microbispora* |
| | (e) | Spores mostly in chains of 4 on aerial mycelium | *Microtetraspora* |
| | (f) | Chains of spores (5–15) on aerial mycelium, sometimes in loops or spirals (similar to *Streptomyces*) | *Actinomadura* |
| | (g) | Long chains of spores (usually >50) on aerial mycelium. Chain shape either straight, flexous, hooked or spiral | *Streptomyces* |
| | (h) | Long chains of spores (usually >50) on aerial mycelium. Chains in whorls | *Streptoverticillium* |
| | (i) | Substrate mycelium in zig-zag form; aerial mycelium with spores of various length, some long (2.5–5.0 μm). Some thermophilic | *Pseudonocardia* |
| | (j) | Globose to irregular vesicles (sporangia) (7–40 μm diameter) with many spores on aerial mycelium. May also be spore chains similar to those of *Streptomyces* | *Streptosporangium* |

## V. Methods for assessing ecological importance

Traditional methods for assessing the importance of microbial groups have been based on the isolation and enumeration of viable organisms, as largely discussed above. Thus, the recovery of a species in significant numbers implies but does not demonstrate growth in that environment. Furthermore, subsequent testing of isolates for physiological and biochemical characters only indicates their metabolic capacity and not necessarily their ecological activity. The proportion of the viable count which is due to spores or other resting forms and the serious underestimation of the microbial population by these techniques (for discussion see Parkes, 1987) are the major considerations. This section is concerned with methods which can be applied to actinomycetes in order to circumvent such problems. It is too early to consider whether recent developments in molecular methods for detecting microorganisms and their activity in the environment in general will alter our perception of actinomycete ecology.

### A. Direct detection

Microscopic observation is the most obvious direct method of detection. Actinomycete hyphae and spores can be easily distinguished from those of fungi, and the unicellular bacteria. One advantage of microscopy is the simple differentiation of hyphae and spores so that observation of the former indicates at least that actinomycetes are growing in the environment. Light microscopy is of limited value but has been used to observe directly outgrowth of thermoactinomycete spores recovered from water samples by membrane filtration (Attwell and Colwell, 1982, 1984). These workers stained samples with acridine orange and collected spores on membrane filters which were then examined by epifluorescence microscopy. A further development is the use of immunofluorescent staining on soil suspensions, a technique applied by Efremenkova *et al.* (1978) specifically to observe *Streptomyces olivocinereus* hyphae as green structures in a red background produced by counterstaining with rhodamine-labelled albumen. Techniques such as fluorescent staining also have the potential to generate more quantitative information by linking them to automated cell sorting devices, but again this is only at the stage of emergence with respect to applications in microbial ecology.

Scanning electron microscopy has been more commonly used for direct observation of actinomycete growth in the environment and has proved a powerful technique for demonstrating *in situ* activity. In this way, the role of streptomycetes as early colonizers of wood (Blanchette *et al.*, 1981), as causative agents of soft rot (Baecker and King, 1981) and as degraders of phloem cell walls (Sutherland *et al.*, 1979) have all been demonstrated. This approach has also been extended to provide direct evidence for extensive wheat straw degradation by actinomycetes during the composting process (Atkey and Wood, 1983). Other noteworthy applications of scanning electron microscopy in this context include the demonstration of vulcanized rubber degradation by actinomycetes (Cundell and Mulcock, 1975), their ability to colonize, probably as symbionts, the termite gut (Bignell *et al.*, 1981) and the more general demonstration of the localized nature of streptomycete growth in soil (Mayfield *et al.*, 1972).

The current surge of research activity on development of molecular methods for detecting microorganisms in the environment derives almost entirely from concern over our ability to monitor and track released genetically engineered microorganisms (GEMS). For a general discussion of such detection methods see Ford and Olson (1988). Specifically, molecular methods aim to improve detection sensitivity, new limits for which are claimed regularly, and also to enable detection of non-viable or non-culturable cells. There has consequently been renewed interest in techniques such as fluorescent antibody labelling, and Wellington *et al.* (1988) are developing methods for ELISA detection of monoclonal and polyclonal antibodies specific to natural and engineered *Streptomyces* strains. The most rapid progress, however, is in nucleic acid-based techniques although these are invariably developed using Gram-negative bacteria as models. The heterogeneity of the actinomycete growth form would add additional factors to what are already complex developmental projects. Efficient recovery of DNA from environmental samples followed by detection of specific microorganisms by hybridization to labelled DNA probes, the sensitivity of which can be dramatically improved by utilizing the polymerase chain reaction (PCR), is the basic approach. The state of the art in this area is exemplified by a number of recent reports (Holben *et al.*, 1988; Steffan and Atlas, 1988). Analysis of 16S rRNA, which has contributed so much to our appreciation of microbial phylogeny, is also emerging as a potential tool for detecting microorganisms in the environment. The ribosomal RNA approach has much to commend it: 16S rRNA genes are present in multiple copies which should result in sensitive detection; the presence of both conserved and variable regions within the molecule could be exploited for determining the taxonomic diversity of microbial populations; the activity status of populations could be quantitatively determined by differential hybridization to rRNA which can only result from active growth in the environment. The RNA approach to the study of microbial ecology has been reviewed by Olsen *et al.* (1986) and a recent methodological development is reported by Weller and Ward (1989), but again actinomycete ecology has yet to be specifically studied using these general techniques.

## B. Estimating activity

Most evidence for the activity of actinomycetes in the environment is circumstantial, i.e. derived from laboratory experiments on isolates obtained in pure culture. Although heavily criticized, this approach to microbial ecology is not without some value. For example, the fact that cellulolytic activity is a common feature of *Micromonospora* isolates (McCarthy and Broda, 1984; Sandrak, 1977) is clear indication that they contribute to cellulose degradation in the environment. On the contrary, antibiotic production by actinomycetes in laboratory culture has eluded numerous attempts at detection in soil, despite observations suggesting a protective role for actinomycetes in the plant rhizosphere (see Goodfellow and Williams, 1983). Methods for detection and quantification of actinomycete products or activity in the environment are few, and include the observation of biodegradation reactions by scanning electron microscopy as discussed above.

Further evidence for the biodegradation of lignocellulose in its native form can be provided by *in vivo* $^{14}C$-labelling techniques. Radiolabelled precursors of lignin or cellulose synthesis are taken up, assimilated into polymers and the plant material harvested and exhaustively treated to remove non-specific radiolabel (Crawford and Crawford, 1978; McCarthy *et al.*, 1984). Such substrates can be used to monitor mineralization of lignocellulose to $CO_2$ by environmental samples (e.g. Benner *et al.*, 1984) or to obtain evidence that actinomycete isolates can degrade intact plant material (e.g. Crawford, 1978; McCarthy and Broda, 1984). Analysis of the water-soluble product of lignocellulose degradation by one species, *Thermomonospora mesophila*, has provided evidence for the involvement of actinomycetes in humification. This material was found to resemble the humic acid fraction of soil in its elemental composition and acid insolubility (McCarthy *et al.*, 1986). Another approach to measuring the contribution of actinomycetes to biodegradative activity in the environment is to correlate enzyme activities with the development and composition of the actinomycete population. This has been attempted with composts, and an early notable success was the detection of a cellulase with pH and temperature relationships similar to those of a cellulase produced by the compost isolate *Thermomonospora curvata* (Stutzenberger, 1971). Penninckx and co-workers have developed this theme by monitoring actinomycete populations and the activity of a range of enzymes during the composting of cattle manure (Godden *et al.*, 1983; Godden and Penninckx, 1984). Enzyme preparations were in the form of glycerol extracts obtained by grinding compost samples with sea sand, and although the evolution of microbial populations and enzyme activities were difficult to correlate, good evidence was obtained that cellulolysis depended on the activity of several actinomycete species rather than fungi. More recently Wang *et al.* (1989) introduced a strain of *Streptomyces lividans*, which exhibited enhanced expression of a peroxidase, into test tubes containing soil, and monitored carbon mineralization by trapping $CO_2$. A transient increase in $CO_2$ evolution, as determined by titration, was detected in non-sterile but not in sterile soil suggesting that the effect was one of stimulation on the degradative activity of the indigenous microflora. The best example, however, of *in situ* measurement of actinomycete activity in the environment concerns the nitrogen-fixing endophyte *Frankia*. Schwintzer and Tjepkema (1983) collected nodulated roots over a 1-year period and used the well-established acetylene reduction test to measure *Frankia* nitrogenase activity directly, providing valuable information on seasonal patterns of metabolic activity in intact roots.

Tastes and odours in potable water can be caused by the products of microbial growth, amongst which the actinomycete metabolites geosmin and methyl isoborneol are mainly responsible for so-called earthy odours. These compounds are secondary metabolites but unlike antibiotics, reliable methods for their direct detection in the environment have been developed. These are based on gas chromatographic analysis of extracts and distillates prepared from environmental samples. Although actinomycetes appear to be the most important source of these compounds in the environment, they are also produced by some species of cyanobacteria, and so their detection is not necessarily evidence for actinomycete growth. Our knowledge of these metabolites derives largely from the work of Gerber (for detailed review, see Gerber, 1979).

### C. Gene exchange

In the current climate, any review on the subject of microbial ecology requires some treatment of gene exchange in the environment. In the case of actinomycetes, a number of studies directed towards determining the frequency of gene transfer between donor and recipient streptomycetes introduced into soil samples are in train. Natural gene exchange has been little studied in other actinomycetes, and in streptomycetes is largely confined to plasmid-mediated conjugation by an unknown mechanism (Chater and Hopwood, 1984), although transmissible plasmids would appear to be common in natural populations of these organisms. Streptomycetes do not exhibit competence, and gene cloning experiments rely heavily on protoplast transformation procedures. Evidence for transduction is confined to a single report in *S. venezuelae* (Stuttard, 1979). A recent description of intergeneric conjugation between *E. coli* and several

*Streptomyces* spp. (Mazodier *et al.*, 1989) has important implications for ecological aspects of gene exchange and will undoubtedly stimulate appropriate experiments.

Published accounts of gene exchange in model environments deal exclusively with plasmid transfer between streptomycetes. Recipient and donor *Streptomyces* strains, the latter containing pIJ101-derived plasmids carrying selectable antibiotic resistance markers, are added to test tubes or pots of soil followed by the isolation and enumeration of transconjugants on plates containing the appropriate combinations of antibiotics (Wellington *et al.*, 1988; Rafii and Crawford, 1988; Bleakley and Crawford, 1989). In general, plasmid transfer events were confirmed by standard restriction analysis and colony or Southern blot hybridization techniques. Wellington *et al.* (1988) compared sterile and non-sterile soils and also investigated the effects of chitin and starch amendments to enhance hyphal development and increase the probability of gene exchange. Rafii and Crawford (1988) were also able to demonstrate mobilization of a non-conjugative plasmid (pIJ702) by a self-transmissible plasmid (pIJ101) in triparental crosses, but only in sterile soil. More recently, Bleakley and Crawford (1989) extended their experiments with sterile soil in test tubes to determine the effects of moisture content and amendments on plasmid transfer between *S. lividans* and *S. parvulus*. The results suggest that nutrient-rich, dry soils somehow enhance plasmid transfer frequencies, but conclusions beyond the demonstration that plasmid transfer between streptomycetes occurs in soil are not advisable at this stage. Preliminary studies in relation to gene transfer in compost have also been reported, with an emphasis on assessing detection limits prior to release and recovery of genetically manipulated actinomycetes (Amner *et al.*, 1988) and designing large-scale contained compost systems for such experiments (Brooks *et al.*, 1990).

## VI. Concluding remarks

With the application of molecular techniques to microbial classification, the boundary between actinomycetes and other Gram-positive bacteria has been eroded. This is of little relevance to the study of actinomycete ecology, where it is the mycelial nature of actinomycete growth and the clear adaptation to colonization of solid substrates which dictates methodology. Their ability to produce spores can also be exploited for obtaining improved recovery of actinomycetes from the environment, but is a complicating factor when attempting to assess their ecological role. In fact it is the application of new, usually molecular, techniques initially for detecting bacteria and hopefully leading to the means of determining their *in situ* activity that is the research area of most rapid proliferation in microbial ecology. Actinomycetes are very much a part of this, particularly as evidence for their involvement in natural biodegradation processes continues to accumulate. However, the importance of actinomycetes as a source of novel and commercially useful metabolites demands that environmental isolates continue to be generated using traditional methods which are improved and targeted on a rational basis.

## References


Agre, N. S. (1964). *Mikrobiol.* **33,** 913–917.

Al-Diwany, L. J. and Cross, T. (1978). *In* "Nocardia and Streptomyces" (M. Mordarski, W. Kurylowicz and J. Jeljaszewicz, Eds), pp. 153–160, Gustav Fischer Verlag, Stuttgart.

Al-Diwany, L. J., Unsworth, B. A. and Cross, T. (1978). *J. Appl. Bacteriol.* **45,** 249–258.

Amner, W., McCarthy, A. J. and Edwards, C. (1988). *Appl. Environ. Microbiol.* **54,** 3107–3112.

Amner, W., Edwards, C. and McCarthy, A. J. (1989). *Appl. Environ. Microbiol.* **55,** 2669–2674.

Athalye, M., Lacey, J. and Goodfellow, M. (1981). *J. Appl. Bacteriol.* **51,** 289–297.

Atkey, P. T. and Wood, D. A. (1983). *J. Appl. Bacteriol.* **55,** 293–304.

Attwell, R. W. and Colwell, R. R. (1982). *Appl. Environ. Microbiol.* **43,** 478–482.

Attwell, R. W. and Colwell, R. R. (1984). *In* "Biological, Biochemical, and Biomedical Aspects of Actinomycetes" (L. Ortiz-Ortiz, L. F. Bojalil and V. Yakoleff, Eds), pp. 441–452, Academic Press, San Diego.

Baecker, A. A. W. and King, B. (1981). *J. Inst. Wood. Sci.* **9,** 65–71.

Baecker, A. A. W. and Ryan, K. C. (1987). *S. Afr. J. Plant Soil* **4,** 165–170.

Baecker, A. A. W., Casim, H. A and Ryan, K. C. (1988). *The Actinomycetes* **20,** 227–252.

Baker, D. D. (1988). *In* "Biology of Actinomycetes '88" (Y. Okami, T. Beppu and H. Ogawar, Eds), pp. 271–276, Japan Scientific Societies Press, Tokyo.

Baker, D. and Torrey, J. E. (1979) *In* "Symbiotic Nitrogen Fixation in the Management of Temperate Forests" (J. C. Gordon, C. T. Wheeler and D. A. Perry, Eds), pp. 38–56, Oregon State University, Corvallis.

Baker, D. and Torrey, J. E. (1980). *Can. J. Microbiol.* **26,** 1066–1071.

Baker, D. Torrey, J. G. and Kidd, G. H. (1979). *Nature (London)* **182,** 76–78.

Baker, D., Newcomb, W. and Torrey, J. E. (1980). *Can. J. Microbiol.* **26,** 1072–1089.

Barcina, I., Iriberri, J. and Egea, L. (1987). *System. Appl. Microbiol.* **10,** 85–91.

Benner, R., Newell, S. Y., Maccubbin, A. E. and Hodson, R. E. (1984). *Appl. Environ. Microbiol.* **48,** 36–40.

Bignell, D. E., Oskarsson, H. and Anderson, J. M. (1981). *Zbl. Bakt. Suppl.* **11,** 201–206.

Blanchette, R. A., Sutherland, J. B. and Crawford, D. L. (1981). *Can. J. Bot.* **59,** 1–7.

Bleakley, B. H. and Crawford, D. L. (1989). *Can. J. Microbiol.* **35,** 544–549.

Brooks, R. C., Fermor, T. R. and McCarthy, A. J. (1990). *J. Sci. Fd. Agric.* **50,** 132–133.
Callaham, D., del Tredici, P. and Torrey, J. G. (1978). *Science* **199,** 899–902.
Chater, K. F. and Hopwood, D. A. (1984). *In* "The Biology of Actinomycetes" (M. Goodfellow, M. Mordarski and S. T. Williams, Eds), pp. 229–286, Academic Press, London.
Crawford, D. L. (1978). *Appl. Environ. Microbiol.* **35,** 1041–1045.
Crawford, R. L. and Crawford, D. L. (1978). *Dev. Ind. Microbiol.* **19,** 35–49.
Cross, T. (1968). *J. Appl. Bacteriol.* **31,** 36–53.
Cross, T. (1981). *J. Appl. Bacteriol.* **50,** 397–424.
Cross, T. (1982). *Dev. Ind. Microbiol.* **23,** 1–18.
Cross, T. (1989). *In* "Bergey's Manual of Systematic Bacteriology", 9th Edn, Vol. 4 (S. T. Williams, M. E. Sharpe and J. G. Holt, Eds), pp. 2340–2343, Williams and Wilkins, Baltimore.
Cundell, A. M. and Mulcock, A. P. (1975). *Dev. Ind. Microbiol.* **16,** 88–96.
Dzingov, A., Marialigeti, K., Jager, K., Contreras, E., Kandics, L. and Szabo, I. M. (1982). *Pedobiol.* **24,** 1–7.
Efremenkova, L. M., Kozhevia, P. A. and Zvyagintsev D. G. (1978). *Mikrobiol.* **47,** 1122–1124.
Ford, S. and Olson, B. H. (1988). *Adv. Microb. Ecol.* **10,** 45–79.
Gerber, N. N. (1979). *CRC Crit. Rev. Microbiol.* **9,** 191–214.
Gochnauer, M. B., Johnson, K. G. and Kushner, D. J. (1989). *In* "Bergey's Manual of Systematic Bacteriology", 9th Edn, Vol. 4 (S. T. Williams, M. E. Sharpe and J. G. Holt, Eds), pp. 2398–2401, Williams and Wilkins, Baltimore.
Godden, B. and Penninckx, M. J. (1984). *Ann. Microbiol. (Inst. Past).* **135B,** 69–78.
Godden, B., Penninckx, M., Pierard, A. and Lannoye, R. (1983). *Eur. J. Appl. Microbiol. Biotechnol.* **17,** 306–310.
Goodfellow, M. (1989a). *In* "Bergey's Manual of Systematic Bacteriology", 9th Edn, Vol. 4 (S. T. Williams, M. E. Sharpe and J. G. Holt, Eds), pp. 2333–2339, Williams and Wilkins, Baltimore.
Goodfellow, M. (1989b). *In* "Bergey's Manual of Systematic Bacteriology" 9th Edn, Vol. 4 (S. T. Williams, M. E. Sharpe and J. G. Holt, Eds), pp. 2362–2371, Williams and Wilkins, Baltimore.
Goodfellow, M. and Haynes, J. A. (1984). *In* "Biological, Biochemical and Biomedical Aspects of Actinomycetes" (L. Ortiz-Ortiz, L. F. Bojalil and V. Yakoleff, Eds), pp. 453–472, Academic Press, Orlando.
Goodfellow, M. and Lechevalier, M. P. (1989). *In* "Bergey's Manual of Systematic Bacteriology", 9th Edn, Vol. 4 (S. T. Williams, M. E. Sharpe and J. E. Holt, Eds), pp. 2350–2361, Williams and Wilkins, Baltimore.
Goodfellow, M. and O'Donnell, A. G. (1989). *In* "Microbial Products: New Approaches" (S. Baumberg, I. S. Hunter and P. M. Rhodes, Eds), pp. 343–383, University Press, Cambridge.
Goodfellow, M. and Williams, S. T. (1983). *Ann. Rev. Microbiol.* **39,** 189–216.
Goodfellow, M. and Williams, E. (1986). *Biotechnol. Genet. Eng. Rev.* **4,** 213–262.
Gregory, P. H. and Lacey, M. E. (1963). *J. Gen. Microbiol.* **30,** 75–88.
Gregory, P. H., Lacey, M. E., Festenstein, G. N. and Skinner, F. A. (1963). *J. Gen. Microbiol.* **33,** 147–174.
Hagedorn, C. (1976). *Appl. Environ. Microbiol.* **32,** 368–375.
Hanka, L. J., Rueckert, P. W. and Cross, T. (1985). *FEMS Microbiol. Lett.* **30,** 365–368.

Haraguchi, T. (1975). *Trans Mycol. Soc. Japan* **16,** 103–105.
Harmsen, L. and Nissen, T. (1965). *Nature (London)* **206,** 319.
Holben, W. E., Jansson, J. K., Chelm, B. K. and Tiedje, J. M. (1988). *Appl. Environ. Microbiol.* **54,** 703–711.
Hsu, S. C. and Lockwood, J. L. (1975). *Appl. Microbiol.* **29,** 422–426.
Hunter, J. C., Eveleigh, D. E. and Casella, G. (1981). *Zbl. Bakt. Suppl.* **11,** 195–200.
Hutchinson, M., Ridgway, J. W. and Cross, T. (1975). *In* "Microbial Aspects of the Biodeterioration of Materials" (R. J. Gilbert and D. W. Lovelock, Eds), pp. 187–202, Academic Press, London.
Ivshina, I. B., Oborin, A. A., Nesterenko, O. A. and Kasumova, S. A. (1981). *Mikrobiol.* **50,** 709–716.
Johnston, D. W. and Cross, T. (1976). *Freshwater Biol.* **6,** 464–469.
Kawamoto, I. (1989). *In* "Bergey's Manual of Systematic Bacteriology" (S. T. Williams, M. E. Sharpe and J. G. Holt, Eds), pp. 2442–2450, Williams and Wilkins, Baltimore.
Khan, M. R. and Williams, S. T. (1975). *Soil Biol. Biochem.* **7,** 345–348.
Küster, E. and Williams, S. T. (1964). *Nature (London)* **202,** 928–929.
Labeda, D. P. (1986). *In* "Perspectives in Microbial Ecology" (F. Megnsar and M. Gantar, Eds), pp. 271–276, Slovene Society for Microbiology, Ljubljana.
Labeda, D. P. (1989). *In* "Bergey's Manual of Systematic Bacteriology", 9th Edn, Vol. 4 (S. T. Williams, M. E. Sharpe and J. G. Holt, Eds), pp. 2586–2589, Williams and Wilkins, Baltimore.
Labonde, M., Calvert, H. E. and Pine, S. (1981). *In* "Current Perspectives in Nitrogen Fixation" (A. H. Gibson and W. E. Newton, Eds), pp. 296–299, Australian Academy of Science, Canberra.
Lacey, J. (1973). *In* "Actinomycetales: Characteristics and Practical Importance" (G. Sykes and F. A. Skinner, Eds), pp. 231–251, Academic Press, London.
Lacey, J. (1974). *Annal. Appl. Biol.* **76,** 63–76.
Lacey, J. (1975). *Trans. Br. Mycol. Soc.* **64,** 265–281.
Lacey, J. (1977). *Lancet* 27th Aug., 455–456.
Lacey, J. (1978). *In* "Nocardia and Streptomyces" (M. Mordarski, W. Kurylowicz and J. Jeljaszewicz, Eds), pp. 161–170, Gustav Fischer Verlag, Stuttgart.
Lacey, J. (1980). *In* "Microbial Growth and Survival in Extremes of Environment" (G. W. Gould and J. E. L. Corry, Eds), pp. 53–70, Academic Press, London.
Lacey, J. (1988). *In* "Actinomycetes in Biotechnology" (M. Goodfellow, S. T. Williams and M. Mordarski, Eds), pp. 359–432, Academic Press, London.
Lacey, J. and Dutkiewicz, J. (1976a). *J. Appl. Bacteriol.* **41,** 315–319.
Lacey, J. and Dutkiewicz, J. (1976b). *J. Appl. Bacteriol.* **41,** 13–27.
Lacey, J. and Lacey, M. E. (1964). *Trans. Br. Mycol. Soc.* **47,** 547–552.
Lechevalier, H. A. (1989). *In* "Bergey's Manual of Systematic Bacteriology", 9th Edn, Vol. 4 (S. T. Williams, M. E. Sharpe and J. G. Holt, Eds), pp. 2344–2347, Williams and Wilkins, Baltimore.
.6.12Lechevalier, M. P. and Lechevalier, H. A. (1974). *Int. J. Syst. Bacteriol.* **24,** 278–288.
.6.12Lechevalier, M. P. and Lechevalier, H. A. (1989). *In* "Bergey's Manual of Systematic Bacteriology", 9th Edn, Vol. 4 (S. T. Williams, M. E. Sharpe and J. G. Holt, Eds), pp. 2410–2417, Williams and Wilkins, Baltimore.
Lechevalier, M. P., Baker, D. and Horriere, F. (1983). *Can. J. Bot.* **61,** 2826–2833.
Lechevalier, M. P., Prauser, H., Labeda, D. P. and Ruan, J-S. (1986). *Int. J. Syst. Bact.* **36,** 29–37.

Lemmer, H. and Kroppenstedt, R. M. (1984). *Syst. Appl. Microbiol.* **5,** 124–135.
Locci, R. and Schofield, G. M. (1989). *In* "Bergey's Manual of Systematic Bacteriology", 9th Edn, Vol. 4 (S. T. Williams, M. E. Sharpe and J. G. Holt, Eds), pp. 2492–2504, Williams and Wilkins, Baltimore.
Luedemann, G. M. and Fonesca, A. F. (1989). *In* "Bergey's Manual of Systematic Bacteriology", 9th Edn, Vol. 4 (S. T. Williams, M. E. Sharpe and J. G. Holt, Eds), pp. 2406–2409, Williams and Wilkins, Baltimore.
Makkar, N. S. and Cross, T. (1982). *J. Appl. Bacteriol.* **52,** 209–218.
Mackay, S. J. (1977). *Appl. Environ. Microbiol.* **33,** 227–230.
Mayfield, C. I., Williams, S. T., Ruddick, S. M. and Hatfield, H. L. (1972). *Soil Biol. Biochem.* **4,** 79–91.
Mazodier, P., Peiter, R. and Thompson, C. (1989). *J. Bacteriol.* **171,** 3583–3585.
McCarthy, A. J. and Broda, P. M. A. (1984). *J. Gen. Microbiol.* **130,** 2905–2913.
McCarthy, A. J. and Cross, T. (1981). *J. Appl. Bacteriol.* **51,** 299–302.
McCarthy, A. J., Macdonald, M. J., Paterson, A. and Broda, P. (1984). *J. Gen. Microbiol.* **130,** 1023–1030.
McCarthy, A. J., Paterson, A. and Broda, P. (1986). *Appl. Microbiol. Biotechnol.* **24,** 347–352.
McClung, N. M. (1960). *Mycologia* **52,** 154–156.
Mertz, F. P. and Yao, R. C. (1988). *Int. J. Syst. Bact.* **38,** 282–286.
Meyer, J. (1989). *In* "Bergey's Manual of Systematic Bacteriology", 9th Edn, Vol. 4 (S. T. Williams, M. E. Sharpe and J. G. Holt, Eds), pp. 2511–2526, Williams and Wilkins, Baltimore.
Millner, P. D. (1982). *Dev. Ind. Microbiol.* **23,** 61–78.
Millner, P. D., Bassett, D. A. and Marsh, P. B. (1980). *Appl. Environ. Microbiol.* **39,** 1000–1009.
Mulkins-Phillips, G. J. and Stewart, J. E. (1974). *Appl. Microbiol.* **28,** 915–922.
Nolan, R. D. and Cross, T. (1988). *In* "Actinomycetes in Biotechnology" (M. Goodfellow, S. T. Williams and M. Mordarski, Eds), pp. 1–32, Academic Press, London.
Nonomura, H. (1989a). *In* "Bergey's Manual of Systematic Bacteriology", 9th Edn, Vol. 4 (S. T. Williams, M. E. Sharpe and J. G. Holt, Eds), pp. 2526–2531, Williams and Wilkins, Baltimore.
Nonomura, H. (1989b). *In* "Bergey's Manual of Systematic Bacteriology", 9th Edn, Vol. 4 (S. T. Williams, M. E. Sharpe and J. G. Holt, Eds), pp. 2545–2551, Williams and Wilkins, Baltimore.
Nonomura, H. and Ohara, Y. (1971). *Hakku Kogaku Zasshi* **49,** 889–894.
Okami, Y. (1972). *J. Antibiotics* **25,** 467–488.
Okami, Y. and Okazaki, T. (1972). *J. Antibiotics* **25,** 456–460.
Okazaki, T. and Okami, Y. (1972). *J. Antibiotics* **25,** 461–466.
Olsen, G. J. and Lane, D. J., Giovannoni, S. J. and Pace, N. R. (1986). *Ann. Rev. Microbiol.* **40,** 337–365.
Olsen, R. A. and Bakken, L. R. (1987). *Microb. Ecol.* **13,** 59–74.
Omura, S., Takahashi, Y. and Iwai, Y. (1989). *In* "Bergey's Manual of Systematic Bacteriology", 9th Edn, Vol. 4 (S. T. Williams, M. E. Sharpe and J. G. Holt, Eds), pp. 2594–2598, Williams and Wilkins, Baltimore.
Palleroni, N. J. (1980). *Arch. Microbiol.* **128,** 53–55.
Palleroni, N. J. (1989). *In* "Bergey's Manual of Systematic Bacteriology", 9th Edn, Vol. 4 (S. T. Williams, M. E. Sharpe and J. G. Holt, Eds), pp. 2419–2428, Williams and Wilkins, Baltimore.

Parkes, R. J. (1987). *In* "Ecology of Microbial Communities" (M. Fletcher, T. R. G. Gray and J. G. Jones, Eds), pp. 147–177, University Press, Cambridge.
Pasti, M. B. and Belli, M. L. (1985). *FEMS Microbiol. Lett.* **26,** 107–112.
Preobrazhenskaya, T. P., Lavrova, N. V., Ukholina, R. S. and Nechaeva, N. P. (1975). *Antibiotiki* **20,** 404–408.
Pridham, T. G. and Lyons, A. J. (1961). *J. Bacteriol.* **81,** 431–441.
Quispel, A., Burggraaf, A. J. P., Borsje, H. and Tak, T. (1983). *Can. J. Bot.* **61,** 2801–2806.
Rafii, F. and Crawford, D. L. (1988). *Appl. Environ. Microbiol.* **54,** 1334–1340.
Rowbotham, T. J. and Cross, T. (1977). *J. Gen. Microbiol.* **100,** 231–240.
Sandrak, N. Y. (1977). *Mikrobiol.* **46,** 478–481.
Schwintzer, C. R. and Tjepkema, J. D. (1983). *Can. J. Bot.* **61,** 2937–2942.
Shearer, M. C., Colman, P. M. and Ferrin, R. M. (1989). *In* "Bergey's Manual of Systematic Bacteriology", 9th Edn, Vol. 4 (S. T. Williams, M. E. Sharpe and J. G. Holt, Eds), pp. 2590–2594, Williams and Wilkins, Baltimore.
Sneath, P. H. A. (1980). *Comp. Geosci.* **6,** 21–26.
Steffan, R. J. and Atlas, R. M. (1988). *Appl. Environ. Microbiol.* **54,** 2185–2191.
Stuttard, C. (1979). *J. Gen. Microbiol.* **110,** 479–482.
Stutzenberger, F. J. (1971). *Appl. Microbiol.* **22,** 147–152.
Sutherland, J. B., Blanchette, R. A., Crawford, D. L. and Pometto, A. L. (1979). *Curr. Microbiol.* **2,** 123–126.
Treuhaft, M. W. and Arden-Jones, M. P. (1982). *J. Clin. Microbiol.* **16,** 995–999.
Trolldenier, G. (1966). *Zentbl. Bakt. Parasitkde Abt. II* **120,** 496–508.
Vickers, J. C. and Williams, S. T. (1987). *Microbios. Lett.* **35,** 113–117.
Vickers, J. C., Williams, S. T. and Ross, G. W. (1984). *In* "Biological, Biochemical and Biomedical Aspects of Actinomycetes" (L. Ortiz-Ortiz, J. P. Bojalil and V. Yakoleff, Eds), pp. 553–561, Academic Press, Orlando.
Vobis, G. (1989). *In* "Bergey's Manual of Systematic Bacteriology", 9th Edn, Vol. 4 (S. T. Williams, M. E. Sharpe and J. G. Holt, Eds), pp. 2437–2442, Williams and Wilkins, Baltimore.
Wakisaka, Y., Kawamura, Y., Yasuda, Y., Koizuma, K. and Nishimoto, Y. (1982). *J. Antibiotics* **35,** 822–836.
Wang, Z., Crawford, D. L., Pometto, A. L. and Rafii, F. (1989). *Can. J. Microbiol.* **35,** 535–543.
Watson, E. T. and Williams, S. T. (1974). *Soil Biol. Biochem.* **6,** 43–52.
Weller, R. and Ward, D. M. (1989). *Appl. Environ. Microbiol.* **55,** 1818–1822.
Wellington, E. M. H., Saunders, V. A., Cresswell, N. and Wipat, A. (1988). *In* "Biology of Actinomycetes '88" (Y. Okami, T. Beppu and H. Ogawara, Eds), pp. 300–305, Japan Scientific Societies Press, Tokyo.
Weyland, H. and Helmke, E. (1988). *In* "Biology of Actinomycetes '88" (Y. Okami, T. Beppu and H. Ogawara, Eds), pp. 294–299, Japan Scientific Societies Press, Tokyo.
Williams, S. T. and Vickers, J. C. (1986). *Microb. Ecol.* **12,** 43–52.
Williams, S. T. and Vickers, J. C. (1988). *In* "Biology of Actinomycetes '88" (Y. Okami, T. Beppu and H. Ogawara, Eds), pp. 265–270, Japan Scientific Societies Press, Tokyo.
William, S. T. and Wellington, E. M. H. (1982). *In* "Methods of Soil Analysis", 2nd Edn, Part 2 (A. L. Page, R. H. Miller and D. R. Keeney, Eds), pp. 969–987, American Society of Agronomy, Madison, Wisconsin.

Williams, S. T., Davies, F. L., Mayfield, C. I. and Khan, M. R. (1971). *Soil Biol. Biochem.* **3,** 187–195.

Williams, S. T., Goodfellow, M., Alderson, G., Wellington, E. M. H., Sneath, P. H. A. and Sackin, M. J. (1983). *J. Gen. Microbiol.* **129,** 1743–1813.

Williams, S. T., Lanning, S. and Wellington, E. M. H. (1984). *In* "The Biology of the Actinomycetes" (M. Goodfellow, M. Mordarski and S. T. Williams, Eds), pp. 481–528, Academic Press, London.

Williams, S. T., Goodfellow, M. and Alderson, G. (1989). *In* "Bergey's Manual of Systematic Bacteriology", 9th Edn, Vol. 4 (S. T. Williams, M. E. Sharpe and J. G. Holt, Eds), pp. 2452–2492, Williams and Wilkins, Baltimore.

# 18

# Methods for Studying the Ecology of Endospore-forming Bacteria

FERGUS G. PRIEST*

*Department of Biological Sciences, Heriot Watt University, Riccarton, Edinburgh EH14 4AS, UK*

AND ROSA GRIGOROVA

*Institute of Microbiology, Bulgarian Academy of Sciences, Sofia 1113, Bulgaria*



METHODS IN MICROBIOLOGY
VOLUME 22 ISBN 0-12-521522-3

## I. Introduction

The capacity to differentiate into a remarkably resistant and dormant structure, the endospore, unifies an otherwise heterogeneous collection of prokaryotes. Endospores and their components show such considerable biochemical and genetic conservation that it is often forgotten that the parent cells are not restricted to rod-shaped bacteria but also occur as cocci (*Sporosarcina*) and filamentous bacteria reminiscent of the actinomycetes in morphology (*Thermoactinomyces*). The major genera of endospore formers are shown in Table I and it is clear that *Bacillus* and *Clostridium* dominate the group in the number of species they represent. This is however a misleading classification since both phenetic and nucleic acid studies show that *Clostridium* and *Bacillus* should be reclassified into several taxa of generic rank (Stackebrandt and Woese, 1981; Priest *et al.*, 1988). Each 'genus' would presumably contain bacteria with similar physiological attributes and this would greatly simplify the isolation, classification and identification of these important organisms as well as distributing the species into more manageable taxa. Such proposals, although first suggested at the turn of the century (see Gibson and Gordon, 1974, for review) require extensive chemotaxonomic and phenotypic data before they can be undertaken with confidence.

The rRNA sequences have also raised the interesting point that non-endospore-forming species such as *Caryophanon latum* and *Planococcus citreus* are phylogenetically interspersed within the endospore-formers (Stackebrandt *et al.*, 1987). This suggests that such bacteria may have lost the ability to make endospores or alternatively, that endospore formation has evolved more than once. If the former explanation is correct, endospore formation must be a very ancient and highly conserved form of differentiation since homology amongst spore structural genes (the small, acid-soluble spore proteins) spreads across existing generic boundaries from *Bacillus* to *Thermoactinomyces* and *Clostridium* (Setlow, 1988). Alternatively, evolution of the spore within separate lineages indicates remarkable convergence to a common form.

The range of endospore-forming genera covers a variety of physiological types. Every form of oxygen metabolism is represented (Table I) as well as chemolithotrophs (sulphate reducers and hydrogen oxidizers such as *B. schlegelii*). Although most are heterotrophs some clostridia and bacilli are autotrophic. Associated with this variety in carbon and energy metabolism is an impressive diversity in adaptation to exotic environments. Few other genera contain bacteria which have adapted to such extremes of pH, temperature and salinity (Table I). The ability to sporulate coupled with adaptation of many species to environmental extremes has enabled

**TABLE I**
Diversity of endospore-forming bacteria

| Feature | Genus | | | | | |
|---|---|---|---|---|---|---|
| | *Bacillus* | *Clostridium* | *Desulfotomaculum* | *Sporolactobacillus* | *Sporosarcina* | *Thermoactinomyces* |
| Cellular morphology | Rods | Rods | Rods/curved rods | Rods | Cocci | Branched filaments |
| Oxygen relationship | Aerobic/ facultative anaerobic | Anaerobic | Anaerobic | Microaerophilic | Aerobic | Aerobic |
| Sulphate reduced to sulphide | – | – | + | – | – | – |
| Nitrate reduced to nitrite | V | V | – | – | V | – |
| Some species: | | | | | | |
| Thermophilic/tolerant | + | + | + | – | – | + |
| Psychrophilic/tolerant | + | – | – | – | – | – |
| Alkaliphilic/tolerant | + | + | – | – | – | – |
| Acidophilic/tolerant | + | + | – | + | – | – |
| Halophilic | + | – | – | – | + | – |
| Approx. No. species | 58 | 88 | 5 | 1 | 2 | 7 |
| Mol. % G + C | 32–69 | 22–55 | 37–50 | 38–39 | 40–42 | 52–55 |

endospore-formers to become amongst the most widespread of all bacteria. Indeed they are ubiquitous and can be isolated from virtually all habitats within the biosphere (Slepecky and Leadbetter, 1983). This raises an important ecological question concerning the status of these organisms since in perhaps a majority of examples they are simply present as dormant spores in a hostile environment and are not growing. For example the recovery of spores of thermoactinomycetes from marine sediments (Cross, 1980) or of the thermophile *B. schlegelii* from lake sediments (Bonjour *et al.*, 1988) suggests that these bacteria are essentially irrelevant to their habitat. It is therefore difficult to consider spore-formers in terms of ecosystems. High numbers of spores of a particular type in a given habitat may indicate earlier or continuing growth of the parental bacterium in that habitat but in other circumstances it may reflect accumulation of spores in an environment that is conducive to spore dornancy (i.e. constant cool temperature) but inhibitory to growth (as in the case of the *B. schlegelii* and thermoactinomycete spores). In other circumstances bacilli have been recovered almost entirely from vegetative cells (Noeth *et al.*, 1988) and it is then simpler to draw valid conclusions as to their contribution to the ecology of that site.

The ecology of the endospore-formers is thus poorly understood (see reviews by Slepecky, 1972; Slepecky and Leadbetter, 1983). This stems largely from difficulties in establishing their role in the environment as noted above but also identification of environmental isolates is problematic. Identification of *Bacillus* (Gordon *et al.*, 1973; Claus and Berkeley, 1986), *Clostridium* (Cato *et al.*, 1986), *Sporosarcina* (Claus and Fahmy, 1986) and *Thermoactinomyces* (Lacey and Cross, 1989) is still largely based on traditional phenotypic tests but is improving with the introduction of miniaturized test systems such as those marketed by API (Logan and Berkeley, 1984) and Minitech (Sullivan *et al.*, 1987). Moreover, a computerized system has been developed for *Bacillus* based on traditional phenotypic tests (Priest and Alexander, 1988). These schemes have been reviewed in previous volumes of this series (Berkeley *et al.*, 1987) and elsewhere (Priest, 1989).

In this article, we shall concentrate on the spore as a unifying and unique feature and deal with methods for studying the ecology of these bacteria largely from this viewpoint. In particular, we shall highlight methods for determining the populations of spore formers in the environment and the proportions that are present as spores. Methods for the direct estimation of spores and vegetative cells in soils are covered in Chapter 10.

## II. Estimation of populations of endospore-formers

Assessment of total populations of endospore-forming bacteria in environmental samples is frustrated by the lack of a single medium capable of supporting the diversity of bacteria. It is therefore not surprising that most studies concern only the easily cultivated bacilli and there is a serious lack of data for clostridia, sporosarcina and other genera. The general way to obtain an indication of the proportion of bacteria present in a sample as spores, compared with vegetative cells, is to conduct two viable counts; one of the total population and a second of a treated sample in which vegetative cells are killed by heat or some other treatment (see Section III). This procedure is reasonably accurate but complication arises from the difficulty of obtaining an effective assessment of the total population. There are no media that are selective for endospore-formers and therefore plates always support growth of other bacteria, both Gram positive and Gram negative. Populations of spore-formers are generally determined on such plates by microscopic examination of random colonies for the presence of spores. Alternatively, replica plating onto velvet pads, heating the pads to destroy vegetative cells (wrap pads in foil suitably marked for orientation and steam for 15 min) and using them to reinoculate fresh plates can give a good estimation if conducted carefully. This method works best if fewer than 50 colonies are present per plate and has the advantage of speed. Moreover, strains that sporulate poorly will be detected but might be missed using microscopic examination.

Total counts are usually prepared on nutritive media such as yeast extract, peptone agar (Parkinson *et al.*, 1971), modified Thornton's medium (Fægri *et al.*, 1977), yeast extract, peptone, soil extract agar (Bunt and Rovira, 1956) or soil extract agar (Gordon *et al.*, 1973). The compositions of these media are given in Table II. Such media are incubated aerobically (30°C) for 7 days for bacilli and anaerobically for clostridia. Although richer media such as blood agar have been used to recover clostridia (Matches and Liston, 1974), virtually any medium rich in carbohydrate with peptone, yeast extract and reducing agents is suitable for saccharolytic clostridia. *Thermoactinomyces* and *Sporosarcina* require their own specialized media (see below). It has been suggested that such nutritively rich media recover only a small percentage of the soil flora (0.1–1%). Olsen and Bakken (1987) compared such media with cold extracted soil extract agar and found 3–5 times higher counts on the latter medium compared with peptone-based media. However the numbers of endospore-formers were similar on both media substantiating the use of fairly rich media for these heterotrophs. It is however always useful to include soil extract since it promotes sporulation in many strains that sporulate poorly *in vitro*.

**TABLE II**
Some typical media used for 'total' counts of endospore-forming bacteria

| Ingredient | YP | YPS | Th[a] | SEA |
|---|---|---|---|---|
| $KH_2PO_4$ | | | 1 | |
| $K_2HPO_4$ | | 0.4 | | |
| $(NH_4)_2HPO_4$ | | 0.5 | | |
| $KNO_3$ | | | 0.5 | |
| $MgSO_4.7H_2O$ | | 0.05 | 0.2 | |
| $MgCl_2$ | | 0.1 | | |
| $CaCl_2.2H_2O$ | | 0.1 | 0.1 | |
| NaCl | | | 0.1 | |
| $FeCl_3.6H_2O$ | | 0.01 | 0.01 | |
| $FePO_4.2H_2O$ | 0.01 | | | |
| Asparagine | | | 0.5 | |
| Peptone | 5 | 1 | | 5 |
| Beef extract | | | | 3 |
| Yeast extract | 1 | 1 | | |
| Mannitol | | | 1 | |
| Soil extract[a] | | 250 ml | 1000 ml | 250 ml |
| Tap water | 1000 ml | 750 ml | 900 ml | 750 ml |
| Agar | | | To solidify | |

YP, yeast extract, peptone agar; YPS, yeast extract, peptone, soil extract agar; Th, modified Thornton's agar; SEA, soil extract agar.

[a] Autoclaved soil extract prepared by autoclaving a 1 : 1 mixture of air-dried soil and water for 20 min at 121°C, filtering and reautoclaving.

Although the media in Table II will recover common endospore-formers, there are many types that will not grow under these conditions. It is for this reason that our knowledge of the distribution of some of the 'rarer' types is so poor. Indeed these 'rarer' types might be common, it is our reluctance to use unusual isolation conditions that perhaps assigns them to the rare category. Claus and Berkeley (1986) have recently provided an excellent review of selecttive methods for the isolation of individdual *Bacillus* species, and for the non-pathogenic clostridia, the article by Gottschalk *et al.* (1981) is very useful. Rather than repeat this information, we decided to provide methods for the isolation of specific physiological types of endospore-formers so that by combining these methods with the standard plate counts mentioned above, a more complete picture of the distribution of endospore-formers can be obtained. In most instances the methods have been developed for bacilli, but anaerobic incubation should often be sufficient to isolate clostridia with similar properties.

## A. Acidophiles

At present, descriptions of obligately acidophilic, heterotrophic endospore-formers are restricted to three thermophilic/thermotolerant bacilli (Darland and Brock, 1971; Deinhard *et al.*, 1987a, b) and one mesophilic type (Jensen and Norman, 1984). These bacteria do not grow at neutral pH and would not be isolated on normal media and yet are common in soils that are not acidic. *Bacillus acidoterrestris* was first isolated using enrichment cultures (Hippchen *et al.*, 1981) but more recently *B. cycloheptanicus* was isolated from soils by direct plating on a medium containing yeast extract (later found to be esssential). Media for the isolation of acidophiles are given in Table III. It should be noted that most acidophilic bacilli are thermophilic or thermotolerant and should be incubated at 45–50°C.

**TABLE III**
Media for the isolation of acidophilic, endospore-forming bacteria

| Ingredient | *B. acidocaldarius* medium[a] ($g\ l^{-1}$) | *B. cycloheptanicus* medium[b] ($g\ l^{-1}$) |
|---|---|---|
| Glucose | 1 | 5 |
| Soluble starch | 2 | – |
| Yeast extract | 2 | 3 |
| $(NH_4)_2SO_4$ | 0.2 | 0.2 |
| $MgSO_4.7H_2O$ | 0.5 | 0.5 |
| $CaCl_2.2H_2O$ | 0.25 | 0.25 |
| $KH_2PO_4$ | 3 | 3 |
| $FeSO_4.7H_2O$ | 0.28 mg | 1 ml of trace elements[c] |
| $MnCl_2.4H_2O$ | 1.25 mg | |
| $ZnSO_4.7H_2O$ | 0.4 mg | |
| Agar | 30.0 | 30.0 |
| | pH 4.0 with 1 M $H_2SO_4$ | pH 4.3 with 1 M $H_2SO_4$ |

[a] Medium of Darland and Brock (1971).
[b] Medium of Deinhard *et al.* (1987a).
[c] Trace elements contain ($mg\ l^{-1}$); $CaCl_2.2H_2O$, 0.66; $ZnSO_4.7H_2O$, 0.18; $CuSO_4.5H_2O$, 0.16; $MnSO_4.4H_2O$, 0.15; $CoCl_2.6H_2O$, 0.18; $H_3BO_3$, 0.10 and $Na_2MoO_4.2H_2O$, 0.30.

## B. Alkaliphiles

Alkaliphilic bacilli are those that grow optimally at high pH and not at neutrality. Thus they would not be recovered on normal media. They are found in both acidic and alkaline soils but are more common in the latter. Numbers range from 10 to $10^6$ per gram soil and generally represent 1–10% of the neutrophilic population (Horikoshi and Akiba, 1982). Some media for the isolation of alkaliphilic bacilli are given in Table IV. It is important to include a sodium salt, generally 0.5–2%, in media for alkaliphiles since they have a sodium requirement. The sodium carbonate in the media in Table IV could be replaced by other sodium salts such as bicarbonate, phosphate, pyrophosphate, borate, sesquicarbonate or sodium hydroxide. Potassium salts are not recommended because of the sodium requirement of the bacteria. The pH should be adjusted to 8.5–11.0; some strains can even grow at pH 12 but it is difficult to maintain media at such a high pH particularly as the bacteria are effective at changing the pH to their optimum requirement.

**TABLE IV**
Media for the isolation of alkaliphilic endospore-forming bacteria

| Ingredient | Medium I[a] ($g\ l^{-1}$) | Medium II[a] ($g\ l^{-1}$) |
|---|---|---|
| Glucose | 10 | – |
| Soluble starch | – | 10 |
| Peptone | 5 | 5 |
| Yeast extract | 5 | 5 |
| $KH_2PO_4$ | 1 | 1 |
| $MgSO_4.7H_2O$ | 0.2 | 0.2 |
| $Na_2CO_3$ | 10 | 10 |
| Agar | 20 | 20 |

pH 8.5 to 11.0 depending on requirements.
[a] Media from Horikoshi and Akiba (1982).

*Sporosarcina ureae* is usually isolated from environments of high pH and rich in urea (Claus and Fahmy, 1986). A suitable medium comprises ($g\ l^{-1}$); tryptic soy broth, 2.7; yeast extract, 5.0; glucose, 5.0; urea, 10.0; adjusted to pH 8.5 with NaOH and solidified with agar.

## C. Thermophiles

Thermophilic microorganisms can be defined as those that grow above 55°C and of the endospore-forming bacteria this includes about 15 *Bacillus*, 11 *Clostridium* and 7 *Desulfotomaculum* species as well as the thermoactinomycetes (Brock, 1986). As with acidophiles and alkaliphiles, thermophiles occur commonly in soils from temperate zones although their numbers will obviously be higher in thermal sites. Techniques for the study of the ecology of thermophiles are given in detail by Brock (1978).

To isolate thermophiles in the temperature range 40–75°C (which includes all endospore-formers), water baths are satisfactory. At temperatures above 80°C water should be replaced by glycerol. Treatment of samples is important (see Weigel, 1986), in particular samples from hot springs and other thermal sites will cool rapidly affecting the $pO_2$, $pCO_2$, PH, etc. and should therefore be examined as soon as possible. Ward (1978) concluded that there was a major reduction in the population of thermophiles in a sample during cooling. With regard to media, agar (often 2–4% is needed to obtain a reasonable gel) can be used up to 70°C but *C. thermoaceticum* does not grow well at high agar concentrations (Weigel, 1986) and this may apply to other types.

Spores of most thermophiles need a substantial heat shock before germination will occur. For example spores of *C. thermohydrosulfuricum* when heat shocked at 100°C for 20–60 min germinated efficiently, but without the heat shock showed a lag phase of 14 days when incubated at 60°C (Weigel, 1986).

Thermoactinomycetes do not grow readily, if at all, on nutrient agar and the most suitable general medium for their isolation and maintenance is CYC agar (Hobbs and Cross, 1983) comprising ($g\ l^{-1}$); Czapek-Dox agar granules (Oxoid), 45; yeast extract, 2; vitamin-free casamino acids, 6; adjusted to pH 7. This can be made highly selective by the inclusion of novobiocin ($25\ \mu g\ ml^{-1}$) which eliminates thermophilic bacilli.

### D. Psychrophiles

Most endospore-formers that grow at low temperature are best described as psychrotrophic since they grow in the range 0–25°C. This includes several bacilli (Larkin and Stokes, 1966; Laine, 1970) but few clostridia. *C. arcticum* apparently flourishes in arctic soils but the paucity of psychrotrophic clostridia probably reflects the few studies in this area. Since many mesophilic endospore-formers grow, albeit slowly, at 10°C and sometimes lower, a temperature below 5°C is recommended for the isolation of psychrotrophs. Normal media such as tryptone soy agar or broths are sufficient (Larkin and Stokes, 1966).

### E. Nitrogen-fixing and denitrifying species

Free-living, nitrogen-fixing clostridia and bacilli are common in soils and include most of the saccharolytic clostridia, for example *C. butyricum, C. butylicum* and *C. pasteurianum*. Some of the facultatively anaerobic bacilli, notably *B. azotofixans, B. macerans*, and *B. polymyxa* and *Desulfotomaculum* species, are also able to fix nitrogen. Selective isolation of these bacteria can be achieved on nitrogen-free media incubated anaerobically. Some typical formulations are given in Table V.

Unusual denitrifying bacilli have been isolated by Pichinoty *et al.* (1979)

**TABLE V**
Media for isolation of nitrogen-fixing, endospore-forming bacteria

| Ingredient (g) | Modified TB medium[a] | *Clostridium* medium[b] |
|---|---|---|
| Glucose[c] | 5 | 10[d] |
| $K_2HPO_4$ | | 0.8 |
| $KH_2PO_4$ | 0.8 | 0.2 |
| $MgSO_4.7H_2O$ | 0.2 | 0.2 |
| NaCl | | 0.2 |
| $FeSO_4.7H_2O$ | | 0.01 |
| $FeCl_3.6H_2O$ | 0.04 | |
| $MnSO_4.4H_2O$ | | 0.01 |
| $CaCl_2.2H_2O$ | 0.15 | 0.01 |
| $Na_2MbO_42H_2O$ | 0.005 | 0.025 |
| Sodium thioglycollate | 0.5 | 1.0 |
| Thiamin | 0.001 | |
| Biotin | 0.001 | |
| Yeast extract | | 0.01 |
| Trace elements[e] | 1 ml | 1 ml |
| Soil extract | | 10 ml |
| Tap water | To 1 litre pH 7.0 | To 1 litre pH 7.2 |
| Agar | To solidify | |

[a] See Seldin *et al.* (1983).
[b] See Skinner (1971).
[c] Glucose can be replaced by sucrose for clostridia.
[d] Sucrose (15%) is specific for *Cl. pasteurianum*.
[e] Suitable trace element solution contains (each at 50 mg $l^{-1}$); $Na_2B_4O_7.10H_2O$, $CoNO_3.6H_2O$, $CdSO_4.2H_2O$, $CuSO_4.5H_2O$, $ZnSO_4.7H_2O$ and $MnSO_4.H_2O$.

from soils by providing peptone media supplemented with $NO_3$, $NO_2$ or $N_2O$ as electron acceptor and incubating under anaerobic conditions. By using $N_2O$, *B. azotoformans* was isolated (Pichinoty *et al.*, 1976) and a heterogeneous collection of bacilli that were quite different from known species was recovered using NO as electron acceptor (Pichinoty *et al.*, 1979). However these bacteria usually grow on normal media and thus would be recovered in ecological studies.

## III. Recovery of spores

Here we will deal with recovery of vegetative bacteria from spores sampled from the environment. This, when compared with total counts, will give an estimate of the proportion of the population present as vegetative cells. It is unlikely that any one treatment will recover all spores in a sample, so to obtain an estimate of the total spore population will require a variety of treatments. It is also important to consider the media and conditions used for the subsequent germination and growth of the bacteria. These are covered in detail by Norris *et al.* (1981).

## A. Heat treatment

The most common procedure for selectively isolating endospore-formers is heating suspensions in buffer, saline or water for 10 min at 80 or 85°C. This kills all vegetative cells (except some extreme thermophiles) but causes no harm to most endospores (spores from true actinomycetes are usually killed). A common procedure is to transfer 4 g of a soil sample to a beaker, add 20 ml of sterile water and heat in a water bath for 10 min with careful agitation. The soil sample is then ready for inoculation into an enrichment broth or onto a plate. It can be useful to vary the heating conditions for certain groups. Thus spores from thermophiles often have increased thermotolerance and those from psychrophiles lower tolerance. Since no guidelines have been established for optimum pasteurization conditions for different groups much is left to the ingenuity of the investigator.

## B. Air drying

An alternative that has not been used extensively for killing vegetative cells of bacilli or clostridia is dry heat. '*Bacillus xerothermodurans*' was isolated following dry heat treatment of a sandy soil for 48 h at 125°C (Bond and Favero, 1977) and *Thermoactinomyces peptonophilus* was first isolated from dry-heated soils (Nomura and Ohara, 1971). It should be remembered however that prolonged heat treatment of spores is often mutagenic (Hopwood, 1970).

## C. Solvent treatment

Ethanol treatment is a valuable alternative to pasteurization for the selective isolation of spore-formers. Claus has established that exposure to 50% ethanol for at least 1 h effectively kills vegetative cells (see Claus and Berkeley, 1986) and Bond *et al.* (1970) used 95% ethanol for 45 min. Similarly, clostridia have been isolated following treatment of samples 1 : 1 (v/v) with 95% ethanol for 45 min (Koransky *et al.*, 1978). Other solvents have not been investigated in detail but toluene and chloroform could perhaps be useful.

### D. Air sampling

The methods outlined above are suitable for liquid or solid samples. Thermoactinomycete spores however are common in dust and air associated with cereals and other vegetation. Indeed, *T. intermedius* has only been found in such samples (Kurup *et al.*, 1980). An Anderson sampler is therefore an invaluable aid for the selective recovery of thermoactinomycete spores. A common procedure is to air dry a batch of compost or straw and recover the spores from dust preparations using the sampler and plating onto CYC agar containing novobiocin which is selective for most thermoactinomycetes (see Section III.C).

### E. Selective germination

Little attention has been given to selective germination followed by pasteurization as a recovery procedure for specific types of endosporeformers but it may have a lot to offer for the isolation of some of the more obscure species. Wakisaga and Kizumi (1982) described heat treatment of soil samples followed by a brief period of growth and a second pasteurization stage. During the growth period, common and vigorous species such as *B. cereus* and *B. subtilis* germinate and are subsequently killed allowing strains of less rapidly growing species such as *B. polymyxa* and *B. macerans* to predominate. A recent innovation for the selective isolation of *B. thuringiensis* involved pasteurization followed by germination of soil samples in the presence of 0.12 M sodium acetate. This inhibited the germination of *B. thuringiensis* spores, and a second heat treatment eliminated all vegetative cells. The *B. thuringiensis* spores were then allowed to germinate on standard media (Travers *et al.*, 1987). Extrapolation of this methodology using antibiotics or other inhibitory chemicals could provide powerful selective isolation procedures for spore-formers.

## IV. Insect pathogens

The significance of insect pathogens is increasing with the advance of environmental pollution. Many of them can be applied directly against insect pests, others play an important role in the regulation of their numbers in ecosystems. The better we understand the ecology of insect pathogens, the more effectively we should be able to use them for safe control of pest and insect vector populations.This use is based on the host–pathogen relationships. Their antagonism cannot always be assessed as the most effective pathogens die together with the reduction or total disappearance of their host. Because of that, and the vastness of the affected territories, a survey of the intensity of a given disease is of great importance, and this calls for the co-operation of scientists throughout the world.

Beside the widespread pathogens, there exist groups of microorganisms which are found only locally. Their further dissemination is hampered by ecological barriers, but sometimes their apparent rarity is the result of insufficient search. Nevertheless, the potential of such local infections is real and it should not be overlooked as a means of biological control of pest insects. This is another consideration illustrating the importance of ecological studies of insect pathogens. Sometimes, the insect pathogens may have an undesired effect as when, for instance, they attack weed pests. A knowledge of their ecology is crucial for preventing such situations.

Representatives of the genus *Bacillus* are of particular importance among insect pathogens. For more than 20 years, *B. thuringiensis* has been successfully applied against various defoliating insects and more recently mosquitoes, blackflies and beetles. *Bacillus sphaericus* is also a promising candidate as an anti-mosquito preparation. *Bacillus popilliae* is now a classic example for its effective restriction of the scarab larvae outbreaks in the United States since the 1940s. Sometimes, strains of *B. cereus* cause up to 50% mortality of Coleoptera, Hymenoptera and the Lepidoptera larvae. Other representatives of the genus applicable in biological insect control may possibly be found.

The main advantage of these pathogens is their high specificity which makes them safe for human beings and the environment. Indeed, they are a natural component of the environment and their introduction there in large numbers can change some population ratios between microorganisms but it is unlikely to cause a dramatic ecological disbalance similar to that caused by chemical pesticides. From an ecological point of view they are one of the safest control agents. Nevertheless, the sad example of some chemicals whose harmful effect was discovered only after many years should be borne in mind. Scientists have to be on the alert for such unexpected consequences of insect pathogen applications too. In the case of biological preparations such apprehensions are far less justified because of the usually rapid destruction of the spores in the environment and it is this process that needs a more detailed investigation. There are some studies on the persistence of some insect pathogens but data on their concentration and their role in the food chain are scarce.

The ecology of insect pathogens has two aspects: epizootiology (to what extent the host is partially or entirely their environment) and persistence in natural habitats. Both are important for the assessment of a given insect disease. A knowledge of the development and distribution of a disease in the insect population provides an opportunity to forecast epizootic outbreaks thus rendering control treatment of the pest unnecessary.

The methods for studying the ecology of insect pathogens do not differ substantially from routine methods. Only some points will be emphasized here.

### A. Sampling

Sampling is as important for the assessment of a disease as it is for the investigation of the pathogen's persistence. It is described in detail by Atlas and Bartha (1981). In epizootiological studies it is necessary to analyse a series of a sufficient quantity of samples from the whole population. The number of individuals in each sample must be large enough to provide statistically reliable results on the disease development. This can be achieved with a sample of about 100 individuals. As a rule the insect populations are very dense and permit the taking of numerous and large samples, but nevertheless care must be taken not to decrease their number too much. The quantity of the individuals in all samples should not exceed 10% of the whole number of the population. If it does the population will be disturbed and further investigation will not give reliable results. The development of an infectious disease depends on many factors including the population density, the potential ability for the infection to be transferred by contact between a diseased and a healthy insect, the frequency of such contacts, and the degree of contamination of the proper stage of the insect's life cycle. Alterations in the population density change the ways of infection transfer. For a correct estimation of the role of a disease it should be studied for every instar and stage of the insect life cycle.

Samples for assessment of the survival and persistence of a pathogen are taken depending on the goal and insect habitat. They may be parts of food plants (leaves, branches, flowers, etc.), soil samples from different depths (Hertlein *et al.*, 1979), litter, water, stored products, air, etc. Samples must be taken at random from the studied ecosystem in order to ensure full representation. In the case of leaves, circles of equal diameter could be used which are cut from leaves collected at random (Pinnock *et al.*, 1971; Lynch *et al.*, 1976). The most simple method of taking air samples is to expose opened Petri dishes with a convenient medium for some time, or preferably to use an Anderson air sampler. Soil and water samples are taken using special soil corers or water samplers (see the comprehensive review by Atlas and Bartha, 1981).

### B. Detection of pathogens in insects

To assess the distribution and significance of a disease in an insect population, healthy, diseased and dead insects must be studied. Detailed reviews on this topic can be found elsewhere (Weiser, 1966; Bulla *et al.*, 1975; Bucher, 1981). It is necessary to discriminate between normal gut microflora and pathogens. For this reason a number of healthy insects are examined and the type and number of the most common colonies are estimated. Sick and dead insects are also studied and when a new type of colony appears at a high frequency it can be supposed to be the pathogen. For unknown microorganisms, the criteria of Koch's postulates must be met: the organism is always found only in infected insects, can be grown in pure culture, can infect healthy insects with the same symptoms, and can be reisolated from the experimental insects. Individual insects are examined in two steps: a microscopic examination and an attempt at culture. Both should be conducted to yield at least rough estimates of bacterial numbers. Culturing is done by spreading bacteria on the surface of an agar plate so that many will form isolated colonies. Some suitable media for *B. thuringiensis* and *B. sphaericus* are given in Section IV.C.2. This is commonly done using spread plates, or in the case of small insects, after the sterilization of their surface in alcohol the insect is cut and its contents are spread on the agar. In the case of *Bacillus* pathogens an experienced scientist may identify the pathogen with great probability on the basis of colony morphology, their frequency in the material from a diseased or dead insect and microscopic examination of the cells. Phase contrast, dark field and Gram staining may be used for this purpose. Characteristic colonies (large, irregularly round, whitish, like galvanized iron under oblique light) together with free crystals and spores are indicative of *B. thuringiensis*. If a great number of similar round colonies are isolated from dead mosquito larvae and rods forming a spherical spore with terminal swelling of the sporangium are observed, it is almost certainly *B. sphaericus*. *Bacillus popilliae* group can be preliminarily detected by the characteristic 'milky' appearance of the dead larvae. They are obligate pathogens and for their isolation and detection special care must be taken (Bulla *et al.*, 1975; Milner, 1981). The characteristics of the main *Bacillus* pathogens – *thuringiensis, sphaericus* and *popilliae* group – as well as of some others may be found elsewhere (Bucher, 1981).

### C. Detection of the pathogens in the environment

#### *1. Agar plate count*

A given quantity of the sample material is washed or suspended in a volume of sterile saline. The fraction of heavy crude particles is allowed to settle and the supernatant is spread on plates with or without dilution in order to obtain no more than 200 colonies per plate. The medium depends on the organism to be traced. When spores are of main interest, it is best to pasteurize the materials before spreading to eliminate non-sporeformers (see Section II). Selective media in such studies are very helpful. They are based on different antibiotic resistance patterns or the use of unusual nutrients. Several have been developed and are listed below.

### 2. *Selective media for* B. thuringiensis

NPP agar is nutrient agar supplemented with 4 μg $ml^{-1}$ penicillin G and 5 μg $ml^{-1}$ polymyxin. Cultivation at 37°C favours competitive growth of *B. thuringiensis* over the soil organisms which are reduced to 1–10% of their original number (Saleh *et al.*, 1969). Alternatively, nutrient agar can be supplemented with 10 μg $ml^{-1}$ acriflavin. This has been used to select *B. thuringiensis* subsp. *israelensis* from natural waters. A 90% reduction of background bacteria was achieved but such a reduction is insufficient for accurate counts of target populations in typical waters (Kalfon *et al.*, 1986).

A novel method for the isolation of *B. thuringiensis* involves selective germination (Travers *et al.*, 1987) and is claimed to be very effective for detecting low numbers of *B. thuringiensis* in soils (see Section III).

### 3. *Selective media for* B. sphaericus

BATS medium (Yousten *et al.*, 1985) seems to be the most successful among the *B. sphaericus* selective media reported to date. It selects not only against many other *Bacillus* species (8 from 9 tested), but also against many non-pathogenic strains of *B. sphaericus* (68% of the tested strains). BATS medium contains per litre: $Na_2HPO_4$, 5.57 g; $KH_2PO_4$, 2.4 g; $MgSO_4.7H_2O$, 50 mg; $MnCl_2.4H_2O$, 4 mg; $FeSO_4.7H_2O$, 2.8 mg; $CaCl_2.2H_2O$, 1.5 mg; L-arginine, 5 g; thiamine, 20 mg; biotin, 2 μg; streptomycin sulphate, 100 mg; and agar, 20 g. The arginine, biotin, thiamin, and streptomycin are prepared as a filter-sterilized stock solution. The $Mg^{2+}$, $Mn^{2+}$, $Fe^{2+}$ and $Ca^{2+}$ salts prepared as an acidified (0.03% v/v concentration of $H_2SO_4$) autoclaved stock solution, are added to the autoclaved phosphate salts–agar mixture when the latter is cooled to 50°C. Plates are incubated at 30°C for 72 h before counting.

Kalfon *et al.* (1986) found that nutrient agar supplemented with a combination of 100 μg $ml^{-1}$ streptomycin and 25 μg $ml^{-1}$ lincomycin reduced the water microorganisms by 90% and did not affect the *B. sphaericus* population but, as in the case of *B. thuringiensis*, they think that this reduction is insufficient for accurate estimations of *B. sphaericus* in water.

Media based on acetate as the only source of carbon have been developed (White, 1972; Massie *et al.*, 1985); they are less selective for *B. sphaericus*, than BATS medium.

## D. Immunofluorescence

A review of the use of immunofluorescent techniques in microbial ecology was made by Bohool and Schmidt (1980). West *et al.* (1984a) used immunofluorescence as outlined below to detect *B. thuringiensis* in soil.

Microorganisms in soil are stained by reacting with antibody, specific to a desired stage of the life cycle. The antibody is labelled either directly or indirectly (using a secondary antibody) with a fluorescent dye. The method permits detection of both viable and non-viable cells and parasporal crystals in contact with soil, with minimal interference from soil microorganisms.

In a typical procedure (West *et al.*, 1984a), antisera are raised against purified (> 99%) parasporal crystals, autoclaved (121°C, 20 min) spores and heat-killed (95°C, 1 h) vegetative cells. Rabbits are injected intramuscularly with $10^9$ bacterial particles emulsified with Freund's complete adjuvant on day 0 and Freund's incomplete adjuvant on days 7 and 14. Antibody titres are assessed by tube agglutination against homologous and heterologous antigens at weekly intervals from day 28. Spores give a satisfactory response by day 42 but rabbits, immunized with cells and crystals, need an intravenous boost on day 56 to achieve an equivalent response and are bled on day 65.

For incubation in soil, suspensions of purified (99% purity) spores, parasporal crystals, viable or heat-killed log phase vegetative cells in 0.1% soil extract (Kennox and Jurgensen, 1975) are smeared over separate slides and air dried. The dry slides are pressed, treated face down, onto 22 g of air-dried soil, sieved <2 mm particle size and contained in a Petri dish maintained in darkness at 25°C, 100% relative humidity. Slides without *B. thuringiensis* are also set up. At intervals, 2–3 replicate slides are removed from soil and stained. The number of bacterial particles is counted for 50 randomly selected microscope fields.

To generate immunofluorescence, slide-borne antigens are reacted with 20 µl of a 1 in 20 dilution of antiserum in phosphate-buffered saline (PBS; Kawamura, 1977) for 30 min at 30°C, rinsed with PBS and then reacted with 20 µl of 1 in 20 diluted fluorescein isothiocyanate (FITC)-labelled IgG from goat, anti-rabbit IgG for the same period, rerinsed with PBS, and air dried. A mixture of both spore and cell antisera is used to detect any germination and sporulation of *B. thuringiensis* in soil. Slides are mounted with carbonate-buffered glycerol (Kawamura, 1977) and observed at 1000× by epifluorescent illumination.

## E. Detection of the toxic activity

### *1. Biotests*

Both *B. thuringiensis* and *B. sphaericus* produce parasporal bodies consisting of toxins which kill insects. The viable count results are not always correlated with the toxic activity, and in many cases it is advisable to test it separately. In the case of samples from the environment, the material is taken as for viable counts but, instead of plating, the washed or suspended material after sedimentation is added directly to the food of the tested insect. When mortality is observed it should be compared with symptoms caused by the agent to ensure that the correct pathogen is responsible. To study the action of soil on insect pathogen toxins, often inoculation of soil samples with the agent being studied is carried out. In this case at least two samples are investigated in parallel, one of them being autoclaved previously in order to facilitate the estimation of the possible action of other soil organisms on the insect (Lynch *et al.*, 1976; West *et al.*, 1984b).

### *2. Methods based on hybridization probes*

Laborious and expensive biotests have encouraged the development of more direct methods to estimate insect pathogen activity (Cheung and Hammock, 1982; Smith and Ulrich, 1983). The use of ELISA has been studied for standardization of biological preparations of *B. thuringiensis* subsp. *israelensis* and it was found to be 10 times less sensitive than biotests (Thiery, 1986). For its application very pure material is needed which at present hampers its use for ecological studies.

An alternative is to detect the toxin genes themselves in the environment using hybridization probes. If the target is to monitor the perseverence of an applied strain, probe formulation is simple since the cloned toxin gene will probably be available or could be readily prepared. If however, the aim is detection of the total *B. thuringiensis* population in the environment probe design is more complicated. Currently, four toxin genes from *B. thuringiensis* are recognized; *cryA*, the lepidopteran toxin, *cryB*, which specifies a crystal protein toxic for Lepidoptera and Diptera (flies), *cryC*, which codes for the Coleopteran (beetle) toxin and *cryD*, from *B. thuringiensis* var. *israelensis* which is toxic to some mosquitoes and blackflies. Although these genes show some conserved sequences, there is considerable divergence even within single genes. Prefontaine *et al.* (1987) have prepared oligonucleotide probes based on conserved sequences in *cryA* genes. With these probes they could detect homologous sequences in DNA from 13 different serotypes; only subsp. *thompsonii* showed no homology. So the approach shows promise and as more genes are sequenced the specificity of the probes should improve.

Although the toxin gene is the obvious probe target, other sequences can be used with perhaps some advantages. Some rRNA sequences are species specific and have been used as probe targets (Saunders *et al.*, 1988) but the problem here is the likely cross reaction with *B. cereus*.

Assuming a suitable probe, several methods for the detection of microorganisms in the environment have been developed recently. Colwell *et al.* (1988) concentrate the microorganisms from aquatic samples and then extract nucleic acids. Litre quantities of sample are first blended to dissociate organisms from particulates, passed through a series of prefilters and concentrated on a single, cylindrical filter membrane (Millipore SWGS01015). Cell lysis and proteolysis are carried out within the filter housing. Crude, high molecular weight nucleic acid solutions are then drawn off the filter. These solutions can be immediately analysed, concentrated or purified depending on intended application. The method was said to be simple and economical and provided DNA/RNA of concentration and purity equal to or exceeding that of previous methods. In cases where greater sensitivity is needed the polymerase chain reaction (using *taq* polymerase) is a powerful procedure for amplifying target sequences by a factor of more than 10 million within a few hours (Atlas and Stefan, 1988).

A most probable number/hybridization procedure has been described by Fredricksen *et al.* (1988) in which *Tn*5 was placed in soil bacteria and antibiotic resistance used to isolate the bacteria (through an MPN system). Hybridization to the transposon was used to identify the bacteria. Such an approach may be useful for monitoring populations of control agents although the widespread dissemination of antibiotic resistance transposons is ill advised.

## V. Methods for screening for extracellular enzymes

The major habitat of endospore-forming bacteria is the soil where they are responsible for the turnover of organic materials derived from plant and animal tissues. Indeed the degradation of these polymers seems to be the major ecological function of these bacteria. Since the bulk of this material will be macromolecular and therefore too large to enter the cell, these bacteria rely on secreted, extracellular enzymes to solubilize the polymers. It is thus not surprising that the bacilli (Priest, 1977), clostridia (Rogers, 1986; Lovitt *et al.*, 1988) and thermoactinomycetes (Lacey and Cross, 1989) are all well endowed with proteases, carbohydrases and nucleic acid-degrading enzymes. Indeed many of these enzymes are produced commercially for use in industry. It therefore seemed appropriate to include in this chapter a survey of simple methods that are used to detect extracellular enzymes.These methods are based on agar plate cultures and therefore can be used to enumerate, for example, amylase-producing bacteria from an environmental sample (Pretorius *et al.*, 1986).

Methods for the detection of extracellular enzymes rely on (1) the hydrolysis of an insoluble substrate or (2) the staining of a soluble substrate after growth, to reveal a zone of hydrolysis (Table VI). Examples in the first category include skimmed milk-containing agar for the detection of proteases and cell walls (or even autoclaved cells) included in a solidified medium for the detection of lytic enzymes. Sometimes soluble substrates can be rendered insoluble or stained with a suitable dye for the detection of appropriate enzymes. Starch azure is such an example in which amylose or starch is combined with Remazol Brilliant Blue R. When it is hydrolysed the oligosaccharide–azure molecules diffuse from the colony area giving rise to a pale blue or colourless halo. This approach has the twin advantages that colonies are not killed by iodine staining and the method is selective for α-amylases since the covalently bound dye molecules inhibit the action of exo-attacking enzymes such as β-amylase or glucoamylase. We have used the basic method described by Rinderknecht *et al.* (1967) to dye various α-glucans such as pullulan with Reactone Red for the preparation of indicator media. Alternatively, Mahasneh and Stewart (1980) autoclaved the β-1,3-glucan pachyman with aniline blue to prepare a dyed substrate and this method also works with the β-1,4-glucan carboxymethyl cellulose.

The alternative procedure is to stain or precipitate the medium after growth of the bacteria. Starch hydrolysis revealed by flooding with iodine solution or sublimation from solid iodine is a common example. Iodine solutions can also be used for pentosans, in particular xylan (Williams, 1983). Congo red washed with sodium chloride is effective for revealing hydrolysis zones with various β-glucans (Williams, 1983). Precipitation with ethanol is useful for pullulan (Morgan *et al.*, 1979) and some other polysaccharides.

An interesting observation with some marine Gram-negative bacteria was that chitin degradation is associated with the ability to hydrolyse chitobiose (O'Brian and Colwell, 1987). The latter can be detected very simply using 4-methylumbelliferyl-*N*-acetyl-β-D-glucosaminide which is much more convenient than preparing colloidal chitin for inclusion in a solid medium. Perhaps this observation will be extended to other problematic substrates since it is always likely that the hydrolysis products from a polymeric substrate will be further metabolized by the bacterium in question.

**TABLE VI**
Methods for screening for extracellular enzymes

| Enzyme | Substrate ($mg\ ml^{-1}$) | Reagent ($mg\ ml^{-1}$) | Comments | Relevant reference |
|---|---|---|---|---|
| Polysaccharides | | | | |
| Agarose | Agar (15) | – | Depressions around colonies in agar-based media | Hunger and Claus (1978) |
| Alginate lyase | Sodium alginate (10) | – | Clearing of opaque medium | Hansen and Nakemura (1985) |
| Amylase | Starch/glycogen (15) | $I_2$ (1)/KI (2) or $I_2$ vapour | Clear zone indicates α-, red zone β-amylase | Hyun and Zeikus (1985) |
| α-Amylase | Starch azure (5) | – | Clear zone indicates endoacting amylase | Madi *et al.* (1987) |
| Cellulase | Cellulose azure (5) | – | Clear zone indicates hydrolysis of crystalline cellulose | Smith (1977) |
| Endocellulase | Carboxymethyl cellulose (4) | Aqueous congo red (1) or trypan blue (1) washed with 1 M NaCl | Clear zones in stained background | Williams (1983) |
| Chitinase | Colloidal chitin (4) | – | Zone of clearing in opaque background | Hsu and Lockwood (1975) |
| Dextranase | 1,3-α-Glucan<br>1,6-α-Glucan | –<br>– | Zones of clearing in opaque background | Ebisu *et al.* (1975) |
| 1,3-β-Glucanse | Pachyman (4), Lichenen (4), or Laminarin (4) | Aqueous congo red (1), washed with 1 M NaCl | Zones of clearing | Alexander and Priest (unpublished) |
| | Pachyman (5) | Aniline blue (1) | Autoclaved substrate and reagent produce complex that is cleared by the enzyme | Mahasneh and Stewart (1980) |

*continued*

**TABLE VI (cont.)**
Methods for screening for extracellular enzymes

| Enzyme | Substrate (mg ml$^{-1}$) | Reagent (mg ml$^{-1}$) | Comments | Relevant reference |
|---|---|---|---|---|
| 1,6-β-Glucanase | Pustulan (4) | Ethanol or aqueous congo red (1) | Precipitation or staining of native pustulan | Martin *et al.* (1980), Priest (unpublished) |
| Pectin/Poly-galacturonate lyase | Apple pectin (5) or polygalacturonate (5) | Aqueous hexadecyl methylammonium bromide (10) | Clear zone in precipitated background | Hankin *et al.* (1971) |
| Pullalanase | Pullulan (4) | Ethanol at 4°C | Clear zone in precipitated background | Morgan *et al.* (1979) |
| | | Dye pullulan with Reactone red or Mikacion brilliant red | Clear zone | Kanno and Tomimura (1985), Priest (unpublished) |
| Xylanase | Xylan (1) | Gram's Iodine | Clear zone in stained background | Williams (1983) |
| | | Ethanol | Clear zone in precipitated background | Flannigan and Gilmour (1980) |
| Proteolytic enzymes | | | | |
| Proteinases | Skimmed milk (10) | – | Clear zone | Cowan (1974) |
| | Elastin (3) | – | Clear zone | – |
| | Solubilized elastin (10) | 30% trichloracetic acid | Clear zone in precipitated background | Williams *et al.* (1988) |
| | Keratin (3) | – | Clear zone | |
| | Gelatin (4) | Acidic mercuric chloride (120) | Clear zone in precipitated background | Frazier (1926) Cowan (1974) |

*continued*

**TABLE VI (cont.)**
Methods for screening for extracellular enzymes

| Enzyme | Substrate ($mg\ ml^{-1}$) | Reagent ($mg\ ml^{-1}$) | Comments | Relevant reference |
|---|---|---|---|---|
| Nucleases and phosphatases | | | | |
| DNase | DNA (3) | 1 M HCl | Clear zone in precipitated background | Cowan (1974) |
| RNase | RNA (3) | 1 M HCl | Clear zone in precipitated background | Cowan (1974) |
| | RNA (15) | Acridine orange (0.01) | Autoclave together to produce complex, view under U.V. light | Lanyi and Lederberg (1966) |
| Phosphatase | Phenolphthalein diphosphate (0.1) | $NH_4OH$ | Invert over $NH_4OH$ to reveal pink/red colonies | Cowan (1974) |
| Lipases | | | | |
| Lipase | Tributyrin (10) | – | Clear zone in opaque background | Cowan (1974) |
| | Tween (10) | – | Opaque halo around growth | Sierra (1958) |
| Phospholipase | Egg yolk emulsion (5% w/v in medium) | – | Precipitation around colonies | Cowan (1974) |
| Phospholipase | Egg yolk emulsion | Aqueous copper sulphate | Precipitates free fatty acids | Cowan (1974) |
| Lytic enzymes | | | | |
| Cell wall lytic enzymes | Cell wall preparations or whole cells | – | Zone of clearing in opaque background | Priest (unpublished) |

## VI. Conclusions

Enodospore-forming bacteria are important ecologically and commercially. In the case of the insect pathogens these two factors overlap and the result is that the ecology of *B. thuringiensis* and *B. sphaericus* is reasonably well known. Similarly the alkaliphiles occupy a unique niche in the enzyme industry and the pressure for new strains prompted studies of their ecology. For the other representatives (with the exception of the human and animal pathogens) the ecology is less well understood but it is to be hoped that as an appreciation of these bacteria in industry grows, new and unusual isolates will be required and this will encourage further studies of the ecology of these fascinating bacteria.

## References

Atlas, R. M. and Bartha, R. (1981). "Microbial Ecology. Fundamentals and Applications", pp. 81–91, Addison-Wesley Publishing Company, Reading, Mass.

Atlas, R. M. and Steffan, R. J. (1988). *In* "The Release of Genetically Engineered Microorganisms" (M. Sussman, C. H. Collins, F. A. Skinner and D. E. Stewart-Tull, Eds), pp. 224–226, Academic Press, London.

Berkeley, R. C. W., Logan, N. A., Shute, L. A. and Capey, A. G. (1987). *Meth. Microbiol.* **16,** 291–328.

Bond, W. W. and Favero, M. S. (1977). *Int. J. Syst. Bacteriol.* **26,** 427–441.

Bond, W. W., Favero, M. S, Peterson, N. J. and Marshall, J. M. (1970). *Appl. Microbiol.* **20,** 573–578.

Bohool, B. B. and Schmidt, E. L. (1980). *Adv. Microb. Ecol.* **4,** 203–241.

Bonjour, F., Graber, A. and Aragno, M. (1988). *Microbiol. Ecol.* **16,** 331–337.

Brock, T. D. (1978). "Thermophilic Microorganisms and Life at High Temperatures", Springer Verlag, New York.

Brock, T. D. (1986). *In* "Thermophiles: General Molecular and Applied Microbiology" (T. D. Brock, Ed.), pp. 1–16, Wiley, New York.

Bucher, G. E. (1981). *In* "Microbial Control of Pests and Plant Diseases 1970–1980" (H. D. Burges, Ed.), pp. 7–33, Academic Press, London.

Bulla, L. A. Jr., Rhodes, R. A. and Julian, G. St. (1975). *Ann. Rev. Microbiol.* **29,** 163–190.

Bunt, J. S. and Rovira, A. D. (1956). *J. Soil Sci.* **6,** 119–128.

Cato, E. P., George, W. L. and Finegold, S. M. (1986). *In* "Bergey's Manual of Systematic Bacteriology", Vol. 2 (P. H. A. Sneath, N. S. Mair and M. E. Sharpe, Eds.), pp. 1141–1200, Williams & Wilkins, Baltimore, Maryland.

Cheung, P. Y. K. and Hammock, B. D. (1982). Annual Report Univ. California, Mosquito Control Res., pp. 66–67.

Claus, D. and Berkeley, R. C. W. (1986). *In* "Bergey's Manual of Systematic Bacteriology", Vol. 2 (P. H. A. Sneath, N. S. Mair and M. E. Sharpe, Eds), pp. 1105–1140, Williams & Wilkins, Baltimore, Maryland.

Claus, D. and Fahmy, F. (1986). *In* "Bergey's Manual of Systematic Bacteriology", Vol. 2 (P. H. A. Sneath, N. S. Mair and M. E. Sharpe, Eds), pp. 1202–1206, Williams and Wilkins, Baltimore, Maryland.

Colwell, R. R., Somerville, C., Knight, I. and Straube, W. (1988). *In* "The Release of Genetically Engineered Microorganisms" (M. Sussman, C. H. Collins, F. A. Skinner and D. E. Stewart-Tull, Eds), pp. 47–60, Academic Press, London.

Cowan, S. T. (1974). "Cowan and Steele's Manual for the Identification of Medical Bacteria", Cambridge University Press, Cambridge.

Cross, T.A. (1980). *J. Appl. Bacteriol.* **50,** 397–423.

Darland, G. and Brock, T. D. (1971). *J. Gen. Microbiol.* **67,** 9–15.

Deinhard, G., Blanz, P., Poralla, K. and Altan, E. (1987a). *System Appl. Microbiol.* **10,** 47–53.

Deinhard, G., Saar, J., Krischke, W. and Poralla, K. (1987b). *System Appl. Microbiol.* **10,** 68–73.

Ebisu, S., Kato, K., Kotani, S. and Misaki, A. (1975). *J. Bacteriol.* **124,** 1489–1501.

Fægri, A., Torsrik, L. and Goksoyr, J. (1977). *Soil Biol. Biochem.* **9,** 105–112.

Flannigan, B. and Gilmour, L. A. (1980). *Mycologia* **72,** 1219–1221.

Frazier, N. C. (1926). *J. Infect. Dis.* **39,** 302–309.

Fredricksen, J. K., Bezdicek, D. F., Brockman, F. J. and Li, S. W. (1988). *Appl. Environ. Microbiol.* **54,** 446–453.

Gibson, T. and Gordon, R. E. (1974). *In* "Bergey's Manual of Determinative Bacteriology", 8th Edn (R. E. Buchanan and N. E. Gibbons, Eds), pp. 529–550, Williams & Wilkins, Baltimore, Maryland.

Gordon, R. E., Haynes, W. C. and Pang, C. H-N. (1973). "The Genus Bacillus", United States Department of Agriculture, Washington D.C.

Gottschalk, G., Andreeson, J. R. and Hippe, H. (1981). *In* "The Prokaryotes; a Handbook on Habitats, Isolation and Identification" Vol. 2 (M. P. Starr, H. Stolp, H. G. Trüper, A. Balows and H. G. Schlegel, Eds), pp. 1767–1780, Springer Verlag, Berlin.

Hankin, L., Zucker, M. and Sands, D. C. (1971). *Appl. Microbiol.* **22,** 205–209.

Hansen, J. B. and Nakemura, L. K. (1985). *Appl. Environ. Microbiol.* **49,** 1019–1021.

Hertlein, B. C., Levy, R. and Miller, T. W. Jr. (1979). *J. Invertbr. Pathol.* **33,** 217–221.

Hippchen, B., Roll, A. and Poralla, K. (1981). *Arch. Microbiol.* **129,** 53–55.

Hobbs, G. and Cross, T. (1983). *In* "The Bacterial Spore", Vol. 2 (A.Hurst and G. W. Gould, Eds), pp. 49–78, Academic Press, London.

Hopwood, D. A. (1970). *Meth. Mircobiol.* **3A,** 364–434.

Horikoshi, K. and Akiba, T. (1982). "Alkalophilic Microorganisms, A New Microbial World", Springer, Berlin.

Hsu, S.C. and Lockwood, J. L. (1975). *Appl. Microbiol.* **29,** 422–426.

Hunger, W. and Claus, D. (1978). *Ant. Leuwenhoek.* **44,** 105–112.

Hyun, H. H. and Zeikus, J. G. (1985). *Appl. Environ. Microbiol.* **19,** 1162–1167.

Jensen, B. F. and Norman, B. E. (1984). *Process Biochem.* **9,** 129–134.

Kalfon, A., Lugten, M. and Margalit, J. (1986). *Appl. Microbiol. Biotechnol.* **24,** 240–243.

Kanno, M. and Tomimura, E. (1985). *Agric. Biol. Chem.* **49,** 1529–1530.

Kawamura, A. (1977). "Fluorescent Antibody Techniques and Their Applications", 2nd Edn, Univ. of Tokio Press, Japan and Univ. Park Press, Baltimore, Md.

Kennox, L. K. and Jurgensen, M. F. (1975). *Biol. Sol.* **21,** 9–13.
Koransky, J. R., Allen, S. D. and Dowell, Jr., V. R. (1978). *Appl. Environ. Microbiol.* **35,** 762–765.
Kurup, V. P., Hollick, G. E. and Pagan, E. F. (1980). *Science-Cierna Boletin Cientifico del Sur* **7,** 104–108.
Lacey, J. and Cross, T. (1989). *In* "Bergey's Manual of Systematic Bacteriology", Vol. 3 (S. T. William, Ed.), Williams & Wilkins, Baltimore, Maryland. In press.
Laine, J. J. (1970). *Ann. Acad. Sci. Fenn. Ser. A, IV Biol.* **169,** 1–36.
Lanyi, J. K. and Lederberg, J. (1966). *J. Bacteriol.* **92,** 1469–1472.
Larkin, J. M. and Stokes, J. L. (1966). *J. Bacteriol.* **91,** 1667–1671.
Logan, N. A. and Berkeley, R. C. W. (1984). *J. Gen. Microbiol.* **130,** 1871–1882.
Lovitt, R. W., Kim, B. H., Shen, G-J. and Zeikus, J. G. (1988). *CRC Crit. Rev. Biotechnol.* **7,** 107–186.
Lynch, R. S., Lewis, L. C. and Brindley, T. A. (1976). *J. Invertebr. Pathol.* **27,** 325–331.
Madi, E., Antranikau, G., Ohmiya, K. and Gottschalk, G. (1987). *Appl. Environ. Microbiol.* **53,** 1661–1667.
Mahasneh, A. M. and Stewart, D. J. (1980). *J. Appl. Bacteriol.* **48,** 457–458.
Martin, D. F., Priest, F. G., Todd, C. and Goodfellow, M. (1980). *Appl. Environ. Microbiol.* **40,** 1136–1138.
Massie, J., Roberts, G. and White, P. J. (1985). *Appl. Environ. Microbiol.* **49,** 1478–1481.
Matches, J. R. and Liston, J. (1974). *Can. J. Microbiol.* **20,** 1–12.
Milner, R.J. (1981). *In* "Microbial Control of Pests and Plant Diseases. 1970–1980" (W. D. Burges, Ed.), pp. 45–60, Academic Press, London.
Morgan, F. J., Adams, K. R. and Priest, F. G. (1979). *J. Appl. Bacteriol.* **46,** 291–294.
Noeth, C., Britz, T. J. and Joubert, W. A. (1988). *Microbiol. Ecol.* **16,** 233–240.
Nomura, H. and Ohara, Y. (1971). *J. Ferment. Technol.* **49,** 895–903.
Norris, J. R., Berkeley, R. C. W., Logan, N. and O'Donnell, A. G. (1981). *In* "The Prokaryotes; a Handbook on Habitats, Isolation and Identification", Vol. 2 (M. P. Starr, H. Stolp, H. G. Trüper, A. Balows and H.G. Schlegel, Eds), pp. 1711–1742, Springer Verlag, Berlin.
O'Brian, M. and Colwell, R. R. (1987). *Appl. Environ. Microbiol.* **53,** 1718–1720.
Olsen, R. A. and Bakken, L. R. (1987). *Microbiol. Ecol.* **13,** 75–87.
Parkinson, D., Gray, T. R. G. and Williams, S. T. (1971). "Methods for Studying the Ecology of Soil Bacteria", IBP Handbook No. 19, Blackwells, Oxford.
Pichinoty, F., de Barjac, H., Mandel, M., Greenway, B. and Garcia, J-L. (1976). *Ann. Microbiol.* **127B,** 351–361.
Pichinoty, F., Mandel, M. and Garcia, J-L. (1979). *J. Gen. Microbiol.* **115,** 419–430.
Pinnock, D. E., Brand, R. J. and Milstead, J. E. (1971). *J. Invertebr. Pathol.* **18,** 405–411.
Prefontaine, G., Fast, P., Lau, P. C. K., Hefford, M. A., Manna, Z. and Brouseau, R. (1987). *Appl. Environ. Microbiol.* **53,** 2808–2814.
Pretorius, I. S., DeKock, M. J., Britz, T. J., Potgeiter, H. J. and Lategan, P. M. (1986). *J. Appl. Bacteriol.* **60,** 351–360.
Priest, F. G. (1977). *Bacteriol. Rev.* **41,** 711–753.
Priest, F. G. (1989). *In* "Handbooks of Biotechnology", Vol. 2 *Bacillus* (C. R. Harwood, Ed.), Plenum Press, New York. In press.

Priest, F. G. and Alexander, B. (1988). *J. Gen. Microbiol.* **134,** 3011–3018.
Priest, F. G., Goodfellow, M. and Todd, C. (1988). *J. Gen. Microbiol.* **134,** 1847–1882.
Rinderknecht, M., Wilding, P. and Haverbuck, B. J. (1967). *Experientia* **23,** 805–810.
Rogers, P. (1986). *Adv. Appl. Microbiol.* **31,** 1–60.
Saleh, S. M., Harris, R. T. and Allen, O. N. (1969). *Can. J. Microbiol.* **15,** 1101–1104.
Saunders, N. A., Harrison, T. G., Kachwalla, N. and Taylor, A. G. (1988). *J. Gen. Microbiol.* **134,** 2363–2374.
Seldin, L., Van Elsas, J. D. and Penido, G. C. (1983). *Plant Soil* **70,** 243–255.
Setlow, P. (1988). *Ann. Rev. Microbiol.* **42,** 319–338.
Sierra, G. (1958). *Ant. Leuvenhoek.* **23,** 15–22.
Skinner, F. A. (1971). *In* "Isolation of Anaerobes" (D. A. Shapton and R. G. Board, Eds), pp. 57–78, Academic Press, London.
Slepecky, R.A. (1972). *In* "Spores V" (H. O. Halvorsen, R. Hanses and L. L. Campbell, Eds), pp. 297–309, American Society for Microbiology, Washington, D.C.
Slepecky, R. A. and Leadbetter, E. R. (1983). *In* "The Bacterial Spore", Vol. 2 (A. Hurst and G. W. Gould, Eds), pp. 79–101, Academic Press, London.
Smith, R. A. (1977). *Appl. Environ. Microbiol.* **83,** 480–481.
Smith, R. A. and Ulrich, J. I. (1983). *Appl. Environ. Microbiol.* **45,** 586–590.
Stackebrandt, E. and Woese, C. R. (1981). *Symp. Soc. Gen. Microbiol.* **32,** 1–31.
Stackebrandt, E., Lundwig, W., Weizenegger, M., Dorn, S., McGill T. J., Fox, G. E., Woese, C. R., Schubert, W. and Schleifer, K. H. (1987). *J. Gen. Microbiol.* **133,** 2523–2529.
Sullivan, N. M., Mills, D. C., Riemann, H. P. and Arnon, S. S. (1987). *Appl. Environ. Microbiol.* **53,** 2680–2682.
Thiery, I. (1986). *In* "Fundamental and Applied Aspects of Invertebrate Pathology" (R .A. Samson, J. M. Vlak and D. Peters, Eds), pp. 677–681, Foundation of the Fourth Int. Colloq. of Invertebr. Pathol., Wageningen. Netherlands.
Travers, R. S., Martin, P. A. W. and Reichelderfer, C. F. (1987). *Appl. Environ. Microbiol.* **53,** 1263–1266.
Wakisaga, Y. and Kizumi, K. (1982). *J. Antibiotics* **35,** 450–457.
Ward, D. M. (1978). *Appl. Environ. Microbiol.* **35,** 1019–1026.
Wiegel, J. (1986). *In* "Thermophiles: General Molecular and Applied Microbiology" (T. D. Brock, Ed.), pp. 17–38, Wiley, New York.
Weiser, J. (1966). "Nemoci hmyzu", Academia, Praha.
West, A. W., Burges, H. D., White, R. J. and Wyborn, C. H. (1984a). *J. Invertebr. Pathol.* **44,** 128–133.
West, A. W., Crook, N. E. and Burges, H. D. (1984b). *J. Invertebr. Pathol.* **43,** 150–155.
White, R. J. (1972). *J. Gen. Microbiol.* **71,** 505–515.
Williams, A. G. (1983). *FEMS Letts.* **20,** 253–258.
Williams, K., Phillipi, K. D. and Willis, A. T. (1988). *Letts. Appl. Microbiol.* **7,** 173–176.
Yousten, A. A., Fretz, S. B. and Jelley, S. A. (1985). *Appl. Environ. Microbiol.* **49,** 1532–1533.

# Index

**Note:** Page numbers for figures and tables are in *italic*.

## N

# Contents of published volumes

**Volume 3A**

*S. P. Lapage, Jean E. Shelton* and *T. G. Mitchell.* Media for the maintenance and preservation of bacteria
*S. P. Lapage, Jean E. Shelton, T. G. Mitchell* and *A. R. Mackenzie.* Culture collections and the preservation of bacteria
*E. Y. Bridson* and *A. Brecker.* Design and formulation of microbial culture media
*D. W. Ribbons.* Quantitative relationships between growth media constituents and cellular yields and composition
*H. Veldkamp.* Enrichment cultures of prokaryotic organisms
*David A. Hopwood.* The isolation of mutants
*C. T. Calam.* Improvement of micro-organisms by mutation, hybridization and selection

**Volume 3B**

*Vera G. Collins.* Isolation, cultivation and maintenance of autotrophs
*N. G. Carr.* Growth of phototrophic bacteria and blue-green algae
*A. T. Willis.* Techniques for the study of anaerobic, spore-forming bacteria
*R. E. Hungate.* A roll tube method for cultivation of strict anaerobes
*P. N. Hobson.* Rumen bacteria
*Ella M. Barnes.* Methods for the gram-negative non-sporing anaerobes
*T. D. Brock* and *A. H. Rose.* Psychrophiles and thermophiles
*N. E. Gibbons.* Isolation, growth and requirements of halophilic bacteria
*John E. Peterson.* Isolation, cultivation and maintenance of the myxobacteria
*R. J. Fallon* and *P. Whittlestone.* Isolation. cultivation and maintenance of mycoplasmas
*M. R. Droop.* Algae
*Eve Billing.* Isolation. growth and preservation of hacteriophages

**Volume 4**

*C. Booth.* Introduction to general methods
*C. Booth.* Fungal culture media
*D. M. Dring.* Techniques for microscopic preparation
*Agnes H. S. Onions.* Preservation of fungi
*F. W. Beech* and *R. R. Davenport.* Isolation, purification and maintenance of yeasts
*G. M. Waterhouse.* Phycomycetes
*E. Punithalingham.* Basidiomycetes: Heterobasidiomycetidae
*Roy Watling.* Basidiomycetes: Homobasidiomycetidae
*M. J. Carlile.* Myxomycetes and other slime moulds
*D. H. S. Richardson.* Lichens
*S. T. Williams* and *T. Cross.* Actinomycetes
*E. B. Gareth Jones.* Aquatic fungi
*R. R. Davies.* Air sampling for fungi, pollens and bacteria
*George L. Barron.* Soil fungi
*Phyllis M. Stockdale.* Fungi pathogenic for man and animals: 1. Diseases of the keratinized tissues
*Helen R. Buckley.* Fungi pathogenic for man and animals: 2. The subcutaneous and deep-seated mycoses
*J. L. Jinks* and *J. Croft.* Methods used for genetical studies in mycology
*R. L. Lucas.* Autoradiographic techniques in mycology

*T. F. Preece*. Fluorescent techniques in mycology
*G. N. Greenhalgh* and *L. V. Evans*. Electron microscopy
*Roy Watling*. Chemical tests in agaricology
*T. F. Preece*. Immunological techniques in mycology
*Charles M. Leach*. A practical guide to the effects of visible and ultraviolet light on fungi
*Julio R. Villanueva* and *Isabel Garcia Acha*. Production and use of fungi protoplasts

**Volume 5A**
*L. B. Quesnel*. Microscopy and micrometry
*J. R. Norris* and *Helen Swain*. Staining bacteria
*A. M. Paton* and *Susan M. Jones*. Techniques involving optical brightening agents
*T. Iino* and *M. Enomoto*. Motility
*R. W. Smith* and *H. Koffler*. Production and isolation of flagella
*C. L. Oakley*. Antigen-antibody reactions in microbiology
*P. D. Walker, Irene Batty* and *R. O. Thomson*. The localization of bacterial antigens by the use of fluorescent and ferritin labelled antibody techniques
*Irene Batty*. Toxin-antitoxin assay
*W. H. Kingham*. Techniques for handling animals
*J. De Ley*. The determination of the molecular weight of DNA per bacterial nucloid
*J. De Ley*. Hybridization of DNA
*J. E. M. Midgley*. Hybridization of microbial RNA and DNA
*Elizabeth Work*. Cell walls

**Volume 5B**
*D. E. Hughes. J. W. T. Wimpenny* and *D. Lloyd*. The disintegration of micro-organisms
*J. Sykes*. Centrifugal techniques for the isolation and characterization of sub-cellular components from bacteria
*D. Herbert, P. J. Phipps* and *R. E. Strange*. Chemical analysis of microbial cells
*I. W. Sutherland* and *J. F. Wilkinson*. Chemical extraction methods of microbial cells
*Per-Åke-Albertsson*. Biphasic separation of microbial particles
*Mitsuhiro Nozaki* and *Osamu Hayaishi*. Separation and purification of proteins
*J. R. Sargent*. Zone electrophoresis for the separation of microbial cell components
*K. Hannig*. Free-flow electrophoresis
*W. Manson*. Preparative zonal electrophoresis
*K. E. Cooksey*. Disc electrophoresis
*O. Vesterberg*. Isolectric focusing and separation of proteins
*F. J. Moss, Pamela A. D. Rickard* and *G. H. Roper*. Reflectance spectrophotometry
*W. D. Skidmore* and *E. L. Duggan*. Base composition of nucleic acids

**Volume 6A**
*A. J. Holding* and *J. G. Collee*. Routine biochemical tests
*K. Kersters* and *J. de Ley*. Enzymic tests with resting cells and cell-free extracts
*E. A. Dawes, D. J. McGill* and *M. Midgley*. Analysis of fermentation products
*S. Dagley* and *P. J. Chapman*. Evaluation of methods to determine metabolic pathways
*Patricia H, Clarke*. Methods for studying enzyme regulation

*G. W. Gould.* Methods for studying bacterial spores
*W. Heinen.* Inhibitors of electron transport and oxidative phosphorylation
*Elizabeth Work.* Some applications and uses of metabolite analogues in microbiology
*W. A. Wood.* Assay of enzymes representative of metabolic pathways
*H. C. Reeves, R. Rabin, W. S. Wegener* and *S. J. Ajl.* Assays of enzymes of the tricarboxylic acid and glyoxylate cycles
*D. T. Gibson.* Assay of enzymes of aromatic metabolism
*Michael C. Scrutton.* Assays of enzymes of $CO_2$ metabolism

**Volume 6B**
*J. L. Peel.* The use of electron acceptors, donors and carriers
*R. B. Beechley* and *D. W. Ribbons.* Oxygen electrode measurements
*D. G. Nicholls* and *P. B. Garland.* Electrode measurements of carbon dioxide
*G. W. Crosbie.* Ionization methods of counting radio-isotopes
*J. H. Hash.* Liquid scintillation counting in microbiology
*J. R. Quayle.* The use of isotopes in tracing metabolic pathways
*C. H. Wang.* Radiorespirometric methods
*N. R. Eaton.* Pulse labelling of micro-organisms
*M. J. Allen.* Cellular electrophysiology
*W. W. Forrest.* Microcalorimetry
*J. Marten.* Automatic and continuous assessment of fermentation parameters
*A. Ferrari* and *J. Marten.* Automated microbiological assay
*J. R. Postgate.* The acetylene reduction test for nitrogen fixation

**Volume 7A**
*G. C. Ware.* Computer use in microbiology
*P. H. A. Sneath.* Computer taxonomy
*H. F. Dammers.* Data handling and information retrieval by computer
*M. Roberts* and *C. B. C. Boyce.* Principles of biological assay
*D. Kay.* Methods for studying the infectious properties and multiplication of bacteriophage
*D. Kay.* Methods for the determination of the chemical and physical structure of bacteriophages
*Anna Mayr-Harting, A. J. Hedges* and *R. C. W. Berkeley.* Methods for studying bacteriocins
*W. R. Maxted.* Specific procedures and requirements for the isolation, growth and maintenance of the L-phase of some microbial groups

**Volume 7B**
*M. T. Parker.* Phage-typing of *Staphylococcus aureus*
*D. A. Hopwood.* Genetic analysis in micro-organisms
*J. Meyrath* and *Gerda Suchanek.* Inoculation techniques—effects due to quality and quantity of inoculum
*D. F. Spooner* and *G. Sykes.* Laboratory assessment of antibacterial activity
*L. B. Quesnel.* Photomicrography and macrophotography

**Volume 8**
*L. A. Bulla Jr, G. St Julian, C. W. Hesseltine* and *F. L. Baker.* Scanning electron microscopy

*H. H. Topiwala.* Mathematical models in microbiology
*E. Canale-Parola.* Isolation, growth and maintenance of anaerobic free-living spirochetes
*O. Felsenfeld, Borrelia*
*A. D. Russell, A. Morris* and *M. C. Allwood.* Methods for assessing damage to bacteria induced by chemical and physical agents
*P. J. Wyatt.* Differential light scattering techniques for microbiology

**Volume 9**
*R. R. Watson.* Substrate specificities of aminopeptidases: a specific method for microbial differentiation
*C.-G. Heden, T. Illeni* and *I. Kuhn.* Mechanized identification of micro-organisms
*D. B. Drucker.* Gas-liquid chromatographic chemotaxonomy
*K. G. Lickfield.* Transmission electron microscopy of bacteria
*P. Kay.* Electron microscopy of small particles, macromolecular structures and nucleic acids
*M. P. Starr* and *H. Stolp. Bdellovibrio* methodology

**Volume 10**
*T. Meitert* and *Eugenia Meitert.* Usefulness. applications and limitations of epidemiological typing methods to elucidate nosocomial infections and the spread of communicable diseases
*C. A. J. Ayliffe.* The application of typing methods to nosocomial infections
*J. R. W. Govan.* Pyocin typing of *Pseudomonas aeruginosa*
*B. Lányi* and *T. Bergan.* Serological characterization of *Pseudomonas aeruginosa*
*T. Bergan.* Phage typing of *Pseudomonas aeruginosa*
*N. B. McCullough.* Identification of the species and biotypes within the genus *Brucella*
*Karl-Axel Karlsson.* Identification of *Francisella tularensis*
*T. Omland.* Serotyping of *Haemophilus influenzae*
*E. L. Biberslein.* Biotyping and serotyping of *Pasteurella haemolytica*
*Shigeo Namioka. Pasteurella multocida*—Biochemical characteristics and serotypes
*Neylan A. Vedros.* Serology of the meningococcus
*Dan Danielsson* and *Johan Maeland.* Serotyping and antigenic studies of *Neisseria gonorrhoeae*
*B. Wesley Catlin.* Characteristics and auxotyping of *Neisseria gonorrhoeae*

**Volume 11**
*Frits Ørskov* and *Ida Ørskov.* Serotyping of *Enterobacteriaceae*, with special emphasis on K-antigen determination
*R. R. Gillies.* Bacteriocin typing of *Enterobacteriaceae*
*H. Milch.* Phage typing of *Escherichia coli*
*P. A. M. Guinée* and *W. J. van Leeuwen.* Phage typing of Salmonella
*S. Ślopek.* Phage typing of *Klebsiella*
*W. H. Traub.* Bacteriocin typing of clinical isolates of *Serratia marcescens*
*T. Bergan.* Phage typing of *Proteus*
*H. Dikken* and *E. Kmety.* Serological typing methods of leptospires

**Volume 12**
*S. D. Henriksen.* Serotyping of bacteria

*E. Thal*. The identification of *Yersinia pseudotuberculosis*
*T. Bergan*. Bacteriophage typing of *Yersinia enterocolitica*
*S. Winblad*. *Yersinia enterocolitica* (synonyms *"Paxteurella X" Bacterium enterocoliticum* for serotype O–8)
*S. Mukerjee*. Principles and practice of typing *Vibrio cholerae*
*H. Brandis*. Vibriocin typing
*P. Oeding*. Genus *Staphylococcus*
*J. Rotta*. Group and type (groups A and B) identification of haemolytic streptococci
*H. Brandis*. Bacteriocins of streptococci and bacteriocin typing
*Erna Lund* and *J. Henricksen*. Laboratory diagnosis, serology and epidemiology of *Streptococcus pneumoniae*

**Volume 13**
*D. E. Mahony*. Bacteriocin, bacteriophage, and other epidemiological typing methods for the genus *Clostridium*
*H. P. R. Seeliger* and *K. Höhne*. Serotyping of *Listeria monocytogenes* and related species
*I. Stoev*. Methods of typing *Erysipelothrix insidiosa*
*A. Saragea, P. Maximescu* and *E. Meitert*. *Corynebacterium diphtheriae*. Microbiological methods used in clinical and epidemiological investigations
*T. Bergan*. Bacteriophage typing of *Shigella*
*M. A. Gerencser*. The application of fluorescent antibody techniques to the identification of *Actinomyces* and *Arachnia*
*W. B. Schaefer*. Serological identification of atypical mycobacteria
*W. B. Redmond, J. H. Bates* and *H. W. Engel*. Methods of bacteriophage typing of mycobacteria
*E. A. Freund. H. Ernø* and *R. M. Lemecke*. Identification of mycoplasmas
*U. Ullmann*. Methods in *Campylobacter*

**Volume 14**
*T. Bergan*. Classification of Enterobacteriacea
*F. Ørskov* and *I. Ørskov*. Serotyping of *Escherichia coli*
*W. H. Ewing* and *A. A. Lindberg*. Serology of the *Shigella*
*I. Ørskov* and *F. Ørskov*. Serotyping of *Klebsiella*
*R. Sakazaki*. Seriology of *Enterobacter* and *Hafnia*
*P. Larsson*. Serology of *Proteus mirabilis* and *Proteus vulgaris*

**Volume 15**
*A. A. Lindberg* and *L. Le Minor*. Serology of *Salmonella*
*B. Lanyi*. Biochemical and serological characterization of *Citrobacter*
*T. L. Pitt* and *Y. J. Erdman*. Serological typing of *Serratia marcescens*
*R. Sakazaki*. Serological typing of *Edwardsiella tarda*
*M. B. Slade* and *A. I. Tiffen*. Biochemical and serological characterization of *Erwinia*
*G. Kapperud* and *T. Bergan*. Biochemical and serological characterization of *Yersinia enterocolitica*
*T. Bergan* and *K. Sirheim*. Gas-liquid chromatography for the assay of fatty acid composition in Gram-negative bacilli as an aid to classification

**Volume 16**

*J. Stringer*. Phage typing of *Streptococcus agalactiae*
*M. J. Corbel*. Phage typing of *Brucella*
*T. Tsuchiya, M. Taguchi, Y. Fukazawa* and *T. Shinoda*. Serological characterization of yeasts as an aid in identification and classification
*M. Popoff* and *R. Lallier*. Biochemical and serological characteristics of *Aeromonas*
*E. I. Garvie*. The separation of species of the Genus Leuconostoc and the differentiation of the Leuconostocs from other lactic acid bacteria
*T. Bergan, R. Solberg* and *O. Solberg*. Fatty acid and carbohydrate cell composition in Pediococci and Aerobocci identification of related species
*A. Bauernfeind*. Epidemiological typing of *Klebsiella* by Bacteriocins
*U. Berger*. Serology of Non-Gonococcal. Non-Meningococcal *Neisseria*
*C. R. Carter*. Serotyping of *Pasteurella Multocida*
*R. Sakazaki*. Serology and epidemiology of *Plesiomonas shigelloid*
*R. Sakazaki* and *T. J. Donovan*. Serology and epidemiology of *Vibrio cholerae* and *fibrio mimicus*
*R. C. W. Berkekley. N. A. Logan, L. A. Shute* and *A. C. Capey*. Identification of *Bacillus* species
*H. Gyllenberg*. Automated identification of bacteria. An overview and examples
*K. Holmberg* and *C. E. Nord*. Application of numerical taxonomy to the classification and identification of microaerophilic Actinomycetes
*K. B. Doving*. Stimulus space in olfaction

**Volume 17**

*P. M. Bennett* and *J. Grinsted*. Introduction
*V. A. Stanisich*. Identification and analysis of plasmids at the genetic level
*N. Willetts*. Conjugation
*J. R. Saunders, A. Docherty* and *G. O. Humphreys*. Transformation of bacteria by plasmid DNA
*L. Caro, G. Churchward* and *M. Chandler*. Study of plasmid replication *in vivo*
*J. Grinsted* and *P. M. Bennett*. Isolation and purification of plasmid DNA
*H. J. Burkardt* and *A. Pühler*. Electron microscopy of plasmid DNA
*W. P. Diver* and *J. Grinsted*. Use of restriction endonucleases
*C. M. Thomas*. Analysis of clones
*T. J. Foster*. Analysis of plasmids with transposons
*P. M. Bennett*. Detection of transposable elements on plasmids
*G. Doughan* and *M. Kehoe*. The minicell system as a method for studying expression from plasmid DNA
*N. L. Brown*. DNA sequencing

**Volume 18**

*John L. Johnston*. Determination of DNA base composition
*John L. Johnston*. DNA reassociation and RNA hybridisation of bacterial nucleic acids
*E. Stackebrandt, W. Ludwig* and *G. E. Fox*. 16S ribosomal RNA oligonucleotide cataloguing
*August Böck*. Analysis of ribosomal proteins by two-dimensional gel electrophoresis
*Karl Heinz Schleifer*. Analysis of the chemical composition and primary structure of murein

**Volume 19**



**Volume 20**

*Horst-Dietmar Tauschel*. Localization of bacterial enzymes by electron microscopic cytochemistry as demonstrated for the polar organelle
*Jochen R. Golecki*. Electron microscopy of isolated microbial membranes
*Frank Mayer* and *Manfred Rohde*. Analysis of dimensions and structural organization of proteoliposomes
*Eberhardt Spiess* and *Rudi Lurz*. Electron microscopic analysis of nucleic acids and nucleic acid-protein complexes
*Walther Johannssen*. Interaction of restriction endonucleases with DNA as revealed by electron microscopy
*Andreas Holzenburg*. Preparation of two-dimensional arrays of soluble protein as demonstrated for bacterial D-nbose-1,5-bisphosphate carboxylase/oxygenase
*Harald Engelhardt*. Correlation averaging and 3-D reconstruction of 2-D crystalline membranes and macromolecules

**Volume 21**
*J. Grinsted* and *P. M. Bennet*. Introduction
*Vilma A. Stanisich*. Identification and analysis of plasmids at the genetic level
*N. Willetts*. Conjugation
*J. R. Saunders* and *Venetia A. Saunders*. Bacterial transformation with plasmid DNA
*J. Grinsted* and *P. M. Bennett*. Preparation and electrophoresis of plasmid DNA
*J. Grinsted* and *P. M. Bennett*. Analysis of plasmid DNA with restriction endonucleases
*H. J. Burkhardt* and *A. Pühler*. Electron microscopy of plasmid DNA
*Russell Thompson*. Plasmid cloning vectors
*P. M. Bennett, J. Grinsted* and *T. J. Foster*. Detection and use of transposons
*G. Dougan* and *N. F. Fairweather*. Detection of Gene products expressed from plasmids
*Nigel L. Brown* and *Peter A. Lund*. DNA sequencing